Predator–Prey Interactions
in the Fossil Record

TOPICS IN GEOBIOLOGY

Series Editors: **Neil H. Landman,** American Museum of Natural History, New York, New York
Douglas S. Jones, University of Florida, Gainesville, Florida

Current volumes in this series

Volume 7	HETEROCHRONY IN EVOLUTION A Multidisciplinary Approach Edited by Michael L. McKinney
Volume 8	GALÁPAGOS MARINE INVERTEBRATES Taxonomy, Biogeography, and Evolution in Darwin's Islands Edited by Matthew J. James
Volume 9	TAPHONOMY Releasing the Data Locked in the Fossil Record Edited by Peter A. Allison and Derek E. G. Briggs
Volume 10	ORIGIN AND EARLY EVOLUTION OF THE METAZOA Edited by Jere H. Lipps and Philip W. Signor
Volume 11	ORGANIC GEOCHEMISTRY Principles and Applications Edited by Michael H. Engel and Stephen A. Macko
Volume 12	THE TERTIARY RECORD OF RODENTS IN NORTH AMERICA William Korth
Volume 13	AMMONOID PALEOBIOGRAPHY Edited by Neil H. Landman, Kazushige Tanabe, and Richard Arnold Davis
Volume 14	NEOGENE PALEONTOLOGY OF THE MANONGA VALLEY, TANZANIA A Window into the Evolutionary History of East Africa Edited by Terry Harrison
Volume 15	ENVIRONMENTAL MICROPALEONTOLOGY The Application of Microfossils to Environmental Geology Edited by Ronald E. Martin
Volume 16	PALEOBIOGEOGRAPHY Bruce S. Lieberman
Volume 17	THE HISTORY AND SEDIMENTOLOGY OF ANCIENT REEF SYSTEMS Edited by George D. Stanley, Jr.
Volume 18	EOCENE BIODIVERSITY Unusual Occurrences and Rarely Sampled Habitats Edited by Gregg F. Gunnell
Volume 19	FOSSILS, PHYLOGENY, AND FORM An Analytical Approach Edited by Jonathan M. Adrain, Gregory D. Edgecombe, and Bruce S. Lieberman
Volume 20	PREDATOR–PREY INTERACTIONS IN THE FOSSIL RECORD Edited by Patricia Kelley, Michał Kowalewski, and Thor Hansen

A Continuation Order Plan is available for this series. A continuation order will bring delivery of each new volume immediately upon publication. Volumes are billed only upon actual shipment. For further information please contact the publisher.

Predator–Prey Interactions in the Fossil Record

Edited by

Patricia H. Kelley
University of North Carolina at Wilmington
Wilmington, North Carolina

Michał Kowalewski
Virginia Polytechnic Institute and State University
Blacksburg, Virginia

and

Thor A. Hansen
Western Washington University
Bellingham, Washington

Kluwer Academic / Plenum Publishers
New York, Boston, Dordrecht, London, Moscow

Cover illustrations:
Malformation due to sublethal predation on trilobite *Elrathia kingii* (Cambrian of Utah) (photograph by Loren E. Babcock).
Skull of predatory dinosaur *Suchomimus tenerensis* from the Cretaceous of Niger (photograph by Thomas R. Holtz, Jr.).
Bivalve *Chione erosa* from the Pliocene of Florida, drilled by muricid gastropod predator (photograph by Gregory P. Dietl).

Top cover illustrations:
Predatory hole in the echinoid *Fibularia ovulum* from the Northern Bay of Safaga, Red Sea (photograph by James H. Nebelsick).
Skull of predatory dinosaur *Allosaurus fragili* (illustration by Tracy Lee Ford).

ISBN: 0-306-47489-1

© 2003 Kluwer Academic / Plenum Publishers, New York
233 Spring Street, New York, New York 10013

http://www.wkap.nl/

10 9 8 7 6 5 4 3 2 1

A C.I.P. record for this book is available from the Library of Congress.

All rights reserved

No part of this book may be reproduced, stored in a retrieval system, or transmitted in any form
or by any means, electronic, mechanical, photocopying, microfilming, recording, or otherwise,
without written permission from the Publisher, with the exception of any material supplied specifically for
the purpose of being entered and executed on a computer system, for exclusive use by the purchaser of the
work.

Printed in the United States of America

Contributors

Richard R. Alexander Department of Geological and Marine Sciences, Rider University, Lawrenceville, New Jersey, 08648-3099

Loren E. Babcock Department of Geological Sciences, The Ohio State University, Columbus, Ohio 43210

Tomasz K. Baumiller Museum of Paleontology and Department of Geological Sciences, University of Michigan, Ann Arbor, Michigan 48109-1079

Carlton E. Brett Department of Geology, University of Cincinnati, Cincinnati, Ohio 45221-0013

David T. Chaffin Department of Geological Sciences, Ohio University, Athens, Ohio 45701

Stephen J. Culver Department of Geology, East Carolina University, Greenville, North Carolina 27858

Gregory P. Dietl Department of Zoology, North Carolina State University, Raleigh, North Carolina, 27695-7617

Ashraf M. T. Elewa Department of Geology, Minia University, Egypt

Forest J. Gahn Museum of Paleontology and Department of Geological Sciences, University of Michigan, Ann Arbor, Michigan 48109-1079

Thor A. Hansen Department of Geology, Western Washington University, Bellingham, Washington 98225

Elizabeth M. Harper Department of Earth Sciences, Cambridge University, Cambridge CB2 3EQ, United Kingdom

Thomas R. Holtz Department of Geology, University of Maryland, College Park, Maryland 20742

Patricia H. Kelley Department of Earth Sciences, University of North Carolina at Wilmington, Wilmington, North Carolina 28403-5944

Michal Kowalewski Department of Geological Sciences, Virginia Polytechnic Institute and State University, Blacksburg, Virginia 24061

Lindsey R. Leighton Department of Geological Sciences, San Diego State University, San Diego, California 92182

Scott Lidgard Department of Geology, The Field Museum, Roosevelt Road at Lake Shore Drive, Chicago, Illinois 60605

Jere H. Lipps Department of Integrative Biology and Museum of Paleontology, University of California at Berkeley, Berkeley, California 94720

Royal H. Mapes Department of Geological Sciences, Ohio University, Athens, Ohio 45701

James McAllister Department of Natural Sciences, Dickinson State University, Dickinson, North Dakota 58601

Frank K. McKinney Department of Geology, Appalachian State University, Boone, North Carolina, 28608-2067, and Honorary Research Fellow, Department of Palaeontology, The Natural History Museum, Cromwell Road, London SW7 5 BD, United Kingdom

Mark McMenamin Department of Earth and Environment, Mount Holyoke College, South Hadley, Massachusetts 01075

James H. Nebelsick Institut und Museum für Geologie und Paläontologie, Universität Tübingen, Sigwartstrasse 10, D-72076 Tübingen, Germany

Richard Potts Human Origins Program, National Museum of Natural History, Smithsonian Institution, Washington, DC 20560-0112

Richard A. Reyment Naturhistoriska Riksmuséet, Avdelningen för paleozoologi, Box 50007, 10405 Stockholm, Sweden

Paul D. Taylor Department of Palaeontology, The Natural History Museum, Cromwell Road, London SW7 5 BD, United Kingdom

J. P. Williams English Heritage, 44 Derngate, Northampton, NN1 1UH, United Kingdom

Rachel Wood Schlumberger Cambridge Research, High Cross, Madingley Road, Cambridge CB2 OEL, UK, and Department of Earth Sciences, University of Cambridge, Downing Street, Cambridge CB2 3EQ, United Kingdom

Foreword

Developing full knowledge of the history of life will require deeper understanding of the biology and interactions of fossil organisms as well as following their phylogenetic history. Cladistic analyses are crucial for developing testable arguments for phylogenetic relationships, but, since all organisms spend their entire lives in organism-physical environment and organism-organism interactions, all evolutionary change occurs in an ecological context. Knowing how organisms interact in life (now or in the past) is as important for understanding the history of life as knowing the evolutionary relationships between lineages. To explain what has happened in the history of life we will need to know the natural history of our fossil organisms. We often say "more work is needed," but in this case it is not just a truism.

Predator-prey interactions are among the most significant of all organism-organism interactions. One of the major aspects of community organization, the context in which every species population exists, is trophic structure. Predator-prey interactions determine the path of energy flow in food chains and the very nature of many community types. Keystone predator species may be responsible for the regulation of diversity and abundance of entire ecosystems. Avoidance of predation, a major component of predator-prey interaction from the viewpoint of the prey, is crucial to survival of most animal species. Species usually can afford some loss (juvenile mortality, accidents, and falling prey to predators), but the degree of loss a population can tolerate is dependent on the rate of reproductive replacement by the population. Selection operates both to increase predator efficiency and predation avoidance. Co-evolutionary balance between predator efficiency and success of predator avoidance thus becomes a major factor in the evolutionary equation. Yet determining the dynamics of predator-prey interaction is difficult when the subjects at hand have been dead for thousands to millions of years!

The maturation of the discipline of taphonomy has given us new and more reliable ways of interpreting the origin and fate of organisms preserved as fossils. Drill holes, chip patterns, and other features reveal both successful and, when continued skeletal growth is seen, unsuccessful predatory attacks. Predators utilize various modes of attack. Their success is dependent on the type of prey available. Analysis of form and function in predators suggests the nature of targeted prey, just as similar analyses of prey give us insight into their strategies for predator avoidance. Statistical analysis of evidence for predation provides a sense of the intensity of predation and how it may vary.

One of the major concepts of a driving force in evolution is Vermeij's idea of escalation. Changes in the diversity and recorded effects of predators over time make it an attractive hypothesis. The early appearance of predators such as *Anomalocaris*, as well as the diversity of priapulids in the Burgess Shale, demonstrates that food chains were already evolving by the Middle Cambrian. Change in shell ornamentation and shell coiling that would enhance resistance to predation occur in the Devonian, just as jawed fishes and other active predators diversified, and the frequency of drilling predation increased in the later Mesozoic at the same time that teleost fish, neogastropods, and crabs also diversified. However, it has been difficult to evaluate the hypothesis in detail. The multitude of potential interactions and the competition for selective effectiveness among various modes of attack, plus the potential for prey to deal with increased predation pressure by change in reproductive capacity rather than change in morphology, make predictions about the evolutionary expression of predator-prey interactions uncertain. For example, different survival strategies can produce different selective effects (thicker shells to frustrate drilling versus lighter shells to permit more rapid movement for escape), and this is further complicated by uncertainty about the abundance of different predators and, therefore, predation intensities from different sources, in most habitats. It will only be by compiling and evaluating data on predator-prey relations as they are recorded in the fossil record that we can hope to tease apart their role in the tangled web of evolutionary interaction over time.

This volume, compiled by a group of expert specialists on the evidence of predator-prey interactions in the fossil record, is a pioneering effort to collate the information now accumulating in this important field. It will be a standard reference on which future study of one of the central dynamics of ecology as seen in the fossil record will be built.

Richard K. Bambach

Professor Emeritus, Virginia Tech
Associate of the Botanical Museum, Harvard University
1998 Raymond C. Moore Medallist, SEPM (Society for Sedimentary Geology)

Preface

Although predation is regarded as a primary process affecting the structure of communities, its role in evolution remains controversial. How important are biotic interactions, such as predation, in shaping the history of life on a macroevolutionary scale? This question can only be answered through knowledge of predator-prey interactions in the fossil record.

Fossil data are essential for developing and testing hypotheses concerning the role of ecological factors in evolution. In recent years, predator-prey interactions have been documented from the fossil record for various microfossil, invertebrate, and vertebrate taxa. Such studies have benefited from improved understanding of taphonomic processes, new analytical approaches (e.g., geochemical and statistical), and the manipulation of large spatial-temporal databases. Rigorous testing of ecological and evolutionary hypotheses has become possible for a variety of taxonomic groups.

This book presents the evidence for predator-prey interactions in the fossil record and the implications of this record for evolutionary paleoecology. The first section of this book, which comprises the bulk of the volume, discusses the fossil record of predation for specific taxonomic groups. These chapters provide a synthesis of the current knowledge of specific predator-prey systems within the context of the broader question of the role of predation in ecology and evolution. Depending on the taxonomic group, authors describe the approaches to inferring predator-prey relations (often based on knowledge of extant representatives), fossil evidence for predator-prey interactions, temporal and spatial predation patterns in the fossil record, and the implications of these results for testing ecological and evolutionary hypotheses. Future directions for research are also identified. Chapter 1 examines predation on and by Foraminifera, and Chapters 2 through 11 review the fossil record of predation for major invertebrate phyla. Vertebrate predator-prey interactions (including fish, dinosaurs, small mammals, and hominids) are discussed in Chapters 12 through 15.

The second section of the book analyzes major macroevolutionary episodes in the history of predation, particularly radiation events. These episodes include the Cambrian radiation of metazoan predators (Chapter 16), the mid-Paleozoic marine revolution (Chapter 17), and the Mesozoic Marine Revolution (Chapter 18).

We thank the contributors of this volume for providing high-quality manuscripts and responding promptly to our editorial demands. We are also grateful to the many reviewers of chapters in this book, including individuals who also authored chapters. Reviewers who agreed to be identified are: Richard Alexander, William Ausich, Loren Babcock, Richard Bambach, Tomasz Baumiller, Carlton Brett, Alan Cheetham, Laurel Collins, Stephen Culver, Philip Currie, James Farlow, Richard Fortey, Michael Gibson,

Elizabeth Harper, Alan Hoffmeister, Roger Kaesler, Alan Kohn, Royal Mapes, James McAllister, Frank McKinney, Mark McMenamin, James Nebelsick, Eleanora Reber, Peter Roopnarine, Kaustuv Roy, J. D. Stewart, Carl Stock, Mark V. H. Wilson, Rachel Wood.

The following individuals at the University of North Carolina at Wilmington provided assistance in preparing the camera-ready copy for the printer: Catherine Morris, Heyward Key, Gregory Dietl, Beth Reimer, and John Huntley. Jonathan Kelley also assisted his frequently frantic wife with reformatting of manuscripts and Timothy Kelley provided help with graphics, and Katherine Kelley helped with the index. Ken Howell, Editor, and Deborah Doherty, Electronic Production Manager, at Kluwer Academic Publishers patiently answered questions about format specifications. Our thanks go to all.

Patricia Kelley
Michał Kowalewski
Thor Hansen

Contents

Introduction

Part I • Taxonomic Review of the Fossil Record of Predation

Chapter 1 • Predation on and by Foraminifera

Stephen J. Culver and Jere H. Lipps

1. Introduction 7
2. Predation on Living Foraminifera..8
3. Predation in Fossil Foraminifera..21
4. Paleobiological Implications of Foraminiferal Predation 24
5. Future Research 27
6. Summary...27
   References 28

Chapter 2 • Predation in Ancient Reef-Builders

Rachel Wood

1. Introduction 33
2. Predation on Modern Coral Reefs...34
3. Potential Antipredatory Traits in Reef-Builders 37
4. The Rise of Biological Disturbance in Reef Ecosystems 43
5. Difficulties of Identification of Antipredatory Traits in Ancient Reef Organisms....45
6. Evolution of Antipredatory Characteristics in Reef Builders...................................46
7. Discussion and Future Work...49
   References 50

Chapter 3 • Trilobites in Paleozoic Predator-Prey Systems, and Their Role in Reorganization of Early Paleozoic Ecosystems

Loren E. Babcock

1. Introduction 55
2. Predation on Trilobites 57

3. Predation by Trilobites 75
4. Final Comments: Predation on and by Trilobites in the Context of Ecosystem
   Reorganization During the Early Paleozoic 81
   References 85

Chapter 4 • Predation by Drills on Ostracoda

Richard A. Reyment and Ashraf M. T. Elewa

1. Introduction 93
2. Gastropod Drill-holes 94
3. Paleoethology and Paleoecology of the Predators 96
4. Size, Predation Pressure and Selectional Effects 103
5. Ornamentation Complexity and Predation Intensity 106
6. Holes of Non-molluscan Origin 107
7. Future Directions for Research 108
   References 109

Chapter 5 • The Fossil Record of Drilling Predation on Bivalves and Gastropods

Patricia H. Kelley and Thor A. Hansen

1. Introduction 113
2. Data on Drilling Predation Extractable from the Fossil Record 115
3. Testable Hypotheses in Evolutionary Paleoecology 121
4. Arms Races and Drilling Frequencies 122
5. Changes in Drilling Behavior through Time 126
6. Relative Effectiveness of Predator and Prey 129
7. Summary and Discussion 130
   References 133

Chapter 6 • The Fossil Record of Shell-Breaking Predation on Marine Bivalves and
   Gastropods

Richard R. Alexander and Gregory P. Dietl

1. Introduction 141
2. Durophages of Bivalves and Gastropods 142
3. Trends in Antipredatory Morphology in Space and Time 145
4. Predatory and Non-Predatory Sublethal Shell Breakage 155
5. Calculation of Repair Frequencies and Prey Effectiveness 160
6. Prey Species-, Size-, and Site-Selectivity by Durophages 164
7. Repair Frequencies by Time, Latitude, and Habitat 166
8. Concluding Remarks 170
   References 170

Contents xiii

Chapter 7 • Predation on Cephalopods: A General Overview with a Case Study From the Upper Carboniferous of Texas

Royal H. Mapes and David T. Chaffin

1. Introduction 177
2. Background..179
3. Case Study: Lethal Predation on Upper Carboniferous Coiled Nautiloids and Ammonoids...187
4. Studies of Predation and Cephalopods Through Time..206
5. Conclusions and Future Studies..208
   Appendix 209
   References 210

Chapter 8 • Predation on Brachiopods

Lindsey R. Leighton

1. Introduction 215
2. Methods 222
3. Results 224
4. Discussion..225
5. After the Paleozoic 229
6. Final Thoughts and Future Directions 230
   Appendix – Ornament and Population Data 230
   References 234

Chapter 9 • Predation on Bryozoans and its Reflection in the Fossil Record

Frank K. McKinney, Paul D. Taylor, and Scott Lidgard

1. Introduction 239
2. Predation on Living Bryozoans 240
3. Fossil Record of Predation..245
4. Possible Evolutionary Responses to Predation 250
5. Summary and Conclusions 256
   References 257

Chapter 10 • Predation on Crinoids

Tomasz K. Baumiller And Forest J. Gahn

1. Introduction 263
2. A Paradigm Shift?..264
3. Neontological Studies of Predation 265
4. Paleontological Studies of Predation 268
5. Conclusion...275
   References 276

Chapter 11 • Predation on Recent and Fossil Echinoids

Michał Kowalewski and James H. Nebelsick

1. Introduction 279
2. Predators of Living Echinoids 280
3. The Fossil Record of Predation and Parasitism on Echinoids 291
4. Future Research Directions 295
5. Summary 297
   References 297

Chapter 12 • Predation of Fishes in the Fossil Record

James McAllister

1. Introduction 303
2. Origin of Fish Predation 304
3. Evidence of Fish Predation from the Fossil Record 311
4. Summary 322
   References 323

Chapter 13 • Dinosaur Predation: Evidence and Ecomorphology

Thomas R. Holtz, Jr.

1. Theropods as Predators 325
2. Fossil Evidence of Theropod-Prey Interaction 327
3. Theropod Ecomorphology 330
4. Prey Defenses 337
5. Conclusions 337
   References 337

Chapter 14 • Bones of comprehension: The Analysis of Small Mammal Predator–Prey Interactions

J. P. Williams

1. Introduction 341
2. Identifying the Predator 342
3. Fossil Evidence 350
4. Summary 356
   References 356

Chapter 15 • Early Human Predation

Richard Potts

1. Introduction 359
2. Interpreting Early Human Predation 360
3. Primate Background 361
4. Evolutionary Basis of Human Predatory Behavior 362
5. Ecological Overlap and Interaction between Hominins and Other Predator/Scavengers 364
6. Predator-Prey Interactions during Human Evolutionary History 365
7. Conclusion 371
   References 373

Part II • Major Macroevolutionary Episodes in the History of Predation

Chapter 16 • Origin and Early Evolution of Predators:
The Ecotone Model and Early Evidence for Macropredation

Mark A. S. Mcmenamin

1. Introduction 379
2. The First Bite 383
3. The Ecotone Model 385
4. Summary 396
   References 398

Chapter 17 • Durophagous Predation in Paleozoic Marine Benthic Assemblages

Carlton E. Brett

1. Introduction 401
2. Paleozoic Predators 402
3. Trace Fossil Evidence of Predatory Attack 406
4. Possible Consequences of Predator Escalation 418
5. Discussion: Tests of the Mid Paleozoic Escalation Hypothesis and a Preliminary Model of Escalation 426
6. Summary 428
   References 429

Chapter 18 • The Mesozoic Marine Revolution

Elizabeth M. Harper

1. Introduction 433
2. Establishing a Chronology for the MMR 435
3. Prey Responses to Increasing Threat of the MMR 440

4. Discussion and Future Directions 447
   References 451

Index 457

Introduction

Predation and competition are the primary ecological processes that control the structure and function of communities. Predation affects the distribution and abundance of organisms, the flow of energy through systems, and the diversity and structure of communities. At the level of individuals, predator-prey interactions provide a major arena in which natural selection takes place.

Despite the importance accorded to predation by ecologists, the role of predator-prey interactions in evolution remains controversial (Allmon, 1994; Jackson, 1988; Jablonski and Sepkoski, 1996; Vermeij, 1996, 1999; Knoll and Bambach, 2000, Allmon and Bottjer, 2001). The evolutionary role of interactions between organisms is a prominent theme of works by Van Valen (1973, 1983), Vermeij (1982, 1987, 1994, 1996, 1999), and Bambach (1993, 1999), among many others. Predation has been claimed to influence rates of evolution (Stanley, 1974) and has been cited as a causal factor in the rise of biomineralization (Stanley, 1976; Bengtson, 1994; Conway Morris, 1998) and in diversification and extinction (Jablonski and Sepkoski, 1996; Vermeij, 1987; Jablonski, 2000). Predator-prey interactions have been claimed to drive long-term morphological and behavioral trends in various clades (Signor and Brett, 1984; Vermeij 1977, 1987).

The importance of biotic interactions such as predation in evolution is disputed by those who emphasize a decoupling of within-species adaptation and macroevolutionary processes, who consider physical factors more important selective agents than biotic factors, or who reject long-term directionality in evolution. This perspective was articulated most forcefully by Gould (e.g., 1985, 1988, 1989, 1990; Gould and Eldredge 1993); see also Knoll and Bambach (2000, p. 2). Gould (1985, 1990) minimized the role of ecological interactions, arguing for a hierarchy of processes that control evolution (i.e., "tiers" of selection). Ecological interactions may be important at the level of natural selection among organisms (Gould's "first tier"). However, according to Gould (1985), adaptation at the level of individuals is less important than, or opposed to, processes that occur at or above the species level (selection or sorting at the species level; mass extinctions); these higher-tier processes may produce trends that reverse, undo, or override the results of first-tier selection (Gould 1985).

Even those who argue that ecological interactions are important in evolution continue to debate which processes are involved (Kitchell, 1990; Vermeij, 1994; Thompson, 1999). The controversy has focused on the roles of two related processes, coevolution and escalation (Dietl and Kelley, in press). Coevolution (Futuyma and Slatkin, 1983) entails reciprocal adaptation (i.e., predator and prey evolve in response to one another). Escalation (Vermeij, 1987, 1994) involves adaptation to enemies, and may not be reciprocal; for instance, prey may respond evolutionarily to predators, but predators are more likely to respond to their enemies (their own predators or competitors). In cases in which the prey are dangerous to their predators, escalation may involve coevolution (producing reciprocal adaptation of predator and prey).

Understanding of the macroevolutionary effects of predation requires knowledge of predator-prey interactions in the fossil record. Documentation of predation for a broad spectrum of microfossil, invertebrate, and vertebrate groups has accelerated in recent years. This work has been enhanced by studies of predation among extant representatives of fossil groups, development of techniques to differentiate taphonomic damage from predation traces, application of new analytical techniques, and the accumulation of large databases. Admittedly, knowledge of the fossil record of predation varies significantly among taxa; for example, our knowledge of predator-prey interactions for bivalves, gastropods, trilobites, and brachiopods is more complete than for cephalopods, bryozoans, and foraminifers. Nevertheless, sufficient data are available for a number of taxa to provide at least a preliminary test of competing ecological and evolutionary hypotheses. Indeed, most of the chapters in this volume argue for the importance of predator-prey interactions in structuring Phanerozoic ecosystems and in influencing biotic evolution. In some cases, data are sufficient to explore the processes involved in such evolution (i.e., coevolution and escalation).

Opportunities are manifold for future research on predator-prey interactions in the fossil record. Virtually every chapter in this book includes suggestions, and sometimes pleas, for areas requiring further study. We hope that this volume will provide a springboard to continued discussion and fruitful testing of the role of predation in shaping the history of life on Earth.

References

Allmon, W. D., 1994, Taxic evolutionary paleoecology and the ecological context of macroevolutionary change, *Evol. Ecol.* **8**:95-112.
Allmon, W. D., and Bottjer, D. J., 2001, *Evolutionary Paleoecology: The Ecological Context of Macroevolutionary Change*, Columbia University Press, New York.
Bambach, R. K., 1993, Seafood through time: changes in biomass, energetics, and productivity in the marine ecosystem, *Paleobiology* **19**:372-397.
Bambach, R. K., 1999, Energetics in the global marine fauna: a connection between terrestrial diversification and change in the marine biosphere, *Geobios* **32**:131-144.
Bengtson, S., 1994, The advent of animal skeletons, in: *Early Life on Earth* (S. Bengtson, ed.), Columbia University Press, New York, pp. 412-425.
Conway Morris, S., 1998, *The Crucible of Creation. The Burgess Shale and the Rise of Animals*, Oxford University Press, Oxford
Dietl, G. P., and Kelley, P. H., 2002, The fossil record of predator-prey arms races: Coevolution and escalation hypotheses, in: *The Fossil Record of Predation* (M. Kowalewski and P. H. Kelley, eds.), The Paleontological Society Papers **8**.
Futuyma, D. J., and Slatkin, M. 1983, Introduction, in: *Coevolution* (D. J. Futuyma and M. Slatkin, eds.), Sinauer Associates, Sunderland, Massachusetts, pp. 1-13.
Gould, S. J., 1985, The paradox of the first tier: an agenda for paleobiology, *Paleobiology* **11**:2-12.
Gould, S. J., 1989, *Wonderful Life: The Burgess Shale and the Nature of History*, Norton, New York.
Gould, S. J., 1990, Speciation and sorting as the source of evolutionary trends, or 'Things are seldom what they seem,' in *Evolutionary Trends* (K. McNamara, ed.), London: Belhaven Press, pp. 3-27.
Gould, S. J., and Eldredge, N., 1977, Punctuated equilibrium comes of age, *Nature* **366**:223-227.
Jackson, J. B. C., 1988, Does ecology matter?, *Paleobiology* **14**:307-312.
Jablonski, D., 2000, Micro- and macroevolution: scale and hierarchy in evolutionary biology and paleobiology, in: *Deep Time: Paleobiology's Perspective* (D. H. Erwin and S. L. Wing, eds.), Allen Press, Lawrence, Kansas, pp. 15-52.
Jablonski, D., and Sepkoski, J. J., Jr., 1996, Paleobiology, community ecology, and scales of ecological pattern, *Ecology* **77**(5):1367-1378.

Kitchell, J. A., 1990, The reciprocal interaction of organism and effective environment: Learning more about "and," in: *Causes of Evolution: A Paleontological Perspective* (R. M. Ross and W. D. Allmon, eds.), University of Chicago Press, Chicago, pp. 151-169.

Knoll, A. H., and Bambach, R. K., 2000, Directionality in the history of life: diffusion from the left wall or repeated scaling of the right, in: *Deep Time: Paleobiology's Perspective* (D. H. Erwin and S. L. Wing, eds.), Allen Press, Lawrence, Kansas, pp. 1-14.

Signor, P. W., III, and Brett, C. E., 1984, The mid-Paleozoic precursor to the Mesozoic marine revolution, *Paleobiology* **10**:229-245.

Stanley, S. M., 1974, Effects of competition on rates of evolution, with special reference to bivalve mollusks and mammals, *Syst. Zool.* **22**:486-506.

Stanley, S. M., 1976, Fossil data and the Precambrian-Cambrian evolutionary transition, *Am. J. Sci.* **276**:56-76.

Thompson, J. N. 1999, Coevolution and escalation: are ongoing coevolutionary meanderings important?, *Am. Nat.* 153:S92-S93.

Van Valen, L. M., 1973, A new evolutionary law, *Evol. Theory* **1**:1-30.

Van Valen, L. 1983, How pervasive is coevolution?, in: *Coevolution* (M. H. Nitecki, ed.), University of Chicago Press, Chicago, pp. 1-19.

Vermeij, G. J., 1977, The Mesozoic marine revolution: evidence from snails, predators, and grazers, *Paleobiology* **3**:245-258.

Vermeij, G. J., 1982, Unsuccessful predation and evolution, *Am. Nat.* **120**:701-720.

Vermeij, G. J., 1987, *Evolution and Escalation: An Ecological History of Life*, Princeton University Press, Princeton, New Jersey

Vermeij, G. J., 1994, The evolutionary interaction among species: selection, interaction, and coevolution, *Ann. Rev. Ecol. Syst.* **25**:219-236.

Vermeij, G. J., 1996, Adaptations of clades: resistance and response, in: *Adaptation* (M. R. Rose and G. V. Lauder, eds.), Academic Press, San Diego, pp. 363-380.

Vermeij, G. J., 1999, Inequality and the directionality of history, *Am. Nat.* **153**:243-252.

I
Taxonomic Review of the
Fossil Record of Predation

Chapter 1

Predation on and by Foraminifera

STEPHEN J. CULVER and JERE H. LIPPS

1. Introduction 7
2. Predation on Living Foraminifera..8
    2.1. Predators of Benthic Foraminifera...9
    2.2. Predators of Planktonic Foraminifera..11
    2.3. The Effects of Predation on Foraminiferal Tests 12
    2.4. The Effects of Predation on Foraminiferal Populations 15
    2.5. Foraminifera as Predators 16
3. Predation in Fossil Foraminifera..21
4. Paleobiological Implications of Foraminiferal Predation 24
    4.1. Predatory Activities Recorded in the Fossil Record...24
    4.2. Taphonomic Effects of Foraminiferal Predation...26
5. Future Research 27
6. Summary..27
    References 28

1. Introduction

>It has been a matter of almost momentary occurrence to see a tiny Copepod blunder against the fully extended pseudopodia of a robust Miliolid and instantaneously fall to the bottom of the tank apparently dead. The Copepod is, however, only stunned, or by some unidentified means terrified, for at the end of, at the most, two minutes, it seems to stretch itself and dart off once more upon its apparently gay and irresponsible career.
>
>Edward Heron-Allen (1915, p. 234-235)

Foraminifera are diverse and numerically important in most marine ecosystems and have been since the early Paleozoic. Their ecology and distribution have been studied

---

STEPHEN J. CULVER • Department of Geology, East Carolina University, Greenville North Carolina 27858. JERE H. LIPPS • Department of Integrative Biology and Museum of Paleontology, University of California Berkeley, Berkeley, California 94720.

Predator-Prey Interactions in the Fossil Record, edited by Patricia H. Kelley, Michał Kowalewski, and Thor A. Hansen. Kluwer Academic/Plenum Publishers, New York, 2003.

extensively across the globe (Murray, 1991). They are also important in the geologic record, being one of the best represented fossil organisms since the Cambrian (Culver, 1991; Lipps, 1992). Foraminifera are widely used in ecologic, paleobiologic, paleoceanographic, and paleoclimatic analyses. Yet we know little about their trophic relationships, a fundamental ecologic feature of any group of organisms.

Despite this huge volume of work, the little knowledge about the trophic relationships of living foraminifera makes the paleobiology of trophic relationships in their fossil counterparts difficult to elucidate. For example, no papers, as far as we are aware, are devoted to predation on foraminifera in the fossil record. The data are few and scattered. Some inferences, however, can be drawn on the effects of predation, on and by foraminifera, on the fossil record from a review of the data on living foraminifera. This paper, therefore, starts with such a review, followed by a summary of the evidence of predation on foraminifera in the fossil record. We conclude with some observations of the paleobiological implications of the current data and some possible avenues of future research.

2. Predation on Living Foraminifera

Given the meiofaunal scale at which foraminifera live, the evidence for predation of foraminifera is often indirect, consisting chiefly of gut-content analyses of invertebrates and vertebrates. A variety of invertebrates ingest foraminifera (Lipps, 1983) as do fish (Daniels and Lipps, 1974; Lipps, 1988) and even shore birds (Lipps, pers. observ., 2001). As long ago as 1702, van Leuwenhoek observed that shrimp feed on foraminifera (Hoole, 1807), and more recently many others have noted foraminifera in the gut contents of a wide variety of organisms. In most instances, however, whether their presence was due to selective or accidental consumption was not evident. Indeed, some have suggested that foraminifera may be ingested not for the food resource of the cytoplasm but for the calcium carbonate or the organic matrix of empty tests. The term foraminiferivory was introduced by Hickman and Lipps (1983) to cover the general phenomenon of foraminiferal ingestion.

More direct evidence of predation on foraminifera may be holes made in tests by some other organism. These holes, which penetrate to the interior of the chambers, are assumed to have given the driller access to protoplasm. Even in these cases, however, predation may not have occurred because actual attacks on the test or eating of protoplasm has not been observed. Other possibilities, such as another organism seeking refuge inside the tests or trying to attach to the test, are equally possible.

Because foraminifera are commonly present in great abundance in the water column, on the sea floor, and in the sediment, they represent an important potential food source for many metazoans and protozoans, including other foraminifera. The density of benthic foraminifera can reach at least 10^6 per square meter of sea floor, and their biomass can range from 0.02 to more than 10 grams per square meter (Saidova, 1967; Wefer and Lutze, 1976; Buzas, 1978; see Gooday et al., 1992, for a review of deep-sea foraminiferal biomass). Planktonic foraminiferal density varies from about 500 (daytime) to 850 (nighttime) specimens/m^3 of near-surface seawater (Holmes, 1982) although Bé and Hamlin (1967) found planktonic foraminifera to be more abundant in daytime tows (up to 10^6 specimens/ 1000 m^3 of water). Thus, foraminifera may well provide a significant food resource for a variety of other organisms.

2.1. Predators of Benthic Foraminifera

Many groups of metazoans probably consume benthic foraminifera, including flatworms, echiuroids, sipunculids, polychaetes, chitons, gastropods, nudibranchs, scaphopods, bivalves, crustaceans, holothuroids, asteroids, ophiuroids, echinoids, crinoids, tunicates, and fish; Boltovskoy and Wright (1976), Lipps (1983) and Gooday *et al.* (1992) provide the primary literature and references. To this list can possibly be added nematodes, which have been reported within the partially emptied chambers of living benthic foraminifera in laboratory culture (Sliter, 1965), although this behavior is less common in natural populations (Sliter, 1971). In the Cretaceous, bottom-feeding ammonites also apparently consumed large numbers of foraminifera (Lehman, 1971).

Representatives from two groups of organisms in particular, gastropods and scaphopods, are known to be selective consumers of benthic foraminifera (Brown, 1934; Bacescu and Caraion, 1956; Hurst, 1965; Burn and Bell, 1974; Shonman and Nybakken, 1978; Hickman and Lipps, 1983; Arnold *et al.*, 1985; Langer *et al.*, 1995) although some deep-sea sipunculans (Heeger, 1990) and assellote isopods (Svavarsson *et al.*, 1991) may also be specialized foraminiferal feeders. The gastropod *Philene* may feed selectively on foraminifera because the proportion of foraminifera to sand grains was higher in the gut than in the sediment where they lived (Brown, 1934). Foraminifera and sand grains also filled the crop and gizzard of the gastropod *Cylichna cylindracea* from the Øresund, Denmark, whereas two species of *Retusa* from the same dredges had also taken other food material (Hurst, 1965). In contrast, Bacescu and Caraion (1956) described two species of *Retusa* from the Black Sea as almost exclusively foraminiferal feeders. *Retusa pelyx* from Victoria, Australia, was selective for certain species of foraminifera (Burn and Bell, 1974). The foraminiferan *Ammonia aoteanus* constituted 48.9% of the gizzard contents but only 3.7% of the foraminiferal fauna. *Elphidium selseyense* was also eaten preferentially; it was twice as frequent in the gizzard as in the sediment. The miliolid *Quinqueloculina seminula*, in comparison, occurred infrequently in the snail's gizzard, despite dominating the ambient population.

Two species of gastropod in Monterey Bay, California, selectively prey on foraminifera (Shonman and Nybakken, 1978). *Cylichna attonsa* fed only on the genus *Nonionella*, one of the more abundant taxa in the sediment. In comparison, *Acteocina culcitella* was more of a generalist, feeding upon 13 species of benthic foraminifera.

The gastropod *Olivella biplicata*, at least as a juvenile, also selectively ingests benthic foraminifera (Hickman and Lipps, 1983); foraminifera in the gut are concentrated by a factor of 1000 compared to the surrounding sediment. Hickman and Lipps (1983), however, pointed out that such ingestion, mainly by juvenile snails, does not necessarily imply predation. Many tests of calcareous foraminifera taken from the stomach of *Olivella* exhibited dissolution beyond that necessary to access cytoplasm. Thus, they suggested that dead foraminifera were being ingested either for the organic matrix of the test wall or for calcium. This speculation has not been confirmed yet, and dissolution merely by stomach action without any purpose seems a likely possibility as well.

Selective drilling by predatory gastropods has also been recorded. Arnold et al. (1985) showed that a presumed juvenile naticid predator fed almost exclusively on the most abundant benthic species, *Siphouvigerina auberiana,* on the Galapagos hydrothermal mounds. Drillings occurred on 27% of the individuals, and most specimens exhibited multiple holes, probably because the predator consumed the cytoplasm from one chamber of the foraminifer and then moved on to drill the next (Fig. 1).

Scaphopods have long been known to feed extensively (in some cases apparently exclusively) on benthic foraminifera (Clark, 1849). Morton (1959) showed that the feeding apparatus, the captacula, composed of filiform prehensile tentacles, is specialized for searching out, locating, and capturing foraminifera. In *Dentalium entalis* from the Celtic Sea, the tentacles terminate in a ciliated flattened area that may function as a suction cup (Morton, 1959).

The captacula of *Dentalium entale stimpsoni,* for example, can distinguish between live foraminifera and empty tests (Bilyard, 1974). Furthermore, *Dentalium* selects calcareous over agglutinated foraminifera and apparently selects smooth rather than rough tests and large (average 150 µm) rather than small ones.

Three scaphopod species from Barkley Sound, Vancouver Island, are selective foraminiferal predators but with differing strategies (Shimek, 1990). *Dentalium rectius* is omnivorous, although foraminifera were the most numerous prey. *Cadulus aberras* preys on the abundant foraminifera *Cribrononion lene* and *Rosalina columbiensis,* whereas *Pulsellum salishorum* preys only on *Cribrononion lene.*

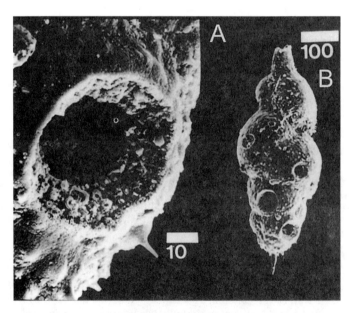

FIGURE 1. Drilled specimen of the benthic foraminifer *Siphouvigerina auberiana* from the Galapagos hydrothermal mounds. Scale bars are in micrometers. A is a closeup of the bottommost hole of specimen in B. Note multiple borings in test. (From Arnold et al., 1985.)

Deep-sea scaphopods also feed selectively on benthic foraminifera (Langer et al., 1995). *Fissidentalium megathyris* from 3000 m depth off central California contained almost exclusively foraminifera (99% of total food items). Seventeen species belonging to agglutinated, porcelaneous and hyaline groups were recorded from dissected scaphopods, but four species (one agglutinated, three hyaline) made up 87% of the ingested foraminifera. Langer et al. (1995) concluded that the primary target for the scaphopods was the foraminiferal cytoplasm.

2.2. Predators of Planktonic Foraminifera

Hemleben et al. (1989) reviewed the very sparse literature on predators of planktonic foraminifera and concluded that there are no known specialist feeders on planktonics. The gut contents of macroplankton and nekton in the Indian Ocean showed that these foraminifera are consumed by tunicates, pteropods, euphausids, and sergestids (Bradbury et al., 1970). They have also been recorded intact in the guts of sergestic prawns in the eastern North Pacific (Judkin and Fleminger, 1972) and in the fecal pellets of salps from the Gulf Stream (Bé, 1977). Crushed tests of planktonic foraminifera were found in the guts of shrimps, salps, and crabs off Baja California (Berger, 1971). Small zooplankton probably feed regularly on planktonic foraminifera.

Planktonic foraminifera may also be preyed on extensively by shallow-water suspension-feeding invertebrates. In shallow water Antarctic environments, the planktonic foraminifer *Neogloboquadrina pachyderma* occurs in the guts of three suspension-feeding polychaete species and four holothurian species, which feed for part of the year on plankton and for part on the benthos (Brand and Lipps, 1982). Hemleben et al. (1989) suggested that *N. pachyderma* had been targeted after dying and settling to the sea bed. Planktonic foraminiferal protoplasm, however, has been found in the gut of two scaphopod species from bathyal depths in the Rockall Trough (Davies, 1987, as reported by Gooday et al., 1992). The question remains as to how considerable numbers of live planktonic foraminifera came to be consumed by a bathyal deposit feeder. A possibility is that the foraminifera are transported to the sea floor in fecal pellets (Lipps, 1979) of zooplankton that fed on them in shallower water.

Hemleben et al. (1989) believed that planktonic foraminifera are probably ingested indiscriminately by filter feeders, whereas Lipps (1979) proposed that raptorial feeders also consume planktonic species. Non-spinose specimens have been observed in cultures cannibalizing smaller non-spinose specimens (Hemleben et al., 1989), but it is not known if this happens under natural conditions. Cannibalism has not been observed in spinose species (Hemleben et al., 1989).

In culture, planktonic foraminifera are attacked and consumed by unidentified ciliates (J. Eraz, pers. comm., 2002). Whether ciliates can do this in the open ocean has not been observed, although ciliates may abound there.

Lipps and Krebs (1974) noted that juvenile planktonic foraminifera inhabit the interstices of sea ice, where they may be fed upon by metazoans grazing the lower ice surfaces and the brine channels where the foraminifera lived. On the other hand, the foraminifera, like many other ice-dwelling organisms, may find the ice a refuge from more intense predation in the open waters where suspension feeding is strong.

2.3. The Effects of Predation on Foraminiferal Tests

Predation on foraminifera affects their tests in three ways: (1) they may be crushed completely, thus removing evidence of their existence; (2) they may be damaged or etched by ingestion; (3) they may be attacked individually by other small organisms that may leave marks or holes. Benthic foraminiferal tests may also pass intact through the guts of bivalves, holothurians, and echinoids, although the cytoplasm may provide nourishment (Boltovskoy and Zapata, 1980). In some cases, at least, foraminifera retract into their tests and can move about once they are released from the gut (Lipps, 1988). Various kinds of damage, both mechanical and chemical, seem to occur commonly during ingestion (Nielsen, 1999). In a study of several groups of invertebrates, Mageau and Walker (1976) described dissolution of calcareous tests, ranging from minor surface etching to total decalcification, caused by *Cancer irroratus* (rock crab), *Littorina littorea* (common periwinkle), *Aporrhais occidentalis* (American Pelican's Foot), *Asterias vulgaris* (starfish), *Strongylocentrotus drobachiensis* (sea urchin), *Molpadia musculus* (sea cucumber), and *Pectinaris hyperborea*, *Praxillella gracilis* and *Amphitrite ornata* (polychaetes). All 88 specimens, belonging to six species of foraminifera, recovered from the gizzard of the gastropod *Retusa pelyx* exhibited dissolution effects, some to the extent that identification to the species level was not possible (Burn and Bell, 1974). Dissolution of foraminiferal chamber walls takes place within the stomachs of the gastropods *Littorina* and *Puncturella* (Walker, 1971; Herbert, 1991), while partial dissolution of tests occurs within the stomach of *Olivella* (Fig. 2A), although this did not

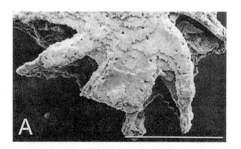
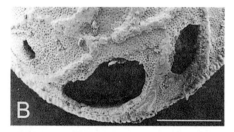

FIGURE 2. Benthic foraminifera from the gut of the neogastropod *Olivella* from Bodega Bay, northern California. (A) damaged and deeply etched test of *Elphidiella hannai*, scale bar=200 μm. (B) Punctured (and etched) test of *Glabratella ornatissima*, scale bar=200 μm. (C) scratched and etched test of *Glabratella ornatissima*, scale bar = 100μm. (From Hickman and Lipps, 1983.)

seem to be required for the snail to obtain access to living protoplasm (Hickman and Lipps, 1983). Ingestion by scaphopods also results in test dissolution and was often most progressed in specimens that previously had been mechanically damaged (Langer et al., 1995).

Mechanical damage to foraminiferal tests results from external attack (borings) or from ingestion. Borings have been attributed to either nematodes or predatory gastropods based on size. Sliter (1965, 1971) reported presumed predaceous nematodes within chambers of *Rosalina globularis* and *Bolivina doniezi* that had been partially emptied of their cytoplasm. Although Sliter did not observe attacks, he argued that "the repeated association of nematode infestation and borings in identical chambers of specimens of *Rosalina globularis* strongly suggests the borings resulted from nematode predation" (Sliter, 1971, p. 21). The borings were round to oval and occasionally roughly but incompletely beveled with a diameter ranging from 5.3 to 14.3 µm (Fig. 3A,B). Sliter (1971) used these characteristics, the oblique orientation, and lack of smooth, beveled edges to distinguish these borings from those putatively made by gastropods. Heeger (1990) and Gooday et al. (1992) also considered nematodes to be predators of benthic foraminifera. Lipps (1983, p.357), however, pointed out that "marine nematodes are not known to possess jaw mechanisms capable of boring holes in calcareous shells or tests and nematodologists G. Pointe and D.J. Raski (pers. comm.) know of no evidence that they do." Lipps (1983) suggested that these tiny borings may be made by naticid gastropods, a marine fungus, or a boring sponge.

Larger borings are made by gastropods, according to several authors. Reyment (1966) described single specimens of *Quinqueloculina* (porcelaneous) and *Spiroplectammina* (agglutinated) from the western Niger Delta, each with a single drillhole. He suggested that the steep-sided, almost cylindrical hole in *Quinqueloculina* was drilled by a muricid, whereas the larger, more conical hole in *Spiroplectammina* was the result of attack by a naticid gastropod.

Sliter (1971) questioned whether round, often multiple borings in both modern and fossil foraminiferal tests were gastropodal in origin. He noted that, although larger than the presumed nematode borings, they are still minute (18 to 58 µm in diameter), much smaller than observed gastropod borings (Fig. 3 C, D). He pointed out that newly hatched *Urosalpinx cinerea* (oyster drill), only 1.5 to 2 mm long, drilled holes 100 to 160 µm in diameter (Carriker, 1957). Further, Sliter (1971) made the pertinent point that no gastropod specimens were recovered from the modern Gulf of Mexico and Cretaceous samples that contained numerous foraminiferal tests with supposed gastropodal borings.

The multiple borings in *Siphouvigerina auberiana* from the Galapagos hydrothermal mounds (Fig. 1) were attributed by Arnold et al. (1985) to juvenile naticid gastropods. The borings averaged 34 µm in diameter and ranged from 10 µm to 125 µm in size. The multiple borings resulted from the predator maximizing the food resource by drilling more than one cytoplasm-containing chamber.

Ingestion of foraminiferal tests by macrofaunal predators results in mechanical damage during capture and processing. Berger (1971) found crushed planktonic foraminiferal tests in the guts of shrimps, salps, and crabs, but most observations are of benthic foraminifera.

The organ that harvests or processes food particles may inflict damage to the test surfaces and chambers (Hickman and Lipps, 1983). These processes occur in marginal

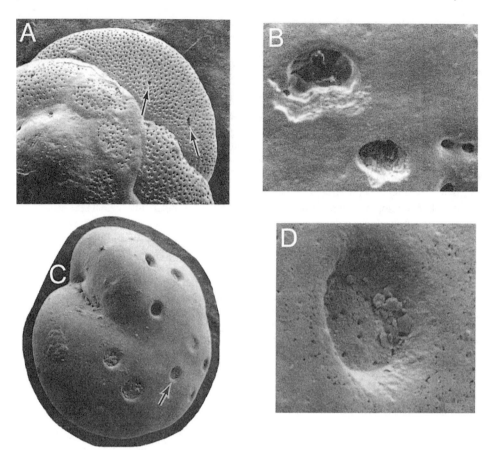

FIGURE 3. Borings in modern benthic foraminifera from Malaga Cove, California, A, B, and 512 m water depth in the Gulf of Mexico, C, D. (A) borings in final chamber of an adult specimen of *Rosalina globularis* which contained a nematode. Length of large boring 15.6 μm, diameter of small boring 8.9 μm. (B) Closeup of boring in spiral side of a juvenile specimen of *Rosalina globularis* showing oblique orientation and poorly beveled edge on one side of each hole. Maximum diameter of larger complete boring 11.1 μm, maximum diameter of smaller complete boring 8.3 μm. (C) Multiple borings in ventral side of adult specimens of *Eponides tumidulus*, maximum test diameter 258 μm. Arrow indicates subject of D. (D) Enlargement of incomplete boring showing flat bottom and slightly beveled edge; outer diameter of boring 20.4 μm, inner diameter 12.3 μm. (From Sliter, 1971.)

to deep-marine environments. In the deep sea, the guts of two asellote isopod species contained many calcareous and agglutinated foraminiferal tests that had been crushed by the predators' mandibles (Svavarsson *et al.*, 1991). On the salt marshes of Georgia, "half tests" of the agglutinated species *Arenoparella mexicana* and *Trochammina* sp. commonly occur (Goldstein and Watkins, 1999). The damage to these tests was attributed to metazoan predation, possibly by fiddler crabs. Mageau and Walker (1976) showed that crustacean predation (by rock crab) in shallow water can result in test punctures or lost chambers. Abundant scratches and deeper, parallel gouges 2 μm wide were probably caused by the grinding process within the gastric mill or by abrasion by other particles ingested with the foraminifera (Mageau and Walker, 1976). Puncture holes and peripheral test breakage are also caused by gastropodal and scaphopodal ingestion. Langer et al. (1995) found foraminifera with punctured tests, broken chambers, and surface scratches in the gut of the deep-sea scaphopod *Fissidentalium megathyris*. The radula of the gastropod *Olivella* punctured tests of *Glabratella ornatissima* (Fig. 2B) but removed segments of the periphery of chambers (often leaving only the septal regions) of *Elphidiella hannai* (Hickman and Lipps, 1983). Again, scratches on the test surface were attributed to abrasion within the gut (Fig. 2C).

Differential effects on foraminiferal tests following gastropodal ingestion were also reported by Shonman and Nybakken (1978). They found that calcareous hyaline and agglutinated tests in the digestive tract of *Acteocina culcitella* showed no sign of physical or chemical damage but calcareous porcelaneous tests were often broken. Foraminifera from the gut of the polychaete worm *Diopatra ornata* are often chipped, presumably by the worm's mandibles (Lipps and Ronan, 1974).

Vertebrates also may be capable of test destruction and damage. Coral reef fish that graze hard substrates may take foraminifera incidentally along with target organisms, but in that process tests may be marked with scratches and breakage (Lipps, 1988). Some larger foraminifera may actually be bitten, with the damaged tests later repaired by the foraminifera (Lipps, 1988).

2.4. The Effects of Predation on Foraminiferal Populations

The effects of predation on planktonic foraminiferal populations are largely unknown. Hemleben et al. (1989) stated that the impact of predation should not, however, be underestimated. The fecal pellets of salps contain mainly the shells of juvenile planktonic foraminifera (Bé, 1977). Such predation must account for a great attrition in numbers following massive gamete release, which itself may be an adaptive mechanism mitigating against the heavy predation pressure on juveniles (Hemleben et al., 1989). The effects of predation are indeed enough to reduce foraminiferal standing crops significantly off Baja California (Berger, 1971; Lipps, 1979).

Bilyard (1974) calculated that in each square meter of sea floor, containing six specimens of *Dentalium entale stimpsoni* and 2,000,000 living foraminifera in the upper 2 cm of sediment, 47,304 foraminifera would be consumed in one year (2.4% of the annual standing crop, not considering the reproductive rate). The predation pressure thus seems quite low, but the most important prey species, *Islandiella islandica,* lost 11% of its standing crop.

Although in the absence of unambiguous experimental results, Shimek (1990) concluded that scaphopod predation caused significant changes to foraminiferal population structure. Both mean size and size-frequency distributions of prey in buccal pouch contents were different from those in the sediment, probably as a result of selective removal of small prey by predators. Furthermore, species most common in the sediment (*Florilus basispinatus* and *Buliminella* spp.) were not as commonly represented in the diet, whereas the prey most commonly eaten, *Cribrononion lene*, was relatively uncommon in the sediment.

Similar significant predation pressure might be exerted by the deep-sea scaphopod *Fissidentalium megathyris* on its foraminiferal prey (Langer et al., 1995). One specimen had more foraminifera in its gut than Shimek (1990) found in shallow-water scaphopods, but data on abundance of both scaphopods and foraminifera were unavailable. Gooday et al. (1992, p. 84), in their review of deep-sea food webs, concluded that benthic foraminifera "represent an important trophic link between bacteria and detritus and higher levels of the deep-sea food chain." They did note, however, that such "speculation [is] constrained by distressingly little real information."

Brand and Lipps (1982) concluded that, in shallow Antarctic waters, ingestion of benthic foraminifera by invertebrates (either accidentally or selectively) may reduce the standing crop, as had been suggested earlier by Said (1951) for the Red Sea and Christiansen (1958) for Norwegian fjords. Brand and Lipps (1982) pointed out that this might affect paleoenvironmental interpretations. Furthermore, they noted that if samples were taken soon after an area of sea floor had been worked over by consumers, then the numbers of foraminifera found would be most likely greatly reduced. This may be an important consideration in ecologic analyses of foraminifera based on test counts.

Douglas (1981) was able to demonstrate the effect of detrital feeders on foraminiferal standing crop in sediments of the deep basins of southern California. In environments where there were large numbers of mud-eaters, such as the irregular echinoid *Brissopsis*, which contained many foraminifera in their guts, benthic foraminiferal standing crop was less than $1/cm^2$. In strong contrast, in environments where oxygen levels fell below 0.2 ml/l, most carnivores and detritus feeders were excluded, and benthic foraminiferal densities exceeded $100/cm^2$.

Caging experiments to test the hypothesis that macrofaunal predators affect foraminiferal densities were run for two years at a subtidal flat in the Indian River, Florida, by Buzas (1978, 1982). He found (1978) that foraminiferal densities were significantly higher in the sediment inside an enclosure with 1 mm openings than in the sediment outside the enclosure, presumably as a result of exclusion of the macrofauna. In an unscreened control cage foraminiferal densities were the same as in the surrounding sediment. The experiments were repeated for a further two years (Buzas, 1982) and reconfirmed that predators play an important role in regulating foraminiferal densities (Fig. 4A,B). Buzas *et al.* (1989) ran similar experiments in deeper waters (125 m) off Fort Pierce Inlet, Florida. They found, for most species, that densities were higher in screened enclosures, again indicating that predation significantly reduces foraminiferal densities.

2.5. Foraminifera as Predators

Foraminifera may be carnivorous, herbivorous, or omnivorous and feed with a wide variety of mechanisms (Lipps, 1983). Some even take up dissolved organic materials from the environment (DeLaca *et al.*, 1981). Most foraminifera, however, seem to graze on bacteria and algae (diatoms in particular), but they also may be suspension feeders or active and passive predators. Since most foraminifera, other than ones kept easily in culture, have not been observed feeding, food preferences and feeding mechanisms should not be assumed. In order to feed, foraminifera employ their pseudopodia to capture their food. Although the pseudopodia probably cannot actively catch food organisms, they are deployed by the foraminifera in arrays that prey may accidentally encounter during their own activities. The foraminifera can then capture, retain, and digest the prey.

Representatives of both the benthic and planktonic foraminifera are known to be carnivorous. Copepods are a common prey of several planktonic species, as noted by

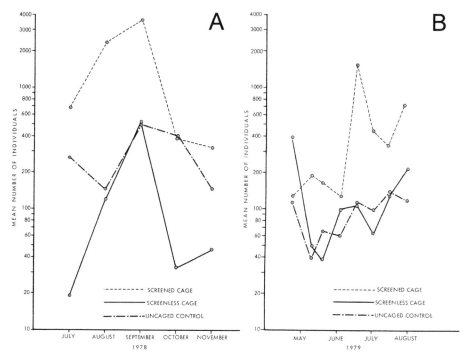

FIGURE 4. Total living benthic foraminifera in 5 ml of sediment at Link Port Florida, A, for 1978 and B, for 1979. Note greater densities in screened cage. (From Buzas, 1982.)

Rhumbler (1911) and later confirmed by laboratory observations (Anderson and Bé, 1976; Bé, 1977). The prey, once captured, is drawn into the bubble capsule surrounding the test of *Hastigerina pelagica* within ten minutes. After six to eight hours the copepod is digested and the carapace is later ejected (Bé, 1977). The entire process is well described by Bé *et al.* (1977, p. 164-165):

> When a crustacean comes into contact with *Hastigerina pelagica*'s spines, the prey is immediately and inextricably snared by the rhizopodia. The great mechanical stress to which the foraminifer is subjected and the extraordinary ability of the rhizopodia to adhere to several actively pulling crustaceans at the same time are worthy of closer observation. Despite strong efforts to escape, the prey is gradually drawn into the bubble capsule, displacing some bubbles in the process. An adhesive substance is released from vacuoles within the rhizopodia which cover the surface of the prey. The rhizopodia penetrate into crevices in the cuticle of the prey, invade the underlying soft tissue and engulf liquid droplets and small particles of tissue that are released by the dying cells. Engulfed particles of food are carried by rhizopodial streaming into the foraminiferal cytoplasm, where food vacuoles containing the engulfed prey tissue become converted into digestive vacuoles.

Although many planktonic species are omnivorous (Hemleben *et al.*, 1989), spinose species tend to take zooplankton more often than non-spinose species that are generally herbivorous (Anderson *et al.*, 1979; Spindler *et al.*, 1984). Spinose species take a wide variety of prey, including copepods (the main food source; Fig. 5), amphipods, ostracodes, crustacean larvae, tintinnids, radiolarians, polychaete and gastropod larvae, heteropods, pteropods, and tunicates (Spindler *et al.*, 1984). Individual spinose specimens capture and digest a zooplanktonic organism approximately every 24 hours. Although the tests of spinose planktonic foraminiferal species are quite small (0.5 to 1.5 mm diameter), they have spines and a rhizopodial network that give an overall diameter of 5 to 15 mm, approximately the same size as their primary prey, the copepods (Bé, 1982).

Given their differing food requirements, the abundance and distribution of non-spinose and spinose planktonic foraminifera are highly dependent on the distribution of zoo- and phytoplankton (Hemleben and Spindler, 1983). Plankton tows in the Red Sea

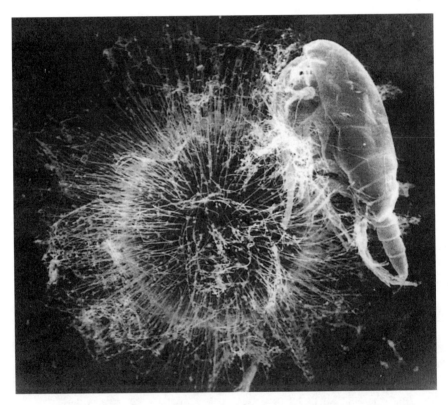

FIGURE 5. The modern spinose planktonic foraminifer *Orbulina universa* feeding on a calanoid copepod. The central calcareous test is surrounded by a network of flexible spines that are covered with sticky rhizopodial strands. Zooplankters such as copepods that inadvertently swim into the spine matrix are ensnared in rhizopodial web and become stuck to the spines. The prey is slowly transported down the spines to the surface of the shell via streaming cytoplasm. Enzymes associated with the rhizopodial web gradually break down the prey tissue, which is then transported through the apertures in the test to the internal body of show a cytoplasm. The food is then phagocytosed into vacuoles and subsequently digested. Magnification unknown. (Courtesy of H. J. Spero, University of California, Davis).

show a boundary near 20°N (which can also be recognized in Pleistocene cores) that coincides with a spring abundance break of phytoplankton (Tardent, 1979). Spinose and non-spinose species occur to the south of the boundary, whereas generally spinose species occur to the north (Zobel, 1973).

Some benthic foraminifera have been known to be carnivores since Schultze (1854) figured and described prey trapped within the reticulopodial network of a foraminiferal predator. The prevalence of carnivory is unclear, however, although it has been reported widely (Sandon, 1932; Boltovskoy and Wright, 1976; Lipps, 1983; Murray, 1991; Goldstein, 1999). A recent review of the mode of prey capture led Langer and Bell (1995) to conclude that carnivores may be more prevalent than previously considered.

Carnivorous benthic foraminifera have not been observed to stalk their prey. Rather, contact with prey appears to be haphazard (Lipps, 1983), with the prey moving into the pseudopodia. Whether these encounters are mediated by the foraminiferan through some attractant or are simply accidental has not been determined. Prey items of various kinds are captured by the pseudopodia, which may have adhesive properties, and the prey soon expires, seemingly from exhaustion (Fig. 6). The prey is then transported near or into the test and digested, leaving a feeding cyst of the undigested material (Winter, 1907).

The strategies involved in capture of prey are many and varied and have been wonderfully described by Christiansen (1971). One undescribed species inhabits empty tests of another benthic foraminifer, *Saccammina*. From its lair, it sends out long, narrow pseudopodia into adjacent foraminiferal tests where protoplasm of the prey is digested in situ. Another species, *Spiculosiphon radiata*, also feeds only on other foraminifera. Its test is composed of sponge spicules but empty tests of calcareous foraminifera are also attached. These are the remains of prey encountered while the predators were moving over the substrate on leg-like deployments of pseudopodial threads. Christiansen (1964) observed that, in one instance, the time from capture to incorporation of the prey into the predator's test was only 20 minutes. As a final example, *Pilulina argentea* constructs a dwelling space just beneath the sediment surface. When a copepod lands in the vicinity, it is captured by pseudopods that draw the victim into the space under the sediment surface, where it is consumed. The undigested carapace is later ejected from the predator's abode (Christiansen, 1971).

Similar predaceous behavior was reported by Buchanan and Hedley (1960; Hedley, 1964) for the large (1 cm diameter) agglutinated foraminifer, *Astrorhiza limicola*, that extends a pseudopodial network over the surface of a muddy substrate. When copepods, nematodes or echinoderms come in contact with the pseudopods, they struggle, weaken, and eventually die. This species also extends pseudopods throughout the interstitial spaces of sediment where they capture interstitial organisms moving between particles (Buchanan and Hedley, 1960). Lipps (1983) considered this to be an example of interstitial suspension feeding. The observations by Buchanan and Hedley (1960) were of specimens in a laboratory setting. Cedhagen (1988) observed the same species in its natural environment and concluded that not only did *A. limicola* orient itself differently on the sediment (vertically rather than horizontally) but it also was sessile and fed on suspended particles and surface deposits. Cedhagen (1988) went further in stating that Thorsen's (1966) idea of great numbers of *A. limicola* constituting a danger zone for settling larvae in northwest European waters may be an exaggeration of their true impact.

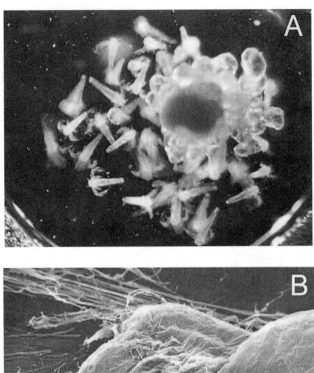

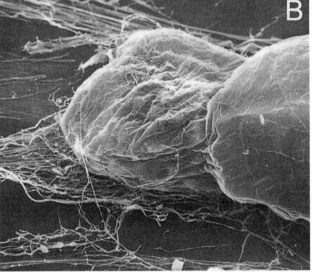

FIGURE 6. (A) Dark field survey micrograph of an isolated *Astrammina rara* cell body surrounded by numerous captured *Artemia*, scale bar = 1 mm. (B) Abdominal region of an *Artemia* (right) and adherent pseudopodial network of *A. rara*, scale bar = 10 μm. (From Bowser *et al.*, 1992.)

In Antarctica shallow waters, the larger (1 cm diameter) agglutinated foraminifera *Astrammina rara*, occurring in abundance of up to 1800 individuals per square meter, not only consumes bacteria but also the larvae, juveniles and adults of small crustacea, molluscs, echinoderms, polychaetes and nematodes (Fig. 6). Some prey items were six times larger than the predator that captured and consumed them (DeLaca, 1985, 1986).

Further observation showed that Antarctic *Astrammina rara* may occasionally abandon its agglutinated test (Bowser *et al.*, 1996). The reason for this is unclear but perhaps related to differing feeding strategies. Bowser *et al.* (1995) speculated that it might be advantageous for *A. rara* to occupy a test while functioning as an infaunal, passive predator. During the Austral summer, however, when algal food is abundant, a bulky test might hinder epifaunal algal grazing. Bowser *et al.* (1986, 1992) demonstrated in laboratory observation and experimentation that *Astrammina rara* is able to capture and hold large, vigorously struggling prey by virtue of the extensive, coiled microtubule cytoskeleton and fibrous extracellular matrix, which enhance the tensile strength or provide elastic properties to the pseudopodia.

How do planktonic and benthic foraminifera manage to capture vigorously mobile metazoan prey, often much larger than the predatory foraminifer? We have noted that at least one species, *Astrammina rara*, is characterized by pseudopods with an extensive coiled microtubule system and a fibrous extracellular matrix (Bowser *et al.*, 1986, 1992). Pseudopods were also suggested to be toxic for smaller crustaceans (Winter, 1907). Langer and Bell (1995) reviewed the evidence and determined that there is minimal evidence for foraminiferal toxicity. They inferred (p. 205) that "foraminiferal prey capture is in large part mediated by molecular cell-surface receptors (e.g., carbohydrates, glycoproteins) present in the extracellular matrix of adhesive pseudopodia."

Recent observations show that a record of foraminiferal predatory behavior, even on other foraminifera, can be preserved. Todd (1965) described a species of *Rosalina* that etched a hole in the shell of a bivalve under its test. She proposed that the *Rosalina* was parasitic and preyed on the bivalve mantle it encountered after penetrating the shell. *Hyrrokkin sarcophaga* is another foraminifer that parasitizes bivalves, sponges and corals off northwest Europe (Cedhagen, 1994). On the bivalve *Arcesta excavata*, *H. sarcophaga* corrodes a pit in the shell, slightly larger than the foraminiferan's test. A canal extends from the pit through the bivalve's shell. Pseudopodia are deployed through the canal, penetrating the mantle of the bivalve. While this parasitic activity is ongoing, *H. sarcophaga* often attacks and feeds on epifaunal organisms (polychaetes and bryozoans) on the surface of the bivalve shell. Bryozoan colonies were observed to reverse their direction of growth to escape the predator's advances (Cedhagen, 1994).

Floresina amphiphaga is predatory on the larger foraminiferan, *Amphistegina gibbosa* (Hallock and Talge, 1994; Hallock *et al.*, 1998). It climbs onto its host, attaches, and drills (the method is unknown) up to 10 holes per site (Fig. 7). Several attachment sites can occur on one host, which is killed over several days, presumably by consumption of cytoplasm through the holes. Adult specimens were preferentially chosen for attack by *F. amphiphaga* (Hallock and Talge, 1994). Morphologically similar borings on a specimen of the planktonic foraminifer *Globorotalia menardii* from Fiji have been attributed to scavenging behavior by *Floresina* (Nielsen *et al.*, 2002).

3. Predation in Fossil Foraminifera

We know from the above review that modern foraminiferal tests often show the effects of predation, such as mechanical breakage, gouges, borings and chemical etching.

 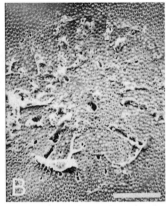 

FIGURE 7. *Floresina amphiphaga* from the Florida Reef Tract, Florida. (A) specimen detached from a larger specimen of *Amphistegina gibbosa* showing holes in the test of prey caused by two separate attachments, scale bar = 100 μm. (B) closeup of the attachment site from which a predator was removed, scale bar = 50 μm. (C) closeup of the central hole in B, drilled into the *A. gibbosa* test by *F. amphiphaga*, scale bar = 5 μm. (From Hallock and Talge, 1994.)

Similarly, foraminiferal predators can leave marks of their activity on the shell of prey material (e.g., pits and holes). Unfortunately, although the literature on fossil foraminifera is vast, there are few mentions of predation and its effects.

The oldest evidence of predation on foraminifera seems to be from the Upper Pennsylvanian of Kansas (Hageman and Kaesler, 2002), where fusulinids exhibit several kinds of predation damage. Some have damage similar to the possible bite mark (Fig. 8) on the Recent foraminifera *Alveolinella quoyii* (Lipps, 1988). Most fusulinid tests were not repaired, suggesting that predation usually resulted in death. Some individuals, however, repaired their tests and continued to grow with malformed tests. Wilde (1965) briefly reviewed reports of teratoid fusulinids; some specimens resulted from probable damage by predators.

 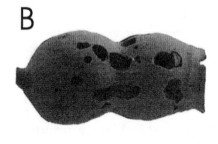

FIGURE 8. *Alveolinella quoyii* test from the Papuan Lagoon, Papua New Guinea, cut by teeth of a fish or the claw of a crab and regenerated. Scale bars = 0.5 mm. (From Lipps, 1988.)

Livan (1937) described and illustrated small, circular to oval, multiple borings in the test of several species of hyaline benthic foraminifera from Lower to Middle Jurassic strata of Germany. She did not know what had made the holes but referred to Rhumbler (1911), who considered similar holes in foraminifera to have been the result, perhaps, of the attentions of juvenile predatory snails. Herrero and Canales (2002) illustrated similar borings as well as test breakage attributed to predators in Lower to Middle Jurassic foraminifera from Spain.

A juvenile gastropod predator (a naticid) was also considered to be responsible for borings (20 µm to 40 µm in diameter) in several genera of benthic foraminifera from Upper Cretaceous strata of Poland (Pozaryska, 1957). Many specimens exhibited multiple drillings, in some cases with one borehole per chamber, closely reminiscent of the pattern of boreholes observed by Arnold *et al.* (1985) in modern deep-sea foraminifera.

Borings in Miocene benthic foraminifera were also attributed to naticid gastropods (Vialov and Kantolinskaja, 1968). Both single and multiple borings were illustrated, with holes ranging from 20 µm to 100 µm in diameter. Smaller borings (11 to 14 µm in diameter) were observed by Collen (1973) in Pliocene and Pleistocene specimens of *Notorotalia*. Most holes were located on sutural pores, had beveled edges, and were attributed to gastropod predation, although Collen (1973) recognized that the small size was unusual for holes drilled by predatory gastropods. Even smaller holes (1.5 to 4 µm in diameter) were common but were not attributed to a specific kind of predator.

Borings in Upper Cretaceous benthic foraminifera from neritic and bathyal paleoenvironments were compared to modern borings (see discussion above) and attributed to nematodes and gastropods (Sliter, 1971). Sliter argued strongly that irregularly round to oval borings from 10 to 31 µm in diameter, oriented at oblique angles to the surface, were attributable to predatory nematodes. Larger (18 to 58 µm), rounded, more numerous borings oriented perpendicular to the test surface and occasionally exhibiting a beveled exterior edge were only tentatively attributed to gastropods due to their small size (cf., Carriker, 1957) and multiple habit, atypical of gastropodal boring, according to Carriker and Yochelson (1968). Perhaps most tellingly, no fossil gastropods were found in the samples with bored foraminiferal tests. Sliter (1971, p. 25) concluded that "evidence of gastropod predation is incomplete." In a later paper, Sliter (1975, p. 901-902) suggested that the larger population of borings was "probably the result of a soft-bodied organism, possibly a nematode or polychaete."

Preferential predation was indicated by the pattern of borings (Sliter, 1975). From 5 to 9% of the total population was bored. Within an assemblage, up to 8 species of a genus are bored and some specimens have as many as 13 borings. Specimens of *Lenticulina muensteri*, *Praebulimina aspera*, *Pleurostomella subnodosa* and *Hoeglundina supracretacea* comprised the major prey items. Sliter (1975) suggested that perhaps infaunal species might be less accessible to predators. In laminated facies there were few bored foraminifera, probably due to the elimination of predators by dysoxic conditions.

Douglas (1973) recorded Miocene deep-water benthic foraminifera with borings similar to those recorded by Sliter (1971) in Upper Cretaceous foraminifera. The size and shape of the borings (Fig. 9) led Douglas (1973) to conclude that predatory nematodes were the culprits. Although, in some samples, most tests were bored or otherwise damaged by predation, in others differential predation was evident. *Bolivina* and *Coryphostoma* species were often bored, as was *Laticarinina*. In contrast,

Cibicidoides was the only calcareous genus that was not bored, perhaps because its large pores make it unnecessary for any predator to drill holes (Douglas, 1973). The benthic agglutinated foraminifera, *Vulvulina*, *Bolivinopsis*, and *Karreriella*, were rarely bored, but the similarly shaped taxa *Textularia* and *Dorothia* were bored as often as most calcareous taxa (Douglas, 1973).

4. Paleobiological Implications of Foraminiferal Predation

4.1. Predatory Activities Recorded in the Fossil Record

As reviewed above, little information on predation on foraminifera in the fossil record exists. The few records range back to the Pennsylvanian and Permian (Wilde, 1965; Hageman and Kaesler, 2002). The great abundance of foraminifera suggests, however, that they were just as important a food source throughout most of the Phanerozoic as they are today. Although foraminifera are recorded in Lower Cambrian rocks (Glaessner, 1978; Culver, 1991; Lipps, 1992; McIlroy *et al.*, 1994; Lipps and Rosanov, 1996), their occurrence is quite patchy (either due to relatively low numbers or lack of preservation or both) until the Silurian and Devonian (e.g., Kircher and Brasier, 1989). Benthic foraminifera were probably an important food resource in peri- and epicontinental sea environments from that time to the end-Paleozoic extinction. We know little about deep-sea occurrences in the Paleozoic. Increasingly through the Mesozoic, foraminifera lived in deeper water environments until, by the Late Cretaceous, benthic foraminifera are found in all continental margin environments and depth zones (Sliter, 1972; Olsson and Nyong, 1984).

Planktonic foraminifera do not appear in the fossil record until the Middle Jurassic (Bajocian) (Simmons *et al.*, 1997). By the Hauterivian they had become holoplanktonic, abundant, and widespread (Banner and Lowry, 1985) and so had become important components of planktonic ecosystems. Indeed, the Mesozoic marks the appearance of increased diversity of microplankton of all kinds. Algal groups such as calcareous nannoplankton, diatoms, and silicoflagellates, first appeared, and preexisting groups, such as dinoflagellates and acritarchs, diversified as well. So did the microzooplankton such as foraminifera, radiolaria, and tintinnids. This diversification may have come

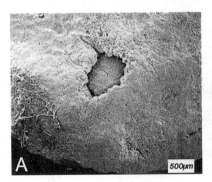 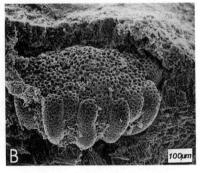

FIGURE 9. Holes bored in the test of Miocene foraminifera. (A) *Siphonodosaria abyssorum*, x 50. (B) boring on upper right of specimen in A, showing partial cutting of a single oval piece, suggesting borings were cut and not ground, x 500. (From Douglas, 1973.)

about by new oceanographic structures, but clearly the trophic relationships in the Mesozoic pelagic environments changed. Planktonic foraminifera were abundant and available as potential food resources to the many larger zooplankton that were likely present but left no fossil record.

Some predators probably were specialized feeders on foraminifera in the past, just as they are today. The large number of metazoan groups that feed today on foraminifera suggests that similar kinds of predation pressure were exerted on foraminifera since they first became abundant. But the presence in the fossil record of abundant metazoans, such as scaphopods, that eat foraminifera would indicate the presence of abundant benthic foraminifera. Similarly, even though some modern gastropods (particularly juveniles of some species) are specialized predators of benthic foraminifera, there is no direct evidence of such trophic interrelationships in the fossil record, other than a few scattered references to drilled foraminiferal tests.

Many benthic foraminifera can be bioeroders (Vénec-Peyré, 1996), and some of their bioeroding activities, preservable in the fossil record, may be related to predation by metazoans. Smyth (1988) described a hyaline foraminifer, *Cymbaloporella tabellaeformis*, from Guam that bores into the outer surface of 25 gastropod species, possibly by chemical dissolution (Fig. 10). Smyth (1988) speculated that, in addition to gaining a feeding advantage by being attached to a mobile substrate, the pit that the foraminifer inhabited may have provided protection from predators. A similar conclusion was reached by Plewes *et al.* (1993), who described small, rosette-shaped borings in skeletal substrates of the Upper Jurassic Oxford Clay and Kimmeridge Clay of England and France. They suggested that the borings, protected by agglutinated collars, were made by foraminifera primarily for protection.

Bioeroding foraminifera are quite well known but the purposes of their activities are not always clear, although they can probably fall into three categories: feeding, protection, and test-building (Vénec-Peyré, 1996). Some foraminifera that make holes or inhabit pits in the shells of other foraminifera or of larger invertebrates have been described as parasites (e.g., Todd, 1965; Banner, 1971; Alexander and DeLaca, 1987; Cherchi and Schroeder, 1991), but it may well be that protection from predation is part of the advantage gained by living on or in a hard and often mobile substrate.

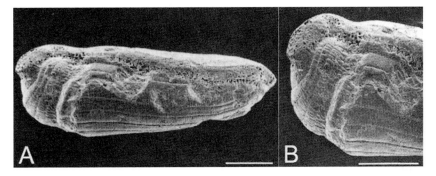

Figure 10. The benthic foraminifer *Cymbaloporella tabellaeformis* in situ in a gastropod shell from the coastal waters of Guam. (A) *C. tabellaeformis* sunken into its boring; channel extending laterally into molluscan shell may represent path of a pseudopodium. (B) molluscan shell broken away revealing peripheral chambers and upper surface of test which appears to have grown in diameter since initial penetration. (From Smyth, 1988.)

4.2. Taphonomic Effects of Foraminiferal Predation

Foraminiferal assemblages undoubtedly have undergone taphonomic change as they fossilize because of predatory activities in life. Three kinds of taphonomic loss are possible: (1) Extreme predation on foraminiferal populations may result, as noted above, in the complete removal through destruction of identifiable tests from the habitat, leaving no or an incomplete fossil record. (2) Selective predation that results in test destruction may bias the fossil record so that some species are relatively more abundant as fossils than they were in the living assemblages. (3) Tests, damaged by drilling or abrasion, may be selectively removed from the fossil record by diagenetic processes.

As noted above, predation pressure can alter the population densities of both benthic and planktonic foraminifera (Buzas, 1982; Hemleben *et al.*, 1989). This kind of total or partial destruction of foraminifera may mean that meaningful paleoecologic analyses of relative abundance cannot be made under these circumstances. An alternative approach might be to examine strata above and below the horizon of interest or laterally along it in order to make some inference about the nature of the assemblage that might have been present.

If predation is selective, for example, on only calcareous benthic foraminifera, then if tests are destroyed during the predation process the resulting fossil assemblage will have a composition that does not reflect accurately the original community. Similar taphonomic loss of information may occur if the predator is selective of particular taxa. Bilyard (1974) estimated the predation pressure of the scaphopod *Dentalium entale stimpsoni*. He found that agglutinated tests with much organic matter (e.g., tests of *Cribostomoides* spp. and *Eggerella advena*) were preferred over those of *Reophax atlantica*, which have tests of coarse sand with little organic matter. Among the calcareous foraminifera, two taxa with smooth walls (*Quinqueloculina* spp. and *Islandiella islandica*) were selected over less smooth taxa.

Damaged tests may be selectively eliminated from the fossil record. Douglas (1973) noted that broken foraminiferal tests increased rapidly down the Neogene section of DSDP boreholes but he believed that below Eocene strata, in firm chalks and limestones, bored shells "had been preferentially destroyed by diagenetic processes" (1973, p. 610). He argued that the number and nature of predatory borings reduced mechanical strength and increased dissolution. Since the bored and damaged tests may also be limited to certain taxa, the resulting fossil record may misrepresent the original ecologic assemblage. This problem may be dealt with partially by presence-absence data, as described above.

The pattern of taphonomic loss undoubtedly varies from environment to environment. Sliter (1971) showed an increased frequency of bored tests in neritic compared to bathyal Upper Cretaceous strata of California. Thus, foraminiferal assemblages in neritic strata, at least in this instance, would have been more affected by taphonomic loss of tests that those of strata deposited in deeper water.

The destructive effects of predation and hence the nature of the fossil record can be related to foraminiferal wall type, but, unfortunately, not in any consistent manner. For example, Boltovskoy and Zapata (1980) observed no damage to foraminiferal tests of different types after passing through the alimentary canals of bivalves, holothurians, and echinoids. But, in contrast, Shonman and Nybakken (1978) showed that tests of miliolid (porcelaneous) foraminifera were often broken, whereas hyaline and agglutinated tests were not broken after ingestion by deposit-feeding gastropods.

Another pattern, described by Mageau and Walker (1976), is the damage of calcareous but not agglutinated wall types. Both porcelaneous and hyaline foraminifera were damaged through ingestion by various invertebrates, but agglutinated foraminifera were not. Agglutinated foraminifera, however, do not escape the taphonomic effects of predation, as clearly indicated by agglutinated tests drilled with boreholes (Sliter, 1975) and cut in half, presumably by fiddler crab claws (Goldstein and Watkins, 1999).

5. Future Research

This review shows that we have few data documenting the interactions of foraminifera and their predators and prey in the fossil record. We can do little more than speculate that, for example, an abundance of fossil scaphopods may indicate the presence of their specialized prey, benthic foraminifera. We can speculate that borings made in fossil foraminiferal tests were made by juvenile naticid gastropods, but there are questions as to whether similar tiny holes in modern bored foraminiferal tests are made by gastropods. We can similarly speculate that even smaller, irregular holes in fossil foraminifer tests were made by predatory nematodes, but we know of no living nematodes that can drill such holes.

We can say with certainty that the actions of predators can and do affect standing crops of planktonic and benthic foraminifera today and must have done so also in the past. But there are few quantitative data on the scale of this predation pressure. Indeed, most comments are anecdotal, except for the predator exclusion experiments in Florida (Buzas, 1978, 1982; Buzas et al., 1989).

We can also say with certainty that the taphonomic effects of predation are considerable but that information loss varies in its nature from time to time and place to place in patterns that are not (as yet) predictable.

Neither is there any evidence about the evolutionary effects of predation on foraminifera. In other groups, the coevolution of predators and prey have driven certain evolutionary trends through time (Vermeij, 1987, 1994), but such trends cannot yet be deciphered for either planktonic or benthic foraminifera. Such trends likely exist but must await detailed study.

In short, we know little of the interrelationships of foraminiferal predators and their prey or of foraminiferal prey and their predators. Perhaps the two most fruitful areas of future research would be a renewed emphasis on experimentation and observation of living systems, and a concerted effort to document the fossil record of predation in foraminifera by the evidence of boring, gouges, breakage, and etching left on their tests and the comparison of co-occurring metazoan fossils to modern foraminiferal predators. The paleobiology of foraminiferal trophic relationships remains a serious topic for additional work.

6. Summary

Foraminifera are important trophic links between the micro- and meiobenthic and planktonic biota and the megabiota simply because of their abundance and diversity in nearly all marine habitats. They are prey for a varied assortment of other protozoans, invertebrates, and vertebrates, both incidentally and purposefully. In particular, grazing invertebrates, such as echinoids and gastropods, and vertebrates, such as fish, take numerous benthic foraminifera. Selective predation on foraminifera occurs chiefly among small gastropods and scaphopods, some of which even select specific test compositions or shapes to consume. Juvenile invertebrates take a larger proportion of foraminifera than their adults do, presumably because of size selectivity. Planktonic foraminifera are preyed on by ciliates and a wide variety of zooplankton. No precise measure of the contribution of foraminifera to the trophic resources of any environment has been recorded, but the consumption of foraminifera surely affects their own reproductive and ecologic strategies.

Foraminifera themselves feed in as many ways as most diverse invertebrate groups do. They are omnivorous, herbivorous, and carnivorous and use feeding mechanisms as varied as simple grazing and complex suspension feeding. They prey on bacteria, on other protozoans including other species and their own species of foraminifera, and on smaller invertebrates that can be encountered and entrapped by their pseudopodia.

Fossil foraminifera must have been preyed on as well, but documentation remains fragmentary. The effects of foraminifera as predators in the geologic past is also unclear. The paleobiologic, paleoecologic, taphonomic, and evolutionary significance of the trophic roles of foraminifera remains largely unknown, although studies of recent foraminiferal relationships suggest their likely major influence in the past.

ACKNOWLEDGMENTS: We thank Dare Merritt and David Lance for their technical help and the various colleagues who kindly allowed us to use their illustrations. This is UC Museum of Paleontology contribution number 1760.

References

Alexander, S. P., and DeLaca, T. E., 1987, Feeding adaptation of the foraminiferan *Cibicides refulgens* living epizoically and parasitically on the Antarctic scallop *Adamussium colbecki*, *Biol. Bull.* **173**:136-159.
Anderson, O. R., and Bé, A. W. H., 1976, A cytochemical fine structure study of phagotrophy in a planktonic foraminifer, *Hastigerina pelagica* (d'Orbigny), *Biol. Bull.* **151**:437-449.
Anderson, O. R., Spindler, M., Bé, A. W. H., and Hemleben, Ch., 1979, Trophic activity of planktonic foraminifera, *J. Mar. Biol. Assoc., U.K.* **59**:791-799.
Arnold, A. J., D'Escrivan, F., and Parker, W. C., 1985, Predation and avoidance responses in the foraminifera of the Galapagos hydrothermal mounds, *J. Foram. Res.* **15**:38-42.
Bacescu, M., and Caraion, F. E., 1956, Animale mincatoare de Foraminifere, *Comm. Acad. Republic P. Rumania* **VI**:551-553.
Banner F. T., 1971, A new genus of the Planorbulinidae an endoparasite of another foraminifer, *Rev. Esp. Micropaleontol.* **3**:113-128.
Banner, F. T., and Lowry, F. D. M., 1985, The stratigraphical record of planktonic foraminifera and its evolutionary implications, *Spec. Papers Palaeontol.* **33**:117-130.
Bé, A. W. H., 1977, An ecological, zoogeographic and taxonomic review of recent planktonic foraminifera, in: *Oceanic Micropaleontology* (A. T. S. Ramsey, ed.), Academic Press, London, pp. 1-100.
Bé, A. W. H., 1982, Biology of planktonic foraminifera, in: *Foraminifera: Notes for a Short Course* (T. W. Broadhead, ed.), University of Tennessee, Department of Geological Sciences, Studies in Geology 6, pp. 51-92.
Bé, A. W. H., and Hamlin, W. H., 1967, Ecology of Recent planktonic foraminifera. Part 3-Distribution in the North Atlantic during the summer of 1962, *Micropaleontology* **13**:87-106.
Bé, A. W. H., Hemleben, Ch., Anderson, O. R., Spindler, M., Hacunda, J., and Tuntivate-Choy, S., 1977, Laboratory and field observations of living planktonic foraminifera, *Micropaleontology* **23**:155-179.

Berger, W. H., 1971, Planktonic foraminifera: sediment production in an oceanic front, *J. Foram. Res.* **1**:95-118.
Bilyard, G. R., 1974, The feeding habits and ecology of *Dentalium entale stimpsoni* Henderson, *Veliger* **17**:126-138.
Boltovskoy, E., and Wright, R., 1976, *Recent Foraminifera*, Dr. W. Junk b.v., The Hague.
Boltovskoy, E., and Zapata, A., 1980, Foraminiferos bentonicos como alimento de otros organismsos, *Rev. Esp. Micropaleontol.* **12**:191-198.
Bowser, S. S., Alexander, S. P., Stockton, W. L., and DeLaca, T. E., 1992, Extracellular matrix augments mechanical properties of pseudopodia in the carnivorous foraminiferan *Astrammina rara*: role in prey capture *J. Protozool.* **39**:724-732.
Bowser, S. S., DeLaca, T. E., and Reider, C. L., 1986, Novel extracellular matrix and microtubule cables associated with pseudopodia of *Astrammina rara*, a carnivorous Antarctic foraminifer, *J. Ultrastructure Molecular Structure Res.* **94**:149-166.
Bowser, S. S., Gooday, A. J., Alexander, S. P., and Bernhard, J. M., 1995, Larger agglutinated foraminifera of McMurdo Sound, Antarctica: are *Astrammina rara* and *Notodendroides antarctikos* allogromiids incognito?, *Mar. Micropaleontol.* **26**:75-88.
Bradbury, M. G., Abbott, D. P., Bovbjerg, R. V., Mariscal, R. N., Fielding, W. C., Barber, R. T., Pearse, V. B., Proctor, S. J., Ogden, J. C., Wourms, L. R., Taylor, L. R., Jr., Christofferson, J. G., Christofferson, J. P., McPhearson, R. M., Wynne, M. J., and Stromberg Tr, P. M., 1970, Studies in the fauna associated with the deep scattering layers in the equatorial Indian Ocean, conducted on R/V Te Vega during October and December 1964, in: *Proceedings of an International Symposium on Biological Sound Scattering in the Ocean* (G.B. Farquhor, ed.), pp. 409-452.
Brand, T. E., and Lipps, J. H., 1982, Foraminifera in the trophic structure of shallow-water Antarctic marine communities, *J. Foram. Res.* **12**: 96-104.
Brown, H. H., 1934, A study of a tectibranch gastropod mollusc, *Philene aperta* (L.), *Trans. R. Soc. Edinburg* **58**:179-210.
Buchanan, J. B., and Hedley, R. H., 1960, A contribution to the biology of *Astrorhiza limicola* (Foraminifera), *J. Mar. Biol. Assoc. U.K.* **39**:549-560.
Burn, R., and Bell, K. N., 1974, Description of *Retusa pelyx* Burn sp. nov. (Opisthobranchia) and its food resources from Swan Bay, Victoria, *J. Malacol. Soc. Antarctica* **3**:37-42.
Buzas, M. A., 1978, Foraminifera as prey for benthic deposit feeders: results of predator exclusion experiments, *J. Mar. Res.* **36**:617-625.
Buzas, M. A., 1982, Regulation of foraminiferal densities by predation in the Indian River, Florida, *J. Foram. Res.* **12**:66-71.
Buzas, M. A., Collins, L. S., Richardson, S. L., and Severin, K. P., 1989, Experiments on predation, substrate preference, and colonization of benthic foraminifera at the shelfbreak off the Ft. Pierce Inlet, Florida, *J. Foram. Res.* **19**:146-152.
Carriker, M. R., 1957, Preliminary study of behavior of newly hatched oyster drills, *Urosalpinx cinerea* (Say), *Elisha Mitchell Scientific Soc. J.* **73**:328-351.
Carriker, M. R., Yochelson, E. L., 1968, Recent gastropod boreholes and Ordovician cylindrical boreholes, *U.S. Geol. Survey Prof. Paper* **593-B**: B1-B26.
Cedhagen, T., 1988, Position in the sediment and feeding of *Astrorhiza limicola* Sandahl, 1857 (Foraminiferida), *Sarsia* **73**:43-47.
Cedhagen, T., 1994, Taxonomy and biology of *Hyrrokkin sarcophaga* gen. et sp. n., a parasitic foraminiferan (Rosalinidae), *Sarsia* **79**:65-82.
Cherchi, A., and Schroeder, R., 1991, Perforations branchues dues à des Foraminiféres cryptobiotiques dans des coquilles actuelles et fossiles, *C. R. Acad. Sci. Paris* **312**:111-115.
Christiansen, B. O., 1958, The foraminifer fauna in the Drobak Sound in the Oslo Fjord (Norway), *Nyutt Mag. Zool.* **6**:5-91.
Christiansen, B. O., 1964, *Spiculosiphon radiata*, a new Foraminifera from Northern Norway, *Astarte* **25**:1-8.
Christiansen, B. O., 1971, Notes on the biology of Foraminifera, *Vie et Milieu, Trois. Symp. Européen Biol. Mar., Suppl.* **22**:465-478.
Clark, W., 1849, On the animal of *Dentalium tarentinum*, *Ann. Mag. Nat. Hist.* **2**:321-330.
Collen, J. D., 1973, Morphology and development of the test surface in some species of *Notorotalia* (Foraminiferida), *Rev. Esp. Micropaleontol.* **5**:113-132.
Culver, S. J., 1991, Early Cambrian Foraminifera from West Africa, *Science* **254**:689-691.
Daniels, R. A., and Lipps, J. H., 1974, Predation on foraminifera by Antarctic fish, *J. Foram. Res.* **8**: 110-113.
Davies, G. J., 1987, Aspects of the biology and ecology of deep-sea Scaphopoda. Unpublished Ph.D. Thesis, Heriot-Watt University.
DeLaca, T. E., 1985, Trophic position of benthic rhizopods in McMurdo Sound, *Antarctic J., U.S.* **20**:147-149.

DeLaca, T. E., 1986, The morphology and ecology of *Astrammina rara*, *J. Foram. Res.* **16**: 216-223.
DeLaca, T. E., Karl, D. M., and Lipps, J. H., 1981, Direct use of dissolved organic carbon by agglutinated benthic foraminifera, *Nature* **289**:287-289.
Douglas, R. G., 1973, Benthonic foraminiferal biostratigraphy in the central North Pacific, Leg 17, Deep Sea Drilling Project, *Init. Rep. Deep Sea Drilling Proj.* **17**:607-671.
Douglas, R. G., 1981, Paleoecology of continental margin basins: a modern case study from the borderland of Southern California, in: *Depositional Systems of Active Continental Margin Basins: Short Course Notes* (R. G. Douglas, I. P. Colburn, and D. S. Gorsline, eds.), Pacific Section, SEPM, pp.121-156.
Glaessner, M. F., 1978, The oldest foraminifera, in: *The Crespin Volume: Essays in Honour of Irene Crespin* (D.J. Belford and V. Scheibernova, eds.), Bureau of Mineral Resources, Geology and Geophysics, **192**:61-65.
Goldstein, S. T., 1999, Foraminifera: a biological overview, in: *Modern Foraminifera* (B. K., Sen Gupta, ed.), Kluwer, Dordrecht, pp. 37-55.
Goldstein, S. T., and Watkins, G. T., 1999, Taphonomy of saltmarsh foraminifera: an example from coastal Georgia, *Palaeogeogr. Palaeoclim. Palaeoecol.* **149**:103-114.
Gooday, A. J., Levin, L. A., Linke, P., and Heeger, T., 1992, The role of benthic foraminifera in deep-sea food webs and carbon cycling, in: *Deep-Sea Food Chains and the Global Carbon Cycle* (G. T. Rowe and V. Pariente, eds.), Kluwer, Dordrecht, pp. 63-91.
Hageman, S. A., and Kaesler, R. L., 2002, Fusulinids: predation damage and repair of tests from the Upper Pennsylvanian of Kansas, *J. Paleontol.* **76**:181-184.
Hallock, P., and Talge, H. K., 1994, A predatory foraminifer, *Floresina amphiphaga*, n. sp., from the Florida Keys, *J. Foram. Res.* **24**:210-213.
Hallock, P., Talge, H. K., Williams, D. E., and Harney, J. N., 1998, Borings in *Amphistegina* (Foraminifera): evidence of predation by *Floresina amphiphaga* (Foraminifera), *Hist. Biol.* **13**:73-76.
Hedley, R. H., 1964, The biology of foraminifera, in: *International Review of General and Experimental Zoology* (W. J. L. Felts and R. J. Harrison, eds.), Academic Press, New York and London, pp. 1-45.
Heeger, T., 1990, Electronenmikroskopische Untersuchungen zur Ernährungsbiologie benthischer Foraminiferen, *Ber. Sonderforschungsbereich* 313, **31**:1-139.
Hemleben, Ch., and Spindler, M., 1983, Recent advances in research on living planktonic foraminifera, *Utrecht Micropaleontol. Bull.* **30**:141-170.
Hemleben, Ch., Spindler, M., and Anderson, O. R., 1989, *Modern Planktonic Foraminifera*, Springer-Verlag, New York.
Herbert, D. G., 1991, Foraminiferivory in a *Puncturella* (Gastropoda; Fissurellidae), *J. Moll. Stud.* **57**:127-129.
Heron-Allen, E., 1915, Contributions to the study of the bionomics and reproductive processes of the Foraminifera, *Phil. Trans. R. Soc. B* **206**:227-279.
Herrero, C., and Canales, M. L., 2002, Taphonomic processes in selected Lower and Middle Jurassic Foraminifera from the Iberian Range and Basque-Cantabrian Basin (Spain), *J. Foram. Res.* **32**:22-44.
Hickman, C. S., and Lipps, J. H., 1983, Foraminiferivory: selective ingestion of foraminifera and test alterations produced by the neogastropod *Olivella*, *J. Foram. Res.* **13**: 108-114.
Holmes, N. A., 1982, Diel vertical variations in abundance of some planktonic foraminifera from the Rockall Trough, northeastern Atlantic Ocean, *J. Foram. Res.* **12**:145-150.
Hoole, S., 1807, *The select works of Antony van Leeuwenhoek containing his microscopical discoveries in many of the works of nature*, Translated from the Dutch and Latin editions, published by the author, Phil. Soc. Lond.
Hurst, A., 1965, Studies on the structure and function of the feeding apparatus of *Philene aperta* with comparative considerations of some other opisthobranchs, *Malacologia* **2**:221-347.
Judkins, D. C., and Fleminger, A., 1972, Comparison of foregut contents of *Sergestes similis* obtained from net collections and Albacore stomachs, *Fisheries Bull.* **70**:217-223.
Kircher, J. M., and Brasier, M. D., 1989, Cambrian to Devonian, in: *Stratigraphical Atlas of Fossil Foraminifera* (D. G. Jenkins and J. W. Murray, eds.), Ellis Horwood, Chichester, pp. 20-31.
Langer, M. R., and Bell, C. J., 1995, Toxic foraminifera: innocent until proved guilty, *Mar. Micropaleontol.* **24**:205-214.
Langer, M. R., Lipps, J. H., and Moreno, G., 1995, Predation on foraminifera by the dentaliid deep-sea scaphopod *Fissidentalium megathyris*, *Deep-Sea Res. I* **42**:849-857.
Lehman, V., 1971, Jaws, radula, and crop of *Arniceras* (Ammonoidea), *Palaeontology* **14**:338-341.
Lipps, J. H., 1979, Ecology and paleoecology of planktic foraminifera, in: *Foraminiferal Ecology and Paleoecology* (J. H. Lipps, W. H. Berger, M. A. Buzas, R. G. Douglas, and C. A. Ross), SEPM Short Course No. 6, pp. 62-104.
Lipps, J. H., 1983, Biotic interactions in benthic foraminifera, in: *Biotic Interactions in Recent and Fossil Benthic Communities* (M. J. S. Tevesz and P. L. McCall, eds.), Plenum Press, New York, pp. 331-376.

Lipps, J. H., 1988, Predation on foraminifera by coral reef fish: taphonomic and evolutionary implications, *Palaios* **3**:315-326.
Lipps, J. H., 1992, Origin and early evolution of Foraminifera, in: *Studies in Benthic Foraminifera* (T. Saito and T. Takayanagi, eds.), Proceedings of the Fourth International Symposium on Benthic Foraminifera "Benthos '90," Sendai, Japan, 1990, Tokai University Press, pp. 3-9.
Lipps, J. H., and Krebs, W. N., 1974, Planktonic foraminifera associated with Antarctic sea ice, *J. Foram. Res.*. **4**:80-85.
Lipps, J. H., and Ronan, T. E., 1974, Predation on foraminifera by the polychaete worm, *Diopatra, J. Foram. Res.* **4**: 139-143.
Lipps, J. H., and Rosanov, A. Yu, 1996, The Late Precambrian-Cambrian agglutinated fossil *Platysolenites*, *Paleontol. J.* **30**:679-687.
Livan, M., 1937, Uber Bohr-Löcher au rezenten und fossilen Invertebraten, *Senckenbergiana* **19**: 138-150.
Mageau, N. C., and Walker, D. A., 1976, Effects of ingestion of foraminifera by larger invertebrates, *Marit. Sed. Spec. Pub.* 1, pp. 89-105.
McIlroy, D., Green, O. R., and Brasier, M. D., 1994, The world's oldest foraminiferans, *Microscopy and Anal.* November 1994:13-15.
Morton, J.E., 1959, The habits and feeding organs of *Dentalium entalis*, *Mar. Biol. Assoc., U.K.* **38**:225-238.
Murray, J. W., 1991, *Ecology and Paleoecology of Benthic Foraminifera*, Longman, New York.
Nielsen, K. S. S., 1999, Foraminiferivory revisited: a preliminary investigation of holes in foraminifera. *Bull. Geol. Soc. Denmark* **45**:139-142.
Nielsen, K. S. S., Collen, I. D., and Ferland, M. A., 2002, *Floresina*: a genus of predators, parasites or scavengers?, *J. Foram. Res.* **32**:93-95.
Olsson, R. K., and Nyong, E. O., 1984, A paleoslope model for Campanian-lower Maastrichtian foraminifera of New Jersey and Delaware, *J. Foram. Res.* **14**:50-68.
Plewes, C. R., Palmer, T. J., and Haynes, J. R., 1993, A boring foramininferan from the Upper Jurassic of England and northern France, *J. Micropaleontol.* **12**:83-89.
Pozaryska, K., 1957, Lagenidae du Cretace Superieur de Pologne, *Acad. Pol. Sci., Paleontol. Polon.* **8**: 1-190.
Reyment, R.A., 1966, Preliminary observations on gastropod predation in the western Niger Delta, *Palaeogeogr. Palaeoclim. Palaeoecol.* **2**:81-102.
Rhumbler, L., 1911, *Die Foraminiferen (Thalamophoren) der Plankton-Expedition; Teil 1. Die allgemeinen Organisation sverhaltnisse der Foraminiferen*, Lipsius and Tisher, Kiel und Leipzig.
Said, R., 1951, Foraminifera of Narragansett Bay, *Contrib. Cushman Found. Foram. Res.* **2**:75-86.
Saidova, K. M., 1967, The biomass and quantitative distribution of live foraminifera in the Kurile-Kamchatka Trench area, *Dok. Akad. Nauk SSSR* **174**:207-209.
Sandon, H., 1932, *The Food of Protozoa*, Misr-Sokkar Press, Cairo.
Shultze, M. S., 1854, *Uber den Organismus der Polythalamien*, Wilhelm Engelmann, Leipzig.
Shimek, R. L., 1990, Diet and habitat utilization in a northeastern Pacific Ocean scaphopod assemblage, *Am. Malacol. Bull.* **7**: 147-169.
Shonman, D., and Nybakken, J. W., 1978, Food preferences, food availability and food resource partitioning in two sympatric species of cephalaspidean opisthobrachs, *Veliger* **21**:120-126.
Simmons, M. D., BouDagher-Fadel, M. K., Banner, F. T., and Whittaker, J. E., 1997, The Jurassic Favusellacea, the earliest Globigerinina, in: *The Early Evolutionary History of Planktonic Foraminifera* (M. K. Boudagher-Fadel, F. T. Banner, and J. E. Whittaker), Chapman and Hall, London, pp.17-51.
Sliter, W. V., 1965, Laboratory experiments on the life cycle and ecologic controls of *Rosalina globularis* d'Orbigny, *J.Protozool.* **12**:210-215.
Sliter, W. V., 1971, Predation on benthic foraminifers, *J. Foram. Res.* **1**:20-29.
Sliter, W. V., 1972, Cretaceous foraminifera depth habitats and their origins, *Nature* **239**:514-515.
Sliter, W. V., 1975, Foraminiferal life and residue assemblages from Cretaceous slope deposits, *Geol. Soc. Am. Bull.* **86**: 897-906.
Smyth, M. J., 1988, The foraminifera *Cymbaloporella tabellaeformis* (Brady) bores into gastropod shells, *J. Foram. Res.* **18**:277-285.
Spindler, M., Hemleben, Ch., Salomons, J. B., and Smit, L. P., 1984, Feeding behaviour of some planktonic foraminifers in laboratory cultures, *J. Foram. Res.* **14**: 237-249.
Svavarsson, J., Gudmundson, G., and Brattegard, T., 1991, Asellote isopods (Crustacea) preying on foraminifera in the deep sea?, *Abstr. 6th Deep-Sea Biol. Symp. Copenhagen*, 1991, p. 63.
Tardent, P., 1979, *Meeresbiologie-eine Einfuhrung*, Thieme, Stuttgart.
Todd, R., 1965, A new *Rosalina* (Foraminifera) parasitic on a bivalve, *Deep-Sea Res.* **12**:831-837.
Thorson, G., 1966, Some factors influencing the recruitment and establishment of marine benthic communities, *Netherlands J. Sea Res.* **3**:267-293.
Vénec-Peyré, M.-T., 1996, Bioeroding foraminifera: a review, *Mar. Micropaleontol.* **28**:19-30.

Vermeij, G. J., 1987, *Evolution and Escalation: An Ecological History of Life*, Princeton University Press, Princeton, N.J.

Vermeij, G. J., 1994, The evolutionary interactions among species: selection, escalation, and coevolution. *Ann. Rev. Ecol. Syst.* **25**:219-236.

Vialov, O. S., and Kantolinskaja, I. I., 1968, Sledy sverlenii Khishchnykh gastropod v rakovinakh miotsenovykh foraminifer [Boring traces by predaceous gastropods in the shells of Miocene foraminifers], *Paleontol. Sbornik, vypusk vtoroi* **5**:88-94.

Walker, D. A., 1971, Etching of the test surface of benthonic foraminifera due to ingestion by the gastropod *Littorina littorea* Linne, *Can. J. Earth Sci.* **8**:1487-1491.

Wefer, G., and Lutze, G. F., 1976, Benthic foraminifera biomass production in the western Baltic, *Kieler Meeresforsch.* **3**:76-81.

Wilde, G., 1965, Abnormal growth conditions in Fusulinids, *Contrib. Cushman Found. Foram. Res.* **16**:121-124.

Winter, F. W., 1907, Zur Kenntnis der Thalamophoren.I. Untersuchung uber *Peneroplis pertusus* Forskal, *Arch Protistenk.* **10**:1-113.

Zobel, B., 1973, Biostratigraphische Untersuchungen au sedimenten des indisch-pakistanischen Kontinentalrandes (Arabisches Meer), *Meteor Forsch. Ergebn. C*, **12**:9-73.

Chapter 2

Predation in Ancient Reef-Builders

RACHEL WOOD

1. Introduction 33
2. Predation on Modern Coral Reefs..34
   2.1. Direct Evidence of Predation...36
   2.2. Indirect Community Effects of Predation 37
3. Potential Antipredatory Traits in Reef-Builders 37
   3.1. Avoidance 37
   3.2. Deterrence 39
   3.3. Tolerance...41
4. The Rise of Biological Disturbance in Reef Ecosystems......................................43
5. Difficulties of Identification of Antipredatory Traits in Ancient Reef Organisms....45
6. Evolution of Antipredatory Characteristics in Reef Builders................................46
   6.1. Attachment to Hard Substrates 46
   6.2. Mobile Life Habit..47
   6.3. Large Size..47
   6.4. Modular Growth Forms...48
   6.5. Regeneration 48
   6.6. Spatial Refuges..49
7. Discussion and Future Work..49
   References 50

1. Introduction

Avoidance of predation is of critical importance to any organism, but reef-building organisms might be considered particularly vulnerable due to their immobile, epifaunal life habit. The need for photosymbiotic metazoans to expose large areas of soft tissue to light further increases the risk of predation, as well as fouling. It has been well established that modern coral reefs grow in particular environments where avoidance of,

RACHEL WOOD • Schlumberger Cambridge Research, High Cross, Madingley Road, Cambridge CB2 OEL, UK, and Department of Earth Sciences, University of Cambridge, Downing Street, Cambridge CB2 3EQ, United Kingdom.

Predator-Prey Interactions in the Fossil Record, edited by Patricia H. Kelley, Michał Kowalewski, and Thor A. Hansen. Kluwer Academic/Plenum Publishers, New York, 2003.

or adaptation to, competition and disturbance are of prime importance (e.g., Connell, 1978; Jackson, 1983; Glynn, 1988). In particular, the control and incidental damage exerted by herbivores, particularly fishes, in limiting the distribution and abundance of algae is probably crucial to the survival of modern coral reefs (e.g., Hay, 1981; Lewis, 1986). As a result, many coral reef organisms are supposed to show a considerable range of anti-predation traits, but unequivocal confirmation of these as adaptations, as well as details of their evolutionary origin and development, remain poorly known.

Reef-builders are restricted from evolving many defensive adaptations by the constraints of their organizational body plans. Sessile organisms have a highly restricted range of antipredatory options at their disposal, as these must be based upon mainly passive constructional defenses. Moreover, susceptibility to partial mortality and reliance upon herbivory and predation to remove competitors or foulers usually entails loss of the prey's own tissues. For this reason, particular anatomies are required that allow resumption of normal growth as quickly as possible, or, even better, create some advantage from this potential adversity.

This contribution explores predator-prey interactions and possible antipredatory traits in living scleractinian corals and ancient reef-building skeletal sponges and cnidarians. These are analyzed with reference to the potential effects of the appearance and radiation of major predator groups in reef ecosystems, particularly the rise of predation after the Paleozoic.

Such an undertaking, however, is highly problematic. Firstly, the basic organization of many reef-building organisms is extremely simple, so that it might be possible to interpret any post-Paleozoic changes in morphology or life habit as defensive. The central question then is to determine the importance of biological disturbance relative to other processes such as physical disturbance, variable recruitment, and competition, so as to evaluate the extent to which escalating predation might be responsible for the many progressive morphological trends documented (e.g., Jackson and Hughes, 1985; Jackson and Coates, 1986; McKinney and Jackson, 1989). Such distinction is in many cases virtually impossible, as traits can be beneficial in a number of unrelated ways or supposed benefits may be in conflict such that the outcome may be a trade-off for many organisms. For example, the cost of investing in increased defense may be a reduction in growth rate or competitive ability.

Secondly, many predators cause disturbance in more than one way: some are also bioturbators; others are capable of significant bioerosion. The rise of new predators with novel feeding methods thus involved coincident developments that might be predicted to have had many common evolutionary consequences.

In this review, I have followed the terminology of Gould and Vrba (1982), where supposed beneficial traits (*aptations*) can be divided into two categories: (1) *adaptations*, which benefit a specific function or effect that has been enhanced by natural selection, and (2) *exaptations*, whose benefits are secondary or incidental to the primary function to which they are adapted.

2. Predation on Modern Coral Reefs

Coral reefs are host to a vast array of predators with greatly varying capabilities and degrees of specialization. The most abundant passive predators on reefs are sessile filter- and suspension-feeders, and planktivorous fish (the so-called "wall of mouths"), which remove up to 60% of the biomass of plankton from the waters streaming across a reef (Glynn, 1973). Coral reefs are also characterized by diverse active predators and herbivores, and borers, which prey upon or otherwise attack sessile organisms (Table 1). Corallivores include crustaceans (hermit crabs), polychaetes (amphinomids), gastropods (prosobranchs and nudibranchs), echinoids (diadematoids), starfish, and numerous fish, which are arguably the most diverse of all reef predators.

TABLE 1. Major Groups of Herbivores, Corallivores and Bioeroders on Modern Coral Reefs

Group	Ecology
Porifera	
Clionidae	Borers
Molluscs	
Polyplacophora	Herbivores (scraping)
Gastropoda	
Archaeogastropods + mesogastropods (excl. limpets)	Herbivores (non-denuding and denuding)
Patellacea	Herbivores (scraping)
Nudibranchs	Corallivores
Prosobranchs	Corallivores
Bivalvia	
Lithophagidae	Borers
Echinoderms	
Echinoidea	
Diadematoida*	Herbivores (excavating) and corallivores
Arbacioida	Herbivores (excavating)
Temnopleuroida	Herbivores (denuding)
Echinoida	Herbivores (excavating)
Asteroidea	Corallivores
Arthropoda	
Isopoda	Herbivores (non-denuding)
Amphipoda	Herbivores (non-denuding)
Decapoda	Corallivores
Polychaetae (Nereidae, Eucicidae and Dorvillidae)	Herbivores (non-denuding) and corallivores
Amphipoda	Herbivores (non-denuding)
Pisces	
Perciformes	
Labridae	Carnivores
Scariidae (79)*	Herbivores (excavating)
Acanthuridae (76)*	Herbivores (denuding)
Pomacentridae (~300)*	Herbivores (non-denuding and denuding) and corallivores
Chaetetodontidae	Herbivores (non-denuding) and corallivores
Blennidae	Herbivores (denuding)
Kyphosidae	Herbivores (denuding)
Siganidae (27)*	Herbivores (denuding)
Tetradontiformes	
Balistidae	Herbivores (non-denuding) and corallivores
Monacanthidae	Herbivores (non-denuding) and corallivores

* indicates most important groups. Species numbers in brackets after Choat 1991. (After Wood, 1999.)

Over 100 families of fishes have coral reef representatives, but of these only a small subset exploits sessile biota (Choat and Bellwood, 1991). Scarids (parrotfish) and acanthuroids (surgeonfish) are largely restricted to coral reefs, but most abundant are the labrids (wrasses) and pomacentrids (damselfish). Estimates suggest that 27-56% of biomass in any reef fish community are benthic invertebrate predators, 7-26% are herbivores, and 4-20% are planktivores (Jones *et al.*, 1991).

Biological disturbance shows marked differences in distribution and intensity across a reef profile. Like physical disturbance, both predation and bioturbation decrease with depth, being usually greatest at shallow depths, from the lower intertidal zone to about 20 m, particularly on reef slopes with substrates of high topographic complexity (Hay, 1984). Herbivory is low above mean low water, often reaches a peak at 1-5 m depth on the forereef, and then declines rapidly with depth (Steneck, 1988). Regardless of depth, however, the effects of biological disturbance may be highly patchy and vary markedly according to local environmental differences.

Reef predators and herbivores can be divided into three main functional categories:

(1) Organisms that consume soft tissues only. These organisms can have either a denuding effect (i.e., are capable of removing substantial quantities of organic matter close to the substrate) or a non-denuding effect. The most important denuders are some molluscs, some echinoids, acanthuroids, siganids (rabbitfish), and pomacentrids. Like herbivore browsers, some corallivores are non-denuding, whereas others (e.g. some polychaetes, gastropods and starfish) are able to denude corals entirely of their soft-parts.

(2) Organisms that crop very close to a substrate, incidentally ingesting substantial portions of living soft tissue, together with associated small invertebrates and underlying skeleton.

(3) Organisms that are capable of removing and ingesting calcareous skeletal material. Epilithic predators feed directly upon sessile invertebrates or algae by etching, rasping or biting, thereby causing incidental skeletal or substrate damage. These predators include excavators, which exert deep bites that result in the removal of large areas of substrate, and scrapers, which have weaker jaw apparatuses that take smaller bite sizes with resultant limited substrate removal. The most important excavators and scrapers are limpets, chitons, some regular echinoids, and acanthuroids and scarids.

2.1. Direct Evidence of Predation

Only the latter two functional groups may leave any recognizable traces in the preservable fossil record. The range of bioerosive trace fossil morphologies is vast, due to the diversity of organisms involved (Bromley, 1992). However, very few of these traces are sufficiently characteristic to allow an unequivocal pairing with a particular predator. For example, Steneck (1983) noted considerable evidence of predatory damage in fossil solenoporacean and corallinacean algal thalli but was unable to determine the origin.

Parrotfish produce distinctive stellate marks on the upper surfaces of scleractinian colonies or algal thalli. Likewise, camerodont echinoderms produce characteristic pentaradiate grazing traces (*Gnathichnus* and *Radulichnus*) due to the action of strengthened teeth in a stirodont lantern (Bromley, 1975). Both these traces are apparent in living and dead modern coral material.

Although there is abundant evidence of damage and subsequent regeneration in scleractinian corals, regeneration from predation has proved difficult to differentiate from that caused by other sources of partial mortality. When the living tissue of a coral is damaged by any means, the surrounding polyps respond by generating new tissue, forming a generic "lesion." One exception to this is the distinctive gall-like structures or "chimneys" known to form in response to the grazing activities of the three-spot damselfish (*Eupomacentrus planifrons*) on the staghorn coral, *Acropora cervicornis* (Kaufman, 1981). Such structures are known from both Pleistocene and Recent specimens. The highly territorial three-spot damselfish grazes upon algal turf by killing all live coral within its territory and by excluding other herbivores. Remnants of *A. cervicornis* can, however, regenerate rapidly from such grazing events. Similar structures are formed by other corals as a result of regeneration of lesions colonized by algae (Bak *et al.*, 1977).

2.2. Indirect Community Effects of Predation

Despite the low apparent standing stocks of algae on reefs, the extremely high growth rates of algal turfs make shallow coral reefs one of the most productive ecosystems known. Most of this production, however, is consumed by herbivores, particularly fish. The abundance of reef fishes is assumed to be of great importance on coral reefs, as evidenced by the dramatic increase of algal growth as a result of their decline on Jamaican reefs (Hughes, 1994). Tropical marine hard substrata are usually sparsely vegetated, but a rich algal flora develops when herbivorous fish are excluded and/or nutrient input increases. Grazers not only allow the dominance of corals and coralline algae on coral reefs, but they also contribute notably to carbonate sediment production and redistribution, algal ridge formation, and the maintenance of overall diversity. Like other predators, they can also ameliorate the effects of competition and may combine with physical controls to produce the characteristic zonation of modern coral reefs. Table 2 outlines the major causes and indirect effects of predation, particularly herbivory, on coral reef communities.

3. Potential Antipredatory Traits in Reef-Builders

Antipredatory traits in immobile epifauna can be divided into three sets: (1) *avoidance* – those that reduce initial accessibility to predators, (2) *deterrence* – traits that discourage predation by increasing its cost, and (3) *tolerance* – traits that reduce, minimize, or even capitalize upon the damaging effects of partial mortality.

3.1. Avoidance

Avoidance includes inhibitive or evasive life habits that reduce accessibility or availability to predators by adoption of restricted spatial or temporal distributions.

Spatial refuges include habitats where predator access is difficult, such as infaunal, endolithic, or cryptic niches, as well as deep sea or intertidal refuges that are beyond the ranges of many predators. Corals that form or occupy topographically complex surfaces (as well as those inhabiting cryptic niches) can preferentially survive predation, including corallivore outbreaks. It has also been well established that some protection

TABLE 2. Causes and Indirect Community Effects of Predation and Herbivory on Coral Reef Ecology and Sedimentology

Effect	Causes
Increase in diversity; retarding dominance	Modification of transitive competitive relationships by specialized predation by reducing or preventing competition
	Increase of biodiversity within herbivorous fish territories, particularly of small, sessile invertebrates, by removal of fast-growing algae
Dominance of coralline algae and corals	Removal of fouling algal turfs and macroalgae by large herbivores
Algal ridge formation	Resilience of coralline algae to excavatory herbivory (parrotfish and limpets) and severe wave exposure
Zonation	Interaction of physical controls with differential effects of damselfish in the survival of different coral species
Sediment production	Intense bioerosion (excavatory herbivory and predation, and boring activity) of reef framework
Sediment grain size reduction	Endolithic infestation and parrotfish ingestion
Reduction in rate of reef progradation	Removal of actively accreting reef carbonate by parrotfish and net transport of sediment from its site of production
Limiting of foraging ranges	Higher predation pressure

from both biological and physical disturbance can be gained by adopting closely aggregating growth (see summary in Jackson, 1983). Dense populations are less susceptible than are isolated individuals to overgrowth from competitors, larval invasion and attack by predation. Many relatively poorly defended living corals (e.g., with small, solitary or low-integration branching organizations with slow rates of regeneration) may be found in protected refuges of the reef framework – under overhangs or within crypts.

Nocturnal, ephemeral or seasonal occurrence represents restricted temporal distributions also based upon a decreased likelihood of encountering predators. The composition of exposed coral-reef organisms at night is very different from daylight hours, and this is thought to be due in part to avoidance of predators that forage only during the daytime. A significant proportion of reef fish, particularly herbivores (Acanthuridae, Labridae, Scaridae and Pomacentridae), "sleep" or undergo torpor (a dormant state of complete physical inactivity) after daylight hours. For example, estimates suggest that the population size of nocturnal swimming coral reef fish is only about 10% that of the diurnal population at One Tree Island, Great Barrier Reef (Talbot and Goldman, 1972), and their feeding habits are more restricted than those of diurnal fish, with a greater consumption of immobile organisms occurring during daylight hours (Hobson, 1968). Numerous crabs and echinoderms congregate under overhangs or within rubble during the day but become active only at night. Zooplankton, which live in cryptic areas during the day, swim freely at night, as do many benthic invertebrates that migrate into the water column at night (Alldredge and King, 1977).

The intertidal habit has reduced predation pressure but presents physiological difficulties, as organisms must overcome the problems of thermal stress, ultraviolet irradiation and desiccation. Some corals can withstand the intertidal habitat by a variety of methods, including the presence of secondary metabolites that act as sunscreens and retraction of polyps (see the summary in Brown, 1997).

Deep waters are successfully occupied by azooxanthellate scleractinians, which can thrive at depths down to 300 m. Some zooxanthellate corals, however, also appear to have evolved additional adaptations for invasion of deeper waters from shallow habitats. *Leptoseris fragilis* can photosynthesize at depths down to 145 m due to the presence of additional photosymbiotic pigments, and this species also possesses specialized skeletal structures that aid heterotrophic feeding in waters of highly reduced illumination (Schlichter, 1991).

3.2. Deterrence

3.2.1. Acquisition of Defenses that Increase the Cost of Successful Predation

Deterrent characteristics include permanent and secure attachment to a stable substrate, heavily armored skeletons, tough rubbery textures, mineralized sclerites, tough fibrous components, and the secretion of unpalatable, noxious or toxic secondary metabolites that inhibit ingestion or digestion.

Organisms without secure attachment to a stable substrate are susceptible to the effects of bioturbation, as inversion and burial will cause death unless individuals are able to re-establish contact with the water column to continue feeding and respiration. Newly settled juveniles will be most susceptible to disturbance, and so initial settlement and growth will be considerably hindered, thereby reducing successful recruitment. In modern shallow shelf seas, many immobile epifauna are excluded from most soft substrates, which are dominated by mobile deposit feeders. While immobile but unattached corals are common today, they are restricted mainly to areas protected from high biological and physical disturbance. Most modern suspension-feeders require a hard substrate, even if these are only isolated patches within areas of unstable, soft substrate ("benthic islands," or in dense aggregations).

Possession of an edge zone in all but the most primitive scleractinian corals allows them to gain permanent attachment to a stable substrate. Not surprisingly, scleractinian corals dominate reef framework environments, especially those in high-energy settings, but they may be scarce in adjacent unconsolidated sediments. However, while secure attachment to a hard substrate allows some scleractinians to grow successfully in such environments, this may be a result of the preponderance of wave-swept, extensive hard substrata for colonization. Permanent attachment allows, however, for the development of very large branching morphologies.

Attachment offers ample defense from bioturbation and reduces considerably the ease of manipulation by potential predators. But because none of the major predators upon corals need to manipulate their prey, this factor is unlikely to have any direct antipredatory value. Indeed, there are no experimental data to show that attached corals are any more susceptible to predation than unattached forms.

Many reef organisms, especially those living on open surfaces, bear heavily calcified skeletons. Most modern immobile reef biota is permanently attached to a hard substrate. Whereas permanent attachment to a stable substrate may reduce pressure from those predators that rely upon prey manipulation, it provides those that do not with prey that are virtual sitting targets (Harper and Skelton, 1993). For example, Steneck (1983) has demonstrated that both the heavily calcified thallus and encrusting habit of coralline algae offer considerable resistance to predatory attack.

Some morphologies are better resistant to predatory breakage than others. For example, colonies with closely spaced branches can make predator access difficult by forming hidden, protected areas. The flattening of branch terminations offers greater resistance to breakage and shearing, and this character is found in erect species of bryozoans, gorgonian corals and stylasterine corals.

Many reef organisms bear toxins, particularly forms that are found on open surfaces (Bakus, 1981), and therefore this has been proposed to be the result of selective predator-prey interaction (e.g. Bakus, 1974). Indeed, there are more toxic sponges, cnidarians, arthropods, holothurians and fishes in the tropics than at higher latitudes. Some have also argued that reef-dwelling, tropical macroalgae generally bear stronger chemical defenses (mainly terpenoids) than do temperate seaweeds (Bolser and Hay, 1997). Although it is likely that some of the most common, exposed invertebrates on reefs are subject to little or no fish predation because they use defenses such as toxicity to prevent even partial mortality, this has not yet been convincingly demonstrated. Indeed, toxins do not confer total immunity to predation; many algae assumed to be protected by antipredatory compounds are still readily consumed by some species. Moreover, compounds with similar structures can differ dramatically in their effects on herbivore feeding, and those that might deter one herbivore may have no effect, or even stimulate feeding, on another (Hay, 1991). The presence of noxious or toxic chemicals is often signalled by warning or cryptic coloration and form. Such a defense relies on correct visual recognition by the potential predator.

The search for, and understanding of, the secondary metabolites present in corals is still in its infancy.

3.2.2. Acquisition of Reduced Nutritional Quality

The potentially deterrent effects of low nutritional characteristics have received little experimental attention, but this strategy would operate best only if alternate food sources of better quality were rarely available (Hay, 1991).

3.2.3. Acquisition of Behavioral Characteristics that Increase the Energetic Cost of Successful Predation

Numerous behavioral aptations have been suggested to occur on coral reefs in response to predation pressure. These aptations include rapid or erratic movement, active escape, assault, migration of demersal zooplankton, and even synchronized spawning (Babcock *et al.,* 1986). Many behavioral mutualistic associations occur in reef organisms, for example the shrimp (*Stenopus*) and the fish *(Amphiprion)* associated with sea anemones, and the aggressive territorial behavior shown by numerous crustaceans (e.g. *Trapezia* and *Alpheus*) that live among the branches of the large coral *Pocillopora* is known to deter some predators.

Mobile, free-living, corals such as fungiids are able to creep, right themselves, and are capable of extraction from burial. Some corals (e.g. *Heteropsammia*) grow around a symbiotic siphunculid, which "tows" its coral host. Although many are solitary, all living mobile corals are photosymbiotic and clonal, suggesting that rapid growth and/or regenerative powers may be important for acquisition of this condition. Again, whereas a free-living habit might provide defense from bioturbation, it is difficult to imagine that such forms have achieved sufficient mobility to flee from potential predators.

3.2.4. Internalization of Soft Tissue or Exposure of Large Areas of Thin Tissue

Some sessile organisms enclose their vulnerable soft parts within heavily armored skeletons. Many sessile reef organisms possess a modular or colonial habit that also reduces soft tissue to a relatively thin veneer over a larger basal skeleton. This defense not only decreases accessibility and the ease of prey manipulation by predators, but also minimizes the tissue biomass while maximizing the cost of collection. For example, in a typical domal colony of *Porites*, only about 0.5% of the colony's radius is occupied by soft tissue (Rosen, 1986). In branching and platy colony forms, the relative proportion of skeleton is even lower.

In coralline algae, a protective outer epithallus overlies the more delicate meristem, fusion cells allow the rapid translocation of photosynthates, and conceptacles that contain reproductive structures are enclosed within the perithallus. These structures have been demonstrated to protect the delicate reproductive anatomy from intensive grazing (Steneck, 1982, 1983). Conceptacles are, however, no match for the deep excavation of parrotfishes, perhaps explaining why such structures are found only on non-tropical, thickened, crusts.

3.3. Tolerance

3.3.1. Growth Form

Growth form is an important determinant of survival in corals. Generally, predators prefer multiserial erect branching or platy corals to encrusting forms (Glynn *et al.*, 1972; Rylaarsdam, 1983). But, as most partial mortality is caused by many processes that occur close to the substrate, "escape in height" by acquisition of a branching habit will increase the chances of survival dramatically (Meesters *et al.*, 1997), and so represents a significant defensive feature.

Damselfish have been demonstrated to promote the growth of *Pocillopora* at shallow depths (Wellington, 1982), suggesting that closely branching colony form in this coral might represent an antipredatory trait.

In coralline algae, thickened crusts are more tolerant to attack than are thin encrusting or branching forms (Steneck, 1985), but, in modern reefs, the dominance of a particular growth form appears to be a trade-off between the cost of investment in increased defense and the reduction in growth rate or competitive ability. As a result, thickened crusts dominate only in areas of high physical and biological disturbance.

3.3.2. Rapid Rates of Growth and Calcification

Rapid rates of growth can give clear competitive superiority to sessile organisms. The autotrophic capabilities of most reef-building corals also removes the need to expose tentacles during the day, when most predators are active. This raises interesting questions as to the possibility that non-zooxanthellate reef-building corals, such as *Tubastraea*, may bear toxic metabolites that counteract predation. The fact that corals – like the alga *Halimeda* – show skeletal extension during the night when herbivorous fish are inactive might also be inferred to be an antipredatory trait. This has yet to be proved experimentally.

3.3.3. Enhanced Ability to Regenerate Soft Tissue and/or Skeleton

Not only do few predators actively excavate underlying skeleton, but their activities also do not normally result in total colony death, only in partial mortality, i.e., sub-lethal damage. In such cases, the capacity of a coral to heal or replace damaged areas of soft tissue becomes critical to its survival. Strategies that rely upon herbivores/predators to remove competition therefore often entail loss of the prey's own tissues. Algal turfs grow very rapidly and so can regenerate from basal portions that have escaped herbivory. Many coralline algae can also tolerate intense herbivory due to their ability to rapidly regenerate removed material (Steneck, 1985, 1988).

Available evidence suggests that in primary (solitary) polyps there is no regeneration when the central part of the polyp (the mouth-stomadaeum) has been destroyed by predation (Bak and Engel, 1979). In clonal/modular organisms, partial predation may remove either individual or a few modules, or large areas may be cleared of living tissue, sometimes together with the excavation of underlying skeleton. Such newly cleared areas of skeletal substratum may be rapidly colonized by fouling or endolithic organisms presenting further threat to the colony. A multiserial modular organization, however, in addition to promoting architectural diversity and flexibility, also allows compartmentalization of damage and enables some colonies to regenerate from fragments (Jackson and Hughes, 1985).

Most significantly, branching corals also show tremendous powers of regeneration: *Acropora palmata* has one of the highest rates recorded (Bak, 1983). Indeed, unlike massive, platy or encrusting forms, damage to branching corals often leads to an immediate increase in growth rate, thus causing an increase in size rather than simply repairing damaged tissue.

Populations of the staghorn coral (*Acropora cervicornis*) frequently form dense, monospecific stands on shallow Caribbean reefs, but there is little evidence of sexual recruitment (Tunnicliffe, 1981). The fragile organization of this species results in easy breakage as a result of both high wave activity and bioerosion, especially by boring sponges that infest the colony bases. However, such corals are able to re-anchor fragments and rapidly regenerate and grow, often fusing with other colonies, at rates up to 150 mm per yr (Tunnicliffe, 1981). Such branching corals have turned adversity into considerable advantage and appear to flourish because, and not in spite, of breakage.

Tissue regeneration as a result of local damage is initially very fast in corals but then slows down exponentially (Bak, 1983). Different coral species show widely varying abilities of regeneration, which are species-specific, related to lesion size, and to environmental factors. Large colonies are able to regenerate lost tissues faster than smaller representatives of the same species (Connell, 1973; Loya, 1976; Bak *et al.*, 1977), and the same species may show different regenerative abilities when growing at varying depths (Bak and Steward-Van Es, 1980). Brooding is associated with small maximum colony size, and this mode of reproduction has been suggested to have evolved in order to overcome the high rates of mortality associated with small size (Meesters *et al.*, 1996).

Different types of lesions may influence regeneration rate. For example, some species (e.g. *Montastraea annularis*) are able to regenerate faster when the skeleton as well as soft tissue is damaged. Such observations suggest that different predators may have differential responses on coral prey. *Montastrea annularis* will recover more slowly from denuding predators than, for example, *Agaricia agaricites*, but will not suffer so greatly from excavating predation (Bak *et al.*, 1977).

3.3.4. Adoption of Large or Small Body Size

Large size offers multiple benefits for sessile reef invertebrates. In most marine invertebrates, larger colonies are less susceptible to the agents of mortality than are smaller ones. Juvenile corals are particularly vulnerable to predation until they can reach an escape in size – that is, become sufficiently large to have gained sufficient powers of regeneration to overcome partial mortality (Birkeland, 1977). Larger colonies not only win more competitive overgrowth interactions and are more fecund than smaller colonies, but also are more likely to survive partial mortality by regenerating tissue following injuries more rapidly than smaller colonies. Competitive ability and sexual reproductive capacity of colonial animals are also strongly correlated to individual colony size (Jackson, 1985). Large body size also allows organisms to achieve a refuge in size.

There is some evidence to suggest that corals direct resources initially to colony growth before the onset of reproductive maturity, i.e., delay reproduction so as to increase the chance of growth to a larger, safer size. Also, older colonies may preferentially allocate resources to the regeneration of damaged tissues rather than for further growth (Bak, 1983). In all reef-building species examined in one study, there was a colony threshold size below which no mature gonads are found (Hall and Hughes, 1996). Interestingly, older colonies will continue gametogenesis even if they become fragmented below the threshold size required for younger colonies to become sexually mature (Kojis and Quinn, 1985). It is therefore likely that the relative investment in growth declines with colony size, and often age. Contrawise, adoption of small body size might allow access to cryptic or otherwise protected refugia.

4. The Rise of Biological Disturbance in Reef Ecosystems

As summarized above, coral reef organisms show highly complex and integrated sets of morphological, spatial, chemical and temporal characteristics, many of which could be considered to be defensive. The probability, however, of any single defensive trait being effective may be decreased as we have seen that much of the predation on coral reefs is non-specific: indeed, different predatory methods may even require the same defensive strategies. In studying supposed defensive aptations, it is therefore necessary to consider the general feeding methods used by predators. These categories are summarized in Table 3, with their inferred first evolutionary appearance.

TABLE 3. General Methods Used by Predators on Sessile Reef Biota and Their Inferred First Appearance in the Fossil Record (after Wood, 1999)

Predatory method	Predatory group and first appearance
Carnivory (non-excavating)	?Polychaetes: Cambrian (Conway Morris, 1979)
Herbivory	
Non-denuding	Archaeogastropods: Cambrian (Steneck and Watling, 1982)
	Polychaetes: Cambrian (Glaessner, 1979)
	Arthropods: Cambrian (Bergstrom, 1979)
Denuding	? Orthothecimorph hyoliths: Lower Cambrian (Edhorn, 1977; Kobluk, 1985)
	Mesogastropods: Devonian (Steneck and Watling, 1982)
	Teleost fish: Eocene (Tyler, 1980)
Excavation of live skeletal material (herbivory and carnivory)	Regular echinoids: Diadematoida: Late Triassic (Smith, 1984)
	Chitons: Chitoninae: Late Cretaceous (van Belle, 1977)
	Molluscs: Patellacea: Late Cretaceous (Lindberg and Dwyer, 1983)
	Teleost fish: Scaridae: Miocene (Bellwood and Schultz, 1991)
Live-boring (non-predatory)	Bivalves: Lithophagidae: Eocene (Savazzi, 1982; Krumm and Jones, 1993)
	Porifera: Clionidae: ?
	Barnacles: ?Eocene (D.S. Jones, pers. comm.)
Substrate Disturbance	Rays and skates: Devonian (Vermeij, 1987)
	Echinoderms: Holothuroids: Devonian (Thayer, 1983)
	Manatees: Eocene
Deep	Polychaetes: Triassic (Thayer, 1983)
	Gastropods: Late Triassic (Thayer, 1983)
	Irregular echinoids: Early Jurassic (Thayer, 1983)
	Decapods: Early Jurassic (Thayer, 1983)

In general, grazers and carnivores throughout the Paleozoic and early Mesozoic were relatively small individuals incapable of excavating calcareous substrates, and with limited foraging ranges. A radiation during the Devonian of durophagous, mobile predators has been proposed by Signor and Brett (1984), but these forms probably relied upon manipulation only in order to crush, or ingest, whole shells (Harper and Skelton, 1993). By the early Mesozoic, sessile organisms had to contend with an increasing battery of far more advanced and novel feeding methods, as well as sediment disruption due to deep bioturbating activity (see summaries in Vermeij, 1987). Most notable was the rise of efficient excavation behaviors. Bioerosion increased in intensity from the Middle to Late Jurassic but appears to have been exclusively post-mortem rather than predatory in reef builders. Biological disturbance increased considerably in intensity from the latest Cretaceous-Early Tertiary when deep-grazing limpets, camerodont sea urchins and especially the reef fishes appeared, and with them the increasing ability for substantial excavation of hard substrata over large areas. A concurrent radiation of endoliths occurred from the Triassic onwards, and deep borings are known from only the Mesozoic and Cenozoic. Clionid sponges – one of the major bioeroders on modern coral reefs – had become abundant by the latest Jurassic. The first live-borers are known from the Eocene (Krumm and Jones, 1993), as are fishes similar to modern reef faunas (50 Ma) (Bellwood, 1996). The complex pharyngeal apparatus of labrids was present at this time, and major labrid clades were already differentiated (Bellwood, 1997). Balistids first appeared in the Oligocene, and the oldest scarid fossil currently known is from the Miocene, 14 Ma (Bellwood and Schultz, 1991). It seems likely that, sometime during the Oligocene - Miocene, reef bioerosion gained a modern cast (Pleydell and Jones, 1988).

How did post-Paleozoic reef communities respond to these new threats? We might be able to make a series of predictions concerning changes in reef community ecology based on their rise to abundance in the fossil record (Table 4). In the sections following, these predictions are tested in order to determine to what extent the evolutionary history of reef-associated herbivores and predators has been bound with the origin of the modern coral reef ecosystem.

5. Difficulties of Identification of Antipredatory Traits in Ancient Reef Organisms

Although it may be possible to demonstrate experimentally that a particular trait confers selective advantage against a given predator, it may be highly problematic to tie its evolutionary appearance to that particular threat. Predictions may be difficult to test, as many agents of mortality are not independent and indeed may occur simultaneously: in the marine environment, all potential threats (biological and physical) broadly decrease with increasing depth. Moreover, the ecological impacts that were important selective influences in the past may no longer operate today (Connell, 1980). Identification of aptations is further complicated by the fact that predator-prey

TABLE 4. Predicted Changes in Reef Community Ecology Based on the Rise to Abundance of New Predatory Methods (and Endoliths) as Evidenced in the Fossil Record (after Wood, 1999)

Event	Prediction	Timing
Rise of macroherbivores	Shift to more conspicuous, well-defended macroalgae on reefs	Late Mesozoic
Rise of excavatory grazers and predators	Shift to organisms with deterrent traits and those which tolerate partial mortality	Late Mesozoic to Miocene
[Increase in abundance of endoliths]	Increase in skeletal sediment production	Late Jurassic
	Formation of sediment aprons	Eocene - Miocene
	Shift from macroalgal-dominated to coral and coralline algal dominated reefs	Late Mesozoic - Eocene
	Increase in the diversity of the Cryptos and other spatial refugia	Jurassic onwards
	Increase in multiserial scleractinian corals	Throughout history of the group
	Increase in multiserial, branching corals	Cretaceous onwards
	Algal ridge formation by branching coralline algae	Cretaceous
	Thick coralline algal crusts	Eocene - Miocene

interactions do not often enter into tight coevolution because predators are usually generalists (see above). Any given predator may also not be universally similar in its effects, and certain prey taxa may be limited to or channelled towards particular defensive strategies by the morphological constraints of their own body plans. For example, whereas a toxic animal might evolve warning coloration in response to a threat from predators, a palatable species would become cryptic or adopt mimicry (Harvey and Greenwood, 1978).

The polyphyletic acquisition of similar traits over a short geological interval can offer compelling evidence for an extrinsic selective force (Skelton, 1991). We can assume that shared anatomical and morphological features, especially of the predator feeding apparatus, might indicate shared feeding (including behavioral) characteristics (see Steneck, 1983). While the pitfalls and generalizations of this methodology are well known (e.g. Choat, 1991), this approach is necessary, given both the poor phylogenetic knowledge of potential prey and the preservational vagaries of the fossil record.

Our detailed knowledge of the first appearance, subsequent diversification and rise to abundance of many predator groups is poor. The evolutionary history of groups with poor preservation potential is dependent upon the finding of exceptionally well preserved material; there will also be a taphonomic bias toward the preferential preservation of organisms with more robust and large skeletons. This is compounded by the fact that many of the feeding habits outlined in Table 3 are not easily recognizable in the fossil record, and moreover may be difficult to differentiate from other causes of skeletal damage. Even the distinctive, but often delicate, grazing or predation traces on reef builders are evident only on original upper surfaces, which often suffer from some degree of post-mortem destruction before final burial and fossilization. All these factors make it extremely difficult to tie the evolution of a perceived defense to the appearance of a particular predatory group. Indeed, very few quantitative data are available to assess the relative importance of different sources of partial and whole mortality in ancient communities.

6. Evolution of Antipredatory Characteristics in Reef Builders

Only skeletal anatomy and morphology, spatial distribution, and skeletal attack or breakage, and regeneration might be detected – or inferred – in the fossil record of reef organisms.

Below, the origin and diversification of such fossilizable traits are considered for Paleozoic reef-building cnidarians and skeletal sponges and scleractinian corals with an analysis of their status as either aptations or exaptations based on the fossil record.

6.1. Attachment to Hard Substrates

Cambrian archaeocyath sponges usually bore small holdfasts that enabled limited attachment to hard substrates.

Many mid- to late Paleozoic reefs were dominated by large, sheet-like invertebrates (stromatoporoid sponges, tabulate and rugose corals, and trepostome and cystoporate bryozoans) that were initially attached to small, ephemeral skeletal debris and then spread over the surrounding sediment, often covering a substantial area. These forms were unspecialized and immobile. Some forms could attach to more stable hard substrates, but in most there is little evidence for any active recruitment onto extensive hard substrates. Small, branching forms (some stromatoporoids and bryozoans) lacking extensive attachment sites were also common, and they were presumably partially rooted in soft sediment.

The late Paleozoic decline of immobile epifauna coincides with the rise of major bulldozing taxa, which passed through the end-Permian extinction unscathed. But whereas it seems possible that the rise of bioturbation is responsible for the decline of the Paleozoic free-lying epifauna, this hypothesis must remain conjectural until tested experimentally.

Thayer (1983) has postulated that the modern deep sea may mimic the same degree of bioturbation as early Paleozoic shelves, and indeed the deep sea appears to harbor an immobile soft-substrate fauna of shallow marine Paleozoic cast. Stalked crinoids, articulate brachiopods, hexactinellid sponges and free-living immobile bryozoans are all concentrated, often in considerable abundance, there. Thayer (1983) further suggested that archaic Cambrian and Paleozoic faunas either migrated to deep-sea environments from shallow shelves, or radiated into both habitats but subsequently suffered selective extinction in shallow waters. Alternatively, rather than being controlled by levels of bioturbation, this migration from near-shore to deeper waters might equally reflect the competitive displacement of these groups by the subsequent appearance of better substrate competitors.

6.2. Mobile Life Habit

Active, free-living microsolenid and caryophyllid corals first appeared in the Middle Jurassic, and Fungiidae and Flabellidae during the Cretaceous (Wells, 1956; Gill and Coates, 1977). The apparently mobile cyclotid corals from the Upper Cretaceous are remarkably convegent to the fungiids in the modern Australian fauna (Gill and Coates, 1977). The first "towed" corals are also known from the Upper Cretaceous. A mobile free-living habit also evolved in bryozoans at this time.

6.3. Large Size

Even though large size offers many clear advantages to reef-builders, only limited evidence is available to assess how colony size has changed through the history of the reef-builders, although we might predict that the average size of shallow-water forms may have increased through geological history.

Archaeocyath sponges, however, show a predicted increase in overall size in branching growth forms (Wood *et al.*, 1992). Many mid-Paleozoic stromatoporoid sponges and tabulate corals, particularly those from the Late Devonian, could reach several meters in diameter (Wood, 1999). Very large scleractinian colonies (up to 10 m in height) – albeit with fragile branches – were certainly present by the Late Triassic (e.g. "*Thecosmillia*"). Anecdotal evidence suggests that large colony size did not become common until the Cenozoic. In an attempt to analyze the causes of accelerated turnover in Caribbean Pliocene/Pleistocene (4 - 1 Ma) reef corals, Johnson *et al.* (1995) analyzed susceptibility to extinction in Miocene to Recent faunas within different ecological groupings based on colony size, colony form, corallite size and reproductive characteristics. Only colony size showed significant differences in evolutionary rate.

Extinction rates were higher for species with small, massive colonies, which tend to live in small, short-lived populations with highly fluctuating recruitment rates and limited dispersal abilities. This differential extinction resulted in an increase in the proportion of species with larger colonies, which tend to have larger population sizes, longer generation times, and more constant rates of population increase.

6.4. Modular Growth Forms

Cambrian archaeocyath sponges show a steady and marked increase in the proportion of complex modular forms during their history (Wood et al., 1992), as do scleractinian corals since the mid-Triassic, which appears to be uninterrupted by the end-Cretaceous extinction event (Coates and Jackson, 1985). In contrast, the percentage of scleractinian erect species (mainly low integration phaceloid-dendroid growth forms) decreased until the Turonian, but increased markedly – particularly in multiserial forms with inferred rates of rapid regeneration – after that time.

This spectacular rise in erect growth forms was coincident with the appearance of new groups of predatory excavators. However, only with increased phylogenetic knowledge will it become clear if this increase is the result of a limited number of differentially proliferating clades, or represents a truly polyphyletic response to changing predation pressure.

Noteworthy is the appearance of most families of modern scleractinian corals that spread throughout Tethys during the Eocene. Groups appearing at this time, such as the poritids, their relatives the actinids, and the favids (which had survived the Cretaceous extinction), dominate most coral reef communities throughout much of the Cenozoic (McCall et al., 1994). By the end of the Eocene, all modern coral families had appeared. Although branching acroporoids appeared in the Eocene, they did not dominate reefs until early Pleistocene. The rise of this group – with its particularly remarkable powers of regeneration from fragmentation and rapid growth – would then seem to be independent of any known changes in predation style.

Mean corallite diameter does not appear to have remained the same through the history of the Scleractinia. In a survey of 1600 species, Coates and Jackson (1985) showed that, while corallite diameter has been remarkably consistent in clonal forms, there was a marked drop in the mean diameter of solitary/aclonal scleractinians in the Eocene. That such a trend is in any way a response to the rise of reef fish at this time remains untested.

After the Eocene, herbivore-susceptible, delicately branched coralline algae decreased in abundance in the tropics, the proportion of thickened encrusting forms increased, and the first algal ridges appeared – all coincident with the rise of herbivorous fish.

6.5. Regeneration

Very few data are available on regeneration from unequivocal predation in ancient reef-builders. However, it is clear that modular representatives (i.e., those with multiple, functional units or individuals) of Cambrian archaeocyath sponges, and stromatoporoid sponges or corals, all show greatly enhanced powers of regeneration in highly integrated growth forms compared to low-integration modular or solitary forms (Wood *et al.*, 1992; Wood, 1999).

6.6. Spatial Refuges

Archaeocyath sponges clearly differentiated into open-surface and crypt-dwellers from the inception of the group (Wood *et al.*, 1992). Scleractinian corals occupied the abyssal plains, cool, high-latitude shelves, deep-water coral banks and tropical-subtropical carbonate banks, platforms and reefs by the end of the Early Jurassic (Toarcian) (Roniewicz and Morycowa, 1993). Low diversity associations of platy corals with skeletal structures similar to the living *Leptoseris fragilis* are known from a variety of Late Jurassic (Oxfordian), Early Cretaceous (Barremian-early Aptian and Albian), and Late Cretaceous localities (see Insalaco, 1996; Rosen, 1998).

7. Discussion and Future Work

There has been a proliferation of traits with proven antipredatory benefits since the Mesozoic, summarized in Fig. 1, and some forms, such as branching corals, appear not only to thrive, but actually require conditions of considerable disturbance for their survival in shallow tropical seas. In addition, many of the functional organizations that proved intolerant of excavatory attack became largely absent from shallow-marine tropical reef biotas during the late Paleozoic to early Mesozoic, e.g. the soft-sediment dwelling benthos. But many of the devices employed to tolerate or resist predation, such as multiserial organizations, aid recovery from all agents of partial mortality. So

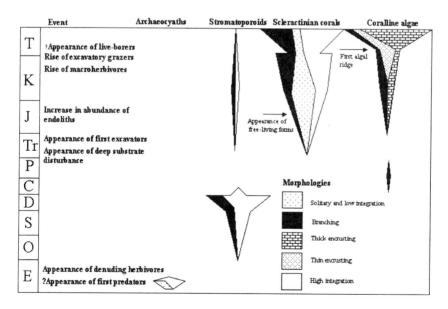

FIGURE 1. Major innovations in the evolution of predatory feeding methods together with broad changes in reef-builder morphology.

whereas the Paleozoic environment might have been characterized by relatively little biological disturbance, it appears that the physical agents of partial mortality that have always been present in shallow marine tropical environments – such as episodic sedimentation – had already selected for the traits that subsequently became beneficial for combating escalating, and excavatory, partial predation. Most of the antipredatory traits surveyed here therefore are considered as exaptations, as they clearly offer multiple benefits.

There are, however, several notable changes in the scleractinian coral biota that all occurred during the Eocene. The increase in the proportion of forms with complex modularity is particularly marked from the Eocene onwards, and, whereas corals displayed the full range of morphological forms and corallite size by the Late Triassic, certain morphological types did not become dominant until the Eocene. Branching coral taxa such as *Acropora, Porites* and *Pocillipora* also appeared during the Eocene. In addition, highly defended, thick crusts in coralline algae become more dominant, and branching forms also become noticeably less conspicuous on reefs, from the Eocene onwards.

This major reorganization of the coral reef ecosystem coincides with the rapid appearance and radiation of reef fish. Much more data, however, are required to assess the incidence of different forms of partial mortality and ability to regenerate damaged skeletons in order to clearly tie these two events in any coevolutionary arms race.

Almost nothing is known as to possible changes in the style of skeletal sediment production and distribution within reefs after the appearance of abundant bioerosion from the Late Jurassic, especially after the appearance of reef fish in the Eocene, and particularly with the rise of the scarids in the Miocene. We might predict that substantial aprons of sediment may not have been present on pre-Eocene reefs, and, likewise in the absence of the grain size reduction activities of reef fish, perhaps the modern style of coral reef lagoon may also not have appeared until that time. Also, barely explored are sedimentological consequences of differences in the geographical distribution of bioeroders – which is especially marked in fish populations due to differential extinction in the Atlantic during the mid-late Cenozoic (Bellwood, 1997). This extinction resulted in the conspicuous loss of large excavating scarids from Caribbean reefs. Such a comparison between the history of Cenozoic reef bioerosion in the Indo-Pacific and the Caribbean realms might therefore be particularly rewarding.

References

Alldredge, A. L., and King, J. M., 1977, Distribution, abundance and substrate preference of demersal zooplankton, *J. Exper. Mar. Biol. Ecol.* **44**:133-156.

Babcock, R. C., Bull, G., Harrison, P. L., Heywood, A. J., Oliver, J. K., Wallace, C. C., and Willis, B. L., 1986, Synchronous multispecific spawnings of 107 scleractinian coral species on the Great Barrier Reef, *Mar. Biol.* **90**:379-394.

Bak, R. P. M., 1983, Neoplasia, regeneration and growth in the reef-building coral *Acropora palmata, Mar. Biol.* **77**:221-227.

Bak, R. P. M., and Engel, M. S., 1979, Distribution, abundance and survival of juvenile hermatypic corals (Scleractinia) and the importance of life history strategies in the parent coral community, *Mar. Biol.* **54**:341-352.

Bak, R. P. M., and Steward-Van Es, Y., 1980, Regeneration of superficial damage in the scleractinian corals *Agaricia agaricites* F. *Purpurea* and *Porites asteroids, Bull. Mar. Sci.* **30**:883-887.

Bak, R. P. M., Brouns, J. J. W. M., and Heys, F. M. L., 1977, Regeneration and aspects of spatial competition in the scleractinian corals, *Proc. 3rd Int. Coral Reef Symp., Miami* **1**:143-148.

Bakus, G. J., 1974, Toxicity in holothurians: a geographic pattern, *Biotropica* **6**:229-236.
Bakus, G. J., 1981, Chemical defense mechanisms on the Great Barrier Reef, Australia, *Science* **211**:497-499.
Bellwood, D. R., 1996, The Eocene fishes of Monte Bolca: the earliest coral reef fish assemblage, *Coral Reefs* **15**:11-19.
Bellwood, D. R., 1997, Reef fish biogeography; habitat associations, fossils and phylogenies, *Proc. 8th Int. Coral Reef Symp., Panama* **2**:1295-1300.
Bellwood, D. R., and Schultz, O., 1991, A review of the fossil record of the parrotfishes (family Scaridae) with a description of a new *Calatomus* species from the middle Miocene (Badenian) of Austria, *Naturhistor. Mus. Wein* **92**:55-71.
Bergstom, J., 1979, Morphology of fossil arthropods as a guide to phylogenetic relationships, in: *Arthropod Phylogeny* (A. P. Gupta, ed.), Van Nostrand Reinhold, New York, pp. 3-58.
Birkeland, C., 1977, The importance of rate of biomass accumulation in early successional stages of benthic communities to the survival of coral recruits, *Proc. 3rd Int. Coral Reef Symp., Miami* **1**:16-21.
Bolser, R. C., and Hay, M. E., 1997, Are tropical plants better defended? Palatability and defenses of temperate vs. tropical seaweeds, *Ecology* **78**:2269-2286.
Bromley, R. G., 1975, Comparative analysis of fossil and Recent echinoid bioerosion. *Palaeontology* **18**:725-739.
Bromley, R. G., 1992, The palaeoecology of bioerosion, in: *The Palaeobiology of Trace Fossils* (S. K. Donovan, ed.), Wiley, Chichester, pp. 134-154.
Brown, B. E., 1997, Adaptations of reef corals to physical environmental stress, *Adv. Mar. Biol.* **31**: 221-299.
Choat, J. H., 1991, The biology of herbivorous fishes on coral reefs, in: *The Ecology of Fishes on Coral Reefs* (P. F. Sale, ed.), Academic Press, London, pp. 120-155.
Choat, J. H., and Bellwood, D. R., 1991, Reef fishes: their history and evolution, in: *The Ecology of Fishes on Coral Reefs* (P. F. Sale, ed.), Academic Press, London, pp. 39-66.
Conway Morris, S., 1979, Middle Cambrian polychaetes from the Burgess Shale of British Columbia. *Phil. Trans. R. Soc. Lond.* **B285**:227-391.
Coates, A. G., and Jackson, J. B. C., 1985, Morphological themes in the evolution of clonal and aclonal marine invertebrates, in: *Population Biology and Evolution of Clonal Organisms* (J. B. C. Jackson, L. W. Buss and R. E. Cook, eds.), Yale University Press, New Haven, pp. 67-106.
Connell, J. H., 1973, Population ecology of reef-building coral, in: *Biology and Geology of Coral Reefs, 2, Biology, 1* (O. A. Jones and R. Endean, eds.), Academic Press, New York, pp. 205-245.
Connell, J. H., 1978, Diversity in tropical rain forests and coral reefs, *Science* **199**:2-10.
Connell, J. H., 1980, Diversity and coevolution of competitors, or the ghost of competition past, *Oikos* **35**:131-138.
Edhorn, A.-S., 1977, Early Cambrian algae croppers, *Can. J. Earth Sci.* **14**:1014-1020.
Gill, G. A., and Coates, A. G., 1977, Mobility, growth patterns and substrate in some fossil and recent corals, *Lethaia* **10**:119-134.
Glynn, P. W., 1973, Ecology of a Caribbean coral reef. The *Porites* reef-flat biotope. II. Plankton community with evidence for depletion, *Mar. Biol.* **20**:297-318.
Glynn, P. W., 1988, Predation on coral reefs: some key processes, concepts and research directions, *Proc. 4th Int. Coral Reef Symp.* **1**:51-62.
Glynn, P. W., Stewart, R. H., and McCosker, J. E., 1972, Pacific coral reefs of Panama: structure, distribution and predators, *Geol. Rundschau* **61**:483-519.
Gould, S. J., and Vrba E. S., 1982, Exaptation – a missing term in the science of form, *Paleobiology* **8**:4-15.
Hall, V. R., and Hughes, T. P., 1996, Reproductive strategies of modular organisms: comparative studies of reef-building corals, *Ecology* **77**:950-963.
Harper, E. M., and Skelton, P. W., 1993, The Mesozoic Marine Revolution and epifaunal bivalves, *Scripta Geol., Special Issue* **2**:127-153.
Harvey, P. H., and Greenwood, P. J., 1978, Anti-predator defense strategies: Some evolutionary problems, in: *Behavioural Ecology: An Evolutionary Approach* (J. R. Krebs and N. B. Davies, eds.), Blackwell, Oxford, UK, pp. 129-151.
Hay, M. E., 1981, Herbivory, algal distribution and the maintenance of between-habitat diversity on a tropical fringing reef, *Am. Natur.* **118**:520-540.
Hay, M. E., 1984, Patterns of fish and urchin grazing on Caribbean coral reefs: are previous results typical?, *Ecology* **65**:446-454.
Hay, M. E., 1991, Fish-seaweed interactions on coral reefs: effects of herbivorous fishes and adaptations of the prey, in: *The Ecology of Coral Reef Fishes* (P. F. Sale, ed.), Academic Press, San Diego, pp. 96-119.
Hobson, E. S., 1968, Predatory behavior of some shore fishes in the Gulf of California, *Res. Report U.S. Fish Wildlife Service* **83**:1-92.

Hughes, T. P., 1994, Catastrophes, phase shifts, and large-scale degradation of a Caribbean coral reef, *Science* **265**:1547-1551.
Insalaco, E., 1996, Upper Jurassic microsolenid biostromes of north and central Europe: fabrics and depositional environment, *Palaeogeogr. Palaeoclim. Palaeoecol.* **121**:169-194.
Jackson, J. B. C., 1983, Biological determinants of present and past sessile animal distributions, in: *Biotic Interactions in Recent and Fossil Benthic Communities* (M. J. S. Tevesz and P. W. McCall, eds.), Plenum Press, New York, pp. 39-120.
Jackson, J. B. C., 1985, Distribution and ecology of clonal and aclonal benthic invertebrates, in: *Population Biology and Evolution of Clonal Organisms* (J. B. C. Jackson, L. W. Buss, and R. E. Cook, eds.), Yale University Press, New Haven, pp. 297-355.
Jackson, J. B. C., and Coates, A. G., 1986, Life cycles and evolution of clonal (modular) animals, *Phil. Trans. R. Soc. Lond.* **B313**:7-22.
Jackson, J. B. C., and Hughes, T.P., 1985, Adaptive strategies of coral-reef Invertebrates, *Am. Sci.* **75**:265-274.
Johnson, K. G., Budd, A. F., and Stemann, T. A., 1995, Extinction selectivity and ecology of Neogene Caribbean reef corals, *Paleobiology* **21**:52-73.
Jones, G. P., Ferrell, D. J., and Sale, P. W., 1991, Fish predation and its impact on the invertebrates of coral reefs and adjacent sediments, in: *The Ecology of Fishes on Coral Reefs* (P. F. Sale, ed.), Academic Press, London, pp.156-179.
Kaufman, L., 1981, There was biological disturbance on Pleistocene reefs, *Paleobiology* **7**:527-532.
Kobluk, D. R., 1985, Biota reserved within cavities in Cambrian *Epiphyton* mounds, Upper Shady Dolomite, South-western Virginia, *J. Paleontol.* **59**:1158-1172.
Kojis, B. L., and Quinn, N. J., 1985, Puberty in *Goniastrea favulus* age or size related?, *Proc. 5th Int. Coral Reef Symp.* **4**:289-293.
Krumm, D. K., and Jones, D. S., 1993, A new coral-bivalve association (*Actinastrea-Lithophaga*) from the Eocene of Florida, *J. Paleontol.* **67**: 945-951.
Lewis, S. M., 1986, The role of herbivorous fishes in the organization of a Caribbean reef community, *Ecol. Monogr.* **56**:183-200.
Lindberg, D. R., and Dwyer, K. R., 1983, The topography, formation and mode of home depression on *Collisella scabra* (Gould) (Gastropod: Acmaeidea), *Veliger* **25**:229-271.
Loya, Y., 1976, Skeletal regenation in a Red Sea coral population, *Nature* **261**:490-491.
McCall, J., Rosen, B. R., and Darrell, J., 1994, Carbonate deposition in accretionary prism settings: early Miocene coral limestones and corals of the Makhran Mountain Range in southern Iran, *Facies* **31**:141-178.
McKinney, F. K., and Jackson, J. B. C., 1989, *Bryozoan Evolution*, Unwin Hyman, Boston.
Meesters, E. H., Wesseling, I., and Bak, R. P. M., 1996, Partial mortality in three species of reef-building corals and the relation with colony morphology, *Bull. Mar. Sci.* **58**: 838-852.
Meesters, E. H., Wesseling, I., and Bak, R. P. M., 1997, Coral colony tissue damage in six species of reef-building corals: partial mortality in relation to depth and surface area, *J. Sea Res.* **37**:131-144.
Pleydell, S. M., and Jones, B., 1988, Boring of various faunal elements in the Oligocene-Miocene Bluff Formation of Grand Cayman, British West Indies, *Palaeontology* **62**:348-367.
Roniewicz, E., and Morycowa, E., 1993, Evolution of the Scleractinina in the light of microstructural data, in: *Proc. 6th Int. Symp. Fossil Cnidaria and Porifera* (P. Oekentorp-Küster, ed.), *Courier Forshungsinstitut Senckenberg* **164**:233-240.
Rosen, B. R., 1986, Modular growth and form of corals: a matter of metamers?, *Phil. Trans. R. Soc. Lond.* **B313**:115-142.
Rosen, B. R., 1998, Corals, reefs, algal symbiosis and global change: the Lazarus factor, in: *Biotic Response to Global Change: The Last 145 Million Years* (S. J. Culver and P. F. Rawson, eds.) Chapman and Hall, London.
Rylaarsdam, K. W., 1983, Life histories and abundance patterns of colonial corals on Jamaican reefs, *Mar. Ecol. Prog. Ser.* **13**:249-260.
Savazzi, E., 1982, Commensalism between boring mytilid bivalves and a soft bottom coral in the Upper Eocene of Northern Italy, *Paläont. Zeit.* **56**:165-175.
Schlichter, D., 1991, A perforated gastrovascular cavity in *Leptoseris fragilis*. A new improved strategy to improve heterotrophic nutrition in corals, *Naturwiss.* **78**:467-469.
Signor, P. W. III, and Brett, C. E., 1984, The mid-Paleozoic precursor to the Mesozoic Marine Revolution. *Paleobiology* **10**:229-245.
Skelton, P. W., 1991, Morphogenetic versus environmental cues for adaptive radiations, in: *Constructional Morphology and Evolution* (N. Schmidt-Kittler and K. Voegel, eds.), Springer-Verlag, Berlin, pp. 375-388.
Smith, A. B., 1984, *Echinoid Palaeobiology*. Allen and Unwin, London.
Steneck, R. S., 1982, Adaptive trends in the ecology and evolution of crustose coralline algae (Rhodophyta, Corallinaceae). Unpublished Ph.D. Dissertation, The John Hopkins University.

Steneck, R. S., 1983, Escalating herbivory and resulting adaptive trends in calcareous algal crusts, *Paleobiology* **9**:44-61.

Steneck, R. S., 1985, Adaptations of crustose coralline algae to herbivory: Patterns in space and time, in: *Paleobiology: Contempory Research and Applications* (D. F. Toomey and M. H. Nitecki, eds.), Springer-Verlag, Berlin, pp. 352-366.

Steneck, R. S., 1988, Herbivory on coral reefs: a synthesis, *Proc. 6th Int. Coral Reef Symp.* **1**:37-49.

Steneck, R. S., and Watling, L., 1982, Feeding capabilities and limitations of herbivorous molluscs: a functional group approach, *Mar. Biol.* **68**: 299-319.

Talbot, F. H., and Goldman, G., 1972, A preliminary report on the diversity and feeding relationships of the reef fishes on One Tree Island, Great Barrier Reef system, *Proc. Symp. Corals and Coral Reefs* **1**: 425-442.

Thayer, C. W., 1983, Sediment-mediated biological disturbance and the evolution of marine benthos, in: *Biotic Interactions in Recent and Fossil Benthic Communities* (M. J. S. Tevesz and P. W. McCall, eds.), Plenum Press, New York, pp. 480-625.

Tunnicliffe, V., 1981, Breakage and propagation of the stony coral *Acropora cervicornis*, *Proc. Nat. Acad. Sci, USA* **78**: 2427-2431.

Tyler, J. C., 1980, Osteology, phylogeny, and higher classification of the fishes of the order Plectognathi (Tetradontiformes), *NOAA Technical Report NMF Circular* **434**:1-422.

Van Belle, R. A., 1977, Sur la classification des Polyplacophora: III. Classification systematique des Subterenochitonidae (Neoloricata: Chitonina), *Inf. Societe Belgique Malacologique* **5**:15-40.

Vermeij, G. J., 1987, *Evolution and Escalation: An Ecological History of Life*, Princeton University Press, Princeton.

Wellington, G. M., 1982, Depth zonation of corals in the Gulf of Panama: control and facilitation by resident reef fish, *Ecol. Monogr.* **52**:223-241.

Wells, J. W., 1956, Scleractinia, in: *Treatise on Invertebrate Paleontology. Coelenterata* (R. C. Moore, ed.), Geological Society of America and University of Kansas Press, Boulder, Colorado and Lawrence, Kansas, USA, pp. 328-400.

Wood, R., 1999, *Reef Evolution*, Oxford University Press, Oxford.

Wood, R., Zhuravlev, A. Yu., and Debrenne, F., 1992, Functional biology and ecology of Archaeocyatha, *Palaios* **7**:131-156.

Chapter 3

Trilobites in Paleozoic Predator-Prey Systems, and Their Role in Reorganization of Early Paleozoic Ecosystems

LOREN E. BABCOCK

---

1. Introduction 55
2. Predation on Trilobites 57
   2.1. Predation Scars 57
   2.2. Gut Contents 67
   2.3. Coprolites..68
   2.4. Broken Sclerites..69
   2.5. Borings..69
   2.6. Predation Resistance or Response..70
3. Predation by Trilobites 75
   3.1. Phylogenetic Evidence for Trilobite Predation 76
   3.2. Functional-Morphological Evidence of Predatory Capability of Trilobites....................76
   3.3. Trace Fossil Evidence of Predatory Behavior by Trilobites.....................80
   3.4. Trilobite Gut Contents 81
4. Final Comments: Predation on and by Trilobites in the Context of Ecosystem
   Reorganization During the Early Paleozoic..81
   References 85

---

1. Introduction

Predation is a fundamental ecological process that has profound effects on the morphology, distribution, abundance, and evolution of metazoans. The earliest verified records of predation date to the Neoproterozoic-Cambrian transition interval (e.g., Conway Morris and Jenkins, 1985; Babcock, 1993a; Bengtson and Yue, 1992;

LOREN E. BABCOCK • Department of Geological Sciences, The Ohio State University, Columbus, Ohio 43210.

Predator-Prey Interactions in the Fossil Record, edited by Patricia H. Kelley, Michał Kowalewski, and Thor A. Hansen. Kluwer Academic/Plenum Publishers, New York, 2003.

Bengtson, 1994; Conway Morris and Bengtson, 1994; Nedin, 1999; Jago and Haines, in press), but the impact of predation almost certainly has a much deeper evolutionary history. Among the earliest and most widespread lines of evidence for the importance of predation in the early Paleozoic comes from the record of trilobites. As biomineralized animals, trilobites have left an excellent fossil record that extends from the latter part of the Early Cambrian (e.g., Zhang, 1987; Geyer, 1996, 1998; Geyer and Palmer, 1995; Luo and Jiang, 1996; Hollingsworth, 1999; Geyer and Shergold, 2000; Peng and Babcock, 2001), c. 520 Ma, to the end of the Permian (e.g., Brezinski, 1992), c. 248 Ma. Predation on and by trilobites evidently exerted influence on the morphological development of metazoans, as well as on ecosystem development, through the Paleozoic.

This paper is a review of the fossil record of predation on and by trilobites. Evidence of lethal and sublethal predation is discussed, together with implications for interpreting the Paleozoic history of predator-prey relationships in which trilobites were prey. Trilobites are a diverse group of arthropods, and they manifested a considerable range in size, life habits, and marine habitats (e.g., Fortey, 1985; Briggs *et al.*, 1994; Whittington, 1997c; Fortey and Owens, 1999). During the Paleozoic, trilobites were evidently prey for a variety of predators, both large and small. Predators exercised a variety of techniques for immobilizing or devouring their trilobite prey; among the techniques are grasping and biting (e.g., Whittington and Briggs, 1985; Nedin, 1999), and boring (e.g., Størmer, 1931; Babcock, 1993a; Babcock and Peng, 2001), of the exoskeleton. As summarized here, the results of predator-prey interactions involving trilobites as prey provide data bearing on important paleobiological problems including the rise and development of Paleozoic marine predators, skeletization of metazoans during the early Paleozoic, wound response, internal organization, and lateralization of the metazoan nervous system.

One significant, but underemphasized, aspect of predator-prey interactions during the Paleozoic is predation by trilobites. Primitively, trilobites are inferred to have been carnivores (Fortey and Owens, 1999), and trace fossils provide a record of hunting behavior in certain trilobites beginning in the Cambrian (e.g., Bergström, 1973; Jensen, 1990; Conway Morris, 1998; Fortey and Owens, 1999) and extending through at least the middle Paleozoic (Brandt *et al.*, 1995; Osgood and Drennen, 1975; Wenndorf, 1990). Indirect support for a predatory habit comes from morphological characteristics and new information concerning gut contents.

Predation on and by trilobites in the Early Cambrian is viewed as part of a much larger reorganization of marine ecosystems in which predation played a major role in guiding the course of biological evolution, and also played an influential role in changing organism-substrate relationships. Trilobites were a major, and well-documented, component of the Cambrian biotic explosion (e.g., Cloud, 1968; Runnegar, 1982, 1989; Conway Morris, 1985, 1998; Lipps and Signor, 1992; Bengtson, 1994; Valentine, 1994; Chen *et al.*, 1997; Hou *et al.*, 1999; Babcock et al., 2001), and were also intimately linked to burrowing and infaunalization of the substrate beginning in the Cambrian (e.g., Seilacher and Pfluger, 1994; Conway Morris, 1998; Bottjer *et al.*, 2000). The early record of trilobites, including aspects of the record relating to predatory activity and predator avoidance, emphasizes that the Cambrian biotic revolution (a series of protracted events in which biomineralized skeletons played only a minor role; Bengtson, 1994) and the Cambrian substrate revolution (Seilacher and Pfluger, 1994; Bottjer *et al.*, 2000) are inseparably linked as components of a larger-scale early Paleozoic marine revolution. Predator-prey systems involving trilobites and other organisms underwent considerable, and dramatic, evolution during the Early Cambrian, although it is perhaps more appropriate to view the Early Cambrian as a starting point of sorts for reorganization of biotic systems that had a long Precambrian history (e.g., Stanley, 1999). Reorganization, some of it apparently driven by predation pressure (e.g., Stanley, 1976; Brasier, 1979), can be interpreted as part of a continuum that extended at least to the Late Cambrian. Information concerning trilobite predator-prey interactions is central to establishing the link between the evolution of marine organisms and changes in substrate conditions, and to illustrating the duration of predation-influenced changes through the early Paleozoic.

2. Predation on Trilobites

Trace fossils, body fossils, and functional-morphological considerations provide an indication that trilobites and their close relatives fell prey to predatory animals. The principal lines of evidence now known for predation on trilobites are: (1) healed, sublethal predation scars; (2) sclerites contained in the alimentary tracts of animals; (3) sclerites, including broken ones, contained in coprolites; (4) broken, disarticulated sclerites independent of recognizable gut tracts and coprolites, particularly when concentrated in stratigraphic intervals containing the body fossils of inferred predators; (5) borings; and (6) predation-resistant ecological strategies inferred from functional morphological considerations and behavioral evidence.

2.1. Predation Scars

Sublethal predation scars on trilobites and their close relatives, the naraoiids, provide strong documentation of predator-prey interactions in Paleozoic marine ecosystems and indicate that predation has been an important forcing factor in evolution since at least the Early Cambrian (e.g., Conway Morris and Jenkins, 1985; Owen, 1985; Babcock and Robison, 1989; Babcock, 1993*a*; Nedin, 1999; Jago and Haines, in press). They also provide information about wound response in trilobites (Ludvigsen, 1977; Conway Morris and Jenkins, 1985; Jell, 1989; Babcock, 1993*a*; Rudkin, 1979, 1985; Owen, 1985) and indicate that cellular repair mechanisms were well advanced in these arthropods by at least the Early Cambrian. Interpretation of predation scars can be somewhat problematic in some examples, however, and, for this reason, criteria have been established for distinguishing these types of injuries from other sublethal malformations. Owen (1985) developed a useful classification of malformations in trilobites based principally on the morphology of malformations and their inferred origin. Babcock (1993*a*) refined this classification, largely through application of information from modern animals concerning cellular reaction to injury. This information, although originally applied to the determination of sublethal predation in trilobites, can be applied to the interpretation of malformations in other skeletonized organisms from the fossil record.

2.1.1. Classification and Origins of Malformations

From modest beginnings as scientific curiosities or monstrosities, malformations (Fig. 1; see Fig. 5.1) have emerged as an important source of paleobiological information concerning trilobites and some coexisting Paleozoic organisms. Recognition of trilobite malformations in paleontologic literature can be traced to the 1840s (Portlock, 1843), but malformations were rarely reported until the middle of the twentieth century.

Most documented examples of malformed trilobites were included in monographic works emphasizing systematics. In one interesting example, an olenellid trilobite having what is now recognized as a healed predation scar was described as a new species, *Olenellus peculiaris* Resser and Howell (1938, p. 223, pl. 6, fig. 10; see Campbell, 1969; Babcock, 1993a; Fig 1.2). Walcott (1883) was the first to provide extensive interpretive details about a malformation in a trilobite. Since that time, a variety of authors have emphasized the paleobiological information provided through study of malformations (e.g., Burling, 1917; Ludvigsen, 1977; Šnajdr, 1979a, 1979b, 1981; Rudkin, 1979, 1985; Vorwald, 1982; Owen, 1985; Briggs and Whittington, 1985b; Conway Morris and Jenkins, 1985; Babcock and Robison, 1989; Babcock, 1993a,b; Jell, 1989; Han and Zhang, 1991; Nedin, 1999). To date, more than 300 malformed trilobites have been reported (excluding Oehlert, 1895, which included approximately 800 teratological specimens of the trinucleid *Onnia pongerardi*).

Malformations in trilobites are categorized into three types (Owen, 1985; Babcock, 1993a), all of which are macroscopic in scale, according to the proximal cause of the abnormality: (1) injuries; (2) teratological conditions; and (3) pathological conditions. Predation scars are one form of injury. For clarity in the interpretation of predation scars, and for distinguishing them from other types of malformations, a brief review of causes and physical manifestations of malformations is provided here.

The root cause of macroscopic injuries, teratologies, and pathologies in all animals is damage or injury at the cellular level followed by a cellular reaction to the damage. Because of potential misinterpretation about use of the term injury at the cellular (or microscopic) and organismic (or macroscopic) levels, I follow historical practice and refer to injuries at a macroscopic level in trilobites as "injuries," whereas injuries at a microscopic level are referred to as "cellular injuries."

Malformations to the exoskeletons of trilobites observed at macroscopic (or organismic) scale resulted from alteration of the structure or function of tissue cells in response to some type of injury at the microscopic (or cellular) scale. Changes in the morphology of once-living epidermis responsible for exoskeletal secretion are faithfully reproduced in the sclerotized exoskeleton during the molt cycle following that in which injury occurred. Certain types of exoskeletal damage (macroscopic injuries) evidently were progressively masked over through successive molt cycles (Ludvigsen, 1977; Rudkin, 1979; Owen, 1983, 1985). Cellular injuries can result from physical agents, macrobiotic agents, microbial agents, chemical agents, hypoxia, immunological sensitivity reactions, malnutrition, inherited defects, aging, and other causes (Babcock, 1993a). Cells of living animals adapt to injury by four processes: (1) hypertrophy; (2) metaplasia; (3) hyperplasia; and (4) atrophy (Purtilo, 1978). Changes to the trilobite exoskeleton resulting from these processes alone or working in concert are identifiable as macroscopic (organismic-scale) malformations, and all of these conditions except atrophy are demonstrably associated with sublethal predaceous injuries on trilobites.

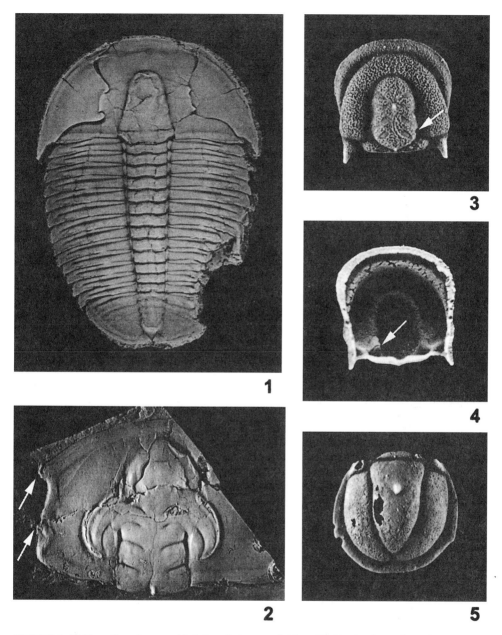

FIGURE 1. Malformations due to sublethal predation on trilobites. (1) *Elrathia kingii* from the Wheeler Formation (Middle Cambrian) of Utah showing arcuate scar on right posterior part of thorax and pygidium; callusing, deformation of the exoskeleton, and an anomalous spine are evident; length = 3.1 cm; University of Kansas Museum of Paleontology (KUMIP) 204773. (2) Cephalon of *Olenellus* (*Olenellus*) *thompsoni* (holotype of *O. peculiaris*) from the Kinzers Formation (Lower Cambrian) of Pennsylvania showing large W-shaped scar on left side; bases of two anomalous spines, broken in preparation, are evident (arrows); length = 4.6 cm; U.S. National Museum of Natural History (USNM) 90809. (3, 4) Cephalon of *Arthrorhachis elspethi* from the Edinburg Limestone (Middle Ordovician) of Virginia in dorsal (3) and ventral (4) views showing small boring along right posterior of axis (arrows); length = 2 mm; KUMIP 204772. (5) Pygidium of *Peronopsis interstricta* showing two small scars on left posterior border; length = 3 mm; KUMIP 204771.

Hypertrophy involves an increase in the size of cells resulting from an increased production of macromolecules; it occurs without cell division (Purtilo, 1978). Hypertrophy occurs in two forms, hormonal and compensatory, but only compensatory hypertrophy has been recognized in trilobites (Babcock, 1993a). Compensatory hypertrophy occurs when cells are either diseased or removed, and the remaining cells compensate for the loss by an increase in mass. Deformation through enlargement around an injured or diseased area of trilobite exoskeleton (Figs. 1.1, 1.2) is interpreted as the result of this process (Babcock, 1993a); it provides a clear indication that the injury occurred during the life of a trilobite, and that it was sublethal.

Hyperplasia involves new cellular growth by an increase in the frequency of mitotic divisions of cells (Purtilo, 1978). Regeneration (e.g., Šnadjr, 1979b; Owen, 1985; Babcock, 1993a), which is the process by which lost tissue is restored, is the result of this process. Scars on exoskeletons, which are key to distinguishing sublethal injuries from lethal injuries (Babcock and Robison, 1989; Babcock, 1993a; Pratt, 1998), are the most common expression of hyperplasia in trilobites. Exoskeletal regeneration through successive molt stages (Ludvigsen, 1977; Rudkin, 1979; Owen, 1983, 1985; Babcock, 1993a), and growth of some neoplasms (gall-like swellings or tumors) through uncontrolled cellular proliferation (e.g., Šnajdr, 1978a; Bergström and Levi-Setti, 1978; Babcock, 1993a, 1994) rather than as a response to parasitic infection (e.g., Šnajdr, 1978b, 1979c, 1981; Conway Morris, 1981; Přibyl and Vaněk, 1981; Owen, 1985), are other expressions of hypertrophy. Growth of anomalous spines, typically in response to predaceous injury (Babcock and Robison, 1989; Babcock, 1993a; Figs. 1.1, 1.2), resulted from either hyperplasia or, less likely, metaplasia.

Atrophy involves a decrease in cell size commensurate with a decrease in the size of the affected tissue and can result from disease or decreased use, workload, blood supply, nutrition, or hormonal stimulation of a tissue (Purtilo, 1978). Atrophication is probably the cause of extreme reduction in the size of certain areas of the trilobite exoskeleton where such reduction is unrelated to scarring (e.g., Ludvigsen, 1979, p. 77, fig. 56), but, to date, has not been associated with sublethal injuries in trilobites.

Metaplasia involves the transformation of one cell type into another type. Generally a highly specialized cell transforms to a less specialized cell (Purtilo, 1978). Metaplasia has not been recognized unequivocally from trilobite exoskeletons. Where anomalous spines have grown on injured surfaces (Babcock and Robison, 1989; Babcock, 1993a; Figs. 1.1, 1.2), the possibility that metaplasia has occurred cannot be ruled out. It is more likely, though, that such spines are the result of hyperplasia (atypical regeneration).

Injuries at the macroscopic (or organismic) scale result from physical breakage of the exoskeleton. Injuries that caused malformation of trilobite exoskeletons were sublethal, as evinced by healed and callused exoskeleton around the broken surface. Regeneration, or hyperplasia, both to close the wound initially, and in other forms (perhaps including anomalous spines) is key evidence of a sublethal injury. Commonly, deformed exoskeleton occurs in the vicinity of the broken area. Deformation probably resulted from compensatory hypertrophy, which would have permitted the mass of cells remaining after injury to compensate for the loss. Evidence of healing, regeneration, and deformation readily distinguishes such sublethal injuries from lethal injuries. Lethal damage to trilobites (Pratt, 1998) and postmortem scavenging (Jago, 1974; Fig. 2.3) is essentially indistinguishable from purely mechanical breakage of postmortem or postecdysial remains (Babcock, 1993a; compare Tshudy et al., 1989).

Sublethal injuries other than borings (see section 2.5) are divided into two types based on inferred origin: (1) sublethal predation; and (2) uncertain origin (Babcock and Robison, 1989; Babcock, 1993a). Injuries attributed with confidence to lacerations incurred during predaceous attack (Figs. 1.1, 1.2, see Fig. 5.1) are: (1) developed on areas of the exoskeleton not likely to have been accidently injured (such as marginal spines, wide and narrow cephalic or pygidial borders, and areas that were operational in molting); (2) developed over a relatively extensive area of the body (often on two or more adjacent sclerites); and (3) generally arcuate to triangular or asymmetrically W-shaped (not simple straight, slightly curved, or jagged breaks as might be expected from accidental damage). Further convincing, but relatively rare, evidence of predation is a large injury (presumably owing to biting) on one side of the trilobite body with minor injury (presumably owing to detainment by a grasping appendage) on the opposite side of the body (Nedin, 1999). Injuries that cannot be attributed with confidence to sublethal predation are considered to be of uncertain origin. In practice, some minor injuries that were the result of unsuccessful predaceous attack probably have been classified as of uncertain origin because they lack strong evidence of attack.

Teratological conditions in trilobites and other animals result from genetic or embryological malfunctions. Disrupted thoracic segments, incomplete separation of segments in the thorax or pygidium, and local atrophy not associated with scarring are examples of teratologies. In practice, it is often difficult to distinguish some teratological conditions from healed injuries that were masked over by regeneration through several molt cycles, and from such pathological conditions as disease-related atrophication.

Pathological conditions result from disease or parasitic infection. Described examples of parasitism in trilobites are small vermiform borings and gall-like swellings (Conway Morris, 1981). Vermiform trace fossils associated with trilobite exoskeletons (Portlock, 1843, p. 360, Pl. 21, fig. 5a; Størmer, 1931; Conway Morris, 1981) are quite common in places but rarely described. From fossils, it is not always easy to determine whether vermiform borings resulted from a parasitic relationship or whether they were formed in an exuvium or carcass (Owen, 1985). Alternatively, in some cases, traces could have been produced in sediment layers slightly subjacent or superjacent to trilobite remains, and, through compaction of sediment, brought into apparent contact with a trilobite exoskeleton. One possible example is the specimen of a putative worm associated with the exoskeleton of an *Olenellus* from the Cambrian of Pennsylvania (Ruedemann and Howell, 1944). This interpretation would apply best to trace fossils that cross the margin of a trilobite exoskeleton into surrounding matrix.

Borings into trilobite exoskeletons can be difficult to interpret, and may in fact result from either parasitism or the activity of a micropredator. Boucot (1990, p. 174) noted that the distinction between parasites and some micropredators may be ambiguous. As noted above, vermiform borings and swellings resembling galls can, in some circumstances, be attributed to parasitism (Conway Morris, 1981). Other borings that more closely resemble drill holes have not been clearly linked to parasitism because of a lack of cellular response beyond simple wound closure by the trilobites, and are therefore more likely the result of sublethal or lethal predation.

Determining the origin of some tumor-like growths on the exoskeletons of trilobites can be difficult. Such growths, which are commonly referred to as neoplasms, have been recorded from several trilobite clades (Bergström and Levi-Setti, 1978; Šnajdr, 1978b, 1979c, 1981; Conway Morris, 1981; Přibyl and Vaněk, 1981; Owen, 1985; Babcock, 1993a, 1994), but the association of such growths with Cambrian paradoxidids

(Bergström and Levi-Setti, 1978; Šnadjr, 1978a; Owen, 1985; Babcock, 1993a, 1994) is striking. The growths are usually interpreted as the result of cancerous, uncontrolled cellular proliferation (Šnadjr, 1978a; Babcock, 1994), but the possibility that some represent cellular response to an invading organism (predaceous or parasitic borer) cannot be ruled out without more detailed study.

2.1.2. Temporal Patterns in Predation Scars

Criteria for distinguishing scars due to sublethal predaceous attack from other forms of malformations are discussed in section 2.1.1. This section provides a review, and some new interpretation, of patterns that have emerged from studies on sublethal predation scars.

Sublethal predation scars in trilobites have a strong tendency to be preserved on the pleural lobes, the posterior part of the body, and the right side of the body (Babcock and Robison, 1989; Babcock, 1993a,b). Despite the addition of more specimens to the data sets published in the late 1980s and early 1990s (Babcock and Robison, 1989; Babcock, 1993a), there is little difference in the percentages of injury occurrences. Updated numbers concerning the positions of injuries due to sublethal predation relative to tagmata are provided in Table 1, and occurrences of healed injuries relative to the right and left sides of the trilobite body are provided in Table 2. Positions of axial occurrences of predation scars relative to those of the margins are not presented in tabular form because all preserved predation scars other than sublethal borings (discussed in section 2.5) on trilobites are on the pleural lobes. Presumably this uniformity of occurrence is because any substantial attack on the axial lobe, which contained most of the vital organs of the nervous, circulatory, and alimentary systems, would have been fatal (Babcock and Robison, 1989; Babcock, 1993a,b), and fatalities due to predation are not included in the calculations.

As indicated in Table 1, predation scars are concentrated on the posterior half of the body. Data are not strictly comparable among species because of differences in the relative length of the pygidium, the number of thoracic segments, taphonomic biases favoring preservation and ultimate discovery of pygidia at a higher rate than other sclerites, taphonomic biases against preservation and ultimate discovery of complete cephalic shields, and other factors. Nevertheless, the overall pattern is clear. Predation scars occur on the posterior half of the trilobite body in 75% of specimens. Such scars occur on the anterior half of the body in 24% of specimens, and in most of these

TABLE 1. Incidence of injuries due to sublethal predation relative to the tagmata of polymeroid trilobites. Percentages are rounded to the nearest integer. The numbers reflect injuries other than borings in trilobite exoskeletons.

Affected region	n (%)
Cephalon	13 (13%)
Cephalon to anterior thorax	1 (1%)
Anterior to middle thorax	10 (10%)
Posterior thorax to pygidium	36 (37%)
Pygidium	36 (37%)
Cephalon to pygidium	1 (1%)
Total sublethal predation scars	97

examples, they are actually on the margins of the thoracic segments or at the posterolateral corners of the cephalon. Numbers for the cephalic and pygidial regions can be disregarded because of taphonomic overprinting, and if so, the frequency of occurrence of scars on the posterior thorax is 78%, compared to 22% on the anterior thorax.

Two hypotheses have been posed to explain the predominance of posteriorly located predation scars: (1) predators pursuing prey from the rear would be most likely to make first contact on the posterior region of the trilobite body (Babcock and Robison, 1989; Babcock, 1993a); and (2) prey were more likely to escape if seized on the posterior part of the body because substantial injury even to marginal areas of the cephalic area or the anterior thorax would have been lethal more commonly than injury to the tail shield or posterior thorax (Babcock, 1993a). Observations on the behavior of modern fishes (Webb, 1975; Eaton et al., 1977) support the first hypothesis. The startle response of trilobites, like that of fishes, may have involved an initial quick turn of the body by unilateral contraction of muscles in the thorax and tail region. This action would rapidly displace the head, which is interpreted as the most vulnerable region of the body, and prepare the animal for escape. In displacing the head, however, the posterior region of the body would become compromised and subject to attack.

As indicated in Table 2, sublethal predation scars are much more commonly preserved on the right side of the trilobite body than on the left side. The number of recorded injuries on the right sides of trilobites exceeds the number expected if the injuries were inflicted purely by chance; in such a case, a random distribution would be expected. Also, there is a relative scarcity of injuries to both sides of the trilobite body (see Nedin, 1999).

The tendency toward lateral asymmetry of sublethal predation scars on polymeroid trilobites may be linked to right-left behavioral asymmetry, or lateralization, in trilobites, their predators, or both trilobites and their predators (Babcock and Robison, 1989; Babcock, 1993a,b). Support for this hypothesis is available in published examples

TABLE 2. Incidence of injuries due to sublethal predation and uncertain causes on right, left, and both sides of 185 polymeroid trilobites, excluding eodiscids. Data for post-Cambrian occurrences have been aggregated because the data for Cambrian trilobites greatly outnumber the data for any other period of the Paleozoic Era. Percentages are rounded to the nearest integer. The numbers reflect injuries other than borings in trilobite exoskeletons. Injuries were analyzed using a binomial test in which the expected distribution of injuries to the right and left sides is 1:1. Injuries due to sublethal predation are statistically significant for Cambrian trilobites ($P < 0.001$, $n = 59$) and post-Cambrian trilobites ($P < 0.05$, $n = 41$), but injuries of uncertain origin are not statistically significant for Cambrian trilobites ($P > 0.10$, $n = 34$), post-Cambrian trilobites ($P > 0.10$, $n = 51$), or pooled Cambrian and post-Cambrian trilobites ($P > 0.10$, $n = 85$).

	Right side only n (%)	Left side only n (%)	Both sides n (%)
Sublethal predation scars:			
Cambrian trilobites	44 (75%)	13 (22%)	2 (3%)
Post-Cambrian trilobites	25 (61%)	12 (29%)	4 (10%)
All trilobites	69 (69%)	25 (25%)	6 (6%)
Injuries of uncertain origin			
Cambrian trilobites	18 (53%)	14 (41%)	2 (6%)
Post-Cambrian trilobites	26 (51%)	23 (45%)	2 (4%)
All trilobites	44 (52%)	37 (44%)	4 (5%)
Total injuries	113	62	10

of behavioral asymmetry in a wide, and continually increasing, array of present-day animals (e.g., Bradshaw, 1988, 1989, 1991, and references therein).

Recent work by Nedin (1999) has elucidated the means by which some predators inflicted injuries to trilobites. Functional analysis indicates that some Cambrian anomalocaridids used their large anterior appendages to restrain trilobite prey and to deliver one edge of the trilobite body to the ventral mouth. Then, together with flexure of the head (Collins, 1996; Nedin, 1999), the trilobites were repeatedly flexed back and forth until the exoskeleton was broken through (producing a bite mark) with compression to the endocuticle and tension to the exocuticle.

Nedin's (1999) interesting conclusions about the technique used by some anomalocaridid species to bite trilobites explain two important aspects of the record of those sublethal predation scars. First, his analysis explains why scars are concentrated on the left and right sides of trilobites, not on the anterior and posterior sides: for back-and-forth flexure to be effective in cracking a trilobite exoskeleton, the stress must be delivered perpendicular to the direction along which sclerites articulate. Repeated flexure of a trilobite from the front or rear would merely cause the body to flop around without breaking. Secondly, Nedin's (1999) analysis explains the relative rarity of bilateral injuries (Conway Morris and Jenkins, 1985; Rudkin, 1985; Nedin, 1999), with at least one of the scars being inflicted by the grasping appendages as the prey tried to escape. According to Table 2, bilateral injuries occur with a frequency of about 3% in Cambrian trilobites, and with a frequency of about 10% in post-Cambrian trilobites. A bilateral predation scar on an Ordovician trilobite, *Pseudogygites*, was ascribed by Rudkin (1985) to the predaceous activity of an endoceratid nautiloid. In the post-Middle Cambrian, following the apparent extinction of anomalocaridids, cephalopods were undoubtedly important predators of trilobites, and it is quite possible they employed a biting-and-flexing attack style (while restraining prey), similar in a general way to that of anomalocaridids.

An increased frequency of both unilateral left-sided predation scars and bilateral predation scars in post-Cambrian trilobites suggests evolution in predatory behavior among post-Cambrian carnivores, or at least a change in the dominant predators of trilobites. Unilateral left-sided predation scars in the post-Cambrian are 29% (compared to 22% in the Cambrian), and bilateral predation scars in the post-Cambrian are 10% (compared to 3% in the Cambrian). Changes in predation strategy on trilobites may have included innovations in capture and restraining techniques, biting style, or breakage through flexing of the exoskeleton. Less strongly lateralized behavior, or fluctuating patterns of behavior mimicking changes in the evasive behavior of prey, may have been a more successful strategy with some post-Cambrian predators (e.g., cephalopods and fishes) than it was among Cambrian predators (e.g., anomalocaridids).

Predation scars can be viewed as a metric of the relative efficiency of predatory activity. Efficient predators should leave little record or, at most, a weak record, in sublethal scars, whereas the record of sublethal scars should be stronger among the prey animals that escaped rapacious carnivores.

Table 2 provides a comparison of known sublethal predation scars on Cambrian (mostly Early and Middle Cambrian) and post-Cambrian trilobites. The raw data of 59 scars on Cambrian trilobites and 41 scars on post-Cambrian trilobites can be better compared by applying a frequency of occurrence of sublethal scars according to a simple formula:

$$F = n / t$$

in which F = frequency of occurrence of scars per million years; n = number of observed specimens having scars; and t = millions of years per time interval. Lower values of F imply greater predatory efficiency. Using the ages provided on the 1999 Geologic Time Scale (Geological Society of America), the frequency of occurrence (F) of Cambrian predation scars is 1.93 per million years, and the frequency of occurrence (F) of post-Cambrian predation scars is 0.17 per million years. The trend toward fewer specimens retaining sublethal predation scars after the Cambrian (represented by an order of magnitude decrease in F) suggests that predatory behavior had become increasingly efficient after the late Middle Cambrian. It is perhaps no coincidence that an increase in predatory efficiency closely parallels both the time of extinction of anomalocaridids and the time during which cephalopods and fish underwent early adaptive radiations. Alternative possibilities for the observed pattern are: (1) fewer attempted attacks on trilobites in the post-Cambrian; and (2) taphonomic biasing against preservation of predation scars on trilobites in the post-Cambrian.

2.1.3. Predators of Trilobites

During the Early and Middle Cambrian, likely predators of trilobites or other arthropods would have mostly included vagile nonmineralizing arthropods. Among many potential predators, anomalocaridids (e.g., *Amplectobelua*, *Anomalocaris*, *Parapeytoia*, *Laggania*; Briggs, 1979, 1994; Briggs and Whittington, 1985a; Whittington and Briggs, 1985; Briggs et al., 1994; Chen et al., 1994, 1997; Hou et al., 1995, 1999; Collins, 1996; Chen and Zhou, 1997; Hou and Bergström, 1997; Nedin, 1999; Fig. 2.4), phyllocarids (e.g., *Canadaspis*; Briggs, 1978; Briggs et al., 1994; Chen et al., 1994, 1997; Hou et al., 1999), *Sidneyia* (e.g., Bruton, 1981; Briggs et al., 1994), *Utahcaris* (Conway Morris and Robison, 1988; Fig. 2.1), *Sanctacaris* (Briggs and Collins, 1988; Briggs et al., 1994), *Kuamaia* (Shu and Zhang, 1996), and so-called large appendage arthropods (e.g., *Jianfengia*; Chen et al., 1994, 1997; Hou et al., 1997, 1999) are notable. Other possible predators include mineralizing arthropods (trilobites such as *Olenoides*; Whittington, 1975, 1980, 1992, 1997c), some worms (cf. Conway Morris, 1977; Briggs et al., 1994), some cnidarians (Alpert and Moore, 1975, but see Babcock, 1993a; cf. Chen and Erdtmann, 1991), some chordates (Shu et al., 1999), and, by the Late Cambrian, some cephalopods (e.g., Crick, 1981). The lack of mineralized jaws in some Cambrian arthropods (anomalocaridids) has been cited as evidence against durophagous predation on trilobites (Hou et al., 1995), but Nedin (1999) demonstrated that heavily sclerotized arthropod cuticle can be used to break the calcite-reinforced cuticle of trilobites. Other arthropods and probably some non-arthropods conceivably developed behaviors adequate for breaking mineralized skeletons of trilobites despite lacking mineralized jaw-like elements themselves. A relevant observation is that modern *Limulus* can break clam shells with its gnathobasic jaws (Fortey and Owens, 1999).

Some Cambrian and Ordovician predators were evidently quite small, as indicated by minute bite marks on small trilobites. Babcock (1993a; Fig. 1.5) and Buchholz (2000) illustrated sublethal predation scars on the pygidial margins of agnostoid trilobites; and Zhang (1989) illustrated a similar scar on the pygidial margin of an eodiscid trilobite.

The substantial number of sublethal predation scars on trilobites dating from the Early Cambrian demonstrates that predation must have played a significant role in ecosystems from at least that point in time. A corollary to this hypothesis is that predation played a role in guiding the course of evolution among metazoans through selection against nonskeletized or lightly skeletized animals. The increase of skeletonized animals, and especially biomineralized skeleton-bearing animals, across the Late Neoproterozoic-Early Cambrian time interval can be viewed as a response to both predation pressure (e.g., Bengtson, 1968, 1994; Stanley, 1976; Miller and Sundberg, 1984; Babcock, 1993a; Conway Morris and Bengtson, 1994; Conway Morris, 1998), and to other factors (Bengtson, 1994) such as changes in marine geochemical systems. Biomineralization of the trilobite exoskeleton, along with biomineralization or other skeletization in many clades of metazoans, can be interpreted as an early expression of escalation in predator-prey systems (cf. Vermeij, 1987; Signor and Brett, 1984; Kelley and Hansen, 1993). Some of the earliest metazoan predators (notably the anomalocaridids) used well-sclerotized chitinous cuticle for durophagous activity, and that may have been a contributing factor in their relative inefficiency compared to such Late Cambrian and post-Cambrian predators as cephalopods and fishes. Extinction of anomalocaridids, which probably occurred about the end of the Middle Cambrian, coincides with a marked decline in the record of sublethal predation scars on trilobites, and may have been partly related to natural selection against an inefficient predation mode.

A greater, and seemingly more capable, array of predators appeared in marine ecosystems through the Late Cambrian-Early Ordovician interval to the Devonian; their appearance was associated with the Mid-Paleozoic Marine Revolution (Signor and Brett, 1984). Potential predators during this time include phyllocarids (Middle Cambrian to Carboniferous), eurypterids (Ordovician to Permian), malacostracan crustaceans (Late Devonian to present), some trilobites (Early Cambrian to Permian), jawless fishes (Late Cambrian to the present), jawed fishes (Middle Devonian to the present), cephalopods (Late Cambrian to present), asterozoan and ophiuroid echinoderms (Early Ordovician to present). The rise of numerous fish and cephalopod species bearing shell-breaking or shell-crushing hard parts during the middle Paleozoic suggests that the style of predation on trilobites had reached a new level by that time. A decline in the relative percentage of unilateral right-sided predation scars on trilobites in the post-Cambrian (22%) compared to the Cambrian (74%; Table 2), together with the increase in bilateral predation scars (3% in the Cambrian versus 10% in the post-Cambrian; Table 2) tends to correlate with increases in predator types and capturing techniques through the middle to late Paleozoic.

Another consideration is that durophagy (shell crushing capability) per se is not necessarily a prerequisite for an animal to be considered predaceous. Non-durophagous predators could threaten a variety of Paleozoic animals. Trilobites, which possessed a mineralized dorsal surface, also had a nonmineralized ventral surface and nonmineralized ventral appendages. As such, they must have numbered among those threatened by predators lacking shell-crushing capability. Vulnerability to predators would have increased shortly following molting, before the new dorsal exoskeletons of trilobites had fully hardened. Some arthropod predators of trilobites may have gained access to appetizing soft tissues by slicing or tearing parts of the nonmineralized (chitinous) anatomy in a manner similar to that used by many chelicerates (Brusca and Brusca, 1990). Additionally, some predators may have fed upon trilobite appendages without leaving much evidence of their activity in the fossil record. Trilobites, like many modern crustaceans and other arthropods, almost certainly could autotomize and regenerate appendages that had been bitten or otherwise lost (see Needham, 1952). If such occurred, little evidence of preyed-upon appendages would be likely to enter the fossil record, even in deposits of exceptional preservation.

Non-durophagous predation is difficult to test for because much of the evidence is negative, and the strongest evidence is circumstantial. However, the argument that biomineralization in trilobites and other Cambrian animals developed partly in response to predation pressure fails if non-durophaghous predators are not considered. Likewise, the argument that enrollment in trilobites (section 2.6.2) led to protection of the nonmineralized ventral anatomy fails if only durophagous predators are considered.

Support for non-durophagous predation, particularly by some arthropods, comes from arthropod gut traces (Butterfield, 2002). By comparison with modern chelicerates, consistently sclerite-free guts in fossil arthropods can be considered the result of relatively minor tearing or slicing of chitinous exoskeleton of other arthropods, followed by possible injection of predigestive juices, and ending in the sucking of vital juices or organs (see section 2.2). Arthropods that fed in this chelicerate-like manner would not be expected to leave a fossil record of sclerite-bearing guts. Deposits of exceptional preservation such as the Burgess Shale and Chengjiang, where nonmineralized cuticle is well preserved, should preserve mineralized or nonmineralized sclerites in arthropod gut traces, assuming that: (1) the arthropods were durophagous; and (2) gut contents were not eliminated from the body prior to burial. Indeed, the widespread lack of sclerite-filled guts among many Paleozoic arthropods (for exceptions, see section 2.2) may be a relatively good indicator of the extent to which non-durophagous predators preyed on macroscopic organisms. The possibility that some arthropods lacking sclerite-filled guts were sediment-deposit feeders cannot be ruled out. However, the rather strong preservation potential of sediment-filled guts in animals other than arthropods from deposits of exceptional preservation (e.g., Babcock and Robison, 1988) indicates that there should be a relatively high frequency of sediment-filled gut traces if sediment deposit-feeding were widespread among Paleozoic arthropods.

2.2. Gut Contents

Metazoans having trilobite sclerites preserved in the stomach or intestinal tract are unusual, but a few examples are known from deposits of exceptional preservation ("Konservat-Lagerstätten" or "Burgess Shale-type deposits"). Good examples are of the Middle Cambrian arthropods *Sidneyia* (Burgess Shale of British Columbia; Bruton, 1981; Briggs *et al.*, 1994) and *Utahcaris* (Spence Shale of Utah; Conway Morris and Robison, 1988; Fig. 2.1). Gut contents do not allow us to readily differentiate whether trilobite remains were ingested through predation or scavenging.

Sclerite-free guts of *Leanchoilia* and other nonmineralizing arthropods from Cambrian deposits of exceptional preservation indicate that those arthropods evolved predation styles similar to those of modern chelicerates (Butterfield, 2002). *Leanchoilia* and other non-durophagous arthropods that ingested vital juices or internal organs of their prey through sucking action, rather than crushing the bodies of prey, can be expected to have alimentary tracts that are free of sclerites and, in some cases, guts or midgut glands filled with early diagenetic minerals (see related discussion in section 3.4). Butterfield (2002) also noted that some sediment-filled guts of fossil arthropods may be the result of sediment sweeping into the alimentary tract following death, meaning that a gut with sediment infill does not definitively identify a sediment deposit-feeding arthropod.

2.3. Coprolites

Coprolites (e.g., Sprinkle, 1973; Conway Morris and Robison, 1986, 1988; Robison, 1991; Chen *et al.*, 1997) occasionally contain trilobite sclerites, and generally, the sclerites are fragmented (Fig. 2.2). Echinoderm sclerites and fragmented brachiopod

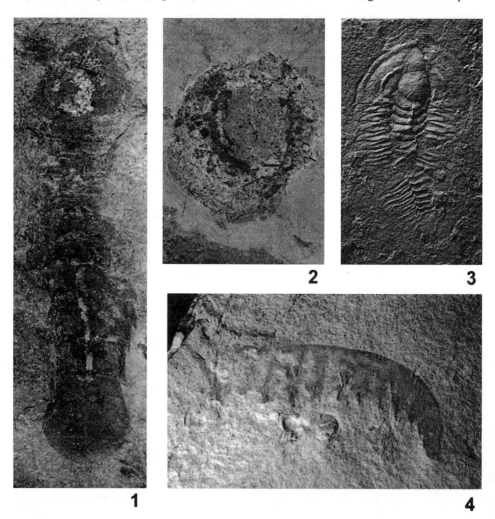

FIGURE 2. Physical evidence of predation or scavenging on trilobites. (1) *Utahcaris orion* from the Spence Shale (Middle Cambrian) of Utah showing disarticulated and broken trilobite sclerites in the stomach cavity under the cephalic shield; length = 9.0 cm; KUMIP 204784; photo courtesy of R. A. Robison. (2) Coprolite from the Spence Shale of Utah showing disarticulated and broken trilobite sclerites, echinoderm plates, and inarticulate brachiopod shells; maximum diameter = 4.4 cm; KUMIP 204369; photo courtesy of R. A. Robison. (3) Exoskeleton of *Buenellus higginsi* from the Buen Formation (Lower Cambrian) of North Greenland showing disarticulation and breakage of thoracic sclerites possibly due to scavenging; length = 3.5 cm; Geologisk Museum, Copenhagen (MGUM) 319571.1640. (4) Grasping appendage of *Anomalocaris canadensis* from the Carrara Formation (Lower Cambrian) of California; length = 4.1 cm; USNM 519542.

shells are also common (Fig. 2.2). Coprolites provide direct evidence that trilobites were targets of carnivores. They also provide proxy evidence of the sizes of some predators on the assumption that coprolite size reflects the terminal diameter of the alimentary tract, which, in turn, is a proxy for length of the predator; Conway Morris and Robison, 1988; Nedin, 1999). As with gut contents, coprolites do not permit us to distinguish trilobite sclerites that an animal ingested through predation from those ingested through scavenging.

2.4. Broken Sclerites

Broken trilobite sclerites provide indirect evidence of predation (Pratt, 1998). Although sclerites could suffer postmortem breakage by such means as transportation or the activity of chitinoclastic bacteria, many broken sclerites in the stratigraphic record, and particularly those stratigraphically associated with remains of predatory animals (see section 2.1.3), are likely the result of successful predation or scavenging. Experiments involving the disarticulation of modern arthropods (Allison, 1986; Babcock and Chang, 1997; Babcock et al., 2000) show that sclerites suffer little observable evidence of the tumbling and transportation to which they have been subjected. Alternatively, durophagous predators normally extract food from arthropods through breakage, crushing, or tearing of their exoskeletons. Trilobite sclerites preserved in the stomach of the durophagous predator *Utahcaris* (Fig. 2.1), and in an illustrated coprolite produced by a large durophagous predator (Fig. 2.2), are broken. Likewise, in a specimen that was likely scavenged (Fig. 2.3), thoracic segments are disrupted and fragmented. The possibility stands that some broken trilobite sclerites are the result of conditions other than durophagy, but current information (e.g., Allison, 1986; Tshudy et al., 1989; Babcock et al., 2000) tends to indicate that many occurrences of broken sclerites resulted from carnivory or scavenging. In such cases, then, concentrations of fragmented sclerites likely represent disaggregrated feces of durophagous predators.

2.5. Borings

Babcock (1993a) and Babcock and Peng (2001) reported borings in the exoskeletons of agnostoid trilobites that evoked a clear, but rather limited, wound response from the living tissues of trilobites. Minute borings on these tiny trilobites are located in the axial furrow adjacent to the glabella, and close to the positions of some vital organs. Small pearl-like projections along the internal surfaces of the cephala apparently sealed the invading organisms from the living tissues of the trilobites (Figs. 1.3, 1.4) and can be attributed to simple wound closure following sublethal attack by micropredators. Except for the internal sealing structures, the borings resemble those inferred to have been produced by nematodes in the tests of foraminiferans (Sliter, 1971; Boucot, 1990; but see Lipps, 1983, p. 357). Borings reported by Babcock (1993a) and Babcock and Peng (2001) date from the Middle Ordovician and Middle Cambrian, respectively. If the borings result from a micropredator-prey relationship between nematodes and agnostoids, they provide circumstantial evidence for the presence of marine nematodes extending to at least the Middle Cambrian. Borings that lack internal sealing structures apparently have not been reported from agnostoid trilobites, but this may be related to non-recognition of such traces.

Peach (1894) and Størmer (1931) illustrated possible borings in the exoskeletons of large polymeroid trilobites. These examples do not show any cellular response to invasion, although they do suggest that trilobites were prey for not only large biting predators, but also for boring micropredators. Weighing against this interpretation is the possibility that the borings were developed in trilobite carcasses or exuviae; borings illustrated by Portlock (1843, Pl. 21, fig. 5a) were a part of the taphonomic history of the trilobites.

2.6. Predation Resistance or Response

A voluminous literature provides details of morphological deterrents to durophagous predation in invertebrate and vertebrate animals. In addition, a rather extensive literature covers chemical deterrents to predation in animals. Good summaries of information concerning one or both of these means of resisting predation among ancient animals include Vermeij (1977, 1987), Signor and Brett (1984), Kelley and Hansen (1993) and Kelley et al. (this volume). In contrast to the extensive work concerning predation-resistant features in such organisms as brachiopods, mollusks, echinoderms, and vertebrates, relatively little published discussion concerns predation-resistant strategies, other than enrollment, in trilobites. Here some morphological features or ecological strategies that may have aided trilobites in deterring predation are briefly considered. If trilobites used any sort of chemical means to avoid predation, which is likely, we have no direct evidence of it. Information provided here is a preliminary attempt to understand the magnitude and complexity of the problem of predation resistance in trilobites, and is not a fully inclusive compilation.

2.6.1. Predation-Resistant Morphological Features

A sclerotized, chitinous exoskeleton, the dorsal surface of which is reinforced by calcite, is the most basic predation-resistant aspect of trilobite anatomy. Calcified trilobites are a clade that descended from a non-calcified arachnomorph ancestor (Fortey and Whittington, 1989; Ramsköld and Edgecombe, 1991) by the latter part of the Early Cambrian (e.g., Zhang, 1987; Geyer, 1993, 1996, 1998; Geyer and Palmer, 1995; Hollingsworth, 1999). Evolution of a calcified exoskeleton from nonmineralized, chitinous cuticle has been interpreted as a response to predation pressure combined with favorable geochemical and biochemical conditions (e.g., Bengtson, 1984; Runnegar, 1989; Conway Morris, 1998; Babcock et al., 2001). Biomineralization in trilobites was part of a sweeping reorganization of marine biotic systems that included increased skeletization, changes in organism-substrate interactions, and general restructuring of ecological relationships (see section 4). In the context of these changing conditions, predation evidently played an important role, one that is recorded by the appearance of biomineralized or other predation-resistant skeletons across a wide spectrum of phylogenetic lines. Hard skeletons certainly conferred some adaptive advantage to animals that developed them, at least with respect to avoiding predators accustomed to feeding upon relatively "soft" animals. As anchoring structures for musculature and other soft anatomy, hard skeletons also indirectly served as predation-resistant devices by permitting growth to larger size (cf. Stanley, 1999), and by increasing mobility and agility. High numerical abundance of Early Cambrian trilobites in certain deposits (e.g.,

Palmer, 1999) attests to the value of calcified skeletons; nevertheless, in the Early Cambrian Chengjiang Biota (Babcock *et al.,* 2001; Babcock and Zhang, 2001) and in the Middle Cambrian Burgess Shale Biota (Conway Morris, 1986), small nonmineralized arthropods (bradoriids and marellids) greatly exceed trilobites in numerical abundance.

Sensory capability, including sight, current-detecting, and chemosensory function, provided trilobites with another line of defense against predation. Primitively, trilobites had eyes (Briggs and Fortey, 1989), with which they could detect, and thereby evade, potential predators. Studies on the visual fields of trilobites are numerous (e.g., Clarkson, 1979; Towe, 1973; Campbell, 1975; Miller, 1976; Miller and Clarkson, 1980). Miller and Clarkson (1980) noted that lenses in the compound eyes of *Phacops* (Devonian) were light concentrators, and beneficial for detecting potential enemies, especially if the trilobites had recently molted under cover of night.

Loss of compound eyes occurred repeatedly in trilobite clades, but light-sensitivity, which could be construed as a form of early warning system, probably remained even in many "blind" species. In agnostoid trilobites, loss of compound eyes was presumably linked to paedomorphic reduction in adult size and a pelagic existence (Robison, 1972, 1984; Müller and Walossek, 1987). Nevertheless, agnostoids likely retained photosensitive structures homologous with the ventral pair of eyes of the frontal eye complex of euarthropods (Müller and Walossek, 1987). A number of illaenid trilobite species have a glabellar tubercle, and it is plausible that its function involved light reception (Fortey and Clarkson, 1976).

Miller (1976) summarized morphological evidence for the sensory fields of the Devonian trilobite *Phacops*. In addition to having remarkable visual coverage of the area surrounding the body, trilobites possessed setal hairs (principally extending from surface tubercles) that provided information about changes in water currents and about chemical signals, both of which should have provided advance warning of predators. Physical and chemical signaling must have been available through the setal hairs to most, if not all, trilobites. Additionally, Miller (1975) suggested that one function of the terrace lines was to monitor water currents.

Sensory appendages provided trilobites with information concerning the approach of predators. Antennae, for example, would have monitored changes in water currents and chemical signals. Conway Morris (1998) suggested that cercae extending from the posterior of the body provided *Olenoides* (Cambrian) with warning against unexpected attack.

Finally, predation-resistant morphological structures in trilobites include a variety of features that functioned, in part, to deter durophagy, or to facilitate predation-resistant behavior. Predation resistance, although adaptive, was most likely a secondary function in many cases. Among them, marginal or other spines (e.g., Levi-Setti, 1993; Whittington, 1997*a*; Fig. 3.1), relatively thickened exoskeleton, and color markings (e.g., Babcock, 1982; Whittington, 1997*b*) that could camouflage trilobites or cause harmless species to mimic poisonous species, quite likely afforded trilobites some protection from predators. Terrace lines, which are characteristic of many trilobites, added mechanical strength to the exoskeleton (Miller, 1975), in addition to functioning in sediment gripping (Schmalfuss, 1978; Seilacher, 1985), mechanoreception (Miller, 1975), and enabling enrollment (see Clarkson and Henry, 1973). Structures enabling trilobite enrollment and predation avoidance are discussed further in section 2.6.2.

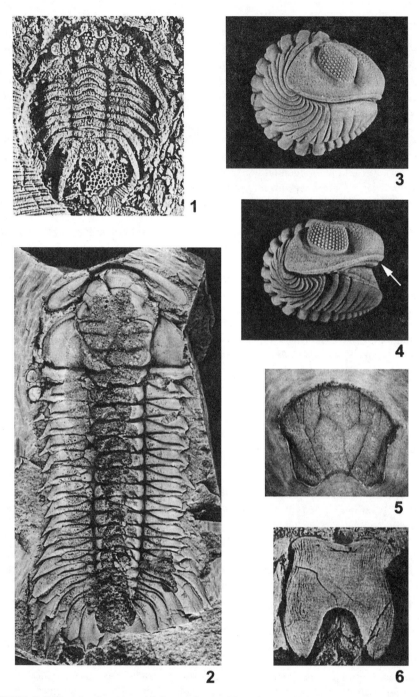

FIGURE 3. Trilobites showing characteristics consistent with predator avoidance (1) or predatory behavior (2-6). (1) *Acidaspis cincinnatiensis* from the Cincinnatian Series (Upper Ordovician) of Ohio showing marginal spines and tuberculate surface; length = 1.3 cm; Cincinnati Museum of Natural History, University of Cincinnati collection (UCGM) 43200. (2) Axial shield of *Centropleura loveni* from the Kap Stanton Formation

2.6.2. Predation-Resistant Behavioral Characteristics

Enrollment (Figs. 3.3, 4.3, 4.4), commonly including self-concealment in sediment, has long been cited as a response to the threat of predation in trilobites (e.g., Pompeckj, 1892; Clarkson and Henry, 1973; Babcock and Speyer, 1987; Speyer, 1990*a*; Fortey, 2000). Based on analogies with modern arthropods, Babcock and Speyer (1987), however, argued that self-burial in mud followed by enrollment was a generalized response to a variety of environmental pressures including predation threats, storms, deoxygenation, sudden lighting changes, and sudden temperature changes (also see Hughes *et al.,* 1999). Arguably, the preservation of most trilobites in an enrolled state is less likely to be a direct response to predation pressure than to other sudden environmental changes, notably episodic sedimentation (Brandt, 1980; Stitt, 1983; Brandt Velbel, 1985; Babcock and Speyer, 1987; Speyer, 1988, 1990*a*). Nevertheless, the capability for enrollment can be reasonably viewed as a type of predation-response mechanism in trilobites.

Enrollment capability in arthropods is evidently a primitive behavioral condition, as it is widespread among groups including trilobites, limulaceans (Fisher, 1977; Anderson and Selden, 1997), decapods (McMahon and Wilkens, 1975), and myriapods (Manton, 1977; Hannibal and Feldmann, 1981). In addition, enrollment is or was part of the behavioral repertoire of modern polychaete worms (Dean *et al.,* 1964), fossil worms (e.g., Robison, 1991; Chen and Zhou, 1997; Chen *et al.,* 1997; Hou *et al.,* 1999), and lobopods (e.g., Chen and Zhou, 1997; Chen *et al.,* 1997; Hou *et al.,* 1999). According to Bergström (1973), enrollment was adopted in all major phylogenetic lines of trilobites except the Cambrian olenellids.

Clarkson and Henry (1973) provided cogent arguments for a coevolutionary relationship between the diversification of major Paleozoic predators (cephalopods and fishes) beginning in the Ordovician and increasing complexity of morphological structures that facilitate enrollment. The ability to enroll was aided by coaptative structures, which are interlocking devices located principally on the anteroventral cephalic rim and the posteroventral pygidial rim; often, they are also developed on the pleural tips. Trilobites used various interlocking strategies (Clarkson and Henry, 1973; Henry and Clarkson, 1974; Speyer, 1988; Clarkson and Whittington, 1997) and enrollment techniques (Bergström, 1973; Levi-Setti, 1993; Clarkson and Whittington, 1997), but the net effect of coaptative structures was to ensure complete closure of a trilobite within a calcified exoskeletal ball (e.g., Bruton and Haas, 1997), and to increase

---

(Middle Cambrian) of North Greenland showing characteristics of a pelagic existence: streamlined shape, enlarged frontal area of glabella, elongate eyes, narrow pleural areas of thorax, and fin-like posterior thorax-pygidial region; the elongate genal spines are missing; length = 11.5 cm; MGUH 21.273. (3) *Phacops rana milleri* from the Silica Shale (Middle Devonian) of Ohio in fully enrolled condition showing enlarged glabella and tight interlocking along the ventral margins of the cephalon and pygidium; length = 3.3 cm; Orton Geological Museum, The Ohio State University (OSU) 16266. (4) *Phacops rana milleri* from the Silica Shale (Middle Devonian) of Ohio in incompletely enrolled condition showing coaptative structures on the cephalon and pygidium, particularly the vincular furrow along the ventral margin of the cephalon (arrow); length = 1.9 cm; OSU 16266. (5) Hypostome of *Centropleura loveni* from the Kap Stanton Formation (Middle Cambrian) of North Greenland showing forked posterior wings; length = 1.5 cm; MGUH 21.286. (6) Hypostome of *Isotelus maximus* from the Cincinnatian Series (Upper Ordovician) of Ohio showing strongly forked posterior wings and pattern of the terrace lines; terrace lines form a rasp-like interior edge along each of the posterior wings; length = 4.8 cm; UCGM 24364A.

the difficulty of outside forces (e.g., predators) from unrolling trilobites. Coaptative structures date from the Cambrian (Stitt, 1983; Clarkson and Whittington, 1997), which suggests that evolution of enrollment-related structures in trilobites began responding to predation pressure prior to the Ordovician.

Use of enrollment as a deterrent to predation must have involved some trade-offs. Non-durophagous predators were probably the least successful predators of enrolled trilobites. For durophagous predators, enrollment is inferred to have been most effective against predators that did not swallow prey whole, and it would have been particularly effective if trilobites first burrowed into sediment before enrolling (Babcock and Speyer, 1987; Speyer, 1988, 1990a). Enrolling at the sediment-water interface likely would have increased the attractiveness of trilobites to predators by eliminating their ability to swim from predators, causing them to project further above the sediment-water interface (making them easier to capture), and diminishing the effective size of the trilobite (making them easier to swallow). Successful predation of a trilobite in an enrolled state, however, would have been relatively unlikely if the predator were small. Borers, for the most part, would have been little deterred by enrollment.

Aside from enrollment, at least two other predation-resistant behavioral patterns have been inferred from the fossil record. The first pattern involves the sheltering of live trilobites in the shells of brachiopods (Brett, 1977), nautiloids (Davis *et al.*, 2001; Fig. 4.5), or elsewhere (Mikulic, 1994). Clearly, trilobites were most susceptible to predation in the intermolt phase, prior to the calcification of a new exoskeleton (Miller and Clarkson, 1980; Mikulic, 1994). Prior to molting, trilobites presumably took shelter to reduce their chances of falling victim to carnivores. Molted exoskeletons (Mikulic, 1994) or putative corpses (Brett, 1977; Davis *et al.*, 2001) found in sheltered areas quite plausibly represent either shed molts or live animals that were overwhelmed by episodic sedimentation following ecdysis. The second pattern of predation-resistant behavior involves the struggling of prey animals following their capture by predators. Nedin (1999) suggested that bilaterally arranged healed injuries on some trilobite exoskeletons (Table 2) record successful struggling of trilobites against their captors.

2.6.3. Reproductive Strategy

Reproductive strategy can be interpreted as one type of response to predation pressure. Intuitively, the most useful ecologic strategy for most trilobites would include a high reproductive rate and rapid maturation (r-strategy), as it would tend to ensure survival of some members of the species to reproductive age in the face of predatory advantage. Large numbers of trilobites representing single species are known from numerous localities (see Speyer, 1990b; Levi-Setti, 1993; Fortey and Owens, 1999; Table 3). These examples suggest that many trilobite species had high reproductive rates. Other than biomineralization of the dorsal exoskeleton, structures facilitating enrollment, and various sensory devices, these trilobites, which are known from thousands to millions of specimens, do not have obvious predation-resistant structures such as large spines. To what extent such trilobites as these may have invested more in reproductive strategy than in evolving elaborate morphological deterrents to predation is unknown.

TABLE 3. Examples of large, monospecific aggregations of trilobites inferred to be related to high reproductive rates. Some examples probably represent episodic burial of molts or live animals clustered as part of their reproductive strategy (Speyer and Brett, 1985; Speyer, 1990b).

Buenellus higginsi, Cambrian of Greenland (Blaker and Peel, 1997)
Olenellus (Olenellus) gilberti, Cambrian of Nevada (Palmer, 1999)
Olenellus (Paedeumias) chiefensis, Cambrian of Nevada (Palmer, 1999)
Peronopsis interstricta, Cambrian of Utah (Levi-Setti, 1993; herein, Fig. 4.1)
Ptychagnostus atavus, Cambrian of Utah, Greenland, and China (Robison, 1984, 1994; Peng and Robison, 2000)
Agnostus pisiformis, Cambrian of Sweden (Westergård, 1946; Müller and Walossek, 1987; Ahlberg, 1998; Fortey and Owens, 1999)
Elrathia kingii, Cambrian of Utah (Bright, 1959; Robison, 1964; Gunther and Gunther, 1981; Levi-Setti, 1993)
Vogdesia bromidensis, Ordovician of Oklahoma (Laudon, 1939; Loeblich, 1940; Shaw, 1974; Levi-Setti, 1993; herein, Fig. 4.2)
Flexicalymene meeki, Ordovician of Ohio, Kentucky, and Indiana (e.g., Osgood, 1970; Ross, 1979; Brandt, 1980)
Flexicalymene aff. *granulosa*, Ordovician of Ohio, Kentucky, and Indiana (Hughes and Cooper, 1999)
Calymene blumenbachii, Silurian of England (Fortey, 2000)
Phacops rana, Devonian of New York and Ohio (Kesling and Chilman, 1975; Speyer and Brett, 1985; Levi-Setti, 1993).

Speyer and Brett (1985) and Speyer (1990b) described clustered monospecific assemblages of adult polymeroid trilobites, including body clusters and molt clusters. These clusters, they hypothesized, record episodically buried aggregations related to swarming and mass-molting in preparation for reproduction. This conclusion is supported by considerable evidence from the reproductive behavior of modern annelids and arthropods (e.g., Thorson, 1950; Alexander, 1964; Speyer, 1990b).

3. Predation by Trilobites

Feeding strategies in trilobites were thoroughly reviewed by Fortey and Owens (1999), who concluded that much of the variation in exoskeletal morphology, particularly in the cephalic region, was a response to feeding mode. Feeding habits of arthropods in general, all of which conceivably apply to trilobites, are: (1) predatory, including hunting and scavenging; (2) particle-feeding; (3) suspension-feeding; (4) filter-feeding; and (5) parasitism (Fortey and Owens, 1999). Hunting is the dominant feeding habit among arachnomorph arthropods (Brusca and Brusca, 1990; Fortey and Owens, 1999), the group to which trilobites belong, and it is common among crustaceans (Schram, 1986). For this reason, and because of functional-morphologic reasons, predation was, in all likelihood, the primitive feeding strategy of trilobites (Fortey and Owens, 1999). Hunting and scavenging are viewed as end-members of a predatory habit, as many carnivorous arthropods that are capable of hunting are also opportunistic scavengers (Schram, 1986; Fortey and Owens, 1999). Phylogenetic, functional-morphologic, and trace-fossil evidence (Fortey and Owens, 1999), together with information from gut contents (section 3.4) supports an interpretation of a predaceous feeding habit for some, if not most, trilobites.

3.1. Phylogenetic Evidence for Trilobite Predation

Comparisons with arthropod relatives support the interpretation that relatively primitive trilobites (e.g., most olenelloids and redlichiids), and some derived forms (e.g., *Olenoides*, *Isotelus*, *Triarthrus*, and *Phacops*), were predatory. Phylogenetic analyses (e.g., Fortey and Whittington, 1989; Briggs and Fortey, 1989; Ramsköld and Edgecombe, 1991; Wills *et al.*, 1995) lead to the conclusion that uncalcified naraoiids are a sister group to the most primitive of calcified trilobites, the olenelloids (Fortey and Owens, 1999). Available evidence indicates that some naraoiids had a predatory habit. Appendage morphology of naraoiids has been well documented (e.g., Whittington, 1977; Zhang and Hou, 1985; Ramsköld and Edgecombe, 1996; Babcock and Chang, 1997; Hou *et al.*, 1999), and the massive spiny basal endites, which resemble gnathobases, in the Cambrian species *Naraoia compacta* and *N. longicaudata* indicate that these species were predatory. Another Cambrian species, *N. spinosa*, however, is commonly preserved with a sediment-filled gut (Babcock and Chang, 1997), which suggests that sediment-ingestion had been adopted by some naraoiids.

Naraoiids and calcified trilobites together comprise a monophyletic group within a more inclusive clade, the Arachnomorpha. The Arachnomorpha includes the diverse chelicerates (spiders, scorpions, whip scorpions, horseshoe crabs, eurypterids, and others). With rare exceptions (principally the mites), nearly all modern chelicerates are predatory; most extinct chelicerates, other than mites, also are inferred to have been predatory.

3.2. Functional-Morphological Evidence of Predatory Capability of Trilobites

Morphological evidence supporting predation by trilobites includes a variety of characters, particularly ones of the limbs and cephala. Whittington (1975, 1980, 1992) argued that the limbs of *Olenoides* (Middle Cambrian of British Columbia) were adapted to hunting and scavenging. Using the spinose endopodites (inner branches of the biramous appendages) to capture prey, *Olenoides* could grind food into small pieces using the gnathobase-like, stout, spinose inner edges of the coxae, after which food particles were passed forward along the ventral-medial line to the backwardly directed mouth. This nektobenthic animal probably preferred worms and other creatures having nonmineralized skeletons (Briggs *et al.*, 1994).

Inner branches of the limbs of *Eoredlichia* (Cambrian of China) are spinose, similar to those of *Naraoia compacta* (Shu *et al.*, 1995; Ramsköld and Edgecombe, 1996; Hou *et al.*, 1999). The spinose limbs of *Eoredlichia* support an interpretation that it was a predator/scavenger (Fortey and Owens, 1999). As the sister group of all trilobites other than olenelloids, a predatory habit in redlichiids means that predation can be viewed as the primitive condition for all the "higher" trilobite groups (Fortey and Owens, 1999).

In Ordovician *Triarthrus* from New York, pyritized appendages show an array of spinose endites on the interior branches of the posterior limbs (the endopodites). This arrangement has been interpreted as a posterior food trap (Whittington, 1997c). Food capture by the posterior limbs would require that food be carried forward to the mouth, situated in the posterior half of the ventral cephalic area, along the midline, by action of the appendages. Trilobites having posterior food traps were hypothesized to have been active predators, scavengers, and deposit-feeders (Whittington, 1997c).

Several lines of evidence suggest predatory behavior in Devonian phacopid trilobites. Stürmer and Bergström (1973) and Bartels *et al.* (1998) reported that the spinose endopodites of pyritized phacopids (*Chotecops*; formerly referred to *Phacops*; Struve and Flick, 1984) from the Devonian of Germany were adapted for capturing prey. Support for that interpretation comes from information about the sensory fields and visual systems of the closely related *Phacops* (Miller, 1976; Stockton and Cowen, 1976). Regarding the lenses as light-gathering devices, Miller (1976) postulated that *Phacops* was a nocturnal hunter. Stockton and Cowen (1976) suggested that *Phacops* had stereoscopic vision; although that interpretation was questioned by Fortey and Owens (1999), the sophisticated visual systems and excellent visual acuity of phacopids is consistent with a predatory feeding habit. Additional support for predatory behavior in *Phacops* comes from the enlarged glabella (Figs. 3.3, 3.4), which Eldredge (1971) inferred housed digestive glands opening to the stomach. Eldredge (1971), however, interpreted *Phacops* as a particle-feeder. Later, Fortey and Owens (1999) suggested that the enlarged glabella of *Phacops* was modified as a gastric mill, an interpretation that is more consistent with a predatory habit.

An exceptionally preserved specimen of *Isotelus maximus* from the Ordovician of Ohio shows an array of appendages under the thoracic and pygidal regions (Fig. 5.4) that suggest a predatory habit. Despite weathering and the consequences of early preparation of the appendages (probably in the late nineteenth century), the bases of rather robust spines still can be observed in places along the inner branches of the endopodites. By analogy with *Olenoides, Redlichia, Triarthrus,* and *Chotecops,* discussed above, the spinose endopodites of *Isotelus* probably functioned in grasping, capturing, and manipulating prey. The illustrated specimen of *Isotelus* also preserves a gut tract filled with calcite spar (Figs. 3.3, 3.5), which, as discussed in section 3.4, is regarded as further support for a predatory feeding habit.

The morphology and attachment style of calcified ventral cephalic plates provide evidence for a predatory feeding habit. According to Fortey and Owens (1999), effective manipulation and breakdown of bulky food using the basal parts of limbs surrounding the mouth in polymeroid trilobites was facilitated by strengthening of the ventral structures. Strengthening the attachment of ventral plates occurred in four basic ways (Fortey and Owens, 1999). The first technique that trilobites used to strengthen ventral structures involved modifications to the hypostomal attachment, including rigid attachment of the hypostome at the cephalic doublure (e.g., *Olenoides*; Whittington, 1988*a*), or bracing the hypostome by either extending the anterior wings of the hypostome (e.g., asaphids and nileids; Whittington, 1988*b*) or ankylosing the hypostome and rostral plate (e.g., paradoxidids; Fortey and Owens, 1999). The second technique involved strengthening of the anterior cephalic doublure by either replacing the rostral plate (Fig. 4.4) with a single median suture (e.g., asaphids such as *Isotelus* and *Vogdesia*; Fig. 4.2), or fusing the cheeks (e.g., nileids and phacopids). The third technique involved rigid attachment of the hypostome to the doublure using either a conterminant condition of attachment (e.g., many paradoxidids, asaphids, encrinurids, and some lichids), or an impendent condition of attachment (e.g., phacopids and many proetids). Commonly, conterminant or impendent conditions of hypostomal attachment were accompanied by expansion of the frontal lobe of the glabella (e.g., paradoxidids [Fig. 3.2] and phacopids [Figs. 3.3, 3.4]), presumably to enlarge the stomach cavity. This would permit the trilobites to accept larger prey, or to store more food for digestion long after ingestion. The fourth technique involved modifications to the posterior hypostomal

FIGURE 4. Trilobites showing various ecological strategies associated with predation pressure. (1) Monospecific assemblage of the agnostoid *Peronopsis interstricta* from the Wheeler Shale (Middle Cambrian) of Utah; length of large specimen at upper left = 7 mm; OSU 46395. (2) Monospecific assemblage of *Vogdesia bromidensis* from the Bromide Formation (Middle Ordovician) of Oklahoma showing specimens in dorsal-up

margin, such as development of a fork (e.g., some paradoxidids [Fig. 3.5], asaphids [Figs. 3.6, 4.2], lichids, calymenids, homalonotids, and proetids), development of roughened tuberculate areas in the grinding areas of the posteromedial border (e.g., some odontopleurids), and development of rasp-like ridges forming serrated inner edges of the posterior wings (e.g., some asaphids such as *Isotelus* [Fig. 3.6] and *Ectenaspis*; Tripp and Evitt, 1986; Rudkin and Tripp, 1989; Sloan, 1992).

Some predatory trilobites developed stalked eyes, making them capable of a semi-infaunal life (Fortey and Owens, 1999). Putative predators having stalked eyes include certain species of homalonotids and cybelids.

Trilobites interpreted as pelagic must have fed in the water column. Pelagic trilobites include both the minute agnostoids (Robison, 1972; Müller and Walossek, 1987; Fig. 4.1) and a variety of polymeroids (Fortey, 1985; Babcock, 1994; Fortey and Owens, 1999; Fig. 3.2) representing a large range in size. Feeding habits in the agnostoids and other minute trilobites have been controversial, as summarized by Müller and Walossek (1987) and Fortey and Owens (1999). Available information indicates that agnostoids and other tiny trilobites (e.g., eodiscids) most likely fed on minute organic particles or microorganisms. Most species probably fed: (1) by collecting suspended detrital food particles, whether from the water column or from a flocculant bottom layer (Müller and Walossek, 1987); or (2) by capturing zooplankton and phytoplankton. Ectoparasitism (Bergström, 1973) is an unlikely possibility for most species (Fortey and Owens, 1999).

Among polymeroid species, some morphologies associated with a pelagic existence (Fortey, 1985; Babcock, 1994; Fortey and Owens, 1999; Fig. 3.2) and consistent with a predaceous feeding habit are: (1) large eyes, which would be useful for detecting prey; (2) streamlined, elongate exoskeletons, including reduced pleural regions, which would reduce frictional response in swimming; (3) powerful thoracic articulations, which would aid in swimming; (4) elongate genal spines, and elongate spines forming a fin-like apparatus in the posterior region, which would aid in stabilization during swimming; and (5) enlarged frontal lobes of the glabella, which would allow the stomach to accept large food items, or which could serve as a reservoir for food. Unfortunately, limbs are unknown for pelagic species other than agnostoids, which means that interpretation of the habits of pelagic species is speculative. Possible food sources for pelagic polymeroid trilobites include phytoplankton and zooplankton, as well as some larger prey.

The foregoing discussion, as well as that in following sections, is based on holaspid (adult) trilobites; however, changes in feeding strategy associated with ontogeny should be considered. Larval trilobites, especially protaspides (see Chatterton and Ludvigsen, 1976; Speyer and Chatterton, 1989), were commonly part of the zooplankton, and, as such, must have fed upon phytoplankton or other zooplankton (Fortey and Owens, 1999). Feeding strategies of these trilobites often must have changed in later ontogeny.

---

and ventral-up position; specimens in ventral-up position show forked hypostome in place and reduction of ventral cephalic sclerites (loss of rostral plate) resulting in single medial suture on the cephalic doublure (compare with Fig. 4.4); length = 5.6 cm; OSU 47616. (3, 4) *Flexicalymene meeki* from the Cincinnatian Series (Upper Ordovician) of Ohio in fully enrolled condition in lateral (3) and frontal (4) views, showing pygidium tucked under the cephalon (compare with Fig. 3.3), and the rostral plate in the medial area of the cephalic doublure (4); length = 2.3 cm, width = 2.7 cm; OSU 46324. (5) Specimens of *Acidaspis cincinnatiensis* from the Cincinnatian Series (Upper Ordovician) of Kentucky sheltered inside mud-filled body chamber of the nautiloid cephalopod ?*Treptoceras duseri*; nautiloid length = 15.2 cm; Cincinnati Museum of Natural History (CiMNH) P2257.

3.3. Trace Fossil Evidence of Predatory Behavior by Trilobites

Traces in sediment provide a record of predatory activity by trilobites. Putative trilobite hunting burrows in which infaunal worm traces are intersected by *Rusophycus* traces have been reported from the Cambrian of Sweden (Martinsson, 1965; Bergström, 1973; Jensen, 1990). Brandt *et al.* (1995) and Osgood and Drennen (1975) described similar occurrences from the Ordovician of Ohio and the Silurian of New York, respectively. Fortey and Owens (1999) inferred predatory behavior from *Rusophycus* traces in the Ordovician of Australia. In all these examples, the *Rusophycus* traces evidently resulted from trilobites burrowing into the sediment to capture worms. Wenndorf (1990), in studying trace fossils associated with Devonian homalonotid trilobites from Germany, suggested that some *Cruziana* traces were the result of grazing and hunting behavior by trilobites.

Burrows from the Lower Cambrian of Sweden (Bergström, 1973; Jensen, 1990) provide interesting insight into predator-prey interactions between trilobites and worms. These traces indicate that the trilobites were capable of locating worms from above and behind their positions in the sediment, then quickly digging to considerable depth to capture the prey. Scratch marks superimposed on some worm burrow casts record manipulation of the worm by the anterior appendages of the trilobite (Bergström, 1973). Rarely, trilobites changed direction while digging, presumably in response to changes in direction of the prey as they fled the trilobites (Jensen, 1990).

A trilobite hunting burrow from the Ordovician of Ohio reported by Brandt *et al.* (1995), together with another specimen (Fig. 5.2), provides evidence for a predatory habit in one species. Large *Rusophyscus carleyi* traces can be attributed with confidence to the activity of *Isotelus* (usually *I. maximus*) because of excellent casting of morphological details of the trilobite in sediment. In both illustrated specimens (Brandt *et al.*, 1995; Fig. 5.2), a worm burrow (*Palaeophycus*) is intersected by the trilobite trace close to the position of the trilobite's mouth. In the specimen illustrated here (Fig. 5.2), scratches produced by anterior endopodites of the trilobite indicate that the worm was forcibly extracted from its burrow by the trilobite. As in other isoteline trilobites, the hypostome is large, robust, and forked; the inside edge of each blade-like posterior wing that extends posteriorly around the mouth is sharp (Fig. 3.6) and seemingly adapted for cutting (Sloan, 1992). In *I. maximus*, the inside edge of each blade is serrated (Fig. 3.6). Combined, these observations provide strong evidence that *Isotelus* hunted worms or other nonmineralized, benthic animals. Further evidence from the gut contents of *Isotelus* (section 3.4) likewise supports the interpretation of a carnivorous food-gathering mode for *Isotelus*.

In contrast to *Rusophycus* traces attributable to *Isotelus*, those attributable to other trilobites from the Ordovician of Ohio and adjacent areas (Osgood, 1970) have not provided convincing evidence that the trilobite tracemakers hunted burrowing worms. *Cryptolithus*, and possibly *Flexicalymene*, are inferred to have stirred bottom sediment and ingested food using a benthic filter chamber (see Schmalfuss, 1978; Seilacher, 1985; Fortey and Owens, 1999). Food extracted from the sediment would have included disseminated organic matter and perhaps live microorganisms.

3.4. Trilobite Gut Contents

Rare instances of preserved gut traces provide valuable information about the diet of trilobites. An exceptionally preserved specimen of *Isotelus maximus* from the Ordovician of Ohio preserves an array of ventral appendages (Fig. 5.4) and a calcite-filled alimentary tract (Figs. 5.3, 5.5). The gut tract extends from the posterior part of the cephalic area (Fig. 5.5) to the end of the axis in the posterior part of the pygidium (Fig. 5.3). The gut is filled with light gray to white calcite spar that contrasts with the gray lime mud, and appears to be entirely mud-free and sclerite-free. This indicates that the animal did not ingest mud during life, nor was mud swept into the gut following death. Lack of any bodily material inside the gut of a fossil that preserves both nonmineralized appendages and soft walls of the internal alimentary tract indicates that this animal was not durophagous, not chitinophagous, and that it had not fully cleared the gut of its contents close to the time of death and burial. Non-compaction of the axial area in this relatively thin-shelled trilobite further suggests that the gut was fluid-filled at the time of burial. Other evidence of a predaceous food-gathering mode in *Isotelus* was discussed in sections 3.2 and 3.3.

A predatory habit can be inferred from the gut contents of few trilobites other than *Isotelus*. Published examples of gut traces or midgut glands in trilobites, notably *Buenellus* (Cambrian of Greenland; Blaker and Peel, 1997) and *Chocetops* (Devonian of Germany; Stürmer and Bergström, 1973; Bartels *et al.*, 1998), lack evidence of either sediment infill or sclerites of ingested animals. Thus, a non-durophagous, non-chitinophagous predatory habit was probable for these forms. Alternatively, sediment filling is present in the gut trace of a ptychopariid (*Pterocephalia* from the Cambrian of British Columbia; Chatterton *et al.*, 1994), and pellets are present in the posterior part of the alimentary trace of a trinucleid (*Deanaspis* from the Ordovician of Bohemia; Šnajdr, 1991). The most parsimonious conclusion is that these two trilobites were sediment deposit-feeders; *Pterocephalia* shows characters typical of particle feeding, and *Deanaspis* shows characters typical of filter-chamber feeding (see Fortey and Owens, 1999). Other trilobites showing gut traces (reviewed by Chatterton *et al.*, 1994) have failed to provide sufficient information for interpretation of feeding strategy.

4. Final Comments: Predation on and by Trilobites in the Context of Ecosystem Reorganization During the Early Paleozoic

This paper provides a brief summary of the role that trilobites played as both predators and prey during the Paleozoic. Having partly mineralized exoskeletons, trilobites have left a rich fossil record spanning much of the Paleozoic. From some of the earliest-known trilobite remains we have clear indications in the form of sublethal predation scars (Babcock, 1993*a*; Nedin, 1999; Babcock and Zhang, 2001) and other evidence that trilobites fell victim to carnivores and, in turn, preyed upon other species, both large and small. There can be little doubt that predation was a major factor in the cycling of energy through Paleozoic ecosystems, and that it was a major forcing factor in animal evolution from at least the beginning of the Phanerozoic (e.g., Stanley, 1976; Brasier, 1979). Borings in cloudiniids from the Late Neoproterozoic (Bengtson and

FIGURE 5. A trilobite that was both predator and prey: body fossils of *Isotelus maximus* (1, 3-5) and trace fossil produced by *I. maximus* (2) from the Cincinnatian Series (Upper Ordovician) of Ohio. (1) Exoskeleton of *I. maximus* showing large sublethal predation scar on right side of cephalon; length = 17.8 cm; OSU 46339. (2)

Yue, 1992; Bengtson, 1994) suggest that predation was also important in evolution prior to the Phanerozoic.

Review of the Phanerozoic fossil record (Sepkoski, 1981) reveals three major radiations of predators accompanied by adoption of predation-resistant strategies in prey species. Vermeij (1977) summarized substantial evidence for a major radiation of durophagous predators that occurred during the Mesozoic, and Signor and Brett (1984) cited extensive evidence for an earlier radiation of durophagous predators that occurred during the middle Paleozoic. Both marine revolutions involved sweeping reorganization of existing marine ecosystems. These two protracted biologic "events" were characterized by widespread changes in the morphology of organisms (notably the evolution of more efficient predation-related structures in predators, and the evolution of predation-resistant structures in prey), in organism-sediment interactions, and in community composition. A comparably important, if not more important, reorganization of marine ecosystems also occurred during the early Paleozoic. Important manifestations of these changes include: (1) the Cambrian radiation of metazoans (e.g., Brasier, 1979; Runnegar, 1982, 1989; Conway Morris, 1985, 1998; McMenamin and McMenamin, 1990; Erwin, 1991; Lipps and Signor, 1992; Bengtson, 1994; Valentine, 1994; Fortey *et al.,* 1996; Chen *et al.,* 1997; Hou *et al.,* 1999; Babcock *et al.,* 2001; Lipps, 2001); (2) evolutionary changes in predator-prey systems (e.g., development of chitinous or biomineralized, predation-resistant skeletons in numerous metazoan clades; and evolution of cephalopods and fishes) continuing through the time of the Ordovician radiation (cf. Sepkoski and Sheehan, 1983; Foote, 1992; Droser *et al.,* 1994, 1996; Signor and Vermeij, 1994; Hughes *et al.,* 1999; Stanley, 1999); (3) a change from mat-dominated communities (Seilacher and Pfluger, 1994; Gehling, 1999; Bottjer *et al.,* 2000) to more highly burrowed sediments (Droser and Bottjer, 1988; Seilacher and Pfluger, 1994; Bottjer *et al.,* 2000); and (4) increased control by organisms over the geochemical cycles of essential (or limiting) nutrients such as carbon, calcium, phosphorous, and silica (Maliva *et al.,* 1989; Vermeij, 1995). Together these biosphere-scale changes can be viewed as part of an Early Paleozoic Marine Revolution (see Vermeij, 1995). Predation resistance, reflecting escalation in predator-prey systems (cf. Vermeij, 1987; Signor and Brett, 1984; Kelley and Hansen, 1993), is characterized in trilobites by increasing morphological complexity and changes in ecological strategy occurring through the Late Cambrian or Early Ordovician (see Clarkson and Henry, 1973; Fortey and Owens, 1990; Foote, 1991).

Evidence for an Early Paleozoic Marine Revolution is compelling (see Vermeij, 1995). Metazoans have a reasonably good fossil record extending from the Neoproterozoic (e.g., Glaessner, 1984; McMenamin, 1986; Seilacher, 1989; Conway Morris, 1990; Knoll, 1992, 1996; Bengtson, 1994; Fedonkin, 1994; Hoffman, 1994; Runnegar, 1994; Sun, 1994; Vidal, 1994; Fortey *et al.,* 1996; Gehling and Rigby, 1996; Narbonne, 1998; Babcock *et al.,* 2001), but their bodies were, with rare exceptions

---

Trace fossil, *Rusophycus carleyi*, intersecting worm burrow at the anterior end (top of photograph; arrow); length = 9.9 cm; Cincinnati Museum of Natural History (CMC-PUC) 50625. (3-5) Exceptionally preserved specimen of *I. maximus* in dorsal (3), ventral (4), and cross sectional (5) views showing: (3) posterior end of calcite-spar-filled gut where exoskeleton of the pygidium has been broken away (light color; arrow) (4) ventral appendages, almost exclusively the endopodites, of the thorax and pygidium (showing spinose inner branches in places); and (5) ovoidal, sparry-calcite-filled, gut tract underlying axial area (as viewed from broken anterior edge of the specimen); calcite spar (light color; arrow) contrasts with lime mud filling (darker color) of much of the body chamber and the matrix; length = 12.5 cm, height at center of axial lobe = 1.7 cm; USNM 33458.

(cloudiniids; Grant, 1990) covered with relatively soft, pliable, nonmineralized surfaces until chitinous and biomineralized coverings evolved in the Early Cambrian. Introduction of biomineralized skeletons during the Cambrian was both dramatic (occurring across a wide range of phylogenetic lines; Runnegar, 1982, 1994; Bengtson, 1994), and rapid (occurring in 10 to 15 million years; Grotzinger et al., 1995; Bowring and Erwin, 1998; Landing et al., 1998). Predation pressure must be considered one of the major reasons for development of resistant skeletons (e.g., Bengtson, 1994; Stanley, 1976; Brasier, 1979; Runnegar, 1994; Conway Morris, 1998), whether chitinous or biomineralized, even though it may not have been the only reason for their development. Changes in marine geochemical conditions must have played a role in biomineralization (Runnegar, 1994), but its possible role in the evolution of chitinous skeletons in many worms, lobopods, and arthropods is less clear. Predation-resistant skeletons arose in small-shelly-fossil-secreters, mollusks, hyoliths, brachiopods, echinoderms, and trilobites roughly contemporaneously with our record of the rise of predaceous arthropods and chordates (section 2.1.3). The record of Early Cambrian shelly fossils slightly precedes the record of these predators (cf. Geyer and Shergold, 2000); however, to some extent, this may be a preservational artifact, as our record of Early and Middle Cambrian predators (e.g., anomalocaridids) is highly influenced by the occurrence of deposits of exceptional preservation (e.g., the Sirius Passet, Chengjiang, and Burgess Shale deposits; see Conway Morris, 1985; Butterfield, 1995). As noted earlier in this paper, predation scars are preserved on some of the earliest-known trilobites, suggesting that a predator-prey "arms race" was in place during the Cambrian. As described more fully by Seilacher and Pfluger (1994) and Bottjer et al. (2000), the Neoproterozoic-Cambrian transition was characterized not only by a reorganization of animal body plans, but also by major changes in substrate conditions. The role of trilobites in facilitating the change from relatively stable, mat-dominated communities at the sediment surface during the Neoproterozoic to increasingly fluidized, animal-burrowed interfaces during the Cambrian was substantial. Trilobites evidently burrowed the substrate in search of food within it, and they used the sediment as a refuge to avoid capture by predators. In addition, species preyed upon by trilobites, including worms and perhaps nematodes or other microorganisms, were themselves churning the sediment, along with other organisms, by the Early Cambrian, and the impact of their activity (Droser and Bottjer, 1988, 1989; Droser et al., 1994) increased in stepwise fashion through the Ordovician.

ACKNOWLEDGMENTS: Over the years, I have had stimulating discussions on topics covered in this paper with colleagues too numerous to mention here. However, I would like to thank in particular W. I. Ausich, S. Bengtson, D. S. Brandt, C. E. Brett, G. Budd, S. Conway Morris, J. S. Hollingsworth, K. C. Hood, J. B. Jago, R. L. Kaesler, J. H. Lipps, C. G. Maples, D. L. Meyer, M. F. Miller, A. R. Palmer, J. S. Peel, S. C. Peng, B. R. Pratt, R. A. Robison, B. Runnegar, S. M. Stanley, E. L. Yochelson, W. T. Zhang (Chang), and M. Y. Zhu for helping to shape my thinking on trilobite predation, and the role of trilobites in Paleozoic biotic systems at key stages. S. Conway Morris, S. C. Peng, and R. A. Robison provided copies of important articles or photographs of specimens. T. Abbott, K. R. Evans, E. Fowler, L. and V. Gunther, J. S. Hollingsworth, and R. Meyer kindly provided specimens of malformed trilobites. Loans of specimens from museum collections were arranged by F. J. Collier, D. M. Gnidovec, A. Kamb, J. S. Peel, and J. Thompson. This manuscript was improved through the reviews of R. A.

Fortey, M. A. S. McMenamin, and the editorial work of P. H. Kelley. This work was supported in part by grants from the National Science Foundation (EAR 9405990, 0073089, and 0106883).

References

Ahlberg, P. (ed.), 1998, Guide to Excursions in Scania and Västergötland, Southern Sweden. IV Field Conference of the Cambrian Stage Subdivision Working Group, International Subcommission on Cambrian Stratigraphy, Lund Publ. Geology 141.
Alexander, R. D., 1964, The evolution of mating behavior in arthropods, in: *Insect Reproduction,* Vol. 2 (K. C. Highnam, ed.), Royal Entomology Society, London, pp. 78-94.
Allison, P. A., 1986, Soft-bodied animals in the fossil record: the role of decay in fragmentation during transport, *Geology* **14**:979-981.
Alpert, S. P., and Moore, J. N., 1975, Lower Cambrian trace fossil evidence for predation on trilobites, *Lethaia* **8**:223-230.
Anderson, L. I., and Selden, P. A., 1997, Opisthosomal fusion and phylogeny of Palaeozoic Xiphosura, *Lethaia* **30**:19-31.
Babcock, L. E., 1982, Original and diagenetic color patterns in two phacopid trilobites from the Devonian of New York, North American Paleontological Convention III, pp. 17-22.
Babcock, L. E., 1993a, Trilobite malformations and the fossil record of behavioral asymmetry, *J. Paleontol.* **67**:217-29.
Babcock, L. E., 1993b, The right and the sinister, *Nat. Hist.* **102**(7):32-39.
Babcock, L. E., 1994, Systematics and phylogenetics of polymeroid trilobites from the Henson Gletscher and Kap Stanton formations (Middle Cambrian), North Greenland, *Grønlands Geol. Under. Bull.* **169**:79-127.
Babcock, L. E., and Chang, W. T. 1997, Comparative taphonomy of two nonmineralized arthropods: *Naraoia* (Nektaspida; Early Cambrian, Chengjiang Biota, China) and *Limulus* (Xiphosurida; Holocene, Atlantic Ocean), *Bull. Nat. Mus. Natur. Hist.* **10**:233-250.
Babcock, L. E., Merriam, D. F., and West, R. R., 2000, *Paleolimulus,* an early limuline (Xiphosurida), from Pennsylvanian-Permian Lagerstätten of Kansas and taphonomic comparison with modern *Limulus, Lethaia* **33**:129-141.
Babcock, L. E., and Peng S. C., 2001, Malformed agnostoid trilobite from the Middle Cambrian of northwestern Hunan, China, in: *Cambrian System of South China* (S. C. Peng, L. E. Babcock, and M. Y. Zhu, eds.), Press of University of Science and Technology of China, Hefei, pp. 250-251.
Babcock, L. E., and Robison, R. A., 1988, Taxonomy and paleobiology of some Middle Cambrian *Scenella* (Cnidaria) and hyolithids (Mollusca) from western North America, *Univ. Kansas Pal. Contrib. Pap.* **121**:1-22.
Babcock, L. E., and Robison, R. A., 1989, Preferences of Palaeozoic predators, *Nature* **337**:695-696.
Babcock, L. E., and Speyer, S. E., 1987, Enrolled trilobites from the Alden Pyrite Bed, Ledyard Shale (Middle Devonian) of western New York, *J. Paleontol.* **61**:539-548.
Babcock, L. E., and Zhang W. T., 2001, Stratigraphy, paleontology, and depositional setting of the Chengjiang Lagerstätte (Lower Cambrian), Yunnan, China, in: *Cambrian System of South China* (S. C. Peng, L. E. Babcock, and M. Y. Zhu, eds.), Press of University of Science and Technology of China, Hefei, pp. 66-86.
Babcock, L. E., Zhang W. T., and Leslie, S. A., 2001, The Chengjiang Biota: record of the Early Cambrian diversification of life and clues to the exceptional preservation of fossils, *GSA Today* **11**(2):4-9.
Bartels, C., Briggs, D. E. G., and Brassel, G., 1998, *The Fossils of the Hunsrück Slate: Marine Life in the Devonian,* Cambridge University Press, Cambridge.
Bengtson, S., 1968, The problematic genus *Mobergella* from the Lower Cambrian of the Baltic area, *Lethaia* **1**:325-351.
Bengtson, S., 1994, The advent of animal skeletons, in: *Early Life on Earth* (S. Bengtson, ed.), Columbia University Press, New York, pp. 412-425.
Bengtson, S., and Yue Z., 1992, Predatorial borings in late Precambrian mineralized exoskeletons, *Science* **257**:367-369.
Bergström, J., 1973, Organisation, life and systematics of trilobites, *Fossils Strata* **21**:1-69.
Bergström, J., and Levi-Setti, R., 1978, Phenotypic variation in the Middle Cambrian trilobite *Paradoxides davidus* Salter at Manuels, S. E. Newfoundland, *Geol. Palaeontol.* **12**:1-40.

Blaker, M. R., and Peel, J. S., 1997, Lower Cambrian trilobites from North Greenland, *Medd. Grønland Geosci.* **35**:1-145.
Bottjer, D. J., Hagadorn, J. W., and Dornbos, S. Q., 2000, The Cambrian substrate revolution, *GSA Today* **10**(9):1-7.
Boucot, A. J., 1990, *Evolutionary Paleobiology of Behavior and Coevolution,* Elsevier Science Publishers, Amsterdam.
Bowring, S. A., and Erwin, D. L., 1998, A new look at evolutionary rates in deep time: uniting paleontology and high-precision geochronology, *GSA Today* **8**(9):1-8.
Bradshaw, J. L., 1988, The evolution of human lateral asymmetries: new evidence and second thoughts, *J. Human Evol.* **17**:615-637.
Bradshaw, J. L., 1989, *Hemispheric Specialization and Psychological Function,* John Wiley & Sons, Chichester, U.K.
Bradshaw, J. L., 1991, Animal asymmetry and human heredity: dextrality, tool use and language in evolution—10 years after Walker (1980), *Brit. J. Psychol.* **82**:39-59.
Brandt, D. S., 1980, Phenotypic Variation and Paleoecology of *Flexicalymene* (Arthropoda: Trilobita) in the Cincinnatian Series (Upper Ordovician) near Cincinnati, Ohio. Unpublished M.S. thesis, University of Cincinnati.
Brandt, D. S., Meyer, D. L., and Lask, P. B., 1995, *Isotelus* (Trilobita) "hunting burrow" from Upper Ordovician strata, Ohio, *J. Paleontol.* **69**:1079-1083.
Brandt Velbel, D. S., 1985, Ichnologic, taphonomic, and sedimentologic clues to the deposition of Cincinnatian shales (Upper Ordovician), Ohio, U.S.A., in: *Biogenic Structures: Their Use in Interpreting Depositional Environment* (A. H. Curran, ed.), Soc. Econ. Paleontol. Mineral. Spec. Publ. 35, pp. 299-307.
Brasier, M. D., 1979, The Cambrian radiation event, in: *The Origin of Major Invertebrate Groups* (M. R. House, ed.), Systematics Association Special Volume 12. Academic Press, London and New York, pp. 103-159.
Brett, C. E., 1977, Entombment of a trilobite within a closed brachiopod shell, *J. Paleontol.* **51**:1041-1045.
Brezinski, D. K., 1992, Permian trilobites from west Texas, *J. Paleontol.* **66**:924-943.
Briggs, D. E. G., 1978, The morphology, mode of life, and affinities of *Canadaspis perfecta* (Crustacea: Phyllocarida), Middle Cambrian, Burgess Shale, British Columbia, *Phil. Trans. R. Soc. London B* **281**:429-487.
Briggs, D. E. G., 1979, *Anomalocaris,* the largest known Cambrian arthropod, *Palaeontology* **22**:631-664.
Briggs, D. E. G., 1994, Giant predators from the Cambrian of China, *Science* **264**:1283-1284.
Briggs, D. E. G., and Collins, D., 1988, A Middle Cambrian chelicerate from Mt. Stephen, British Columbia, *Palaeontology* **31**:779-798.
Briggs, D. E. G., Erwin, D. H., and Collier, F. J., 1994, *The Fossils of the Burgess Shale,* Smithsonian Institution Press, Washington and London.
Briggs, D. E. G., and Fortey, R. A., 1989, The early radiation and relationships of the major arthropod groups, *Science* **246**:241-243.
Briggs, D. E. G., and Whittington, H. B., 1985a, Modes of life of arthropods from the Burgess Shale, British Columbia, *Trans. R. Soc. Edinburgh Earth Sci.* **76**:149-160.
Briggs, D. E. G., and Whittington, H. B., 1985b, Terror of the trilobites, *Nat. Hist.* **94**:34-39.
Bright, R. C., 1959, A paleoecologic and biometric study of the Middle Cambrian trilobite *Elrathia kingii* (Meek), *J. Paleontol.* **33**:83-98.
Brusca, R. C., and Brusca, G. J., 1990, *Invertebrates,* Sinauer Associates, Sunderland, Massachusetts.
Bruton, D. L., 1981, The arthropod *Sidneyia inexpectans,* Middle Cambrian, Burgess Shale, British Columbia, *Phil. Trans R. Soc. London B* **295**:619-656.
Bruton, D. L., and Haas, W., 1997, Functional morphology of Phacopinae (Trilobita) and the mechanics of enrollment, *Paleontographica Abt. A* **245**:1-43.
Buchholz, A., 2000, Die Trilobitenfauna der oberkambrischen Stufen 1 – 3 in Geschieben aus Vorpommern und Mecklenburg (Norddeutschland), *Archiv für Geschiebekun.* **2**:697-776.
Burling, L. D., 1917, Was the Lower Cambrian trilobite supreme?, *Ottawa Natural.* **31**:77-79.
Butterfield, N. J., 1995, Secular distribution of Burgess Shale-type preservation, *Lethaia* **28**:1-13.
Butterfield, N. J., 2002, *Leanchoilia* guts and the interpretation of three dimensional structures in Burgess Shale-type fossils, *Paleobiology* **28**:155-171.
Campbell, L. D., 1969, Stratigraphy and Paleontology of the Kinzers Formation, Southeastern Pennsylvania. Unpublished M.S. thesis, Franklin and Marshall College.
Campbell, K. S. W., 1975, The functional anatomy of trilobites: musculature and eyes, *J. Proc. R. Soc. N. S. Wales* **108**:168-188.
Chatterton, B. D. E., Johanson, Z., and Sutherland, G., 1994, Form of the trilobite digestive system: alimentary structures in *Pterocephalia, J. Paleontol.* **68**:294-305.

Chatterton, B. D. E., and Ludvigsen, R., 1976, Silicified Middle Ordovician trilobites from the South Nahanni River area, District of Mackenzie, Canada, *Palaeontographica, Abt. A* **154**:1-106.
Chen J. Y., and Erdtmann, B.-D., 1991, Lower Cambrian Lagerstätte from Chengjiang, Yunnan, China: insights for reconstructing early metazoan life, in: *The Early Evolution of Metazoa and the Significance of Problematic Taxa* (A. M. Simonetta and S. Conway Morris, eds.), Cambridge University Press, Cambridge, pp. 57-76.
Chen J. Y., Ramsköld, L., and Zhou G. Q., 1994, Evidence for monophyly and arthropod affinity of Cambrian giant predators, *Science* **264**:1304-1308.
Chen J. Y., and Zhou G. Q., 1997, Biology of the Chengjiang fauna, *Nation. Mus. Nat. Hist. Bull.* **10**:11-105.
Chen J. Y., Zhou G. Q., Zhu M. Y., and Yeh K Y., 1997, *The Chengjiang Biota: A Unique Window of the Cambrian Explosion,* National Museum of Natural Science, Taichung, Taiwan.
Clarkson, E. N. K., 1979, The visual system of trilobites, *Palaeontology* **22**:1-22.
Clarkson, E. N. K., and Henry, J.-L., 1973, Structures coaptative et enroulement chez quelques trilobites ordoviciens et siluriens, *Lethaia* **6**:105-132.
Clarkson, E. N. K., and Whittington, H. B., 1997, Enrollment and coaptative structures, in: *Treatise on Invertebrate Paleontology,* Part O, Arthropoda 1, Trilobita, Revised. Volume 1: Introduction, Order Agnostida, Order Redlichiida (R. L. Kaesler, ed.), Geological Society of America and University of Kansas, Boulder, Colorado, and Lawrence, Kansas, pp. 67-74.
Cloud, P. E., 1968, Pre-metazoan evolution and the origins of the Metazoa, in: *Evolution and Environment* (E. T. Drake, ed.), Yale University Press, New Haven, pp. 1-72.
Collins, D., 1996, The "evolution" of *Anomalocaris* and its classification in the arthropod class Dinocarida (nov.) and order Radiodonta (nov.), *J. Paleontol.* **70**:280-293.
Conway Morris, S., 1977, Fossil priapulid worms, Spec. Pap. Palaeontol. 20.
Conway Morris, S., 1981, Parasites and the fossil record, *Parasitology,* **82**:489-509.
Conway Morris, S., 1985, Cambrian Lagerstätten: their distribution and significance, *Phil. Trans. R. Soc. London B* **311**:49-65.
Conway Morris, S., 1986, The community structure of the Middle Cambrian phyllopod bed (Burgess Shale), *Palaeontology* **29**:423-467.
Conway Morris, S., 1990, Late Precambrian and Cambrian soft-bodied faunas, *Ann. Rev. Earth Planet. Sci.* **18**:101-122.
Conway Morris, S., 1998, *The Crucible of Creation. The Burgess Shale and the Rise of Animals,* Oxford University Press, Oxford.
Conway Morris, S., and Bengtson, S., 1994, Cambrian predators: possible evidence from boreholes, *J. Paleontol.* **68**:1-23.
Conway Morris, S., and Jenkins, R. J. F., 1985, Healed injuries in Early Cambrian trilobites from South Australia, *Alcheringa* **9**:167-177.
Conway Morris, S., and Robison, R. A., 1986, Middle Cambrian priapulids and other soft-bodied fossils from Utah and Spain, *Univ. Kansas Paleontol. Contrib. Pap.* **117**:1-22.
Conway Morris, S., and Robison, R. A., 1988, More soft-bodied animals and algae from the Middle Cambrian of Utah and British Columbia, *Univ. Kansas Paleontol. Contrib. Pap.* **122**:1-48.
Crick, R. E., 1981, Diversity and evolutionary rates of Cambrian-Ordovician nautiloids, *Paleobiology* **7**:216-229.
Davis, R. A., Fraaye, R. H. B., and Holland, C. H., 2001, Trilobites within nautiloid cephalopods, *Lethaia,* **34**:37-45.
Dean, D., Rankin, J. S., Jr., and Hoffman, E., 1964, A note on the survival of polychaetes and amphipods in stored jars of sediment, *J. Paleontol.* **38**:608-609.
Droser, M. L., and Bottjer, D. J., 1988, Trends in depth and extent of bioturbation in Cambrian carbonate marine environments, western United States, *Geology* **16**:233-236.
Droser, M. L., and Bottjer, D. J., 1989, Ordovician increase in extent and depth of bioturbation: implications for understanding early Paleozoic ecospace utilization, *Geology* **17**:850-852.
Droser, M. L., Hughes, N. C., and Jell, P., 1994, Infaunal communities and tiering in Lower Palaeozoic nearshore clastic environments: trace fossil evidence from the Cambro-Ordovician of New South Wales, *Lethaia* **27**:273-283.
Droser, M. L., Fortey, R. A., and Xing, L., 1996, The Ordovician radiation, *Am. Sci.* **84**:122-131.
Eaton, R. C., Bombardieri, R. A., and Meyer, D. L., 1977, The Mauthner-initiated startle response in teleost fish, *J. Exp. Biol.* **66**:65-81.
Eldredge, N., 1971, Patterns of cephalic musculature in the Phacopina (Trilobita) and their phylogenetic significance, *J. Paleontol.* **45**:52-67.
Erwin, D. H., 1991, Metazoan phylogeny and the Cambrian radiation, *Trends Ecol. Evol.* **6**:131-134.

Fedonkin, M. A., 1994, Early multicellular fossils, in: *Early Life on Earth* (S. Bengtson, ed.), Columbia University Press, New York, pp. 370-388.
Fisher, D. C., 1977, Mechanism and significance of enrollment in xiphosurans (Chelicerata, Merostomata), *Geol. Soc. Amer. Abstr. Progr.* **9**:264-265.
Foote, M., 1991, Morphologic patterns of diversification: examples from trilobites, *Palaeontology* **34**:41-485.
Foote, M., 1992, Paleozoic record of morphological diversity in blastozoan echinoderms, *Proc. Nat. Acad. Sci. USA* **89**:7325-7329.
Fortey, R. A., 1985, Pelagic trilobites as an example of deducing the life habits of extinct arthropods, *Trans. R. Soc. Edinburgh Earth Sci.* **76**:219-230.
Fortey, R. A., 2000, *Trilobite! Eyewitness to Evolution,* Alfred A. Knopf, New York.
Fortey, R. A., Briggs, D. E. G., and Wills, M. A., 1996, The Cambrian evolutionary 'explosion': decoupling cladogenesis from morphological disparity, *Biol. J. Linn. Soc.* **57**:13-33.
Fortey, R. A., and Clarkson, E. N. K., 1976, The function of the glabellar 'tubercle' in *Nileus* and other trilobites, *Lethaia* **9**:101-106.
Fortey, R. A., and Owens, R. M., 1990, Trilobites, in: *Evolutionary Trends* (K. J. McNamara, ed.), Belhaven Press, London, pp. 121-142.
Fortey, R. A., and Owens, R. M., 1999, Feeding habits in trilobites, *Palaeontology* **42**:429-465.
Fortey, R. A., and Whittington, H. B., 1989, The Trilobita as a natural group, *Hist. Biol.* **2**:125-138.
Gehling, J. G., 1999, Microbial mats in Proterozoic siliciclastics: Ediacaran death masks, *Palaios* **14**:40-57.
Gehling, J. G., and Rigby, J. K., 1996, Long expected sponges from the Neoproterozoic Ediacaran fauna of South Australia, *J. Paleontol.* **70**:185-195.
Geyer, G., 1993, The giant Cambrian trilobites of Morocco, *Beringia* **8**:71-107.
Geyer, G., 1996, The Moroccan fallotaspidid trilobites revisited, *Beringia* **18**:89-199.
Geyer, G., 1998, Intercontinental, trilobite-based correlation of the Moroccan early Middle Cambrian, *Can. J. Earth Sci.* **35**:374-401.
Geyer, G., and Palmer, A. R., 1995, Neltneriidae and Holmiidae (Trilobita) from Morocco and the problem of Early Cambrian intercontinental correlation, *J. Paleontol.* **69**:459-474.
Geyer, G., and Shergold, J. S., 2000, The quest for internationally recognized divisions of Cambrian time, *Episodes* **23**: 188-195.
Glaessner, M. F., 1984, *The Dawn of Animal Life. A Biohistorical Study,* Cambridge University Press, Cambridge.
Grant, S. W. F., 1990, Shell structure and distribution of *Cloudina,* a potential index fossil for the terminal Proterozoic, *Amer. J. Sci.* **290-A**:261-294.
Grotzinger, J. P., Bowring, S. A., Saylor, B. Z., and Kaufman, A. J., 1995, Biostratigraphic and geochronologic constraints on early animal evolution, *Science* **270**:598-604.
Gunther, L. F., and Gunther, V. G., 1981, Some Middle Cambrian fossils of Utah, *Brig. Young Univ. Geol. Stud.* **28**:1-87.
Han N. R., and Zhang J. L., 1991, Malformed thoracic pleurae of *Redlichia (Redlichia) hupehensis* Hsu, *Acta Palaeontol. Sinica* **30**:126-128.
Hannibal, J. T., and Feldmann, R. M., 1981, Systematics and functional morphology of oniscomorph millipedes (Arthropoda: Diplopoda) from the Carboniferous of North America, *J. Paleontol.* **55**:730-746.
Henry, J. L., and Clarkson, E. N. K., 1974, Enrollment and coaptation in some species of the Ordovician trilobite genus *Placoparia, Fossils Strata* **4**:87-95.
Hoffman, H. J., 1994, Proterozoic carbonaceous compressions ("metaphytes" and "worms"), in: *Early Life on Earth* (S. Bengtson, ed.), Columbia University Press, New York, pp. 342-357.
Hollingsworth, J. S., 1999, The problematical base of the Montezuman Stage: should the Laurentian fallotaspidids be in a non-trilobite series?, in: *Laurentia 99. V Field Conference of the Cambrian Stage Subdivision Working Group, International Subcommission on Cambrian Stratigraphy* (A. R. Palmer, ed.), Institute for Cambrian Studies, Boulder, Colorado, pp. 5-9.
Hou X. G., and Bergström, J., 1997, Arthropods of the Lower Cambrian Chengjiang fauna, Southwest China, *Fossils Strata* **45**:1-116.
Hou X. G., Bergström, J., and Ahlberg, P., 1995, *Anomalocaris* and other large animals in the Lower Cambrian Chengjiang fauna of Southwest China, *Geol. Fören. Stockholm Förhand.* **117**:162-183.
Hou X. G., Bergström, J., Wang H. F., Feng X. H., and Chen A. L., 1999, *The Chengjiang Fauna: Exceptionally Well-Preserved Animals from 530 Million Years Ago,* Science and Technology Press, Yunnan.
Hughes, N. C., Chapman, R. E., and Adrain, J. M., 1999, The stability of thoracic segmentation in trilobites: a case study in developmental and ecological constraints, *Evol. Develop.* **1**:24-35.
Hughes, N. C., and Cooper, D. L., 1999, Paleobiologic and taphonomic aspects of the "*granulosa*" trilobite cluster, Kope Formation (Upper Ordovician, Cincinnati region), *J. Paleontol.* **73**:306-319.

Jago, J. B., 1974, Evidence for scavengers from Middle Cambrian sediments in Tasmania, *Neues Jahrb. Geol. Paläontol., Monatsch.* **1974**:13-17.
Jago, J. B., and Haines, P. W., in press, Repairs to an injured early Middle Cambrian trilobite, Elkedra area, Northern Territory, *Alcheringa*.
Jell, P. A., 1989, Some aberrant exoskeletons from fossil and living arthropods, *Queensland Mus. Mem.* **27**:491-498.
Jensen, S., 1990, Predation by Early Cambrian trilobites on infaunal worms—evidence from the Swedish Mickwitzia Sandstone, *Lethaia* **23**:29-42.
Kelley, P. H., and Hansen, T. A., 1993, Evolution of the naticid gastropod predator-prey system: an evaluation of the hypothesis of escalation, *Palaios* **8**:358-375.
Kelley, P. H., Kowalewski, M., and Hansen, T. A. (eds.), this volume, *Predator-Prey Interactions in the Fossil Record*. Kluwer Academic/Plenum Publishers, New York.
Kesling, R. V., and Chilman, R. B., 1975, Strata and Megafossils of the Middle Devonian Silica Formation, *Univ. Michigan Mus. Paleontol. Pap. Paleontol.* **8**:1-408.
Knoll, A. H., 1992, The early evolution of eukaryotes: a geological perspective, *Science* **256**:622-627.
Knoll, A. H., 1996, Daughter of time, *Paleobiology* **22**:1-7.
Landing, E., Bowring, S. A., Davidek, K. L., Westrop, S. R., Geyer, G., and Heldmaier, W., 1998, Duration of the Early Cambrian: U-Pb ages of volcanic ashes from Avalon and Gondwana, *Can. J. Earth Sci.* **35**:329-338.
Laudon, L. R., 1939, Unusual occurrence of *Isotelus gigas* DeKay in the Bromide Formation (Ordovician) of southern Oklahoma, *J. Paleontol.* **13**:211-213.
Levi-Setti, R., 1993, *Trilobites*, second edition, University of Chicago Press, Chicago and London.
Lipps, J. H., 1983, Biotic interactions in benthic Foraminifera ecosystems, in: *Biotic Interactions in Recent and Fossil Benthic Communities* (M. J. S. Tevesz and P. L. McCall, eds.), Plenum Press, New York and London, pp. 331-376.
Lipps, J. H., 2001, Protists and the Precambrian-Cambrian skeletonization event, in: *Cambrian System of South China* (S. C. Peng, L. E. Babcock and M. Y. Zhu, eds.), Press of University of Science and Technology of China, Hefei, p. 280.
Lipps, J. H., and Signor, P. W. (eds.), 1992, *Origin and Early Evolution of the Metazoa*, Plenum Press, New York.
Loeblich, A. R., Jr., 1940, An occurrence of *Isotelus gigas* DeKay in the Arbuckle Mountains, Oklahoma, *J. Paleontol.* **14**:161-162.
Luo H. L., and Jiang Z. W., 1996, The Sinian-Cambrian boundary section and the Meishucun and Chengjiang faunas in Yunnan, 30th Int. Geol. Congr. Field Trip T118/381, Geological Publishing House, Beijing, pp. 1-23.
Ludvigsen, R., 1977, Rapid repair of traumatic injury by an Ordovician trilobite, *Lethaia* **10**:205-207.
Ludvigsen, R., 1979, *Fossils of Ontario. Part 1: The Trilobites*, R. Ontario Mus. Life Sci. Misc. Publ., 96 pp.
Maliva, R. G., Knoll, A. H., and Siever, R., 1989, Secular change in chert distribution: a reflection of evolving biological participation in the silica cycle, *Palaios* **5**:519-532.
Manton, S. M., 1977, *The Arthropoda*, Clarendon Press, Oxford.
Martinsson, A., 1965, Aspects of a Middle Cambrian thanatotope on Öland, *Geol. Fören. Stockholm Forhand.* **87**:181-230.
McMahon, B. R., and Wilkens, J. L., 1975, Respiratory and circulatory responses to hypoxia in the lobster *Homarus americanus*, *J. Exp. Biol.* **62**:637-655.
McMenamin, M. A. S., 1986, The garden of Ediacara, *Palaios* **1**:178-182.
McMenamin, M. A. S., and McMenamin, D. L. S., 1990, *The Emergence of Animals: The Cambrian Breakthrough*, Columbia University Press, New York.
Mikulic, D. G., 1994, Sheltered molting by trilobites, *Geol. Soc. Amer. Abstr. Prog.* **26**(5):55.
Miller, J., 1975, Structure and function of trilobite terrace lines, *Fossils Strata* **4**:155-178.
Miller, J., 1976, The sensory fields and life mode of *Phacops rana* (Green, 1832) (Trilobita), *Trans. R. Soc. Edinburgh* **69**:337-367.
Miller, J., and Clarkson, E. N. K., 1980, The post-ecdysial development of the cuticle and the eye of the Devonian trilobite *Phacops rana milleri* Stewart 1927, *Phil. Trans. R. Soc. London B* **288**:461-480.
Miller, R. H., and Sundberg, F. A., 1984, Boring Late Cambrian organisms, *Lethaia* **17**:185-190.
Müller, K. J., and Walossek, D., 1987, Morphology, ontogeny, and life habit of *Agnostus pisiformis* from the Upper Cambrian of Sweden, *Fossils Strata* **19**:1-124.
Narbonne, G. M., 1998, The Ediacara Biota: a terminal Neoproterozoic experiment in the evolution of life, *GSA Today* **8**(2):1-6.
Needham, A. E., 1952, *Regeneration and Wound-Healing*, Methuen, London.
Nedin, C., 1999, *Anomalocaris* predation on nonmineralized and mineralized trilobites, *Geology* **27**:987-990.

Oehlert, D.-P., 1895, Sur les *Trinucleus* de l'Ouest de la France, *Bull. Soc. Géol. France ser. 3*, **23**:299-336.
Osgood, R. G., Jr., 1970, Trace fossils of the Cincinnati area, *Palaeontogr. Amer.* **6**:281-444.
Osgood, R. G., Jr., and Drennen, W. T., 1975, Trilobite trace fossils from the Clinton Group (Silurian) of east-central New York, *Bull. Am. Paleontol.* **67**:300-348.
Owen, A. W., 1983, Abnormal cephalic fringes in the Trinucleidae and Harpetidae (Trilobita), *Spec. Pap. Palaeontol.* **30**:241-247.
Owen, A. W., 1985, Trilobite abnormalities, *Trans. R. Soc. Edinburgh Earth Sci.* **76**:255-272.
Palmer, A. R., 1999, Terminal Early Cambrian extinction of the Olenellina: documentation from the Pioche Formation, Nevada, *J. Paleontol.* **72**:650-672.
Peach, B. N., 1894, Additions to the fauna of the *Olenellus*-zone of the Northwest Highlands, *Quart. J. Geol. Soc. London* **50**:661-676.
Peng S. C., and Babcock, L. E., 2001, Cambrian of the Hunan-Guizhou region, South China, in: *Cambrian System of South China* (S. C. Peng, L. E. Babcock and M. Y. Zhu, eds.), Press of University of Science and Technology of China, Hefei, pp. 3-51.
Peng S. C., and Robison, R. A., 2000, Agnostoid biostratigraphy across the Middle-Upper Cambrian boundary in Hunan, China, *Paleontol. Soc. Mem.* **53** (supplement to *J. Paleontol.* **74** (4)), 104 pp.
Pompeckj, J., 1892, Bemerkungen über das Einrollungsvermögen der Trilobiten, *Gesellsch. Natur. Württemberg Stuttgart Jahr.* **48**:93-101.
Portlock, J. E., 1843, *Report on the Geology of the County of Londonderry, and of Parts of Tyrone and Fermanagh. Examined and Described Under the Authority of the Master General and Board of Ordnance,* Andrew Milliken, Dublin, and Longman, Brown, Green, and Longmans, London.
Pratt, B. R., 1998, Probable predation on Upper Cambrian trilobites and its relevance for the extinction of soft-bodied Burgess Shale-type animals, *Lethaia* **31**:73-88.
Přibyl, A., and Vaněk, J., 1981, Preliminary report on some new trilobites of the family Harpetidae Hawle and Corda (Trilobita), *Cas. Pro. Min. Geol.* **26**:187-193.
Purtilo, D. T, 1978, *A Survey of Human Diseases,* Addison-Wesley, Menlo Park, California.
Ramsköld, L., and Edgecombe, G. D., 1991, Trilobite monophyly revisited, *Hist. Biol.* **4**:267-283.
Ramsköld, L., and Edgecombe, G. D., 1996, Trilobite appendage structure of *Redlichia* reconsidered, *Alcheringa* **20**:269-276.
Resser, C. E., and Howell, B. F., 1938, Lower Cambrian *Olenellus* Zone of the Appalachians, *Geol. Soc. Am. Bull.* **49**:195-248.
Robison, R. A., 1964, Late Middle Cambrian faunas from western Utah, *J. Paleontol.* **38**:510-566.
Robison, R. A., 1972, Mode of life of agnostid trilobites, *Proc. 24th Internat. Geol. Congr.* **7**:33-40.
Robison, R. A., 1984, Cambrian Agnostida of North America and Greenland, Part 1, Ptychagnostidae, *Univ. Kansas Paleontol. Contrib. Pap.* **109**:1-59.
Robison, R. A., 1991, Middle Cambrian biotic diversity: examples from four Utah Lagerstätten, in: *The Early Evolution of Metazoa and the Significance of Problematic Taxa* (A. M. Simonetta and S. Conway Morris, eds.), Cambridge University Press, Cambridge, pp. 77-98.
Robison, R. A., 1994, Agnostoid trilobites from the Henson Gletscher and Kap Stanton formations (Middle Cambrian), North Greenland, *Grønlands Geol. Under. Bull.* **169**:25-77.
Ross, R. J., Jr., 1979, Additional trilobites from the Ordovician of Kentucky, *U.S. Geol. Surv. Prof. Pap.* **1066-D**:1-13.
Rudkin, D. M., 1979, Healed injuries in *Ogygopsis klotzi* (Trilobita) from the Middle Cambrian of British Columbia, *R. Ontario Mus. Misc. Coll. Pap.* **32**:1-8.
Rudkin, D. M., 1985, Exoskeleton abnormalities in four trilobites, *Can. J. Earth Sci.* **22**:479-483.
Rudkin, D. M., and Tripp, R. P., 1989, The type species of the Ordovician trilobite genus *Isotelus*: *I. gigas* Dekay, 1824, *R. Ontario Mus. Life Sci. Contrib.* **152**:1-19.
Ruedemann, R., and Howell, B. F., 1944, Impression of a worm on the test of a Cambrian trilobite, *J. Paleontol.* **18**:96.
Runnegar, B., 1982, The Cambrian explosion: animals or fossils?, *J. Geol. Soc. Australia* **29**:395-411.
Runnegar, B., 1989, The evolution of mineral skeletons, in: *Origin, Evolution, and Modern Aspects of Biomineralization in Plants and Animals* (R. E. Crick, ed.), Plenum Press, New York, pp. 75-94.
Runnegar, B., 1994, Proterozoic eukaryotes: evidence from biology and geology, in: *Early Life on Earth* (S. Bengtson, ed.), Columbia University Press, New York, pp. 287-297.
Schmalfuss, H., 1978, Structure, patterns, and function of cuticular terraces in Recent and fossil arthropods, *Zoomorphologie* **90**:19-40.
Schram, F. R., 1986, *Crustacea,* Oxford University Press, Oxford.
Seilacher, A., 1985, Trilobite paleobiology and substrate relationships, *Trans. R. Soc. Edinburgh Earth Sci.* **76**:231-237.
Seilacher, A., 1989, Vendozoa: organismic construction in the Proterozoic biosphere, *Lethaia* **22**:229-239.

Seilacher, A., and Pfluger, F., 1994, From biomates to benthic agriculture: a biohistoric revolution, in: *Biostabilization of Sediments* (W. S. Krumbein et al., eds.), Bibliotheks und Informationsystem der Universität Oldenberg, Oldenberg, Germany, pp. 97-105.
Sepkoski, J. J., 1981, A factor analytic description of the Phanerozoic marine fossil record, *Paleobiology* **7**:36-53.
Sepkoski, J. J., and Sheehan, P. M., 1983, Diversification, faunal change, and community replacement during the Ordovician radiations, in: *Biotic Interactions in Recent and Fossil Benthic Communities* (M. J. S. Tevesz and P. L. McCall, eds.), Plenum, New York, pp. 673-717.
Shaw, F. C., 1974, Simpson Group (Middle Ordovician) trilobites of Oklahoma. *Paleontol. Soc. Mem.* **6** (supplement to *J. Paleontol.* **48**(5)), 54 pp.
Shu D. G., Geyer, G., Chen, L., and Zhang, X. L., 1995, Redlichiacean trilobites with soft-parts from the Lower Cambrian Chengjiang fauna (South China), *Beringia Spec. Issue* **2**:203-241.
Shu D. G., Luo H. L., Conway Morris, S., Zhang X. L., Hu S. X., Chen L., Han J., Zhu M., Li Y., and Chen L. Z., 1999, Lower Cambrian vertebrates from South China, *Nature* **402**:42-46.
Shu D. G., and Zhang X. L., 1996, *Kuamaia*, an Early Cambrian predator from the Chengjiang Fossil Lagerstätte, *J. Northwest Univ.* **1996**:27-33.
Signor, P. W., III, and Brett, C. E., 1984, The mid-Paleozoic precursor to the Mesozoic marine revolution, *Paleobiology* **10**:229-245.
Signor, P. W., and Vermeij, G. J., 1994, The plankton and the benthos: origins and early history of an evolving relationship, *Paleobiology* **20**:297-319.
Sliter, W. V., 1971, Predation on benthic foraminifers, *J. Foram. Res.* **1**:20-29.
Sloan, R. E., 1992, Functional anatomy of *Ectenaspis* and the isoteline hypostome, *Geol. Soc. Am. Abstr. Progr.* **24**(4):65.
Šnajdr, M., 1978a, Anomalous carapaces of Bohemian paradoxid trilobites, *Sb. Geol. Ved. Paleont.* **20**:1-31.
Šnajdr, M., 1978b, Pathological neoplasms in the fringe of *Bohemoharpes* (Trilobita), *Věstn. Ústřed. ústavu geolog.* **53**:49-50.
Šnajdr, M., 1979a, Two trinucleid trilobites with repair of traumatic injury, *Věstn. Ústřed. ústavu geolog.* **54**:49-51.
Šnajdr, M., 1979b, Note on the regenerative ability of injured trilobites, *Věstn. Ústřed. ústavu geolog.* **54**:171-173.
Šnajdr, M., 1979c, Patologické exoskeletony dvou Ordovických trilobitů Barrandienu, *Cas. Národ. Muz.* **148**:173-176.
Šnajdr, M., 1981, Bohemian Proetidae with malformed exoskeletons, *Sb. Geol. Ved. Paleont.* **24**:37-61.
Šnajdr, M., 1991, On the digestive system of *Deanaspis goldfussi* (Barrande), *Cas. Národ. Muz.* **156**:8-16.
Speyer, S. E., 1988, Biostratinomy and functional morphology of enrollment in two Middle Devonian trilobites, *Lethaia* **21**:121-138.
Speyer, S. E., 1990a, Enrollment in trilobites, in: *Evolutionary Paleobiology of Behavior and Coevolution* (by A. J. Boucot), Elsevier Science Publishers, Amsterdam, pp. 450-455.
Speyer, S. E., 1990b, Gregarious behavior and reproduction in trilobites, in: *Evolutionary Paleobiology of Behavior and Coevolution* (by A. J. Boucot), Elsevier Science Publishers, Amsterdam, pp. 405-409.
Speyer, S. E., and Brett, C. E., 1985, Clustered trilobite assemblages in the Middle Devonian Hamilton Group, *Lethaia* **18**:85-103.
Speyer, S. E., and Chatterton, B. D. E., 1989, Trilobite larvae and larval ecology, *Hist. Biol.* **3**:27-60.
Sprinkle, J., 1973, Morphology and evolution of blastozoan echinoderms, *Mus. Comp. Zool., Harvard Univ., Spec. Publ.*, pp. 1-283.
Stanley, S. M., 1976, Fossil data and the Precambrian-Cambrian evolutionary transition, *Amer. J. Sci.* **276**:56-76.
Stanley, S. M., 1999, *Earth System History*, W. H. Freeman and Company, New York.
Stitt, J. H., 1983, Enrolled Late Cambrian trilobites from the Davis Formation, southeast Missouri, *J. Paleontol.* **57**:93-105.
Stockton, W. L., and Cowen, R., 1976, Stereoscopic vision in one eye: paleophysiology of the schizochroal eye of trilobites, *Paleobiology* **2**:304-315.
Størmer, L., 1931, Boring organisms in trilobite shells, *Norsk Geol. Tidsskr.* **12**:533-539.
Struve, W., and Flick, H., 1984, *Chotecops sollei* und *Chotecops ferdinandi* aus den devonischen Schiefern des rheinischen Gebirges, *Senck. leth.* **65**:137-163.
Stürmer, W., and Bergström, J., 1973, New discoveries on trilobites by x-ray, *Paläontol. Zeit.* **47**:104-141.
Sun W. G., 1994, Early multicellular fossils, in: *Early Life on Earth* (S. Bengtson, ed.), Columbia University Press, New York, pp. 358-369.
Thorson, G., 1950, Reproductive and larval ecology of marine bottom invertebrates, *Biol. Rev.* **25**:1-45.
Towe, K. M., 1973, Trilobite eyes: calcified lenses in vivo, *Science* **179**:1007-1009.

Tripp, R. P., and Evitt, W. R., 1986, Silicified trilobites of the family Asaphidae from the Middle Ordovician of Virginia, *Palaeontology* **29**:705-724.
Tshudy, D. M., Feldmann, R. M., and Ward, P. D., 1989, Cephalopods: biasing agents in the preservation of lobsters, *J. Paleontol.* **63**:621-626.
Valentine, J. W., 1994, The Cambrian explosion, in: *Early Life on Earth* (S. Bengtson, ed.), Columbia University Press, New York, pp. 401-411.
Vermeij, G. J., 1977, The Mesozoic faunal revolution: evidence from snails, predators and grazers, *Paleobiology* **3**:245-258.
Vermeij, G. J., 1987, *Evolution and Escalation: An Ecological History of Life*, Princeton University Press, Princeton, New Jersey.
Vermeij, G. J., 1995, Economics, volcanoes, and Phanerozoic revolutions, *Paleobiology* **21**:125-152.
Vidal, G., 1994, Early ecosystems: limitations imposed by the fossil record, in: *Early Life on Earth* (S. Bengtson, ed.), Columbia University Press, New York, pp. 298-311.
Vorwald, G. R., 1982, Healed injuries in trilobites -- evidence for a large Cambrian predator, *Geol. Soc. Am. Abstr. Progr.* **14**:639.
Walcott, C. D., 1883, Injury sustained by the eye of a trilobite at the time of the moulting of the shell, *Amer. J. Sci. ser. 3*, **26**:302 [reprinted, 1884, Annals Mag. Nat. Hist. 15:69].
Webb, P. W., 1975, Acceleration performance of rainbow trout *Salmo gairneri* and green sunfish *Lepomis cyanellus*, *J. Exp. Biol.* **63**:451-465.
Wenndorf, K.-W., 1990, Homalonotinae (Trilobita) aus dem Rheinischen Unter-Devon, *Palaeontographica Abt. A* **211**:1-184.
Westergård, A. H., 1946, Agnostidea of the Middle Cambrian of Sweden, *Sver. Geol. Unders. C* **477**:1-141.
Whittington, H. B., 1975, Trilobites with appendages from the Middle Cambrian Burgess Shale, British Columbia, *Fossils Strata* **4**:97-136.
Whittington, H. B., 1977, The Middle Cambrian trilobite *Naraoia*, Burgess Shale, British Columbia, *Phil. Trans. R. Soc. London B* **280**:409-443.
Whittington, H. B., 1980, Exoskeleton, moult stage, appendage morphology and habits of the Middle Cambrian trilobite *Olenoides serratus*, *Palaeontology* **23**:171-204.
Whittington, H. B., 1988a, Hypostomes and ventral cephalic sutures in Cambrian trilobites, *Palaeontology* **31**:577-610.
Whittington, H. B., 1988b, Hypostomes of post-Cambrian trilobites, *Mem. New Mexico Bur. Mines Min. Resour.* **44**:321-39.
Whittington, H. B., 1992, *Trilobites. Fossils Illustrated*, Vol. 2, Boydell Press, Woodbridge and Suffolk, U.K.
Whittington, H. B., 1997a, Morphology of the exoskeleton, in: *Treatise on Invertebrate Paleontology, Part O, Arthropoda 1, Trilobita, Revised. Volume 1: Introduction, Order Agnostida, Order Redlichiida* (R. L. Kaesler, ed.), Geological Society of America and University of Kansas, Boulder, Colorado, and Lawrence, Kansas, pp. 1-67.
Whittington, H. B., 1997b, Supposed color markings in: *Treatise on Invertebrate Paleontology, Part O, Arthropoda 1, Trilobita, Revised. Volume 1: Introduction, Order Agnostida, Order Redlichiida* (R. L. Kaesler, ed.), Geological Society of America and University of Kansas, Boulder, Colorado, and Lawrence, Kansas, pp. 84-85.
Whittington, H. B., 1997c, Mode of life, habits, and occurrence, in: *Treatise on Invertebrate Paleontology, Part O, Arthropoda 1, Trilobita, Revised. Volume 1: Introduction, Order Agnostida, Order Redlichiida* (R. L. Kaesler, ed.), Geological Society of America and University of Kansas, Boulder, Colorado, and Lawrence, Kansas, pp. 137-169.
Whittington, H. B., and Briggs, D. E. G., 1985, The largest Cambrian animal, *Anomalocaris*, Burgess Shale, British Columbia, *Phil. Trans. R. Soc. London B* **306**:569-609.
Wills, M. A., Briggs, D. E. G., Fortey, R. A., and Wilkinson, M., 1995, The significance of fossils in understanding arthropod evolution, *Verhand. Deutsch. Zoolog. Gesellsch.* **88**:203-215.
Zhang W. T., 1987, World's oldest Cambrian trilobites from eastern Yunnan, in: *Stratigraphy and Palaeontology of Systemic Boundaries in China, Precambrian-Cambrian Boundary 1*, Nanjing University Publishing House, Nanjing, pp. 1-18.
Zhang W. T., and Hou X. G., 1985, Preliminary notes on the occurrence of the unusual trilobite *Naraoia* in Asia, *Acta Palaeontol. Sinica* **24**:591-595.
Zhang X. G., 1989, Ontogeny of an Early Cambrian eodiscid trilobite from Henan, China, *Lethaia* **22**:13-29.

Chapter 4

Predation by Drills on Ostracoda

RICHARD A. REYMENT and ASHRAF M. T. ELEWA

1. Introduction 93
2. Gastropod Drill-holes 94
   2.1. Drilling Process in Muricids and Naticids 94
   2.2. Muricid versus Naticid Holes 95
3. Paleoethology and Paleoecology of the Predators 96
   3.1. Multiple Boreholes and Failed Borings 99
   3.2. Site Chosen for Drilling 103
4. Size, Predation Pressure and Selectional Effects 103
5. Ornamentation Complexity and Predation Intensity 106
   5.1. Ornamentation and Borehole Dimensions 106
6. Holes of Non-molluscan Origin 107
7. Future Directions for Research 108
   References 109

1. Introduction

Identifiable evidence for predation on fossil ostracods is limited to the drilling gastropods of the families Naticidae and Muricidae. Ostracods are admittedly swallowed by fish, as has been reported for finds in the stomach contents of freshwater fish. This is perhaps more the result of chance events rather than the fish in question having sought out ostracods with the definite intent of preying upon them. The stomach contents of water-birds have yielded ostracod remains and the eggs of some freshwater ostracod species are distributed in the feces of wading birds. There are also isolated reports of echinoids having consumed ostracods, but whether this is by design or by accident has never been resolved.

Most work on drilling gastropods has been concerned with their importance as predators of edible molluscs, with the greatest attention having been given to oysters because of their commercial importance. A great amount of data is available for the oyster whelk *Urosalpinx cinerea* Say, not only on the morphology of the holes drilled

---

RICHARD A. REYMENT • Naturhistoriska Riksmuséet, Avdelningen för paleozoologi, Box 50007, 10405 Stockholm, Sweden.   ASHRAF M. T. ELEWA • Department of Geology, Minia University, Egypt.

Predator-Prey Interactions in the Fossil Record, edited by Patricia H. Kelley, Michał Kowalewski, and Thor A. Hansen. Kluwer Academic/Plenum Publishers, New York, 2003.

but also on the length of rasping periods (Carriker, 1961, 1969, 1981; Carriker *et al.*, 1974). Analogously well-documented information for naticids has yet to accrue. The most authoritative treatment of the predational biology of naticids is still the monograph of Ziegelmeier (1954).

The first figured record of a drilled ostracod seems to be that of Livan (1937, p. 147, figure 25) for a species from the Oligocene of the Mainz Basin. Remarkably, she also figured a drilled foraminifer. Drilled tests of *Quinqueloculina* and *Spiroplectammina* were recorded, and figured, by Reyment (1966a) in sediments from the present-day Niger Delta. The first detailed studies of gastropod predation on ostracods are Reyment (1963, 1966a, 1966b, 1967). Reyment *et al.* (1987) provided a detailed account of gastropod predation on Late Cretaceous and early Paleocene ostracods. Kabat's (1990, pp. 167, 179) claim that he could find no evidence of gastropod predation for the Paleocene has perforce a hollow ring to it; the Paleocene was in effect a time of intense gastropod predation and, indeed, the first well-documented records of drilled ostracods are of that age (Reyment, 1963). Ostracods have not entered into the general discussion on modern gastropod predational webs, not even as an alternative food source, despite the fundamental results of Fischer-Piette (1935) on lag in food-source switching in muricids, justly lauded by Moore (1958). In recognition of Edouard Fischer-Piette's outstanding research, Reyment (1999) designated the process of prey substitution as the "Fischer-Piette effect."

The mode of predation of drills on ostracods cannot be equated directly to their mode of feeding on barnacles, bivalves and gastropods because the terminal growth stage of ostracods lies within the lower range of growth of the predator. Hence, the theoretical concept of the "size-refuge" model does not arise as a possible interactive mechanism (Colbath, 1985, p. 858). Drills do not attack younger instars, which indicates that there is a size for prey below which the attack-reaction of juvenile drills does not seem to be triggered. In an environment rich in bivalves and ostracods of all age classes, ostracods are not the main target for predation (Reyment, 1971, p. 135).

The material used for illustrating this chapter comes from our own work in the Santonian of Israel, the Paleocene of Nigeria, the Eocene of Spain, the Miocene of Egypt and the present-day Bight of Benin (Niger Delta) and therefore cannot be claimed to be absolutely general in scope. The Israeli material was collected in 1986 (cf. Honigstein, 1984); the fossil Nigerian material is referenced in Reyment (1966b). The Recent data from the Niger Delta were obtained by Reyment in 1966 (Reyment, 1966a). The reference for the Egyptian Miocene fossils is Bassiouni and Elewa (2000). The quantitative appraisal of the material was made using various methods of multivariate statistical analysis (cf. Reyment and Savazzi, 1999).

2. Gastropod Drill-holes

2.1. Drilling Process in Muricids and Naticids

It is widely accepted that holes drilled by muricid predators can usually be distinguished from those made by naticids. Ziegelmeier (1954) established the criterion of counter-sinking of naticid holes and the lack of this feature in muricids. At an early stage in the study of gastropod predation it was suggested this difference in behavioral mode could possibly be due to the naticids drilling by use of the radula, and hence a purely mechanical procedure, whereas muricids were thought to combine chemical softening with rasping. However, it has been shown that both kinds of holes can present bevelled outer margins, so that the identification of countersinking in thin shells can be confounded (Carriker, 1981, p. 404; Hingston, 1985, p. 49; Reyment, 1967, p. 34). Carriker (1957, p. 340) reported that newly hatched muricids drill irregular holes. Inasmuch as all holes in ostracods must perforce derive from juvenile drills, it is noteworthy that most holes in our material are regular in form. The radula scratches made by muricids during drilling can be expected to be erased by the secretion produced by the Accessory Boring Organ (acronym ABO), as noted by Carriker (1969). This is, however, not a general rule, and muricid holes with preserved radula markings do occur. The likelihood of there being a biochemical agent involved in naticid drilling was first fully realized by Ankel (1937), who surmised that some chemical agent could be secreted in order to soften the shell of the prey (there had been earlier speculation on this, dating back to the nineteenth century). It has been shown by Carriker (1981) that muricids are unable to drill if the ABO has been excised. There is also a sexual dimorphic factor of location of the ABO to be considered for muricids, a feature that seems to be lacking in naticids (Carriker, 1981). Carriker (1955, p. 23) noted that the ABO is disproportionately large in newly hatched muricid drills, being about one-third the length of the foot. This factor must be kept in mind when extrapolating from hole-diameter to an estimate of predator size. Carriker and Yochelson (1968, p. 8) reported that newly hatched drills bore holes less than 0.1 mm in diameter. Wiltse (1980, pp. 187, 198) is a further useful reference for predation by juvenile snails.

2.2. Muricid versus Naticid Holes

Maddocks (1988) attempted to classify all types of holes she observed in Texan Cretaceous to Recent ostracod shells, thus initiating a line of research worthy of future development. The question of whether a particular drill-hole derives from a naticid or a muricid is usually easy to resolve in thick-shelled fossils, applying the criteria of Ziegelmeier (1954, 1957) and Carriker (1969, 1981). Holes left in thin-shelled individuals may be more difficult to identify. This is sometimes a problem in ostracods but, in many cases, a correct assignment can be made. Muricids drill steep-sided, tapering to cylindrical holes that may be narrowly "frosted" around the distal (in relation to the prey) opening and that may be gently countersunk. Holes drilled by naticids are parabolic, frequently with a broad zone of "frosting" around the distal orifice and usually, though not invariably, markedly countersunk (Guerrero and Reyment, 1988). Both categories of drill-holes frequently display bevelled outer margins (Reyment, 1967, p. 34; Carriker, 1981, p. 404; Hingston, 1985, p. 49) and, in some cases, this can be the cause of a misleading identification.

Reyment *et al.* (1987) analyzed data for Santonian and Paleocene drill-holes by the multivariate statistical method of discriminant functions to resolve whether undoubted naticid holes differ statistically from identifiable muricid holes. The analysis of 21 naticid holes and 12 undoubted muricid holes from the Nigerian Paleocene (Table 1) indicates that there is no strong difference in size between holes made by juvenile naticids and juvenile muricids, which is manifested, for example, in the high probability of misidentification of holes. The linear discriminant function misidentifies 33.3% of the naticid holes, placing them among muricids, and 16.7% of the muricid holes, placing them among naticids. A preliminary examination of bivariate relationships pointed to a slight possibility of curvilinearity in the data, notwithstanding that the covariance matrices are homogeneous, with chi-squared = 20.6 for Box's criterion of homogeneity. The corresponding quadratic discriminant function (which is more appropriate for moderate curvilinearity in a multivariate data-set) yields an improvement in the levels of misidentification; thus, the error level for allocating naticid holes was reduced to 19% and that for the incorrect classification of muricid holes fell to zero. Examples of naticid holes drilled in shells of varying ornamental complexity are illustrated in Fig. 1, from reticulated (1a, 1b) to smooth (1d, 1f).

Reyment (1971) reported that more than 99.5% of the holes drilled in molluscs and ostracods in the Niger Delta are due to naticids. By way of comparison, 98.5% of the holes in the Spanish Eocene species of *Echinocythereis* were determined as having been drilled by naticids. This value is the same for all three species studied. The Egyptian Miocene presents a somewhat different picture in that 30% of three species occurring were drilled, with *Keijella* heading the list.

3. Paleoethology and Paleoecology of the Predators

Although it is not possible to make categorical statements about the originators of the boreholes, it is within the realm of feasibility to provide certain deductions of paleoethological and paleoecological interest. Muricids hunt the epifauna and also eat carrion, whereas naticids usually consume their prey within the sediment. They are known to attack on the surface, but usually they drag their prey into the sediment to consume it. Savazzi and Reyment (1989) recorded a case of a living naticid species (*Natica gualteriana*) attacking, drilling and consuming its bivalve prey on the sediment surface.

Ostracods are relatively agile creatures and it is therefore surprising that the slow-moving drill is able to overpower them and to drill successfully. It is logical to assume that most attempts are doomed to failure. However, a behavioral property of many ostracods may explain how the seemingly impossible becomes possible. Reyment found in laboratory experiments on (freshwater) ostracods that many individuals, when lightly touched, ceased to move and became inert. If this behavior carries over to marine ostracods it would give the predator time to envelop the carapace with mucous. There is the added possibility that the mucous itself contains an anaesthetizing substance (Richter, 1962), an opinion to which Carriker (1981) subscribed. Lying inert would seem to be some kind of self-preservation reaction. In relation to predatory gastropods this strategy is not a wise move.

Table 1. Discriminant Analysis for Nigerian Naticid and Muricid Boreholes (Paleocene)

	L	H	d	D
Difference mean vector	0.011	0.011	0.079	0.094
Linear discriminant coefficients	2.01	-11.35	8.48	19.90

Unbiased generalized distance = 0.465
Probability of misidentification = 0.37
L = difference for length-dimension; H = difference for height-dimension of specimen
d = diameter of interior of borehole; D = external diameter of borehole

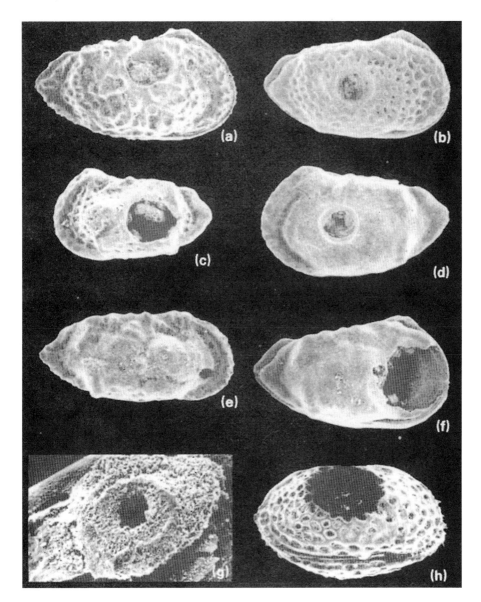

FIGURE 1. (a) *Cythereis cretaria* van den Bold. A naticid hole drilled in a reticulated surface; the hole is breached dorsally, x 45. Santonian, Israel; (b) *Cythereis cretaria* van den Bold. Naticid hole drilled in a reticulated shell surface. x 45. Santonian, Israel; (c) *Cythereis diversereticulata* Honigstein Borehole possibly made by a naticid. x 45. Santonian, Israel; (d) *Veenia fawwarensis* Honigstein. Borehole probably made by a naticid. x 45. Santonian, Israel; (e) *Cythereis diversereticulata* Honigstein. Muricid (?) hole in an anterior position. x 40. Santonian, Israel; (f) *Veenia fawwarensis* Honigstein. Entire anterior zone excavated by a naticid borehole. x 45. Santonian, Israel, (g) *Leguminocythereis lagarhirobensis* Apostolescu. Muricid hole drilled through secondary calcite, deposited presumably post-mortem. x 220. Paleocene, Nigeria, borehole sample Gbekebo 2288 ft; (h) *Anticythereis? bopaensis* Apostolescu. Large borehole drilled in a strongly reticulated surface. x 40.Paleocene, Nigeria, borehole sample Gbekebo 2039 ft. Reprinted from *Cretaceous Research*, 8 (Fig. 4, Reyment *et al.*, 1987).

Ostracods are not to be compared with bivalves or gastropods as a source of easily digested food and it is not likely that they would be hunted in preference to molluscs. In the present-day Niger Delta, providing the number of spat is great enough, molluscs are always attacked in preference to ostracods (Reyment, 1966a, p. 61). The Fischer-Piette effect may come into play if and when the numbers of a favored prey species have been reduced to virtual local extinction. The predator population then overshoots the carrying capacity of the prey, leading to a "crash" and ensuing starvation of the drills. Fischer-Piette (1935) studied the effects of overshooting with respect to barnacles that were being preyed upon heavily by *Nucella lapillus*. In the case of predation on ostracods, it seems unlikely that juvenile drills would be affected in this manner. Rather, on hatching, if they entered an environment in which the normal dietary preferences were depleted, they would be constrained to select ostracods as a means of survival, these being the only available source of nourishment that might be expected to unleash the attack response, not least because of the morphological similarity shared by ostracods and bivalves.

Fig. 2 is a schematic graphical representation of fluctuations in the number of ostracods and the relative abundance of juvenile predaceous gastropods in samples from the borehole Araromi, Western Niger Delta (Paleocene age). The lower line marks the predator frequencies, the upper line denotes the total number of individuals, and the dashed line the total numbers of adults, level by sampled level. This illustrates the "averaged" correlation between abundance of prey and numbers of predaceous gastropods. For obvious reasons, this is not a Volterra-D'Ancona portrayal of the classical predator-prey model of population dynamics, but it does indicate how the two components have interacted (cf. Reyment, 1971, pp. 107-150).

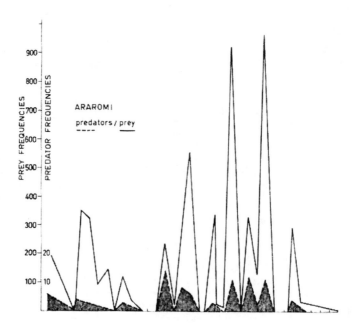

FIGURE 2. Graphical stylized representation of fluctuations in number of ostracods and the relative abundance of predaceous gastropods for the Gkebebo I borehole, Nigeria, Reprinted from *Stockholm Contributions in Geology* xiv (Fig. IV:7, Reyment, 1966b).

3.1. Multiple Boreholes and Failed Borings

Incomplete perforations and failed gastropod borings (i.e. holes abandoned shortly after having been started) are by no means uncommon (Hofmann *et al.* 1974; Kitchell *et al.,* 1986). The latter authors found that completed holes made by the naticid *Polinices duplicatus* have an internal (d) to external (D) hole-diameter ratio of > 0.5, whereas "non-functional" holes have d/D < 0.5. In the present material, the d/D ratio is about 0.8, i.e., the inner diameter is relatively greater in relation to the outer diameter than in the species of bivalves studied by Kitchell and coworkers. This seems to be a natural outcome of the relatively thin shell of the ostracod. Hingston (1985, p. 52) reported for molluscs that stronger ornament tends to be associated with drilling failures.

Partly completed holes in the present collection sometimes show a kidney-shaped break-through, said by Ziegelmeier (1954, p. 17) to always be located in the anterior sector of the naticid hole (see also Carriker and Yochelson, 1968, p. 5). These examples of holes have not been included in the statistical analyses, but are figured here for their paleoethological interest (Fig. 3(c), (g)). In Fig. 4(c), (e), two equally large partial perforations are shown, deriving presumably from a naticid. Further examples of incomplete holes are illustrated in Fig. 3(b) enlarged view of anterior zone of Figure 5(d)), Figure 4(f) and Figure 5(h), the latter being strongly elongated.

The Santonian material is rich in failed borings. Examples figured here are Fig. 3(d), (e), Fig. 4 (b), (d), (h) and Figure 5 (b). All of these failed holes (Fig. 3(c)) but one display features concordant with drilling by muricids. Thus, the twin holes in Fig. 3(b) are conical and with a concave base. The failed hole depicted in Fig. 3(d) has a raised central zone, such as often occurs in naticid holes. An early record of the boss visible in some incomplete naticid holes is that of Livan (1937, p. 13).

Shell surfaces seldom show drilling marks and solution effects side-by-side (Carriker, 1969; Carriker *et al.,* 1974). Such are rare in Recent material and exceptional in fossils. It is therefore of interest to record the presence of parallel radula markings, as well as evidence of dissolution of the shell surface by the ABO (Fig. 3(a)), in a Nigerian Paleocene *Buntonia* (cf. figure 8 in Carriker et al., 1974, p. 69). Not a few specimens were found to be drilled at both ends. The *Bythocypris* material suggests that the naticid predators adopt a stereotyped mode of attack and that, where holes occur at the anterior and posterior ends of the same carapace, two gastropods have probably been engaged in drilling at the same time. Where two holes are located side by side, it would seem that this is due to the rounding of the shell interfering with the normal rasping function of the radula. Ziegelmeier (1954, p.17), who made extensive use of cinematography in his laboratory work, has depicted diagrammatically the way in which the naticid drilling operation is carried out. The hole is scraped sectionally in quadrants of a circle, passing from left to right in the lower part of the approximately circular field and then from left to right in the upper part thereof, the rasping being carried out roughly equally in each quadrant and consecutively from quadrant to quadrant, so that progress is roughly equal over the circular field. In the case of a strongly arched shell, the regular action of the radula is hindered, which leads to an atypically formed hole.

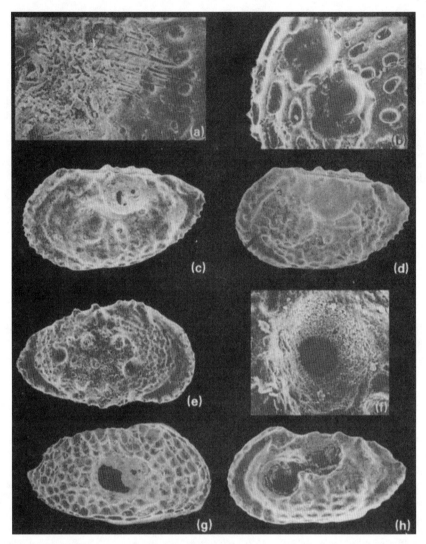

FIGURE 3. (a) *Buntonia* sp. Early phase of drilling displaying parallel radula scratches and the effects of dissolution of the shell by the ABO (the porous surface). x 165. Paleocene, Nigeria, borehole sample Gbekebo 2288 ft; *(b) Leguminocythereis lagaghiroboensis* Apostolescu. The partially completed holes of the specimen in Figure 5(d). The surface texture of the pits shows evidence of solution. x 110. Paleocene, Nigeria, borehole sample Gbekebo 2290ft; (c) *Veenia fawwarensis fauwarensis* Honigstein. Abandoned hole. The broken-through parts of the hole, in particular the crescent-shaped perforation, point to a naticid predator (cf. Ziegelmeier, 1954). x 45. Santonian, Israel; (d) *Veenia fawwarensis dividua* Honigstein. Incompletely bored muricid hole (cf. Hofmann *et al.* 1974, p. 251), x 45. Santonian, Israel; (e) *Trachyleberis teiskotensis* (Apostolescu). Specimen showing two muricid? boreholes, the anterior of which is complete. x 60. Paleocene, Nigeria, borehole sample Gbekebo 2290 ft. See also Figure 5(b); (f) *Ovocytheridea* cf. *pulchra* Reyment. Muricid borehole. x 200. Paleocene, Nigeria, borehole sample Gbekebo 2284 ft; *(g) Anticythereis* (?) *bopaensis* Apostolescu. Almost complete naticid borehole. x 50. Paleocene, Nigeria, borehole sample Gbekebo 2042 ft; (h) *Veenia fawwarensis* Honigstein. Twin, completed naticid holes x 45. Santonian, Israel. Reprinted from *Cretaceous Research*, 8 (Fig. 2, Reyment *et al.*, 1987).

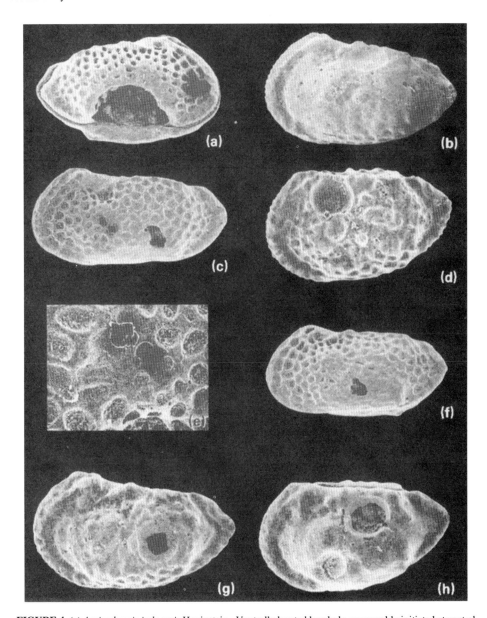

FIGURE 4. (a) *Anticythereis judaensis* Honigstein. Ventrally located borehole, presumably initiated at ventral margin contact and drilled obliquely. x 45. Santonian, Israel. (b) *Veenia fawwarensis* Honigstein. Incomplete muricid borehole. x 45. Santonian, Israel. (c) *Cythereis cretaria dorsocaudata* Honigstein. Two partially completed boreholes; the one with two small perforations was made by a naticid (see 4(e). x 50. Santonian, Israel. (d) *Veenia fawwarensis* Honigstein. Incomplete muricid? borehole. x 40. Santonian, Israel; (e) *Cythereis cretaria dorsocauda* Honigstein. One of the incomplete borings of (c) at higher magnification. The broken-through parts of the excavation present one of the criteria of naticid drilling (Ziegelmeier, 1954). x 145. Santonian, Israel. (f) *Cythereis cretaria dorsocaudata* Honigstein. Incomplete naticid borehole showing a broad annular zone over which the reticular ornament was scraped away. x 45. Santonian, Israel; (g) *Veenia fawwarensis* Honigstein. Almost complete borehole with a broad zone of "frosting". x 45. Santonian, Israel. (h) *Veenia fawwarensis* Honigstein. Two holes, one complete and one incomplete, both probably deriving from muricids. x 50. Santonian, Israel. Reprinted from *Cretaceous Research*, 8 (Fig. 3, Reyment *et al.*, 1987).

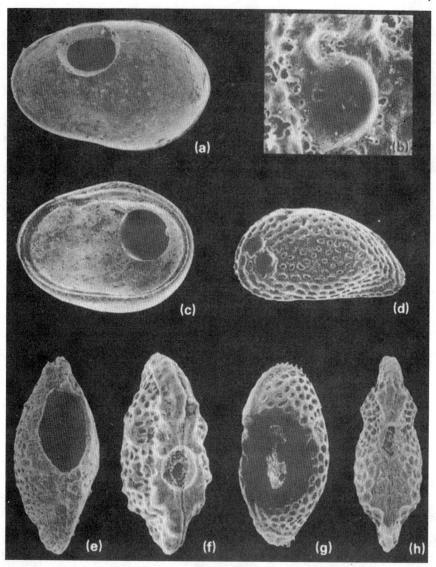

FIGURE 5. (a) *Cytherella sylvesterbradleyi* Reyment. A thick shell with the characteristically sloping sides of well developed muricid borehole x 50. Paleocene, Nigeria, borehole sample Gbekebo 2042 ft; *(b) Trachyleberis teiskotensis* (Apostolescu), Magnification of the incomplete posterior borehole in the specimen shown in Figure 2(e). x 300. Paleocene, Nigeria, borehole sample Gbekebo 2290 ft; *(c) Cytherella sylvesterbradleyi* Reyment. Internal aspect of a muricid hole. x 50. Paleocene, Nigeria, borehole sample Gbekebo . 2042 ft; *(d) Leguminocythereis lagaghiroboensis* Apostolescu. Twin, incomplete muricid holes (see Figure 2(b)). x 40. Paleocene, Nigeria, borehole sample Gbekebo 2290 ft; (e) *Buntonia livida* Apostolescu. Dorsally located naticid borehole. x 65. Paleocene, Nigeria, borehole sample Gbekebo 2042 ft; (f) *Cythereis cretaria* van den Bold. Naticid hole drilled on the smooth dorsal surface at the line of contact of the two valves. x 45. Santonia,n, Israel; (g) *Anticythereis ? bopaensis* Apostolescu. Large naticid? borehole. x 45. Paleocene, Nigeria, borehole sample Gbekebo 2039 ft; (h) *Cythereis cretaria dorsocaudata* Honigstein. Irregularly shaped (elongated) hole drilled at the dorsal marginal contact; surface weakly reticulated. x 50. Santonian, Israel. Reprinted from *Cretaceous Research*, 8 (Fig. 5, Reyment *et al.*, 1987).

3.2. Site Chosen for Drilling

Reyment (1966a, p.64, 1971) found that the holes drilled in the Recent bivalves of the Niger Delta are not randomly distributed, but are located in accord with a well-defined preference for the site of attack. It was also shown, conversely, in that paper that Nigerian Paleocene ostracods were not drilled in accordance with a definite pattern, but that the borings were distributed randomly (supported by a positive Poisson test for randomness). A similar test for *Bythocypris* indicates non-randomness in the material studied but it was not possible to obtain a decisive result for *Cytherella*.

"Pope diagrams" (cf. Carriker, 1955, in reference to work by T. E. B. Pope in an unpublished U. S. Fisheries Report from 1910-1911) showing the sites selected for attack for the Santonian ostracods are shown in Fig. 6, with respect to three categories of ornamentation (almost smooth, moderately strongly ornamented, and strongly ornamented). Most holes can be seen to lie in the middle zone of the shell with some spreading towards the dorsal part of the valve. Dorsally drilled holes are always located around the line of contact of the shells (Figs. 5e, 5f , 5h). The nature of the surface of the shell does not seem to have had an influence on the siting of the borehole, apart from a tendency for reticulated shells to be entered along the dorsal contact (cf. Carriker, 1981, pp. 408-409). It is noteworthy that the anterior and posterior zones of the species concerned are not of interest to the predators. The locations of holes drilled in *Venus striatula* by *Natica alderi,* reported by Ansell (1960, Fig. 3), tend to be concentrated towards the ventral marginal zone and hence in contrast with what was found for the Santonian ostracods from Israel. Ansell (1960, p. 161) thought this to reflect a stereotyped behavioral pattern involving recognition of the prey and the adoption of a particular orientation while boring, but he also noted that, in some cases, the dorsal zone is preferred.

4. Size, Predation Pressure and Selectional Effects

Colbath (1985) reported a predation percentage of 11% on Miocene bivalves. Confidence intervals computed for predation on the Santonian ostracods gives a much higher value than for the Paleocene material. The confidence intervals for the total level of predation as registered by the percentage of drilled holes is 0.0637 - 0.0913 for the Santonian and 0.0157 - 0.0253 for the Paleocene. Hence, the level of predation registered in the Israeli Santonian is four times that observed for the Nigerian Paleocene. This result is of no more than approximate significance but we include it *faute de mieux*. We hope that the lists for Texan data published by Maddocks (1988) will form the basis for quantitative appraisal. It is worth keeping in mind that seasonal fluctuations in feeding maxima, which are seldom identifiable in fossil material (but see Reyment, 1971, for an attempted reconstruction), can deliver a confusing picture of predation intensity (Fischer Piette, 1935, pp. 167, 176; Ziegelmeier, 1954, p. 20; 1957, p. 386). An approximate determination of intensity of natural selection (cf. Reyment, 1982, 1991 for an outline of the methodology) gave one selective death per million per generation for the Santonian ostracods. This is insufficient to bring about significant evolutionary change in morphology; this conclusion seems to offer support for Kitchell's (1986) opinion concerning behavioral stasis in drills. Similar opinions have been voiced by Thomas (1976, p. 498) and Hofmann *et al.* (1974, p. 257).

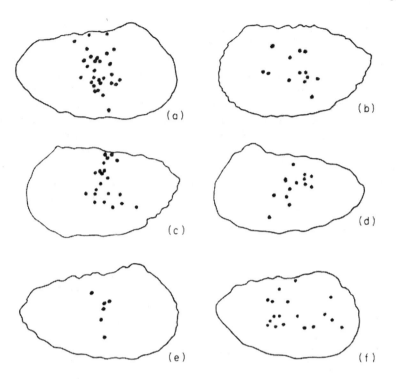

FIGURE 6. Pope diagrams for the location of boreholes in the shells of Santonian ostracods (Israel). Strongly ornamented (a,c), moderately strongly ornamented (b,e), smooth surface (d,f). Reprinted from *Cretaceous Research*, 8 (Fig. 6, Reyment *et al.*, 1987).

Reyment *et al.* (1987) reconstructed juvenile predator sizes by extrapolating from drillhole dimensions in fossil ostracods (see also Kitchell *et al.*, 1981). The computed range was from 0.6 to 3.9 mm for the Santonian examples and 0.7 to 3.9 mm for Paleocene ostracods. In arriving at these results, Reyment *et al.* (1987) used the few measurements on ABOs in relation to shell height available in the literature, as well as the observation that the diameter of the borehole is closely correlated with the size of the predator (cf. Kitchell *et al.*, 1981, p. 549) in order to estimate the size of the predator that gave rise to a particular borehole. Although the values listed in Table 2 are perforce rough estimates, it is nonetheless noteworthy that the size-range of predators runs from 0.6 mm to a maximum size of 3.9 mm, with the mode lying at 1.8 mm. These values accord closely with published observations and fall into the expected size range for juvenile drills likely to obtain a reasonable energy return from adult ostracods. Where juvenile drills have been studied, it is reported that they are able to bore immediately on hatching; the size of hatchlings of *Urosalpinx cinerea* ranges over 1.0-1.5 mm (Carriker, 1957, p. 40; Wiltse, 1980, p. 197). Fischer-Piette (1935, p. 159), on the grounds of a detailed suite of observations made over many years, reported that the muricid *Nucella lapillus* does not drill juvenile barnacles. Moore (1958, p. 372), referring to French work, noted that newly hatched *N. lapillus* feed on unfertilized eggs, after which they eat small polychaetes, before switching to bivalves or barnacles on attaining a shell height of about 8 to 10 mm.

TABLE 2. Reconstructed Shell Heights for Predators (All Figures in mm)

Predator	Diameter of borehole	Predator height	Prey
			Paleocene species
Muricid	0.288	2.3	*Cytherella sylvesterbradleyi*
Muricid	0.120	1.0	*Leguminocythereis lagaghiroboensis*
Muricid?	0.070	0.6?	*Trachyleberis teiskotensis*
Muricid	0.090	0.7	*Trachyleberis teiskotensis*
Naticid	0.282	2.3	*Buntonia livida*
Naticid	0.490	3.9	*Anticythereis (?) bopaensis*
Naticid	0.400	3.2	*Anticythereis (?) bopaensis*
Muricid	0.090	0.7	*Ovocytheridea pulchra*
			Santonian species
Naticid	0.226	1.8	*Cythereis cretaria*
Naticid	0.199	1.6	*Cythereis cretaria*
Naticid	0.240	1.9	*Cythereis cretaria*
Muricid?	0.071	0.6	*Cythereis diversereticulata*
Muricid?	0.231	1.8	*Cythereis diversereticulata*
Naticid	0.374	2.7	*Veenia fawwarensis*
Naticid	0.183	1.5	*Veenia fawwarensis*
Failed muricid	0.221	1.8	*Veenia fawwarensis*
Failed muricid	0.213	1.7	*Veenia fawwarensis*
Naticid	0.206	1.7	*Veenia fawwarensis*
Failed naticid	0.225	1.8	*Veenia fawwarensis*
Failed naticid	0.251	2.0	*Veenia fawwarensis*
Muricid	0.273	2.2	*Veenia fawwarensis*
Naticid	0.227	1.8	*Veenia fawwarensis*
Naticid	0.427	3.1	*An ticy thereis judaensis*

For purposes of estimation of size of prey, some kind of areal approximation is required. The length and height of the ostracod carapace provide a good estimate of the lateral area of the shell. It has been noted by several workers that the size of the hole made by drills is correlated with the size of the prey (Ansell, 1960, p. 159 and figure 2; Thomas, 1976; Kitchell *et al.*, 1981, p. 547; DeAngelis *et al.*, 1984, p. 132; DeAngelis *et al.*, 1985). However, Reyment (1966a, pp. 81, 95) found many cases of samples showing poor correlation between the size of the hole drilled and the size of the prey for living forms from the Niger Delta, although he also observed that cases of significant correlation between the two sets may occur. Reyment *et al.* (1987) applied the multivariate statistical procedure of canonical correlation to the Santonian and Paleocene material (cf. Reyment and Savazzi, 1999, for an account of this method). This analysis yielded the unexpected result that, whereas the correlation between hole size and size of prey is highly significant for the Paleocene data, it is not statistically significant for the Santonian fossils. We doubt that this result reflects evolutionary adjustment, although we concede freely that the subject is worthy of detailed consideration.

The poor correlations sometimes observed can be provided with a simple explanation. Many of the samples analyzed by Reyment *et al.* (1987) seem to be examples of upward-truncated predation systems in which the prey population lacked an upper "size-refuge" beyond which freedom from attack would be assured and the predators remained in the same environment as the ostracods. In this case, all adult ostracods, and even some individuals of later instars, are interesting to the predator. Such a situation may have existed for the Santonian ostracods; hence, the size-distributional requirements for producing a significant correlation are not represented in the data. The Paleocene material, with the significant association between sets, must have a different ecological background. A reasonable model is that the drills feeding on ostracods only did so for a short initial period in their life, and then migrated to shallower water. This would be evidenced in the fossil record as a size-oriented predation pattern for juvenile snails. Micro-environmental factors might be decisive. In any event, a relationship such as shown by Ansell (1960, figure 2) for *Natica alderi* and *Venus striatula* is seldom found for ostracods.

Reyment (1966b) found Paleocene ostracods to be differentially drilled. Some species seem never to have been attacked, whereas the predation level for others was very high. Two species of *Brachycythere* top the list with figures of about 25%, thus nearing the level for the Egyptian Miocene data and some of the observations tabulated by Maddocks (1988). For example, she recorded 49% of *Brachycythere* in the Texan Danian to have been drilled. There is a chance that these observations could reflect the role of specific micro-environments. Individuals inhabiting the upper layer of sediment would be more accessible to naticids than those roaming the surface. Ostracods that lived on phytoplankton would, of course, not have been available as a source of nourishment other than under exceptional circumstances.

5. Ornamentation Complexity and Predation Intensity

In Table 3, the order of proneness to attack of Paleocene ostracod species is listed (observed for the Nigerian borehole Gbekebo I). The ordering does not give any clear concept of susceptibility in relation to shape of the carapace, apart from the positions occupied by the first four species. A possible explanation of this situation could lie with the ecological niche occupied by a particular species. There is, nevertheless, an appealing picture of vulnerability to predation that could possibly be related to complexity of ornamentation. There seem to be several main groupings occurring naturally in the hierarchy of liability to attack. The first group encompasses smooth shells (drilled proportion <1:20), the second encompasses lightly ornamented shells (drilled proportion=1:30), and the third more strongly sculptured forms (drilled proportion > 1:40). The category referred to as being anomalously located denotes shells that fall out of the hierarchical sequence. The reason for this in some cases is the small size of adults. There is also the possibility of ecological differentiation due to habitat (epifaunal or endofaunal). A statistical test for random proneness to attack was rejected (Reyment, 1966b).

5.1. Ornamentation and Borehole Dimensions

The investigation of the question of the influence of the surface texture of the ostracod shell as a factor in predation was made by a canonical variate analysis of three arbitrary categories of Santonian shells, namely, smooth, moderately strongly ornamented, and strongly ornamented (reticulated). A general account of the method of the multivariate statistical method of canonical variate analysis is given in Reyment *et al.* (1984) and Reyment and Savazzi (1999). A one-way analysis of variance disclosed that there are no significant differences among the holes drilled in these three sculptural types, at the univariate level. The generalized statistical differences indicate, however, that for the length of the carapace, and the inner and outer hole-diameters, there is a significant difference between the ornamental categories "smooth" and "moderately strongly ornamented" (P=0.02) and between the categories "moderately strongly ornamented" and "strongly ornamented (reticulated)" (P=0.05).

The implications of the foregoing can be obtained from the plot of the first and second canonical variate scores (see Fig. 7); this graph shows a tendency towards segregation of the three arbitrary ornamented categories. Thus, the smooth shells (inverted triangles) tend to lie in the upper part of the graph, moderately ornamented shells (dots) congregate to the left, and strongly ornamented shells fall in the middle field (diamonds). The role of the inner diameter of the holes is surprisingly insignificant. It seems that the texture of the surface of the shell is important for the initial determination of the external dimensions of the area to be scraped, but it has little relevance for the dimensions of the inner diameter of a hole.

6. Holes of Non-molluscan Origin

A final observation of interpretational significance concerns the occurrence of minute holes occasionally found in ostracod shells and which may be confused with holes made by drilling gastropods. These derive from post-mortem degradation by marine fungi and, although largely represented by holes through the shell, they may also occur as a kind of scoring of the surface. Some examples observed on shells of *Keijella punctigibba* (Capeder) from the middle Miocene of Mersa Matruh, Egypt, are illustrated

TABLE 3. Order of Proneness to Attack by Drills on Paleocene Ostracods (Western Nigeria)

Species	Proportion drilled	Species	Proportion drilled
Smooth			
Brachycythere ogboni	1:4	*Cytherella beyrichi ?)*	1:9
B. armata	1:10	*Iorubaella ologuni*	1:4
Buntonia triangulate	1:16	*Cytherella sylvesterbradleyi*	1:17
Bairdia ilaroensis	1:18	*Paracypris nigeriensis*	1:20
Buntonia beninensis	1:20		
Lightly ornamented			
Buntonia pulvinata	1:20	*Leguminocytherereis bopaensis*	1:25
Buntonia ioruba	1:36	*Buntonia apatayeriyerii*	1:27
More strongly ornamented			
Veenia ornatoreticulata	1:36	*Quadracytbere lagaghiroboensis*	1:36
Veenia warriensis	1:40	*Dahomeya alata*	1:42
Leguminocytbereis lagahirobonensis	1:44	*Trachyleberis teiskotensis*	1:47
Actinocythereis asanmamoi	1:57	*Cythereis deltaensis*	1:70
Ruggieria tattami	1:73	*Veenia acuticostata*	1:86
Anomalously located species			
Buntonia livida	1:50	*Buntonia bopaensis*	1:51
Buntonia keiji	1:82	*Buntonia fortunata*	1:84

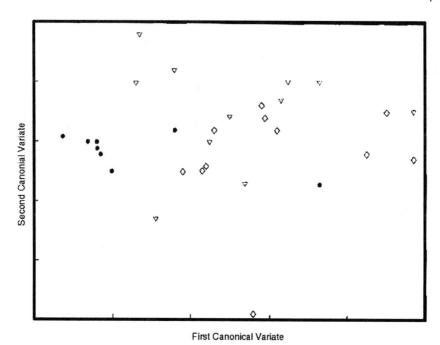

FIGURE 7. Plot of the first and second canonical variate scores for three samples of Santonian ostracods (Israel). Smooth shells are denoted by inverted triangles, moderately strongly ornamented shells are denoted by dots, and reticulated, strongly ornamented shells are marked by diamonds. Reprinted from *Cretaceous Research*, 8 (Fig. 1, Reyment *et al.*, 1987).

in Fig. 8, a-d (cf. Bassiouni and Elewa, 2000). This is an area for future research. So far, there is only one monographic treatment of non-molluscan boreholes, notably that of Maddocks (1988).

7. Future Directions for Research

The experimental study of gastropod predation on marine ostracods is a far more difficult task than for macrofauna. For this reason it seems unrealistic to expect a breakthrough in this area, at least in the near future, and workers in the field will have to continue by way of "paleoforensics." Nonetheless, this is an area for research that could well profit from the engagement of skilled experimental biologists. We believe that the initiated application of modern biometrical techniques can be expected to yield useful information about ecological relationships and evolutionary trends, where such occur. The rapidly developing subject of geometric morphometrics would seem to be eminently suitable for disclosing polymorphic shifts in shape and size in relation to predation pressure (cf. Reyment, 1991). It is, however, necessary to stress the need for conforming to the tenets of statistical design and strategic forethought. All too often, statistical methods are applied to data obtained in a haphazard manner and without initial statistical

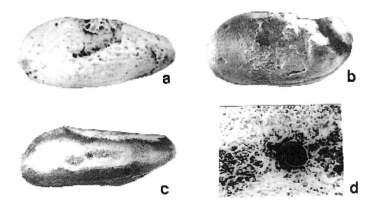

FIGURE 8. Examples of post-mortem holes made in the Middle Miocene species *Keijella punctigibba* (Capeder) from Mersa Matruh, Egypt, by marine fungi. Figs a, b and c x 160. Fig. d is a detail of Fig. c showing the rimmed hole x 800. Original material.

planning -- like calling in a medical practitioner to a wake. Paleontologically applicable methods for studying evolution in the phenotype are now available and, with suitable care and adequate material, they could be tried on predation on ostracods. The topic is considered in Reyment (1991, pp. 168-203).

ACKNOWLEDGMENTS: Thanks to the graciousness of Academic Press and the Editor of *Cretaceous Research*, Professor David Batten, we have been permitted to use the figures that illustrated Reyment *et al.* (1987). We also thank Professor Jan Backman and Stockholm University for permission to reproduce Fig. 2. We also wish to thank the referees for constructive and thoughtful criticism.

References

Ankell, W. E., 1937, Wie bohrt *Natica?*, *Biol. Zentralblatt* **57**:75-82.
Ansell, A. D., 1960, Observations on predation of *Venus striatula* (Da Costa) by *Natica alderi* (Forbes), *Proc. Malacol. Soc. London* **34**:157-164.
Bassiouni, M. A., and Elewa, A. M. T., 2000, Miocene ostracods of the southern Mediterranean: a first record from Wadi Um Ashtan, Mersa Matruh, Western Desert, Egypt, *Monat. Neues Jahrb. Geol. Paläontol.* **2000**(8):449-466.
Carriker, M. R., 1955, Critical review of biology and control of oyster drills *Urosalpinx* and *Eupleura*, *U. S. Department of Interior; Fish and Wildlife Service, Special Report*, 148: 150 typewritten pages. Washington DC.
Carriker, M. R., 1957, Preliminary study of behavior of newly hatched oyster drills, *Urosalpinx cinerea* (Say), *J. Elisha Mitchell Sci. Soc.* **73**:328-351.
Carriker, M. R., 1961, Comparative functional morphology of boring mechanisms in gastropods, *Am. Zool.* **1**:263-266.

Carriker, M. R., 1969, Excavation of boreholes by the gastropod *Urosalpinx:* an analysis by light and scanning electron microscopy, *Am. Zool.* **9**:917 933.
Carriker, M. R., 1981, Shell penetration and feeding by naticacean and muricacean predatory gastropods: a synthesis, *Malacologia* **20**:403-422.
Carriker, M. R., and Yochelson, E. L, 1968, Recent gastropod boreholes and Ordovician cylindrical borings, *U.S. Geol. Surv. Prof. Paper,* 593-B:1-26.
Carriker, M. R., Schaadt, J. G., and Peters, V., 1974, Analysis by slow-motion picture photography and scanning electron microscopy of radular function in *Urosalpinx cinerea follyensis* (Muricidae, Gastropoda) during shell penetration, *Mar. Biol.* **25**:63-76.
Colbath, S. L., 1985, Gastropod predation and depositional environments of two molluscan communities from the Miocene Astoria Formation at Beverly Beach State Park, Oregon, *J. Paleontol.* **59** :849-869.
DeAngelis, D. L., Kitchell, J. A., Post, W. M. and Travis, C. C., 1984, A model of naticid gastropod predator-prey coevolution, *Lect. Notes Biomath.* **54**:120-136.
DeAngelis, D. L., Kitchell, J. A., and Post, W. M., 1985, The influence of naticid predation on evolutionary strategies of bivalve prey: conclusions from a model, *Am. Naturalist* **126**:817 -842.
Fischer-Piette, E., 1935, Histoire d'une moulière. Observations sur une phase déséquilibre faunique, *Bull. Biol. France et Belgique* **69**:153-177.
Guerrero, Alba, S. and Reyment, R. A., 1988, Predation and feeding in the naticid *Naticarius intricatoides* (Hidalgo), *Palaeogeogr. Palaeclim. Palaeoecol.* **68**:49-52.
Hingston, J. P.,1985, Predation patterns among molluscs in the Victorian Tertiary, *Proc. R. Soc Victoria* **97**:49-57.
Hofmann, A., Pisera, A., and Ryskiewicz, M., 1974, Predation by muricid and naticid gastropods on the Lower Tortonian mollusks from the Korytnica clays, *Acta Geol. Polon.* **24**: 249-260.
Honigstein, A., 1984, Senonian ostracods from Israel, *Bull. Geol. Surv. Israel* **78**: 1-48.
Kabat, A. R., 1990, Predatory ecology of naticid gastropods with a review of shell boring predation, *Malacologia* **32**:155-193.
Kitchell, J. A., 1986, The evolution of predator-prey behavior: naticid gastropods and their molluscan prey, in: *Evolution of Animal Behavior: Paleontological and Field Approaches* (M. Nitecki and J.A. Kitchell, eds.), Oxford University Press, Oxford, pp. 88-110.
Kitchell, J. A., Boggs, C. H., Kitchell, J. F., and Rice, J. A., 1981, Prey selection by naticid gastropods: experimental tests and application to the fossil record, *Paleobiology* **7**:533-552.
Kitchell, J. A., Boggs, C. H, Rice, J. A., Kitchell, J. F., Hofmann, A., and Martinell, J., 1986, Anomalies in naticid predatory habits: a critique and experimental observation, *Malacologia* **27**:291-298.
Livan, M., 1937., Über Bohrlöcher an rezenten und fossilen Invertebraten, *Senckenbergiana* **19**:138 150.
Maddocks, R. F., 1988, One hundred million years of predation on ostracodes: the fossil record in *Texas,* in: *Proc. 9th Ostracod Conference, Shizuoka, Japan: Evolutionary Biology of Ostracods* (T. Hanai, N. Ikeya, and K. Ishizaki, eds.), pp. 637-657.
Moore, H. B., 1958, *Marine Ecology,* Wiley & Sons, 493 pp.
Reyment, R. A., 1963, Bohrlöcher bei Ostracoden, *Paläontol. Zeitschrift* **37**:283-291.
Reyment, R. A., 1966a, Preliminary observations on gastropod predation in the western Niger Delta, *Palaeogeogr. Palaeoclim. Palaeoecol.* **2**:81 102.
Reyment, R. A, 1966b, Studies on Nigerian Upper Cretaceous and Lower Tertiary Ostracoda. Part 3. Stratigraphical, paleoecological and biometrical conclusions, *Stockholm Contr. Geol.* **14**:1-151.
Reyment, R. A., 1967, Paleoethology and fossil drilling gastropods, *Trans. Kansas Acad. Sci.***70** :33 50.
Reyment, R. A., 1971, *Introduction to Quantitative Paleoecology,* Elsevier, Amsterdam, 226 pp.
Reyment, R. A., 1982, Analysis of trans-specific variation in Cretaceous ostracods, *Paleobiology* **8**:292-305.
Reyment, R. A., 1991. *Multidimensional Palaeobiology,* Pergamon Press, Oxford, 377 pp.
Reyment, R. A., 1999, Drilling gastropods, in: *Functional Morphology of the Invertebrate Skeleton,* (E. Savazzi, ed), John Wiley & Sons, Ltd, pp. 197-204
Reyment, R.A., and Savazzi, E., 1999, *Aspects of Multivariate Statistical Analysis in Geology,* Elsevier, Amsterdam, 285 pp.
Reyment, R. A., Blackith, R. E., Campbell, N. A., 1984, *Multivariate Morphometrics,* Academic Press, London, 224 pp.
Reyment, R. A., Reyment, E. R., and Honigstein, A., 1987, Predation by boring gastropods on Late Cretaceous and Early Palaeocene ostracods, *Cretaceous Res.* **8**:189-209.
Richter, G., 1962, Beobachtungen zum Beutefang der marinen Bohrschnecke *Lunatia nitida, Natur und Museum* **18**:185-192.
Savazzi, E., and Reyment, R. A., 1989, Subaerial hunting behaviour in *Natica gualteriana* (naticid gastropod), *Palaeogeogr. Palaeoclimatol. Palaeoecol.* **74**:355-364.

Thomas, R. D. K., 1976, Gastropod predation on sympatric Neogene species of *Glycymeris* (Bivalvia) from the eastern United States, *J. Paleontol.* **50**:488-499.
Wiltse, W. I.,1980, Predation by juvenile *Polinices duplicatus* (Say) on *Gemma gemma* (Totten). *J. Exper. Mar. Biol. Ecol.* **42**:187 199.
Ziegelmeier, E., 1954, Beobachtungen über den Nahrungserwerb bei der Naticide *Lunatia nitida* Donovan (Gastropoda Prosobranchia), *Helgoländer Wissenschaftliche Meeresforschung* **5**:1- 33.
Ziegelmeier, E., 1957, Ein kleiner lebender Drillbohrer, *Orion,* **5**:385 392.

Chapter 5

The Fossil Record of Drilling Predation on Bivalves and Gastropods

PATRICIA H. KELLEY and THOR A. HANSEN

1. Introduction 113
2. Data on Drilling Predation Extractable from the Fossil Record 115
3. Testable Hypotheses in Evolutionary Paleoecology 121
4. Arms Races and Drilling Frequencies 122
5. Changes in Drilling Behavior through Time 126
6. Relative Effectiveness of Predator and Prey 129
7. Summary and Discussion 130
   References 133

1. Introduction

The fossil record yields abundant data on the interaction between drilling predators and their shelled prey. Predatory drill holes may date to the late Precambrian (Bengtson and Zhao, 1992) and have been reported from various Paleozoic assemblages (e.g., Sheehan and Lesperance, 1978; Smith *et al.*, 1985; Conway Morris and Bengtson, 1994). Paleozoic drill holes have been reported primarily from brachiopods, although gastropods (Rohr, 1991) and bivalves (Kowalewski *et al.*, 2000; Hoffmeister *et al.*, 2001) also exhibit apparent predatory drill holes. In most cases, the identities of the Paleozoic drilling predators are unknown; platyceratid gastropods documented in association with drilled Paleozoic crinoids and blastoids appear to have been parasitic (Baumiller, 1990, 1996, 2001).

Few drill holes have been reported from the early Mesozoic (Kowalewski *et al.*, 1998), but the primary drillers of modern molluscs did not radiate until the Cretaceous. Although predatory drilling behavior is known within several families of gastropods, and also among octopods and even flatworms and nematodes (Sohl, 1969; Kabat, 1990;

PATRICIA H. KELLEY • Department of Earth Sciences, University of North Carolina at Wilmington, Wilmington, North Carolina 28403-5944.   THOR A. HANSEN • Department of Geology, Western Washington University, Bellingham, Washington 98225.

Predator-Prey Interactions in the Fossil Record, edited by Patricia H. Kelley, Michał Kowalewski, and Thor A. Hansen. Kluwer Academic/Plenum Publishers, New York, 2003.

Bromley, 1981, 1993), most drill holes reported in molluscs resemble those produced by naticid or muricid gastropods. The earliest holes resembling those of naticids are Triassic (Newton, 1983; Fursich and Jablonski, 1984), and Harper et al. (1998) reported muricid-like drill holes from the Jurassic. However, body fossils of these predatory groups are unknown before the Cretaceous, suggesting that other gastropod taxa were most likely responsible for these early Mesozoic holes (Harper et al., 1998). Both naticids and muricids diversified beginning in the Cretaceous (Sohl, 1969; Kabat, 1990).

In this chapter, we review the fossil record of drilling predation on bivalves and gastropods and the implications for testing ecological and evolutionary hypotheses. Pre-Cretaceous data will be reviewed as available. However, clearly predatory drill holes in bivalves and gastropods are relatively rare prior to the Cretaceous, and the perpetrators are more difficult to ascertain; much more work is needed on drilling in the Paleozoic and early Mesozoic. Consequently, we focus on predation by naticid and muricid gastropods from the Cretaceous onward.

Naticids are infaunal marine gastropods that plow through the sediment in search of prey. Hunting on the surface of the substrate has been observed on rare occasions in the laboratory (Guerrero and Reyment, 1988) and field (Grey, 2001); Savazzi and Reyment (1989; see also Reyment and Elewa, Ch. 4, in this volume) observed subaerial hunting at low tide, followed by drilling within the substrate. Epifaunal drilling appears to occur on occasion (Dietl, in press). Details of naticid predation have been determined by Ziegelmeier (1954) and Carriker (1981); Kitchell (1986), Kabat (1990) and Reyment (1999) provided excellent summaries. Prey are detected, apparently using chemoreception and/or mechanoreception, and evaluated. Typically, the predator covers the prey in mucus (which is hypothesized to have an anesthetizing effect; Ansell and Morton, 1987; Carriker, 1981), envelopes it in its large mesopodium, and orients the item in a preferred position for drilling. Drilling involves mechanical rasping with the radula as well as secretions of the accessory boring organ (ABO), including hydrochloric acid as well as enzymes and chelating agents (Carriker and Gruber, 1999). Depending on thickness of the prey, drilling of a single item may take hours to days. The proboscis is then inserted through the completed hole and the tissue ingested.

Muricid gastropods search for and drill prey epifaunally, although they may also dig up shallow infaunal prey. Following detection of prey, apparently by chemical cues (Carriker, 1981), a borehole site is selected (in *Urosalpinx cinerea*, the predator crawls over the surface of the epifaunal bivalve prey, exploring the surface for up to half an hour; Carriker, 1981). The drilling process in muricid gastropods is generally similar to that of naticids, and involves alternating application of the accessory boring organ and the radula. (The ABO occupies a position in the mid-ventral part of the muricid foot, in contrast to its location at the tip of the naticid proboscis, however; Carriker and Gruber, 1999.) Following penetration of the prey shell, some muricids may inject toxins into their victims (Brown and Alexander, 1994).

The prey of naticids are most often infaunal gastropods (including naticids) and bivalves. Naticid boreholes are infrequently found in scaphopods (Yochelson et al., 1983, reported that 1 – 5% of specimens in Cretaceous through Holocene samples had apparent naticid borings, though some scaphopod species occupying coarse sediments exhibited higher frequencies). Juvenile naticids prey upon ostracods (Reyment, 1963, 1966, 1967; Reyment and Elewa, Ch. 4, in this volume) and possibly foraminifers (Reyment, 1966, 1967; Livan, 1937; Arnold et al., 1985). Reyment (1999) reported that naticids are not known to eat carrion, although Grey (2001) observed *Euspira lewisii* feeding on dead crabs in the field. In contrast, although muricids prefer live prey, they also eat carrion, and their diet may include bivalves, gastropods, barnacles, bryozoans, crabs, and ostracods (Carriker, 1981; Reyment, 1999). Naticid- and muricid-like holes have also been reported from egg cases (Ansell, 1961; Cox *et al.*, 1999). Prey switching has been documented for the muricid *Nucella lapillus*, which was found to attack mussel prey only when the favored barnacle prey had been extirpated locally by drilling (Fischer-Piette, 1935; see Reyment and Elewa, Ch. 4, in this volume). Such preferences in *Nucella lapillus* are developed through ingestive conditioning; *Nucella* raised on barnacles but subsequently offered mussels continue to prefer barnacles, and vice versa (Hughes and Dunkin, 1984).

2. Data on Drilling Predation Extractable from the Fossil Record

Deciphering predator-prey interactions from drill holes requires assurance that the holes were indeed the result of predation, rather than some other biotic or taphonomic process. For example, pseudo-borings may develop during diagenesis through pressure dissolution by shell fragments (Lescinsky and Benninger, 1994), though the grooves produced do not resemble predatory drill holes. Hard-substrate borers may excavate holes in shells, which are used as domiciles, or may bore through a lithified substrate, penetrating the enclosed shells (Richards and Shabica, 1969). In addition, Carriker and Yochelson (1968) suggested that Ordovician putative predatory drill holes might be epibiont attachment scars. Several authors have reviewed the criteria for recognizing drill holes as predatory (e.g., Carriker and Yochelson, 1968; Rohr, 1991; Baumiller, 1996; Kaplan and Baumiller, 2000; Leighton, 2001*a*): perpendicular penetration of the shell surface; regular shape (circular or oval); penetration from the shell exterior; failure to pass through both valves of an articulated bivalved specimen; and non-random distribution of holes with respect to valve, position on the valve, and/or size or type of prey. An additional criterion sometimes used is that only one complete drill hole is present on a specimen, but many exceptions to this criterion occur among modern predatory drillers (see Leighton, 2001*a*, and below).

Once a drill hole has been recognized as predatory, the taxonomic affinity of the driller can often be determined based on size and morphology of the hole. Bromley (1981) and Kabat (1990) reviewed the hole morphologies of various groups of drillers. Prosobranch gastropods drill holes that are circular in plan view, whereas the turbellarian flatworm *Pseudostylochus ostreophagus* produces small, irregular oval holes (Kabat, 1990). Octopods also drill irregular or oval holes, although circular holes resembling those drilled by muricid gastropods are produced as well (Bromley, 1981, 1993; Cortez *et al.*, 1998; Harper, 2002). Naticid drill holes usually can be distinguished from those made by muricid gastropods by their size and shape. Muricid holes tend to be smaller and more cylindrical, although Gordillo and Amuchastegui (1998) found that holes drilled by the muricid *Trophon geversianus* may vary from conical to cylindrical depending on the prey species drilled. Naticid holes are beveled and parabolic in cross section such that the "outer borehole" diameter is larger than the "inner borehole" diameter (Bromley, 1981; Carriker, 1981; Reyment, 1963, 1999; Kowalewski, 1993). Bromley (1981) designated the ichnofossil typically produced by naticid drilling as *Oichnus paraboloides* and that characteristic of muricids as *Oichnus simplex*. Muricid and naticid holes are usually distinguishable, with some exceptions (Kowalewski, 1993); characteristic differences may not be evident in edge-drilled holes, holes in thin-shelled prey, some taphonomically altered holes, and cases where shell structure, geometry, or ornamentation affects hole shape (e.g., presence of a conchiolin layer in corbulid valves). Kowalewski (1993), using subfossil assemblages, found multivariate morphometrics useful in discriminating naticid and muricid holes based on size; however, Reyment and Elewa (Ch. 4, in this volume) reported no significant size difference in holes drilled in ostracods by juvenile naticids and muricids). Extant muricids tend to be small; differences in drill hole size may be less definitive in fossil assemblages containing similar-sized muricids and naticids.

A few additional taxa have been discovered to drill holes that resemble those made by naticids or muricids (Harper *et al.*, 1998). For instance, naticid-like holes are made by a few members of the Nassariidae, although the holes drilled by post-larval *Nassarius festivus* (Morton and Chan, 1997) appear to be more irregular than naticid holes. Some Marginellidae are also reported to make naticid-like holes (Ponder and Taylor, 1992), although Carriker and Gruber (1999) likened the marginellid holes to those drilled by octopods. The nudibranch *Okadaia elegans* also drills beveled holes but preys on spirorbid and serpulid polychaetes and not on molluscs (Carriker and Gruber, 1999). Muricid-like holes are sometimes drilled by octopods and also were produced by the buccinid *Cominella* in small, thin *Katelysia* bivalve prey during aquarium experiments (Peterson and Black, 1995). The ectoparasitic Capulidae also drill holes that resemble those of muricids, but these can be recognized by the accompanying attachment scar when preserved (Kabat, 1990). In most molluscan assemblages in which naticids and/or muricids are abundant, they were the most probable drillers of *O. paraboloides* and *O. simplex* holes, respectively. (See also Reyment and Elewa, Ch. 4, in this volume, for further comparison of naticid and muricid drilling.)

The data most readily available for reconstructing the fossil record of drilling predation are drilling frequency data. These data generally are reported as frequency of drilled individuals (only whole or nearly whole specimens must be used). For gastropods, calculation of drilling frequency is straightforward; for disarticulated pelecypod valves, drilling frequency has usually been calculated as twice the number of drilled valves divided by the total number of valves, because drilling of only one of a bivalved individual's two valves causes mortality. Bambach and Kowalewski (2000) have advocated dividing the number of drilled valves by half the total number of valves; the same drilling frequencies result, but statistical tests of differences in drilling frequencies are more conservative and this second approach provides correct estimates of Type I errors (see also Hoffmeister and Kowalewski, 2001, and Kowalewski, in press.) An alternative approach is to calculate separately and sum the proportions of drilled left and right valves (De Cauwer, 1985; Anderson *et al.*, 1991); this approach is useful in cases where left and right valves are not represented in equal numbers and where a preference exists for drilling a particular valve. Drilling frequencies may be reported for individual taxa or for entire assemblages of bivalves and gastropods. Vermeij (1987) used a combined approach that calculated the percent of abundant species (i.e., those represented by more than 10 individuals) in a fauna drilled at frequencies greater than 10%. Some studies have lumped together all forms of drilling; others have used drill hole size and morphology to attempt to separate mortality caused by naticids from that caused by muricids.

Drilling frequencies generally have been assumed to represent accurately the proportion of gastropod and bivalve mortality caused by drilling predation. The following assumptions are implicit: (1) drilling predators do not kill prey except by drilling; (2) predators drill only live prey; (3) drilling frequencies are not altered by taphonomic processes. The first assumption appears to be largely correct in the case of naticids; some instances of naticid predation without boring (through suffocation or insertion of the proboscis through gapes) have been reported (see summary in Kabat, 1990). Behaviors differ among taxa; in aquarium experiments, Ansell and Morton (1987) found that *Natica gualteriana* and *Glossaulax didyma* always bored their prey, but that *Polinices tumida* suffocated and consumed some of its prey without drilling (the frequency of such non-drilling predation varied among prey species). In addition, although drilling appears to be the "primary and probably plesiomorphic method of feeding" of muricids (Vermeij and Carlson, 2000, p. 36), some muricids attack bivalve prey by wedging open or chipping the valve margins until the proboscis can be inserted (Taylor *et al.*, 1980). Some muricids attack gastropod prey through the aperture without drilling as well (Vermeij and Carlson, 2000). Naticids typically do not feed on carrion but attack live prey (Berg, 1978; Kabat, 1990), consistent with the second assumption. (Pek and Mikulas, 1996, attributed apparent cannibalism to boring of dead prey, a conclusion contrary to the results of Kelley, 1991a; but see Dietl and Alexander, 1995, on the occurrence of rare holes apparently bored from the shell interior and therefore in empty shells). On the other hand, muricids are reported to ingest carrion, but are not known to drill empty shells (Carriker, 1981).

The assumption about taphonomy is more controversial. Lever *et al.* (1961) showed that drilled and undrilled valves exhibited different hydrodynamic behavior, with bored valves carried up higher onto the beach, although this factor is probably less important in shelf settings. Roy *et al.* (1994) found that bored *Mulinia* valves are weaker under point-load compression than are non-bored shells, although Hagstrom (1996) showed that such vulnerability would bias drilling frequencies only in high-energy environments. Zuschin and Stanton (2001) confirmed that drilled valves of *Anadara* were weaker under point-load compression, but found that drill holes produced no taphonomic bias during sediment compaction experiments. In fact, drilled *Mulinia* shells had a lower percent breakage than undrilled shells and drill holes did not act as local stress concentrators; few (10%) fractures passed through drill holes. Zuschin and Stanton (2001, p. 167) concluded that, in shell beds, "contact between shells is the crucial factor, and the presence of a borehole is less important." Thus preferential breakage of shells due to the presence of drilling does not appear to be a significant concern for most shelf assemblages. These results are consistent with those of Nebelsick and Kowalewski (1999), who demonstrated a lack of pre-burial taphonomic bias against drilled tests of echinoids. Nevertheless, studies of drilling frequencies should assess the potential for taphonomic bias. Both Roy *et al.* (1994) and Kaplan and Baumiller (2000) provided guidelines for determining the degree of taphonomic bias of drilled assemblages; Kaplan and Baumiller also presented a method for "back-calculating" drilling frequencies on bivalved fossils in cases of between-valve taphonomic biases. In addition, to minimize taphonomic bias, studies that compare drilling frequencies among assemblages should keep environment constant.

Because drilling predators may attack some shelled prey without drilling, and because of possible preferential destruction of drilled shells, drilling frequencies may provide minimum estimates of the impact of drilling predators on fossil bivalves and gastropods (see also Leighton, 2001b). However, the effect of crushing predators on drilling frequencies should also be considered. Obliteration of shells by crushing predation may artificially inflate drilling frequencies, because the frequency of drilling predation is calculated relative to the sample of whole or nearly whole shells (i.e, those that did not experience crushing predation). Shells with incomplete drill holes could also be destroyed by shell crushers, affecting calculations of failed drilling as well (Harper, pers. comm., 2002).

Drill holes in shells provide a variety of additional information about predator-prey interactions, including the size and selectivity of the predator and the success of the predation attempt. Carriker and Gruber (1999, p. 579), in a global interspecific comparison of drilling gastropods (primarily muricids), found that the size and shape of the drill hole "mirror the external morphology of the extended ABO." Calculations based on the data they reported indicate that species mean height, ABO mean diameter, and radula mean width are all highly correlated with one another ($r = 0.7782$ for ABO width vs. H, $r = 0.8167$ for radula width vs. H, and $r = 0.7345$ for ABO and radula width; 57 observations). Thus drill hole size should be indicative of the size of the driller; see also Reyment and Elewa (Ch. 4, in this volume). These results are consistent with the observations of drilling by extant naticids. Kitchell et al. (1981) found a highly significant correlation between size of *Neverita duplicata* predators and both the outer (OBD) and inner (IBD) borehole diameters. Grey (2001) found a significant correlation between predator size and outer borehole diameter for *Euspira lewisii* and *E. heros*; however, the ratio of IBD:OBD differed among *E. lewisii, E. heros*, and *Neverita duplicata*. Both the outer and inner borehole diameters have been used by various authors as a measure of predator size; although Kabat (1990) advocated using IBD, we have used OBD because the inner borehole diameter may not always be widened enough to admit the proboscis in nonfunctional holes (Kitchell et al., 1986).

Various studies have used the position of a drill hole on the prey's shell to investigate the degree of behavioral stereotypy of the predator. For example, drill hole site for naticids has been related to the way the predator manipulates a prey item (Ziegelmeier, 1954), which may be influenced by size and shape of the prey (Ansell, 1960; Stump, 1975; Kitchell, 1986; Reyment, 1999; Roopnarine and Willard, 2001). Drill hole position has been observed to shift with growth of the predator, reflecting ontogenetic change in its ability to manipulate prey, for some naticids (Calvet, 1992; Vignali and Galleni, 1986) and for the muricid *Chorus giganteus* (Urrutia and Navarro, 2001). The position of predatory drill holes has been linked to access to particular soft parts (Hughes and Dunkin, 1984; Arua and Hoque, 1989; Leighton, 2001a), though this may be less important for naticids (and also large muricids; Dietl, pers. comm., 2002), which normally consume the entire soft tissue. Carriker (1981) suggested that chemical cues may influence drill hole site selection by muricids, which preferentially drill through preexisting cracks or holes in shells or, for some species, near valve margins. The muricids *Nucella lapillus* and *Morula musiva* prefer drilling the older parts of mussel shells, where abrasion removes the periostracum, which in experiments was found to deter commencement of drilling (Harper and Skelton, 1993).

A variety of approaches exists for determining drill hole site. Most studies have made use of some sort of grid or sector designation (e.g., Kelley, 1988, 1991a; Allmon et al. 1990, Anderson et al., 1991; Gordillo, 1994; Harper et al., 1998) or have otherwise divided shells into areas (Zlotnik, 2001). Other techniques include landmark analysis (Roopnarine and Beussink, 1999; Dietl and Alexander, 1995, 2000), contour diagrams (Stump, 1975), bivariate scattergrams or relative position along antero-posterior and dorso-ventral axes (e.g., Berg, 1978; Anderson, 1992; Dietl and Alexander, 1997), rose diagrams (Dietl and Alexander, 1997, 2000), and "Pope diagrams" (Reyment, 1999;

Reyment and Elewa, Ch. 4, in this volume). Some investigators have also examined preference for drilling the right vs. left valves of bivalves (e.g., Sohl, 1969; Adegoke and Tevesz, 1974; Stump, 1975; Anderson, 1992; Harper *et al.*, 1998).

Other aspects of predator behavior extractable from the fossil record include selectivity of prey size and prey species. Size selectivity has been examined by comparing the frequency distributions of bored vs. unbored specimens, or bored specimens vs. the total sample (e.g., Ansell, 1960; Allmon *et al.*, 1990). Correlation of prey size and drill hole size have also been used (e.g., Ansell, 1960; Kitchell, 1986; Kelley 1988, 1991*a*; Anderson *et al.*, 1991; Harper *et al.*, 1998). Selectivity of prey species has been examined by comparing frequencies of drilling of species with their relative abundance in the fauna (Hoffman and Martinell, 1984; Kohn and Arua, 1999).

Some studies have attempted to predict preferences of naticid predators by employing cost-benefit analysis. Naticid prey are not chosen at random; work with extant naticids indicates a high degree of predator selectivity with respect to prey size and species. This selectivity appears to be a stereotyped behavior that has developed over evolutionary time, because in the laboratory naticids are unable to learn to select a novel (artificially thinned) prey despite its greater profitability (Boggs *et al.*, 1984; see also Berg, 1978). Kitchell *et al.* (1981) showed that prey selection by extant naticids is consistent with the predictions of cost-benefit analysis. They developed an optimum foraging model in which cost of a predation event was dependent on drilling time and benefit was proportional to prey biomass. (In the fossil record, the cost:benefit ratio can be represented as shell thickness:internal volume of the prey.) Predators were predicted to favor the prey item with the lowest cost-benefit ratio in the size range that can be handled (cost-benefit ratio of a particular prey species decreases with ontogeny, but prey beyond a certain species-specific size cannot be captured and manipulated successfully and are therefore immune to predation). Although some criticisms of the assumptions of cost-benefit modeling have been raised (Anderson *et al.*, 1991; Leighton, 2001*b*), cost-benefit analysis successfully predicted prey preferences of naticids in the laboratory and for several fossil assemblages (Kitchell *et al.*, 1981; Kitchell, 1982; Kelley, 1988, 1991*a*).

The completeness of the drill hole indicates the success of the predation event. In general, a drill hole that does not penetrate to the shell's interior represents a failed predation attempt. (Exceptions may occur; Ansell and Morton, 1987, found that *Glossaulax didyma* in aquarium experiments abandoned drilling and consumed some large prey that suffocated during drilling, yielding incomplete holes despite a successful attack.) An attack may be unsuccessful for any of a number of reasons, including mechanical limits imposed by prey shell thickness (Kitchell *et al.*, 1981), wearing of the radula (Reyment, 1999), or interruption by physical or biological events such as escape attempts of the prey (Kitchell *et al.* 1981, 1986; Dietl and Alexander, 1997) or interference by other predators (Morton, 1985). In addition, Dietl (2000) demonstrated that *Mytilus edulis* is able to repair complete muricid holes; such repaired holes also represent unsuccessful predation. Kitchell *et al.* (1986) also argued that a naticid hole in which the inner diameter was not enlarged enough to allow insertion of the proboscis was not functional. Their data indicate that a ratio of the inner borehole diameter:outer borehole diameter greater than 0.5 indicates successful penetration by the naticid's proboscis and thus a functional hole. This criterion for functionality should be used with caution; Grey (2001) demonstrated variation in IBD:OBD ratios of functional holes drilled by different naticid species. Most large surveys have not distinguished between functional holes and complete but nonfunctional holes but instead record the occurrence of complete vs. incomplete holes (Kelley and Hansen, 1993, 1996a; Kelley et al., 2001). The ratio of incomplete to complete holes has been defined as prey effectiveness (Vermeij, 1987) and used to measure the relative capabilities of naticid predators and prey.

Some prey shells bear multiple drill holes, which, in the case of naticid holes, may indicate failed drilling. Naticid predatory behavior precludes simultaneous attack of a prey item by multiple naticids (Carriker, pers. comm., 1995; Kitchell et al., 1986; but see also Reyment and Elewa, Ch. 4, in this volume); multiple holes thus result from successive predation attempts. If interrupted during drilling, naticids apparently cannot reoccupy the same hole but produce a new hole, often adjacent to the earlier hole (Kojumdjieva, 1974; Kitchell, 1986; Kabat, 1990; although observations by Dietl, pers. comm., 2002, indicate exceptions occur). Although multiple holes in a shell typically include no more than one complete hole, multiple complete holes may occur if a prey item is able to escape after the hole penetrates the shell but before feeding commences (Kitchell et al., 1986) and if the timing of successive attacks prevents repair of holes. Multiple complete holes may also derive from taphonomic abrasion of incomplete holes (Kitchell et al., 1986). Although Reyment (1966) reported that double holes may be caused by a naticid rasping at the suture of a gastropod with inflated whorls, most cases of multiple naticid holes in bivalves and gastropods indicate failed attacks (Kelley and Hansen, 1996a). Some authors (Hoffman et al., 1974; Stanton and Nelson, 1980; Pek and Mikulas, 1996) have argued that multiple drill holes on a prey individual indicate that the predator bored an already dead shell. Such predation is unlikely (Kitchell et al., 1986; Kelley, 1991a), but if it occurs would indicate a highly inefficient predatory behavior. Some multiple holes were postulated to result from attacks on shells occupied by hermit crabs (Kohn and Arua, 1999), although field and laboratory observations suggest that the hermit crab *Pagurus longicarpus* actively avoids shells that have been drilled (Pechenik and Lewis, 2000) if undrilled shells are available. Thus the percent of holes that occur in multiply bored shells is a useful measure of relative efficiency of naticid predators and prey (Kelley and Hansen, 1996a). In the case of bivalved specimens, articulated valves are needed for accurate estimates of the frequency of multiply bored valves (Harper et al., 1998), because some multiple drilling involves attacks on both right and left valves (De Cauwer, 1985; Dietl and Alexander, 1997). Data based on disarticulated valves should be regarded as underestimates of multiple drilling.

Multiple and incomplete holes are less suitable as indicators of muricid efficiency. Multiple holes need not indicate unsuccessful drilling, because multiple predators are able to attack a victim simultaneously. Brown and Alexander (1994) demonstrated that group foraging, especially on large oysters, is common in the muricid *Stramonita haemostoma*. Taylor and Morton (1996) also observed multiple *Thais clavigera* and *T. luteostoma* attacking single prey items simultaneously. Interpretation of incomplete drill holes is also complicated. Because muricid predators can locate, reoccupy and subsequently complete a hole, if interrupted in drilling, the incidence of incomplete muricid holes is a minimum estimate of drilling failure. On the other hand, Dietl (2000) suggested that muricids in a gang attack may abandon a partially completed hole and feed through the gape, so that incomplete holes do not necessarily signify failed attacks.

3. Testable Hypotheses in Evolutionary Paleoecology

The information provided by drill holes, and the abundance of such holes in the fossil record, permits the reconstruction of ancient predator-prey systems and the testing of hypotheses in evolutionary paleoecology. A primary question concerns the role that ecological interactions such as predation play in evolution. The related concepts of coevolution and escalation both assume that biological factors are major agents of natural selection and that organisms are able to respond evolutionarily to selective factors imposed by other organisms in the environment, yielding arms races between evolving taxa. (See Dietl and Kelley, in press, for a more detailed discussion of arms race hypotheses and the fossil record of predation.)

Coevolution is defined as the evolution of two or more species in response to one another. In the strict sense the term has been applied to reciprocal adaptation of species: "a trait of one species has evolved in response to a trait of another species, which trait itself has evolved in response to the trait in the first" (Futuyma and Slatkin, 1983, p. 1). The coevolving species may be predator and prey, competitors, host and guest, or mutualists. The term "diffuse coevolution" has been applied to interactions involving more than two species (for instance, a predator with multiple prey species, such as a drilling gastropod). Coevolution has been demonstrated in some terrestrial ecosystems by ecological and other studies (Futuyma and Slatkin, 1983), though some previously accepted examples of coevolution have been questioned recently (see summary in Jablonski and Sepkoski, 1996). Among marine organisms, examples of coevolution are less obvious. Circumstantial evidence suggests coevolution for some predator-prey or mutualistic relationships, but Vermeij (1983) argued that data are generally insufficient to document convincingly coevolution in the sea.

Vermeij (1987, 1994) argued for the prevalence of escalation rather than coevolution. Vermeij's hypothesis of escalation claims that, over the course of the Phanerozoic, biological hazards (such as predation) have become more severe, and adaptations to those hazards have increased in expression. The hypothesis of escalation considers the most significant selective agent to be an organism's enemies. In the case of predator-prey interactions, Vermeij (1987, 1994) has argued that prey respond to their predators, but that the predators are more likely to respond to their own enemies (for instance, their predators) than they are to their prey. Thus, adaptation need not be reciprocal because the predator exerts stronger selective pressure on the prey than the prey exert on the predator (the consequences of the interaction are more severe for the prey than for the predator). However, if the prey are dangerous (able to inflict damage) to the predator, coevolution may occur because both escalating parties are enemies (Brodie and Brodie, 1999; Dietl and Alexander, 2000; Dietl and Kelley, in press).

Consistent with hypotheses of coevolution and/or escalation, various authors have argued for the importance of biotic interactions in evolution (e.g., Steneck, 1983; Signor and Brett, 1984; Jackson, 1988; Lidgard and Jackson, 1989; Aronson, 1992; McKinney, 1995; Roy, 1996; Bambach, 1999; Knoll and Bambach, 2000). Others (e.g., Boucot, 1983; Guensberg and Sprinkle, 1992) have argued that physical factors are more important as selective agents than are biological factors, or that extrinsic events such as mass extinction "undo and reset any pattern that might accumulate during normal times" through adaptation (Gould, 1985, p. 7). The prevalence of stasis in some lineages has been used to argue against a significant role for ecological interactions in evolution (see discussion by Allmon, 1994). Gould (1990) and Kitchell (1990) also argued against the reality of trends in evolution, including those implicated in arms race scenarios. Adaptation of organisms to one another is also seen to be limited by genetic or developmental constraints, or by tradeoffs among competing selective forces; Kitchell (1990) argued, based on modeling the naticid predator-prey system, that such tradeoffs limit the role of escalation in evolution. See Vermeij (1987, 1994) for a response to these criticisms of the hypothesis of escalation.

Several studies of coevolution and/or escalation have focused on predation by drilling gastropods. For instance, Kitchell (1986, 1990) and coworkers (DeAngelis *et al.*, 1984, 1985) argued that coevolution has characterized the naticid gastropod predator-prey system and developed mathematical models of the coevolutionary response of predators and prey. Vermeij (1987) used data on drilling through the Phanerozoic to support the hypothesis of escalation (see also studies of Dudley and Vermeij, 1978, and Vermeij *et al.*, 1980). Various additional studies of drilling gastropod predation have tested explicitly, or provide data applicable to testing, hypotheses of coevolution and/or escalation. We review these studies and discuss the implications of the results for arms race hypotheses.

4. Arms Races and Drilling Frequencies

Temporal patterns in drilling frequencies have been used as evidence of arms races. For instance, the hypothesis of escalation predicts a temporal increase in the hazard of predation, including that by drilling gastropods (Vermeij, 1987). Testing this hypothesis requires data from a range of stratigraphic units representing similar paleoenvironments; at the least, if environment cannot be kept constant, an understanding of how drilling varies with environment is necessary (see Hagadorn and Bottjer, 1993). Several studies have examined variation in drilling with environment. For example, Sander and Lalli (1982) examined drilling frequencies of Recent molluscs along a depth transect on the west coast of Barbados but no convincing trend was found. Likewise, Hansen and Kelley (1995) reported no significant bathymetric trend in drilling among localities in the Eocene Gulf Coast Moodys Branch Formation, although drilling frequency was higher in the deeper water Yazoo Formation. Hoffmeister and Kowalewski (2001) also documented variation of drilling frequencies among facies in the Miocene of Central Europe. Drilling appears to vary with latitude, but some studies have found an increase in drilling with increasing latitude (Vermeij *et al.*, 1989; Allmon *et al.*, 1990; Hansen and Kelley, 1995; Hoffmeister and Kowalewski, 2001) and others have found a decrease (Dudley and Vermeij, 1978; Alexander and Dietl, 2001) or no difference (Kelley *et al.*, 1997). Ongoing work by Hansen *et al.* (2001) aims to develop a Recent data base on drilling with latitude.

Drilling frequencies for individual mollusc assemblages of various ages have been reported in a number of studies (e.g., Permian: Kowalewski *et al.*, 2000, Hoffmeister *et al.*, 2001; Jurassic: Kowalewski *et al.*, 1998, Harper *et al.*, 1998; Cretaceous: Taylor *et al.*, 1983, Vermeij and Dudley, 1982, Pan Hua-zhang, 1991; Eocene: Taylor, 1970, Adegoke and Tevesz, 1974, Stanton and Nelson, 1980; Miocene: Hoffman *et al.*, 1974, Kojumdjieva, 1974, Colbath, 1985, Kowalewski, 1990; Pliocene: Hoffman and Martinell, 1984, Walker, 2001; Pleistocene: Kabat and Kohn, 1986). An accurate depiction of drilling through time is difficult to determine by comparing data from published studies, however, due to variations in environments of assemblages and methodology used by different researchers.

A few studies have attempted to synthesize temporal patterns of drilling frequencies for individual taxa. The results of these studies suggest that drilling predation may have increased during the Cretaceous and/or early Cenozoic, in accordance with the hypothesis of escalation, and then stabilized thereafter. For instance, De Cauwer (1985) reported drilling frequencies on Cretaceous, Eocene, Miocene, Pliocene and Recent corbulids. Although drilling frequencies varied substantially, the most marked increase in drilling frequencies occurred in the Cretaceous, between the Albian and the Campanian to Maastrichtian. Vermeij et al. (1980) compared drilling on terebrid gastropods from museum samples and reported a possible increase in drilling between the Eocene and Oligocene to Recent samples. Dudley and Vermeij (1978) examined museum specimens and determined that drilling frequencies on turritellid gastropods were significantly lower in the Cretaceous than in the Eocene, Miocene, Pliocene, and Recent. They claimed an increase in drilling predation had taken place between the Cretaceous and Eocene, but could not determine exactly when the increase occurred due to lack of Paleocene samples. Allmon et al. (1990) compiled data on turritelline gastropods that suggested that drilling frequencies in the Cretaceous were lower than for most Cenozoic samples. Hagadorn and Boyajian (1997) found no difference in drilling between Miocene and Pliocene samples of *Turritella* from the U.S. Coastal Plain. Likewise, Thomas (1976) found no temporal trends in drilling frequency on lower Miocene through Pliocene *Glycymeris* from the Coastal Plain. Differences in drilling frequency on Pleistocene and Recent *Turritella* species from California were no greater than among contemporaneous samples (Tull and Bohning-Gaese, 1993).

Vermeij (1987) compiled data from the published literature on drilling predation on Cretaceous through Recent gastropods and on drilling on Phanerozoic bivalved animals. Plots of the percent of abundant species in a fauna drilled at frequencies greater than 10% showed a pattern interpreted by Vermeij to represent escalation. Vermeij (1987) stated that drilling frequencies for both groups were low in the Cretaceous and reached modern levels by the Eocene (Fig. 1). He proposed that a significant episode of escalation of drilling occurred between the Cretaceous and Eocene, although the details of this apparent escalation were unclear due to the small number of assemblages (only six of Cretaceous through Eocene age, with none from the Paleocene).

Results consistent with those of Vermeij (1987) were produced by a literature compilation of 254 occurrences of drilled shells and of drilling frequencies from 32 fossil assemblages (Kowalewski et al., 1998). Three phases in drilling were identified. The initial (Paleozoic) phase included a Devonian peak, in which drilling frequencies reached 12% and brachiopods were the primary victims. A second phase spanned the Permian through Early Cretaceous, with drilling frequencies <1%; Kowalewski et al. (2000) and Hoffmeister et al. (2001) confirmed similar low frequencies for drilling on bivalves from Permian bulk samples (drilling frequencies of 20% on Jurassic astartids reported by Harper et al., 1998, are not directly comparable because they are not assemblage estimates). A Cenozoic phase was also identified, in which drilling frequencies on molluscan assemblages increased from ~ 10% in the Late Cretaceous to values reaching ~ 40% in the Paleogene and Neogene. Thus, significant escalation in drilling appears to have occurred since the Late Cretaceous, although lumping of all Paleogene and all Neogene samples obscured finer details of the pattern.

Kelley and Hansen (1993, 1996a) conducted a study designed to test the hypothesis of escalation for the naticid predator-prey system. They tabulated drilling frequencies on >46,000 bivalve and gastropod specimens from bulk samples of Cretaceous through

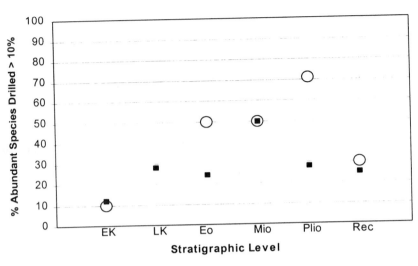

FIGURE 1. Data on drilling of local assemblages of gastropods (squares) and pelecypods (open circles) compiled from the literature by Vermeij (1987). Data presented as percent of abundant species (>10 individuals) in an assemblage drilled >10%. EK, Early Cretaceous; LK, Late Cretaceous; Eo, Eocene,; Mio, Miocene; Plio, Pliocene; Rec, Recent.

Oligocene age (17 formations, 14 stratigraphic levels from the Gulf and Atlantic Coastal Plains). Patterns in drilling were more complex than revealed in earlier studies (Fig. 2). Drilling frequencies were low to moderate in the Cretaceous, declined crossing the Cretaceous-Tertiary boundary, rose dramatically in the early Paleocene, and remained high until the late Eocene. A drastic decrease in drilling in late Eocene samples was followed by a statistically significant increase in the early Oligocene. Kelley and Hansen (1996a) suggested that this episodic pattern of drilling frequencies was linked to mass extinctions, which caused drilling frequencies to rise, possibly because of the elimination of highly escalated prey species (see also Vermeij, 1987).

To test this hypothesis, Hansen et al. (in review) bulk sampled mollusc assemblages of latest Oligocene through Pleistocene age from the U.S. middle Atlantic Coastal Plain (>97,000 specimens). Drilling frequencies again exhibited a fluctuating pattern (Fig. 2), with statistically significant increases in drilling synchronous with the middle Miocene extinction event and in the Recent, following the regional Plio-Pleistocene extinction. However, Hansen et al. (1999) failed to substantiate the proposed causal link between surges in drilling and preferential extinction of highly escalated prey. Of forty tests across four extinction boundaries (Cretaceous-Tertiary, Eocene-Oligocene, middle Miocene, and Plio-Pleistocene), only one supported the hypothesis that morphologically escalated prey were preferentially eliminated. However, the morphological approach to determining level of escalation in bivalves may be oversimplified because it does not account for metabolic rate (i.e., physiological escalation), which can be assessed by determining shell growth rates isotopically. The possibility remains, therefore, that physiologically escalated prey may have been affected preferentially by extinction, in accordance with the hypothesis (Dietl et al., 2002).

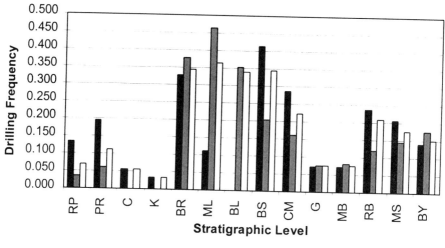

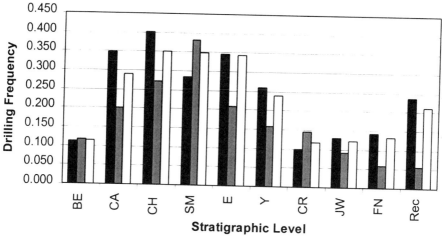

FIGURE 2. Drilling frequencies for bivalves (black), gastropods (diagonal), and the total mollusc fauna (white; bivalves + gastropods) from Kelley-Hansen database on naticid predation in the Gulf and Atlantic Coastal Plain. Top: Cretaceous through Paleogene; bottom, Neogene through Recent. Stratigraphic units: Cretaceous: RP, Ripley Fm; PR, Providence Fm; C, Corsicana Fm. Paleocene: K, Kincaid Fm; BR, Brightseat Fm; ML, Matthews Landing Mbr of Naheola Fm; BL, Bells Landing Mbr of Tuscahoma Fm. Eocene: BS, Bashi Mbr of Hatchetigbee Fm; CM, Cook Mountain, Upper Lisbon, and Piney Point Fms; G, Gosport Fm; MB, Moodys Branch Fm. Oligocene: RB, Red Bluff Fm; MS, Mint Spring Fm; BY, Byram Fm. Latest Oligocene/Early Miocene: BE, Belgrade Fm. Miocene: CA, Calvert Fm; CH, Choptank Fm; SM, St. Marys Fm; E, Eastover Fm. Pliocene: Y, Yorktown Fm; CR, Chowan River Fm. Pleistocene: JW, James City and Waccamaw Fms; FN, Flanner Beach and Neuse Fms. Rec, Recent. (Recent data from Key et al., 2001.)

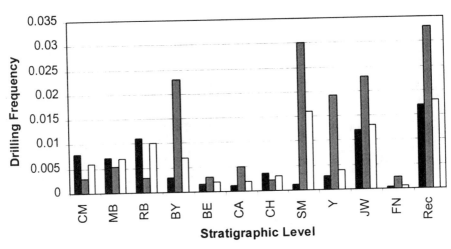

FIGURE 3. Drilling frequencies of bivalves (black), gastropods (diagonal), and the total mollusc fauna (white; bivalves + gastropods) from Kelley-Hansen database on muricid predation in the Gulf and Atlantic Coastal Plain. Recent data from Key et al., 2001. Abbreviations as in Fig. 2.

Similar detailed syntheses of long-term drilling trends by muricid gastropods are lacking. Key *et al.* (2001) compiled data on muricid drilling from the Kelley-Hansen database and found much lower frequencies of drilling by muricids on the largely infaunal assemblages. Drilling by muricids was not significantly correlated with muricid abundance or with naticid drilling frequencies. However, muricid drilling on bivalves exhibited a fluctuating pattern of drilling as well (Fig. 3).

5. Changes in Drilling Behavior through Time

A possible prediction of hypotheses of coevolution or escalation is that behavioral changes of predators have developed over time. For instance, the hypothesis of escalation would predict that the hazard of drilling predation has increased since the Cretaceous, when drilling predators diversified. Behavioral changes, such as an increase in predator selectivity, may have been involved in this intensification of the hazard of drilling predation.

Extant naticids are selective of drill hole position on the prey shell and of prey size and species. Muricids may exhibit similar stereotyped behavior (Gosselin and Chia, 1996; Gordillo and Amuchastegui, 1998; Dietl, 2000), which has been observed to increase with predator experience (Hughes and Dunkin, 1984). Some authors have reported less selectivity by muricids than naticids (Stump, 1975; Thomas, 1976; Roopnarine and Willard, 2001). Stereotypy is exhibited by naticids immediately upon metamorphosis (Berg, 1976); drill hole position may change during ontogeny, however (Calvet, 1992). Stereotyped behavior appears to increase the probability of successful predation. For instance, selectivity of drill hole site fosters drilling the thinnest area of the prey shell (Kitchell, 1986). Hughes and Dunkin (1984) found that drilling the thinnest area of the shell (a learned behavior in *Nucella lapillus*) increased the profitability of mussel prey by 17%. Kelley (1988), in a study of nine species of Miocene Chesapeake Group bivalves, found a positive relationship between degree of selectivity of prey size and percentage of predation attempts that were successful. If naticids and muricids were less capable early in their history, selectivity of drill hole site and prey item may have been less for Cretaceous and Paleogene assemblages than for Neogene and Recent samples.

Various studies have examined the siting of boreholes on prey shells. Work on individual Neogene and Pleistocene assemblages generally has revealed selective siting of boreholes, indicating behavioral stereotypy of the predator (Berg and Nishenko, 1975; Stump, 1975; Martinell and Marquina, 1978; Hoffman and Martinell, 1984; Kelley, 1988; Anderson *et al.*, 1991; Anderson, 1992; Pek and Mikulas, 1996). In accordance with the hypothesis of escalation, several studies of drill hole position have found variable drill hole positions in the Cretaceous or Paleogene. For instance, Kitchell (1986) reported variability of drill hole position in Cretaceous species of the Ripley Formation and suggested that site specificity may have increased since the Cretaceous. A study of drilling of molluscs from the Eocene of Nigeria (Adegoke and Tevesz, 1974) also indicated variability of drill hole position; this variability was not due to the presence of multiple predator species (the assemblages were dominated by one naticid species) but instead indicates lack of predator behavioral stereotypy. (But see Arua and Hoque, 1989, for contrary results). In addition, stereotypy of drill hole position was reported for Jurassic astartids (Harper *et al.*, 1998).

A few studies have examined long-term temporal trends in stereotypy of drill hole position. Berg (1978) described results of a study of naticid drill hole position on gastropod shells from museum collections. Borings were distributed randomly around the shells of Cretaceous prey but, by the Paleocene, boring of the adoral surface was preferred. Longitudinal distribution of drill holes also became more stereotyped between the Cretaceous and Recent (Berg, 1978). Kelley and Hansen (1993, 1996*a*) tabulated drill hole position on all species from bulk samples of 17 Cretaceous through Paleogene U.S. Coastal Plain formations. Only three of 28 gastropod species in which naticid holes were abundant enough for statistical analysis did not show a significant preference for longitudinal drill hole position. Thirty bivalve species had sufficient holes for statistical analysis of their distribution with respect to a nine-sector grid. No general temporal trend in selectivity of drill hole site was observed, although the greatest selectivity of drill hole site occurred in the Cretaceous, opposite expectations (100% of the Cretaceous species exhibited site stereotypy). The least stereotypy occurred above mass extinction boundaries; percent of bivalve species exhibiting site stereotypy was 43%, 69%, and 40% for the Paleocene, Eocene, and Oligocene respectively (Kelley and Hansen, 1996*a*).

Dietl *et al.* (2001) and Alexander *et al.* (in review) examined temporal trends in stereotypy of naticid drill hole position by applying the Shannon-Weaver diversity index (H') to drill hole site data from the Kelley-Hansen database as well as from additional fossil and Recent specimens (120 samples of bivalve species total). In calculating H', each sector in the nine-sector grid (Kelley, 1988) is analogous to a species, and each drill hole within a sector is analogous to an occurrence of that "species." They used J = H'/Hmax (a measure of evenness) as an index of degree of stereotypy; the higher the J value, the more even the distribution of boreholes among sectors, and thus the less stereotyped the drill hole position. Stereotypy increased significantly from the Miocene to Pliocene and Pleistocene to Recent, a time interval over which the species of predator (*Neverita duplicata* and *Euspira heros*) did not change.

Several studies have explored temporal shifts in drill hole position for specific taxa. Hagadorn and Boyajian (1997) found that the preferred site for naticid and muricid drilling on *Turritella* in the U.S. Coastal Plain shifted away from the aperture between the Miocene and Pliocene. They concluded that this shift represented a subtle change in predator behavior consistent with the hypothesis of escalation. Dietl and Alexander (2000) found a shift in borehole position on the body whorl toward the umbilicus from the Miocene to Recent in confamilial predation on *Euspira heros* and especially *Neverita duplicata*. They attributed this behavioral change, which would prohibit egress of the prey's foot, to coevolution between the two naticid predators, which represent dangerous prey to one another. Thomas (1976) also reported a shift in the position of drill holes away from the umbones on *Glycymeris* shells during the Neogene due either to a change in predator behavior or the presence of different predator species. (The latter explanation is consistent with Berg and Porter's 1974 observation that different predators may prefer different drill hole positions on a single prey species.) Roopnarine and Beussink (1999) reported a significant decrease in variance of drill hole position on the bivalve *Chione* from the southeastern U.S. between the Pliocene (*C. erosa*) and Pleistocene (*C. cancellata*), though the mean drill hole position did not change. In contrast, Culotta (1988) reported a ventral shift in drill hole position in *Chione cancellata* and four additional bivalve species in the Plio-Pleistocene of Florida. (But see Herbert and Dietl, in press, for a discussion of problems in differentiating naticid and muricid drill holes in *Chione* that may affect the results of the above studies.)

Selectivity of prey size has been examined in a number of studies by correlating borehole diameters (indicative of predator size) with prey size. Harper *et al.* (1998) found no significant correlation between the diameter of muricid-like boreholes and *Astarte* height in the Jurassic of England, indicating a lack of prey size selectivity. Although Kitchell (1986) demonstrated size-selective predation in the Cretaceous Ripley Formation, size selectivity in the Paleogene appears less developed. Adegoke and Tevesz (1974) found no significant correlation of predator and prey size in the Eocene of Nigeria. Eocene assemblages from the U.S. Coastal Plain showed size selectivity for nine of eleven bivalve species (Kelley and Hansen, 1996*b*), but Eocene bivalves of Seymour Island, Antarctica, exhibited limited size selectivity by naticid predators (Kelley *et al.*, 1997). Sickler (1995) and Sickler *et al.* (1996) found significant correlations between borehole size and prey size for *Astarte* and *Corbula* from the Oligocene Red Bluff Formation of Mississippi, but not for *Scapharca*.

Neogene to Recent assemblages have exhibited correlations between borehole diameter and prey size, indicating prey size selectivity; e.g., Atlantic Coastal Plain Miocene (Kitchell *et al.*, 1981; Kelley, 1988, 1991*a*) and Pliocene (Kitchell *et al.*, 1981), Miocene of Bulgaria, with naticid holes showing a greater correlation than those of muricids (Kojumdjieva, 1974), the Pliocene of Emporda, Spain (Hoffman and Martinell, 1984), Pleistocene of Fiji (Kabat and Kohn, 1986) and California (Tull and Bohning-Gaese, 1993), and Recent U.S. Atlantic Coast (Kitchell *et al.*, 1981; Dietl and Alexander, 1995; Alexander and Dietl, 2001; Dietl, 2000, for muricids). Anderson (1992) reported size-selective predation for three of five corbulid species from the Miocene of the Dominican Republic and Plio-Pleistocene of Florida. Hagadorn and Boyajian (1997) found size-selective drilling predation on Neogene *Turritella* and stated that predator-to-prey size ratios were significantly better correlated in the Pliocene than in the Miocene. However, Dietl and Alexander (2000) reported a decline in the OBD – prey size correlation between the Miocene and Recent; this change corresponded to a decrease in slope of the cost-benefit function, indicating less difference in profitability of small and large prey. The studies by Kelley (1988) and Kelley and Hansen (1996*b*) found that correlations of borehole size and prey size were greater for successful holes than when unsuccessful holes were included, perhaps because unsuccessful holes often result from a mismatch of predator and prey size. In addition, prey survive and continue to grow after an unsuccessful attack, so that actual prey size at the time an incomplete hole was drilled is not clear (Dietl and Alexander, 1997; Dietl, 2000).

Cost-benefit analyses have been applied to several assemblages in order to determine if fossil naticids selected prey consistently with the predictions of an optimum foraging model developed by Kitchell *et al.* (1981). Most Neogene assemblages studied have exhibited predation patterns consistent with the Kitchell *et al.* model (though see Kowalewski, 1990; Anderson *et al.*, 1991; and Jones, 1999, for counterexamples). Kitchell *et al.* (1981) successfully used their cost-benefit model to predict prey preferences in the Neogene Yorktown and Jackson Bluff Formations of the Atlantic Coastal Plain. Selection of prey species was also in accord with cost-benefit modeling for the Miocene of Maryland (Kelley 1988, 1991*a*), Plio-Pleistocene Pinecrest, Caloosahatchee, and Bermont Formations of Florida (Culotta, 1988), and for Recent naticids of New Jersey (Dietl and Alexander, 1995). Among Miocene to Pleistocene assemblages, departures from the model's predictions occurred for specific taxa (primarily corbulids, for which the conchiolin layer of the shell may deter drilling; Kelley, 1988; Culotta, 1988) or when slopes of cost-benefit functions are low (Jones, 1999), so that different prey items have similar cost-benefit ratios. In contrast, predation on assemblages of Eocene and Oligocene age departed from the expectations of the cost-benefit model. For instance, of four Eocene assemblages examined by Kelley and Hansen (1996*b*), only the Bashi Marl Member of the Hatchetigbee Formation (Alabama) yielded results consistent with cost-benefit predictions. In the Oligocene Red Bluff Formation of Mississippi, a cost-benefit analysis that compared predation on the common prey (*Astarte*, *Corbula*, and *Scapharca*) found all prey classes to be drilled at about the same frequency (11 – 20%), despite their different cost-benefit rankings (Sickler, 1995; Sickler *et al.*, 1996).

6. Relative Effectiveness of Predator and Prey

If predator and/or prey responded evolutionarily to one another, through coevolution and/or escalation, changes in their relative effectiveness may have occurred over time. Vermeij (1987) tabulated data from the global literature on prey effectiveness (number of incomplete drill holes divided by number of attempted holes) for 34 bivalve and 47 gastropod species for which at least ten drill holes were reported. He concluded that Cenozoic bivalves were more resistant to predation than were gastropods. No general conclusions were drawn concerning temporal trends in resistance of prey to drilling; Vermeij (1987, p. 237) remarked that, "Too little information is currently available to permit evaluation of the hypothesis that shells have become more effective against drilling since the Late Cretaceous."

Kelley and Hansen's (1993) survey of naticid predation found a statistically significant increase in failed drilling from the Cretaceous through Oligocene of the U.S. Coastal Plain. Prey effectiveness for bivalves ranged from 1.6% in the Cretaceous

Providence Formation to 18.9% in the Oligocene Red Bluff Formation. Consistent with Vermeij's (1987) conclusion, prey effectiveness for gastropods was much lower than for bivalves (5% versus 11% for the Cretaceous through Oligocene) and ranged from 0 (Cretaceous Providence and lower Paleocene Brightseat and Matthews Landing) to 25.7% (upper Eocene Moodys Branch Formation). A statistically significant rank correlation between prey effectiveness and stratigraphic position can be demonstrated for both bivalves ($r = 0.6168$, $p < 0.05$) and gastropods ($r = 0.8531$, $p < 0.01$), indicating a trend of increasing prey effectiveness from the Cretaceous through Oligocene. Likewise, Kelley and Hansen (1993) found a statistically significant rank correlation between stratigraphic position and percent of holes that occurred in multiply bored shells for both bivalves ($r = 0.6964$, $p < 0.01$) and gastropods ($r = 0.6276$, $p < 0.05$). For bivalves, no multiple boreholes occurred in the Cretaceous, but by the Oligocene 15% of all drill holes occurred in multiply bored valves. Similarly, the incidence of multiple boreholes in gastropods increased from 3% to 18% from the Cretaceous to Oligocene.

In contrast, Kelley *et al.* (2001) found no temporal trends in incomplete or multiple naticid drill holes in the Neogene through Pleistocene. Incomplete and multiple drill holes had frequencies of 6% and 3% for bivalves and 4% and 11% for gastropods, and Spearman's rank correlation with stratigraphic position was nonsignificant in all cases. Consistent with these results, Hagadorn and Boyajian (1997) found no significant difference between the Miocene and Pliocene in the occurrence of incomplete and multiple drill holes in *Turritella*. Generalizing from these studies of failed drilling, prey defenses appear to have increased relative to predator efficiency from the Cretaceous through Oligocene; subsequently, predator abilities have stabilized at a higher level relative to prey defenses. This conclusion is supported by the low prey effectiveness documented for the Miocene of Bulgaria (<1% for key species; Kojumdjieva 1974), Plio-Pleistocene of Florida (5% or less for abundant species; Culotta 1988) and the Pleistocene of Fiji (3%; Kohn and Arua 1999). Although Walker (2001) did not report prey effectiveness directly, she found a low frequency (<5%) of individuals with incomplete drill holes in Pliocene deep-water gastropod assemblages of Ecuador.

An additional approach to determining predator and prey effectiveness is through examining the range of prey sizes selected by the predator. Kelley (1989) used the slopes and intercepts of regression lines of borehole size and prey length to demonstrate temporal changes in selectivity of prey size. Predators of a given size attacked smaller individuals of *Marvacrassatella* (formerly *Eucrassatella*), *Astarte*, and *Dallarca* (formerly *Anadara*) at higher stratigraphic levels within the Maryland Miocene compared to lower in the section. These results suggest that the relative effectiveness of the predator declined during the Miocene. In contrast, Roopnarine *et al.* (1997) reported that Pleistocene drillers successfully preyed upon *Chione cancellata* that were twice the size of *Chione erosa* preyed upon in the Pliocene. Likewise, Culotta (1988) observed a temporal decrease in predator-prey size ratios in four taxa (*Anadara, Varicorbula caloosae, Chione cancellata*, and the Naticidae) from the Plio-Pleistocene of Florida. These results suggest an increase in predator capabilities since the Miocene (but see Herbert and Dietl, in press).

7. Summary and Discussion

The debate over arms race hypotheses is part of a broader controversy involving the importance of biological factors in evolution. As several authors have asked (Jackson,

1988; Allmon, 1994; Allmon and Bottjer, 2001), does ecology matter in evolution? Or are physical factors so overwhelming or adaptation so constrained and trends so ephemeral that species interactions can have no significant role in evolution? If ecological interactions can be demonstrated to play a role in evolution, what processes are involved (e.g., coevolution and/or escalation)?

Insight into these questions is provided by studies of drilling gastropod predation in the fossil record. Such studies have examined spatio-temporal variation in intensity of predation and temporal changes in predator behavior and in relative effectiveness of predator and prey. Conclusions emerging from these studies are that biological interactions have been important in the evolution of these predators and prey, but that the situation is more complex than simple arms race scenarios might predict. A similar assessment was provided by Jablonski and Sepkoski (1996) and Jablonski (2000), whose review of the evolutionary paleoecological literature concluded that biotic interactions have affected large-scale ecological and evolutionary patterns, but that the interactions have been diffuse and protracted and the exact mechanisms that caused such patterns are unclear.

Vermeij (1987) supported his hypothesis of escalation with data from the literature on drilling predation that indicated significant increases in drilling between the Cretaceous and Eocene, when modern levels of drilling were reached. Surveys of individual taxa (terebrids, turritellines, glycymerids) also support this conclusion. Likewise, a literature survey by Kowalewski et al. (1998) showed a dramatic increase in drilling frequencies between the Late Cretaceous and Paleogene. More detailed study of entire molluscan assemblages, however, indicated a complex pattern of escalation in naticid drilling frequencies. Kelley and Hansen (1993, 1996a) and Hansen et al. (in review) found a fluctuating pattern of drilling frequency, with significant increases following mass extinctions. We suggested, following Vermeij (1987), that mass extinctions may selectively eliminate highly escalated prey, causing drilling frequencies to rise. This hypothesis has not been supported, however, by subsequent analysis of extinction selectivity (Hansen et al., 1999; Reinhold, 2000; but see Dietl et al., 2002). The cause of this fluctuating pattern is thus unclear but the apparent link to mass extinction suggests an abiotic control on the rate and timing of escalation.

Studies of behavioral stereotypy of drilling gastropods provide further evidence for change in the dynamics of the predator-prey system. Several studies of Cretaceous and Paleogene assemblages reported variability of drill hole position, whereas Neogene assemblages generally show well-developed selectivity of drill hole position. Kelley and Hansen (1993, 1996a), however, found strong selectivity of drill hole position in the Cretaceous and no temporal trend of increasing selectivity through the Paleogene, though stereotypy does appear to have increased since the Miocene (Dietl et al., 2001; Alexander et al., in review). Studies of selectivity of prey size and species appear to be more definitive with regard to temporal patterns. Size selectivity appears to have increased with time, and the predictions of prey choice are generally consistent with optimum foraging models for Neogene and younger assemblages but not for the Eocene and Oligocene.

Along with these apparent changes in behavioral stereotypy, the relative effectiveness of predator and prey varied temporally. Statistically significant temporal trends of increasing percentages of incomplete and multiple borings characterize the Cretaceous through Oligocene faunas examined by Kelley and Hansen (1993). These results suggest that, despite possible increases in predator behavioral stereotypy, the relative offensive capability of predators declined through the Paleogene. In contrast, Neogene and younger assemblages provide evidence for improved predator capabilities relative to the prey (i.e., failed drilling declined after the Oligocene). Comparison of incomplete and multiple drilling of *Turritella* yielded no difference between the Miocene and Pliocene (Hagadorn and Boyajian, 1997). Within the Miocene, however, Kelley (1989) found changes in patterns of prey size selectivity, indicating decreased effectiveness of the predator through time. Roopnarine *et al.* (1997) and Culotta (1988) also reported changes in size selectivity of prey between the Pliocene and Pleistocene, but in the direction of increased predator effectiveness (but see Herbert and Dietl, in press). These results are consistent with the observation that edge-drilling, a more rapid and efficient mode of attack than drilling through the side of a valve (Dietl and Herbert, 2002), has become common only in the latest Neogene to Recent (Vermeij and Roopnarine, 2001).

To determine the processes by which these changes have occurred, several studies have investigated morphological evolution of predator and/or prey. The incidence of well-armored or highly mobile prey has increased through time, presumably in response to predation by drilling gastropods and other predators (Vermeij, 1987). These data do not indicate, however, the process by which predator-resistant taxa have become better represented within faunas (for instance, by selection of individuals or some form of species selection that may or may not have been mediated by biological factors).

Apparent antipredatory increases in shell thickness within several bivalve prey lineages have been documented by Kelley (1989, 1991*b*), Kelley and Hansen (2001), and Kent (as described by Vermeij, 1987). Although some antipredatory adaptation occurred during speciation, directional within-species changes in thickness dominated (Kelley and Hansen, 2001). Such evolution appears to have been in response to predation rather than to changes in the physical environment; Kelley (1991*b*) found that the taxa that experienced the greatest predation exhibited the greatest thickness changes. This evolution was an effective deterrent to predation; thickness increases were accompanied by an increase in incomplete drilling (Kent as cited by Vermeij, 1987) and a decrease in drilling frequency and in size of prey attacked (Kelley, 1989).

Hagadorn and Boyajian (1997) reported size increase between Miocene and Pliocene species of *Turritella*, which they attributed to drilling predation. Concomitant shifts in drill hole site selectivity and prey size selectivity were considered indicative that predator-prey escalation occurred through the Neogene. Their results contrast with those of Allmon *et al.* (1990), who claimed morphological stasis within Cenozoic *Turritella*. Allmon *et al.* (1990) concluded that any morphological antipredatory adaptation that occurred must have done so prior to the Cenozoic. Neither study traced temporal patterns in detail; the differing conclusions of these two studies are difficult to evaluate.

Additional studies have examined the morphologic evolution of naticids (Kitchell, 1986; Kelley, 1992). Traits presumed to affect predatory capabilities of naticids (aperture dimensions and shell globosity) exhibited stasis; in contrast, traits that functioned defensively (shell height and thickness) showed temporal increases. Naticids thus appear to have responded more to their predators (as predicted by escalation) than to their prey (as would be expected in coevolution), at least in terms of morphologic evolution. These results suggest that coevolution was not a significant component in escalation of the system, with the exception of changes in cannibalism, in which confamilial prey are dangerous to the predator (Dietl and Alexander, 2000). Behavioral evolution of naticids could be interpreted as due to coevolution, but the evolutionary response of predator and prey were not balanced. Despite increased selectivity by the predator, several lines of evidence indicate that relative effectiveness of naticid predators declined through the Paleogene and then increased again in the Neogene.

Studies of drilling frequencies, predator selectivity, and relative effectiveness of predator and prey thus indicate significant changes in drilling gastropod predator-prey systems through time. These changes are consistent with the hypothesis of escalation and support the importance of biotic interactions in evolution. As predicted by escalation, the hazard of naticid predation appears to have increased through time. For instance, predator behavioral stereotypy apparently was less developed in the Cretaceous and Paleogene than in younger assemblages. However, a fluctuating pattern of naticid drilling frequencies indicates that escalation proceeded by a complex pathway that was mediated by abiotic events linked to mass extinction. Although some authors have suggested that most escalation was restricted to the Mesozoic (Allmon *et al.*, 1990) or early Cenozoic (Vermeij, 1987, among others), changes in the dynamics of the predator-prey system appear to have occurred throughout the history of the system.

Studies of drilling gastropod predator-prey systems have a high potential for further increasing our understanding of predator-prey evolution. We have a general understanding of the history of drilling predation, but some studies are contradictory and many questions remain. For instance, what factors are responsible for the episodic pattern in naticid drilling frequencies observed by Kelley and Hansen? How do assemblage-level patterns compare to those at the level of lineages? Is the pattern of fluctuations a coincidental by-product of local variability in drilling? In order to assess that question, more data are needed on the distribution of drilling frequencies in the modern marine environment. Are similar patterns present in areas other than the U.S. Coastal Plain? If so, global causes may be implicated. What processes link the surges in drilling frequency to mass extinction? Are similar patterns and dynamics characteristic of other predators, such as muricid gastropods and octopods, which have received much less study? Answers to these questions will help elucidate the contributions of drilling predation to the evolutionary process.

ACKNOWLEDGMENTS: We thank Gregory Dietl, Elizabeth Harper, Michał Kowalewski, and Peter Roopnarine for helpful reviews of this manuscript. This chapter is CMS Contribution Number 272.

References

Adegoke, O. S., and Tevesz, M. J. S., 1974, Gastropod predation patterns in the Eocene of Nigeria, *Lethaia* 7:17-24.
Alexander, R. R., and Dietl, G. P., 2001, Latitudinal trends in naticid predation on *Anadara ovalis* (Bruguière, 1789) and *Divalinga quadrisulcata* (Orbigny, 1842) from New Jersey to the Florida Keys, *Am. Malacol. Bull.* **16**:179-194.
Alexander, R. R., Dietl, G. P., Kelley, P. H., and Hansen, T. A., in review, Increased stereotypy of naticid predation on Atlantic coastal bivalves since the Miocene, *Palaeogeogr. Palaeoclim. Palaeoecol.*
Allmon, W. D., 1994, Taxic evolutionary paleoecology and the ecological context of macroevolutionary change, *Evolutionary Ecol.*, **8**:95-112.
Allmon, W. D., and Bottjer, D. J., 2001, *Evolutionary Paleoecology: The Ecological Context of Macroevolutionary Change*, Columbia University Press, New York.
Allmon, W. D., Nieh, J. C. and Norris, R. D., 1990, Predation in time and space revisited: drilling and peeling in turritelline gastropods, *Palaeontology* **33**:595-611.

Anderson, L. C., 1992, Naticid gastropod predation on corbulid bivalves: Effects of physical factors, morphological features, and statistical artifacts, *Palaios* **7**:602-620.

Anderson, L. C., Geary, D. H., Nehm, R. H., and Allmon, W. D., 1991, A comparative study of naticid gastropod predation on *Varicorbula caloosae* and *Chione cancellata*, Plio-Pleistocene of Florida, USA, *Palaeogeogr. Palaeoclim. Palaeoecol.* **85**:29-46.

Ansell, A. D., 1960, Observations on predation of *Venus striatella* (Da Costa) by *Natica alderi* Forbes, *Proc. Malac. Soc.* **34**:157-164.

Ansell, A. D., 1961, Egg capsules of the dogfish (*Scilliorhinus canicula*, Linn.) bored by *Natica* (Gastropoda, Prosobranchia), *Proc. Malac. Soc.* **34**: 248-249.

Ansell, A. D., and Morton, B., 1987, Alternative predation tactics of a tropical naticid gastropod, *J. Exper. Mar. Biol. Ecol.* **111**:109-120.

Arnold, A. J., D'Escrivan, F., and Parker, W. C., 1985, Predation and avoidance responses in the Foraminifera of the Galapagos hydrothermal mounds, *J. Foram. Res.* **15**:38-42.

Aronson, R. B., 1992, Biology of a scale-independent predator-prey interaction, *Mar. Ecol. Prog. Ser.* **89**:1-13.

Arua, I., and Hoque, M., 1989, Predatory gastropod boreholes in an Eocene molluscan assemblage from Nigeria, *Lethaia* **22**:49-52.

Bambach, R. K., 1999, Energetics in the global marine fauna: A connection between terrestrial diversification and change in the marine biosphere, *Geobios* **32**:131-144.

Bambach, R. K, and Kowalewski, M., 2000, How to count fossils, *Geol. Soc. Am. Abstr. Prog.* **32**:A95.

Baumiller, T. K., 1990, Non-predatory drilling of Mississippian crinoids by platyceratid gastropods, *Palaeontology* **33**: 743-748.

Baumiller, T. K., 1996, Boreholes in the Middle Devonian blastoid *Heteroschisma* and their implications for gastropod drilling, *Palaeogeogr. Palaeoclim. Palaeoecol.* **123**:343-351.

Baumiller, T. K., 2001, Cost-benefit analysis as a guide to the ecology of drilling platyceratids, *PaleoBios* **21** supplement to number 2: 29.

Bengtson, S., and Yue Zhao, 1992. Predatorial borings in late Precambrian mineralized exoskeletons, *Science* **257**:367-369.

Berg, C. J., 1976, Ontogeny of predatory behavior in marine snails (Prosobranchia: Naticidae), *Nautilus* **90**:1-4.

Berg, C. J., 1978. Development and evolution of behavior in molluscs, with emphasis on changes in stereotypy, in: *The Development of Behavior: Comparative and Evolutionary Aspects* (G. M. Burghardt and M. Bekoff, eds.), Garland STPM Press, New York, NY, pp. 3-17.

Berg, C. J., and Nishenko, S., 1975, Stereotypy of predatory boring behavior of Pleistocene naticid gastropods, *Paleobiology* **1**:258-260.

Boggs, C. H., Rice, J. A., Kitchell, J. A., and Kitchell, J. F., 1984, Predation at a snail's pace: what's time to a gastropod?, *Oecologia* **62**:13-17.

Boucot, A. J., 1983, Does evolution take place in an ecological vacuum? II. *J. Paleontol.* **57**:1-30.

Brodie, E. D., III, and Brodie, E. D., Jr., 1999, Predator-prey arms races: Asymmetrical selection on predators and prey may be reduced when prey are dangerous, *Bioscience* **49**: 557-568.

Bromley, R. G., 1981, Concepts in ichnotaxonomy illustrated by small round holes in shells, *Acta Geol. Hisp.* **16**:55-64.

Bromley, R. G., 1993, Predation habits of octopus past and present and a new ichnospecies, *Oichnus ovalis*, *Bull. geol. Soc. Denmark* **40**:167-173.

Brown, K. M., and Alexander Jr., J. E., 1994, Group foraging in a marine gastropod predator: benefits and costs to individuals, *Mar. Ecol. Prog. Ser.* **112**:97-105.

Calvet, C., 1992, Borehole site-selection in *Naticarius hebraeus* (Chemnitz in Karsten, 1769) (Naticidae: Gastropoda)?, *ORSIS* **7**:57-64.

Carriker, M. R., 1981, Shell penetration and feeding by naticacean and muricacean predatory gastropods: a synthesis, *Malacologia* **20**:403-422.

Carriker, M. R., and Gruber, G. L., 1999, Uniqueness of the gastropod accessory boring organ (ABO): comparative biology, an update, *J. Shellfish Res.* **18**:579-595.

Carriker, M. R., and Yochelson, E. L., 1968, Recent gastropod boreholes and Ordovician cylindrical borings, *Contr. Paleontology, Geol. Survey Prof. Paper* **593B**:B1-B26.

Colbath, S. L., 1985, Gastropod predation and depositional environments of two molluscan communities from the Miocene Astoria Formation at Beverly Beach State Park, Oregon, *J. Paleontol.* **59**:849-869.

Conway Morris, S., and Bengtson, S., 1994, Cambrian predators: possible evidence from boreholes, *J. Paleontol.* **68**:1-23.

Cortez, T., Castro, B. G., and Guerra, A., 1998, Drilling behavior of *Octopus mimus* Gould, *J. Exper. Mar. Biol. Ecol.* **224**:193-203.

Cox, D. L., Walker, P., and Koob, T. J., 1999, Predation on eggs of the thorny skate, *Trans. Am. Fisheries Soc.* **128**: 380-384.
Culotta, E., 1988, Predators and Available Prey: Naticid Predation during a Neogene Molluscan Extinction Event. MS thesis, University of Michigan.
DeAngelis, D. L., Kitchell, J. A., and Post, W. M., 1984, A model of naticid gastropod predator-prey coevolution, *Lect. Notes Biomath.* **54**:120-136.
DeAngelis, D. L., Kitchell, J. A., and Post, W. M., 1985, The influence of naticid predation on evolutionary strategies of bivalve prey: conclusions from a model, *Am. Naturalist* **126**:817-842.
De Cauwer, G., 1985, Gastropod predation on corbulid bivalves: paleoecology or taphonomy? *Annals Soc. r. zool. Belg.* **115**:183-196.
Dietl, G. P., 2000, Successful and unsuccessful predation of the gastropod *Nucella lapillus* (Muricidae) on the mussel *Mytilus edulis* from Maine, *Veliger* **43**:319-329.
Dietl, G. P., in press, Traces of naticid predation on the gryphaeid oyster *Pycnodonte dissimilaris*: epifaunal drilling of prey in the Paleocene, *Hist. Biol.*
Dietl, G. P., and Alexander, R. R., 1995, Borehole site and prey size stereotypy in naticid predation on *Euspira (Lunatia) heros* Say and *Neverita (Polinices) duplicata* Say from the southern New Jersey coast, *J. Shellfish Res.* **14**:307-314.
Dietl, G. P., and Alexander, R. R., 1997, Predator-prey interactions between the naticids *Euspira heros* Say and *Neverita duplicata* Say and the Atlantic surfclam *Spisula solidissima* Dillwyn from Long Island to Delaware, *J. Shellfish Res.* **16**:413-422.
Dietl, G. P., and Alexander, R. R., 2000, Post-Miocene shift in stereotypic naticid predation on confamilial prey from the mid-Atlantic shelf: coevolution with dangerous prey, *Palaios* **15**:414-429.
Dietl, G. P., and Herbert, G. S., 2002, Experiments with the predatory muricid *Chicoreus dilectus* and its bivalve prey *Chione elevata*: Does edge drilling decrease prey handling time? *Am. Malacol. Soc. Abstr.*, p. 44.
Dietl, G. P., and Kelley, P. H., 2002, The fossil record of predator-prey arms races: Coevolution and escalation hypotheses, in *The Fossil Record of Predation* (M. Kowalewski and P.H. Kelley, eds.), The Paleontological Society Papers 8.
Dietl, G. P., Alexander, R., Kelley, P. H., and Hansen, T. A., 2001, Stereotypy of naticid predation on bivalves since the Cretaceous: trends, controlling factors, and implications for escalation, *PaleoBios* **21** supplement to number 2:78.
Dietl, G. P., Kelley, P. H., Barrick, R., and Showers, W., 2002, Escalation and extinction selectivity: morphology versus isotopic reconstruction of bivalve metabolism, *Evolution* **56**(2):284-291.
Dudley, E. C., and Vermeij, G. J., 1978, Predation in time and space: drilling in the gastropod *Turritella*, *Paleobiology* **4**:436-441.
Fischer-Piette, E., 1935, Histoire d'une moulière. Observations sure une phase déséquilibre faunique, *Bull. Biol. France et Belgique* **69**:153-177.
Fursich, F. T., and Jablonski, D., 1984, Late Triassic naticid drillholes: carnivorous gastropods gain a major adaptation but fail to radiate, *Science* **224**:78-80.
Futuyma, D. J., and Slatkin, M., 1983. Introduction, in: *Coevolution* (D.J. Futuyma and M. Slatkin, eds.), Sinauer, Sunderland, Mass., pp. 1-13.
Gordillo, S., 1994, Borings on subfossil and modern specimens of two bivalves from Beagle Channel, Tierra del Fuego, *Ameghiniana* **31**: 177-185.
Gordillo, S., and Amuchastegui, S., 1998, Predation strategies of *Trophon geversianus* (Pallas) (Muricoidea: Trophonidae), *Malacologia* **39**:83-91.
Gosselin, L. A., and Chia, F. S., 1996, Prey selection by inexperienced predators: do juvenile snails maximize net energy gains on their first attack?, *J. Exper. Mar. Biol. Ecol.* **199**:45-58.
Gould, S. J., 1985, The paradox of the first tier: an agenda for paleobiology, *Paleobiology* **11**:2-12.
Gould, S. J., 1990, Speciation and sorting as the source of evolutionary trends, or 'things are seldom what they seem', in: *Evolutionary Trends* (K. McNamara, ed.), Belhaven Press, London, pp. 3-27.
Grey, M., 2001, Predator-Prey Relationships of Naticid Gastropods and their Bivalve Prey. Unpublished M.S. Thesis, University of Guelph.
Guensberg, T. E., and Sprinkle, J., 1992, Rise of echinoderms in the Paleozoic evolutionary fauna: significance of paleoenvironmental controls, *Geology* **20**:407-410.
Guerrero, S., and Reyment, R. A., 1988, Predation and feeding in the naticid gastropod *Naticarius intricatoides* (Hidalgo), *Palaeogeogr. Palaeoclim. Palaeoecol.* **68**:49-52.
Hagadorn, J. W., and Bottjer, D. J.,1993, Paleobiological data collection: Outcrop-scale variability in turritelline gastropod populations from the U.S. Gulf Coast, *Geol. Soc. Am. Abstr. Prog.* **25**:A55-A56.
Hagadorn, J. W., and Boyajian, G. E., 1997, Subtle changes in mature predator-prey systems: an example from Neogene *Turritella* (Gastropoda), *Palaios* **12**:372-379.

Hagstrom, K. M., 1996, Effects of Compaction and Wave-Induced Forces on the Preservation and Macroevolutionary Perception on Naticid Predator-Prey Interactions. Unpublished M.S. Thesis, Indiana University.

Hansen, T. A., and Kelley, P. H., 1995, Spatial variation of naticid gastropod predation in the Eocene of North America, *Palaios* **10**:268-278.

Hansen, T. A., Kelley, P. H., Melland, V. D., and Graham, S. E., 1999, Effect of climate-related mass extinctions on escalation in molluscs, *Geology* **27**:1139-1142.

Hansen, T. A., Kelley, P. H., and Hall, J. C., 2001, The Moonsnail Project: a collaborative research partnership between middle schools and universities, *Geol. Soc. Am. Abstr. Prog.* **33**(6):A33.

Hansen, T. A., Kelley, P. H., Graham, S. E., Huntoon, A. G., and Jones, M. A., in review, Episodic escalation in the naticid gastropod predator-prey system, Cretaceous to Recent of the U.S. coastal plain, *Palaios*.

Harper, E. M., 2002, Plio-Pleistocene octopod drilling behavior in scallops from Florida, *Palaios* **17**:292-296.

Harper, E. M., Forsythe, G. T. W., and Palmer, T., 1998, Taphonomy and the Mesozoic Marine Revolution: Preservation state masks the importance of boring predators, *Palaios* **13**:352-360.

Harper, E. M., and Skelton, P. W., 1993, The Mesozoic Marine Revolution and epifaunal bivalves, *Scripta Geol., Spec. Issue* **2**:127-153.

Herbert, G. S., and Dietl, G. D., in press, Tests of the escalation hypothesis: the role of multiple predators, *Geol. Soc. Am. Abstr. Prog.* **34**.

Hoffman, A., and Martinell, J., 1984, Prey selection by naticid gastropods in the Pliocene of Emporda (Northeast Spain), *Neues Jahrb. Geol. Paläontol., Monatschefte* **1984**:393-399.

Hoffman, A., Pisera, A., and Ryszkiewicz, M., 1974, Predation by muricid and naticid gastropods on the Lower Tortonian molluscs from the Korytnica clays, *Acta Geol. Polon.* **24**:249-260.

Hoffmeister, A. P., and Kowalewski, M., 2001, Spatial and environmental variation in the fossil record of drilling predation: A case study from the Miocene of Central Europe, *Palaios* **16**:566-579.

Hoffmeister, A. P., Kowalewski, M., Bambach, R. K., and Baumiller, T. K., 2001, Evidence for predatory drilling in Late Paleozoic brachiopods and bivalve mollusks from west Texas, *PaleoBios* **21** supplement to number **2**:66-67.

Hughes, R. N., and Dunkin, S. D., 1984, Behavioral components of prey selection by dogwhelks *Nucella lapillus* feeding on mussels *Mytilus edulis* in the laboratory, *J. Exper. Mar. Biol. Ecol.*, **77**:45-68.

Jablonski, D., 2000, Micro- and macroevolution: scale and hierarchy in evolutionary biology and paleobiology, in: *Deep Time: Paleobiology's Perspective* (D. H. Erwin and S. L. Wing, eds.), The Paleontological Society, Lawrence, KS, pp. 15-52.

Jablonski, D., and Sepkoski, J. J.,1996, Paleobiology, community ecology, and scales of ecological pattern, *Ecology* **77**:1367-1378.

Jackson, J. B. C., 1988, Review: Evolution and Escalation: An Ecological History of Life, by Geerat J. Vermeij, *Paleobiology* **14**:307-312.

Jones, M. A., 1999, Cost-Benefit analysis of naticid gastropod predation across the Plio-Pleistocene boundary from North Carolina and Virginia. Unpublished M.S. Thesis, University of North Carolina at Wilmington.

Kabat, A. R., 1990, Predatory ecology of naticid gastropods with a review of shell boring predation, *Malacologia* **32**:155-193.

Kabat, A. R., and Kohn, A. J., 1986, Predation on early Pleistocene naticid gastropods in Fiji, *Palaeogeogr. Palaeoclim. Palaeoecol.* **53**:255-269.

Kaplan, P., and Baumiller, T. K., 2000, Taphonomic inferences on boring habit in the Richmondian *Onniella meeki* epibole, *Palaios* **15**:499-510.

Kelley, P. H., 1988, Predation by Miocene gastropods of the Chesapeake Group: stereotyped and predictable, *Palaios* **3**:436-448.

Kelley, P. H., 1989, Evolutionary trends within bivalve prey of Chesapeake Group naticid gastropods, *Hist. Biol.* **2**:139-156.

Kelley, P. H., 1991a, Cannibalism by Chesapeake Group naticid gastropods: a predictable result of stereotyped predation, *J. Paleontol.* **65**:75-79.

Kelley, P. H., 1991b, The effect of predation intensity on rate of evolution of five Miocene bivalves, *Hist. Biol.* **5**:65-78.

Kelley, P. H., 1992, Evolutionary patterns of naticid gastropods of the Chesapeake Group: an example of coevolution?, *J. Paleontol.* **66**:794-800.

Kelley, P. H., and Hansen, T. A., 1993, Evolution of the naticid gastropod predator-prey system: an evaluation of the hypothesis of escalation, *Palaios* **8**:358-375.

Kelley, P. H., and Hansen, T. A., 1996a, Recovery of the naticid gastropod predator-prey system from the Cretaceous-Tertiary and Eocene-Oligocene extinctions, *Geol. Soc. Spec. Publ.* **102**:373-386.

Kelley, P. H., and Hansen, T. A., 1996b, Naticid gastropod prey selectivity through time and the hypothesis of escalation, *Palaios* **1**:437-445.

Kelley, P. H., and Hansen, T. A., 2001, The role of ecological interactions in the evolution of naticid gastropods and their molluscan prey, in: *Evolutionary Paleoecology: The Ecological Context of Macroevolutionary Change* (W. D. Allmon and D. J. Bottjer, eds.), Columbia University Press, New York, pp. 149-170.
Kelley, P. H., Hansen, T. A., Graham, S. E., and Huntoon, A. G., 2001, Temporal patterns in the efficiency of naticid gastropod predators during the Cretaceous and Cenozoic of the United States Coastal Plain, *Palaeogeogr. Palaeoclim. Palaeoecol.* **166**(1/2):165-176.
Kelley, P. H., Thomann, C., Hansen, T. A., Aronson, R. and Blake, D., 1997, A world apart but not so different: predation by naticid gastropods in Antarctica and the U.S. Gulf Coast during the Eocene, *Geol. Soc. Am. Abstr. Progr.* **29**(6):A107.
Key, H. M., Kelley, P. H., Dietl, G. P., and Hansen, T. A., 2001, Muricid vs. naticid gastropod predation on Cretaceous to Recent Coastal Plain mollusc assemblages: drilling cycles of different periodicities, *PaleoBios* **21** supplement to number 2:78.
Kitchell, J. A., 1982, Coevolution in a predator-prey system, *Third North American Paleontological Convention, Proc.* **2**:301-305.
Kitchell, J. A., 1986, The evolution of predator-prey behavior: naticid gastropods and their molluscan prey, in: *Evolution of Animal Behavior: Paleontological and Field Approaches* (M. Nitecki and J.A. Kitchell, eds.), Oxford University Press, Oxford, pp. 88-110.
Kitchell, J. A., 1990, The reciprocal interaction of organism and effective environment: Learning more about "and," in: *Causes of Evolution: A Paleontological Perspective* (R. M. Ross and W. D. Allmon, eds.), The University of Chicago Press, Chicago, pp. 151-169.
Kitchell, J. A., Boggs, C. H., Kitchell, J. F., and Rice, J. A., 1981, Prey selection by naticid gastropods: experimental tests and application to the fossil record, *Paleobiology* **7**:533-552.
Kitchell, J. A., Boggs, C. H., Rice, J. A., Kitchell, J. F., Hoffman, A. and Martinell, J., 1986, Anomalies in naticid predatory behavior: a critique and experimental observations, *Malacologia* **27**:291-298.
Knoll, A. H., and Bambach, R. K., 2000, Directionality in the history of life: diffusion from the left wall or repeated scaling of the right?, in: *Deep Time: Paleobiology's Perspective* (D. H. Erwin and S. L. Wing, eds.), The Paleontological Society, Lawrence, KS, pp. 1-14.
Kohn, A. J., and Arua, I., 1999, An early Pleistocene molluscan assemblage from Fiji: gastropod faunal composition, paleoecology and biogeography, *Palaeogeogr. Palaeoclim. Palaeoecol.* **146**:99-145.
Kojumdjieva, E., 1974, Les gasteropodes perceurs et leurs victimes du Miocene de Bulgarie du Nord-Ouest, *Bulgarian Acad. Sci., Bull. Geol. Inst., Ser. Paleontol.* **23**:5-24.
Kowalewski, M., 1990, A hermeneutic analysis of the shell-drilling gastropod predation on mollusks in the Korytnica Clays (Middle Miocene; Holy Cross Mountains, Central Poland), *Acta Geol. Polon.* **40**:183-213.
Kowalewski, M., 1993, Morphometric analysis of predatory drillholes, *Palaeogeogr. Palaeoclim. Palaeoecol.* **102**:69-88.
Kowalewski, M., in press, The fossil record of predation: An overview of analytical methods, in *The Fossil Record of Predation* (M. Kowalewski and P. H. Kelley, eds.), The Paleontological Society Papers 8.
Kowalewski, M., Dulai, A., and Fursich, F. T., 1998, A fossil record full of holes: the Phanerozoic history of drilling predation, *Geology* **26**: 1091-1094.
Kowalewski, M., Simões, M. G., Torello, F. F., Mello, L. H. C., and Ghilardi, R. P., 2000, Drill holes in shells of Permian benthic invertebrates, *J. Paleontol.*, **74**:532-543.
Leighton, L. R., 2001*a*, New example of Devonian predatory boreholes and the influence of brachiopod spines on predator success, *Palaeogeogr. Palaeoclim. Palaeoecol.* **165**:53-69.
Leighton, L. R., 2001*b*, Evaluating the accuracy of drilling frequency as an estimate of prey preference and predation intensity, *PaleoBios* **21** supplement to number 2:83.
Lescinsky, H. L., and Benninger, L., 1994, Pseudo-borings and predator traces: artifacts of pressure-dissolution in fossiliferous shales, *Palaios* **9**;599-604.
Lever, J., Kessler, A., Van Overbeeke, P., and Thijssen, R., 1961, Quantitative beach research. II. The 'hole effect': a second mode of sorting lamellibranch valves on sandy beaches: *Netherlands J. Sea Res.* **1**:339-358.
Lidgard, S., and Jackson, J. B. C., 1989, Growth in encrusting cheilostome bryozoans: I. Evolutionary trends, *Paleobiology* **15**:255-282.
Livan, M., 1937, Über Bohr-locher an rezenten und fossilen Invertebraten, *Senckenbergiana* **19**:138-150.
Martinell, J., and Marquina, M. J., 1978, Senales de depredacion en los Gastropoda procedentes de un yacimiento pliocenico de Molins de Rei (Barcelona).--Implicaciones paleoecologicas, *Acta Geol, Hisp.* **13**:125-128.
McKinney, F. K., 1995, One hundred million years of competitive interactions between bryozoan clades: asymmetrical but not escalating, *Biol. J. Linnean Soc.* **56**:465-481.

Morton, B., 1985, Prey preference, capture and ration in *Hemifusus tuba* Prosobranchia Melongenidae, *J. Exper. Mar. Biol. Ecol.* **94**:191-210.
Morton, B., and Chan, K., 1997, The first report of shell boring predation by a member of the Nassariidae (Gastropoda), *J. Molluscan Stud.* **63**:476-478.
Nebelsick, J. H., and Kowalewski, M., 1999, Drilling predation on Recent clypeasteroid echinoids from the Red Sea, *Palaios* **14**:127-144.
Newton, C. R., 1983, Triassic origin of shell-boring gastropods, *Geol. Soc. Am. Abstr. Progr.* **15**:652-653.
Pan Hua-zhang, 1991, Lower Turonian gastropod ecology and biotic interaction in *Helicaulax* community from western Tarim Basin, southern Xinjiang, China, in: *Palaeoecology of China,* vol. 1 (Jin Yu-gan, Wang Jun-geng, and Xu Shan-hong, eds.), Nanjing University Press, Nanjing, pp. 266-280.
Pechenik, J. A., and Lewis, S., 2000, Avoidance of drilled gastropod shells by the hermit crab *Pagurus longicarpus* at Nahant, Massachusetts, *J. Exper. Mar. Biol. Ecol.* **253**:17-32.
Pek, I., and Mikulas, R., 1996, The ichnogenus *Oichnus* Bromley, 1981 -- predation traces in gastropod shells from the Badenian in the vicinity of Ceska Trebova (Czech Republic), *Bull. Czech Geol. Survey* **71**(2):107-120.
Peterson, C. H., and Black, R., 1995, Drilling by buccinid gastropods of the genus *Cominella* in Australia, *Veliger* **38**:37-42.
Ponder, W. F., and Taylor, J. D., 1992, Predatory shell drilling by two species of *Austroginella* (Gastropoda: Marginellidae), *J. Zool. London* **228**:317-328.
Reinhold, M. E., 2000, The effect of anti-predatory morphology on mollusc survivorship across the Cretaceous/Tertiary boundary. Unpublished M.S. Thesis, University of North Carolina at Wilmington.
Reyment, R. A., 1963, Bohrlocher bei Ostrakoden, *Paläontol. Zeitschrift* **37**:283-291.
Reyment, R., 1966, Preliminary observations on gastropod predation in the western Niger Delta, *Palaeogeogr. Palaeoclim. Palaeoecol.* **2**:81-102.
Reyment, R., 1967, Paleoethology and fossil drilling gastropods, *Trans. Kansas Acad. Sci.* **70**:33-50.
Reyment, R. 1999, Drilling gastropods, in: *Functional Morphology of the Invertebrate Skeleton* (E. Savazzi, ed.), John Wiley & Sons, Ltd, pp. 197-204.
Richards, R. P., and Shabica, C. W., 1969, Cylindrical living burrows in Ordovician dalmanellid brachiopod beds, *J. Paleontol.* **43**:838-841.
Rohr, D. M., 1991, Borings in the shell of an Ordovician (Whiterockian) gastropod, *J. Paleontol.* **65**:687-688.
Roopnarine, P. D., and Beussink, A., 1999, Extinction and naticid predation of the bivalve *Chione* von Muehlfeld in the late Neogene of Florida, *Palaeontol. Electron.* **2**, 1.
Roopnarine, P. D., and Willard, S., 2001, Bivalve prey morphology as a guide to drilling stereotypy of naticid and muricid gastropods: venerids, naticids and muricids in Tropical America. *PaleoBios* **21** supplement to number 2:109-110.
Roopnarine, P. D., Beussink, A., and Schuff, L., 1997, Late Neogene changes of morphology and naticid drilling in tropical American chionine bivalves, *Geol. Soc. Am. Abstr. Progr.* **29**(6):A101.
Roy, K., 1996, The roles of mass extinction and biotic interaction in large-scale replacements: a reexamination using the fossil record of stromboidean gastropods, *Paleobiology* **22**:436-453.
Roy, K., Miller, D. K., and LaBarbera, M., 1994, Taphonomic bias in analyses of drilling predation: effects of gastropod drill holes on bivalve shell strength, *Palaios* **9**:413-421.
Sander, F., and Lalli, C. M., 1982, A comparative study of the mollusk communities on the shelf-slope margin of Barbados, West Indies, *Veliger* **24**:309-318.
Savazzi, E., and Reyment, R. A., 1989, Subaerial hunting behaviour in *Natica gualteriana* (naticid gastropod), *Palaeogeogr. Palaeoclim. Palaeoecol.* **74**:355-364.
Sheehan, P. M., and Lesperance, P. J., 1978, Effect of predation on the population dynamics of a Devonian brachiopod, *J. Paleontol.* **52**:812-817.
Sickler, R. N., 1995, Prey selectivity of naticid gastropods from Tertiary sediments of the United States Coastal Plain. Unpublished M.A. Thesis, University of North Dakota.
Sickler, R. N., Kelley, P. H. and Hansen, T. A., 1996, Prey selectivity of naticid gastropods from Tertiary sediments of the United States Coastal Plain, *Geol. Soc. Am. Abstr. Progr.* **28**(6):64.
Signor, P. W., and Brett, C. E., 1984, The mid-Paleozoic precursor to the Mesozoic marine revolution, *Paleobiology* **10**:229-245.
Smith, S. A., Thayer, C. W., and Brett, C. E., 1985, Predation in the Paleozoic: gastropod-like drillholes in Devonian brachiopods, *Science* **230**:1033-1035.
Sohl, N. F., 1969, The fossil record of shell boring by snails, *Am. Zool.* **9**:725-734.
Stanton, R. J., Jr., and Nelson, P. C., 1980, Reconstruction of the trophic web in paleontology: community structure in the Stone City Formation (middle Eocene, Texas), *J. Paleontol.* **54**:118-135.
Steneck, R. S., 1983, Escalating herbivory and resulting adaptive trends in calcareous algal crusts, *Paleobiology* **9**:44-61.

Stump, T. E., 1975, Pleistocene molluscan paleoecology and community structure of the Puerto Libertad region, Sonora, Mexico, *Palaeogeogr. Palaeoclimatol. Palaeoecol.* **17**:177-226.

Taylor, J. D., 1970, Feeding habits of predatory gastropods in a Tertiary (Eocene) molluscan assemblage from the Paris basin, *Palaeontology* **13**:254-260.

Taylor, J. D., Cleevely, R. J., and Taylor, C. N., 1980, Food specialization and the evolution of predatory prosobranch gastropods, *Palaeontology* **23**:375-409.

Taylor, J. D., Cleevely, R. J., and Morris, N. J., 1983, Predatory gastropods and their activities in the Blackdown Greensand (Albian) of England, *Palaeontology* **26**:521-553.

Taylor, J. D., and Morton, B.,1996, The diets of predatory gastropods in the Cape D'Aguilar Marine Reserve, Hong Kong, *Asian Mar. Biol.* **13**:141-165.

Thomas, R. D. K., 1976, Gastropod predation on sympatric Neogene species of *Glycymeris* (Bivalvia) from the Eastern United States, *J. Paleontol.* **50**:488-499.

Tull, D. S., and Bohning-Gaese, K., 1993, Patterns of drilling predation on gastropods of the family Turritellidae in the Gulf of California, *Paleobiology* **19**:476-486.

Urrutia, G. X., and Navarro, J. M., 2001, Patterns of shell penetration by *Chorus giganteus* juveniles (Gastropoda: Muricidae) on the mussel *Semimytilus algosus*, *J. Exper. Mar. Biol. Ecol.* **258**:141-153.

Vermeij, G. J., 1983, Intimate associations and coevolution in the sea, in: *Coevolution* (D.J. Futuyma and M. Slatkin, eds.), Sinauer Associates, Inc., Sunderland, MA, pp. 311-327.

Vermeij, G. J., 1987, *Evolution and Escalation: An Ecological History of Life*, Princeton University Press, Princeton, NJ, 527 pp.

Vermeij, G. J., 1994, The evolutionary interaction among species: selection, escalation, and coevolution, *Ann. Rev. Ecol. Syst.* **25**:219-236.

Vermeij, G. J., and Carlson, S. J., 2000, The muricid gastropod subfamily Rapaninae: phylogeny and ecological history, *Paleobiology* **26**:19-46.

Vermeij, G. J., and Dudley, E. C., 1982, Shell repair and drilling in some gastropods from the Ripley Formation (Upper Cretaceous) of the southeastern U.S.A., *Cretaceous Res.* **3**:397-403.

Vermeij, G. J., and Roopnarine, P. D., 2001, Edge-drilling: History and distribution of a novel method of predation, *PaleoBios* **21** supplement to number 2:130.

Vermeij, G. J., Zipser, E., and Dudley, E. C., 1980, Predation in time and space: drilling in terebrid gastropods, *Paleobiology* **6**:352-364.

Vermeij, G. J., Dudley, E. C., and Zipser, E., 1989, Successful and unsuccessful drilling predation in Recent pelecypods, *Veliger* **32**:266-273.

Vignali, R., and Galleni, L., 1986, Naticid predation on soft bottom bivalves: a study on a beach shell assemblage, *Oebalia* **13**:157-178.

Walker, S. E., 2001, Paleoecology of gastropods preserved in turbiditic slope deposits from the Upper Pliocene of Ecuador, *Palaeogeogr. Palaeoclimatol. Palaeoecol.* **166**:141-163.

Yochelson, E. L., Dockery, D., and Wolf, H., 1983, Predation of sub-Holocene scaphopod mollusks form Southern Louisiana, *U.S. Geol. Survey, Prof. Paper* 1282, 13 pp.

Ziegelmeier, E., 1954, Beobachtungen uber den Nahrungserwerb bei der Naticide *Lunatia nitida* Donovan (Gastropoda Prosobranchia), *Helgolander Wissenschaftliche Meeresforschung* **5**:1-33.

Zlotnik, M., 2001, Size-related changes in predatory behaviour on naticid gastropods from the middle Miocene Korytnica Clays, Poland, *Acta Palaeontol. Polon.* **46**:87-97.

Zuschin, M., and Stanton, R. J., Jr., 2001, Experimental measurement of shell strength and its taphonomic interpretation, *Palaios* **16**:161-170.

Chapter 6

The Fossil Record of Shell-Breaking Predation on Marine Bivalves and Gastropods

RICHARD R. ALEXANDER and GREGORY P. DIETL

1. Introduction 141
2. Durophages of Bivalves and Gastropods 142
3. Trends in Antipredatory Morphology in Space and Time 145
4. Predatory and Non-Predatory Sublethal Shell Breakage 155
5. Calculation of Repair Frequencies and Prey Effectiveness 160
6. Prey Species-, Size-, and Site-Selectivity by Durophages 164
7. Repair Frequencies by Time, Latitude, and Habitat 166
8. Concluding Remarks 170
References 170

1. Introduction

Any treatment of durophagous (shell-breaking) predation on bivalves and gastropods through geologic time must address the molluscivore's signature preserved in the victim's skeleton. Pre-ingestive breakage or crushing is only one of four methods of molluscivory (Vermeij, 1987; Harper and Skelton, 1993), the others being whole-organism ingestion, insertion and extraction, and boring. Other authors in this volume treat the last behavior, whereas whole-organism ingestion, and insertion and extraction, however common, are unlikely to leave preservable evidence. Bivalve and gastropod ecologists and paleoecologists reconstruct predator-prey relationships based primarily on two, although not equally useful, categories of pre-ingestive breakage, namely lethal and sublethal (repaired) damage. Peeling crabs may leave incriminating serrated, helical

RICHARD R. ALEXANDER • Department of Geological and Marine Sciences, Rider University, Lawrenceville, New Jersey, 08648-3099. GREGORY P. DIETL • Department of Zoology, North Carolina State University, Raleigh, North Carolina, 27695-7617.

Predator-Prey Interactions in the Fossil Record, edited by Patricia H. Kelley, Michał Kowalewski, and Thor A. Hansen. Kluwer Academic/Plenum Publishers, New York, 2003.

fractures in whorls of high-spired gastropods (Bishop, 1975), but unfortunately most lethal fractures are far less diagnostic of the causal agent and often indistinguishable from abiotically induced, taphonomic agents of shell degradation.

Published research has focused on gastropod and bivalve shell repairs in the fossil record (Vermeij, 1987), although reliance on sublethal damage to reconstruct predator-prey interactions is not without complications. Sublethal, repaired shell damage is commonly inversely related to lethal, unrepaired shell damage (Vermeij, 1982a), a relationship that may lead to ambiguous reconstruction of predation intensity (Kowalewski et al., 1997; Leighton, in press). As strength of the predator's crushing elements increase, or, concurrently, strength of the prey shell decreases, lethal damage increases, whereas sublethal damage decreases. Furthermore, a clam victim may sustain sublethal tissue damage, such as siphon nipping (Peterson and Quammen, 1982) by fish or crabs that may not break the shell. No evidence, other than slowed growth rate (Coen and Heck, 1991) records the predatory attempt. Furthermore, time-averaging (Cadée et al., 1997) and high within-habitat variability (Schmidt, 1989) may complicate inferences on prey evolution derived from the record of shell repairs, particularly if sample sizes are small.

Although predation intensity cannot be inferred directly from bivalve (Dietl et al., 2000; Alexander and Dietl, 2001) or gastropod (Vermeij et al., 1981; Schindel et al., 1982) frequencies of breakage-induced shell repair, injury to the shell is a necessary condition for selection for defensive adaptations (Vermeij, 1982b). Utility of repair frequencies to tests of the escalation hypothesis in marine molluscs is central to this review; escalation involves adaptation to enemies, stressing the importance of predation as a driving force in evolution. Vermeij et al. (1981, p. 1024) postulated that:

> selection favoring the evolution of breakage-resistant shells can occur only if individuals in a population reproduce after they have suffered nonlethal shell-breaking attacks. If all breakage were lethal, there would be no selection between weak and strong shell variants and no shells would show the scars that record nonlethal injury. High frequencies of sublethal damage imply that the shell, together with other defenses, is effective in protecting the gastropod [or bivalve] against locally prevailing, shell-breaking agents. The higher the frequency of sublethal shell damage, the greater is the likelihood that selection will maintain or enhance shell armor.

2. Durophages of Bivalves and Gastropods

Vermeij (1987) traced the evolutionary history of predators capable of shell crushing as one line of evidence to support his hypothesis that breakage became a more important component of natural selection for shallow-water, marine, shelled invertebrates through the Phanerozoic. Vermeij (1987) reasoned that if individuals are exposed to increasingly greater risks from their enemies over geologic time, a temporal trend towards greater reliance on methods of rapid prey subjugation would be expected. The developing emphasis on breakage, a rapid predation method, through the Phanerozoic, agrees with the prediction that high-energy modes of predation are favored as risks to the predators themselves increase (Vermeij, 1987).

The dominant predators responsible for shell damage in marine gastropods and bivalves over geologic time include a varied array of clawed and jawed higher taxa. Marine Phanerozoic life has been apportioned among three "evolutionary faunas," namely the Cambrian, Paleozoic, and Modern (Sepkoski, 1984). Each "fauna" had its shell breakers. Potential shell breakers that dominated the Cambrian fauna were the anomalocarids and *Sidneyia* (Conway Morris and Whittington, 1979; Nedin, 1999). *Sidneyia* is preserved with brachiopod fragments in its stomach contents (Conway Morris and Whittington, 1979) and could have been molluscivorous. The Mid-Paleozoic revolution in durophagy (Signor and Brett, 1984) represented the first large-scale diversification of shell-breaking vertebrate and invertebrate predators. Phyllocarids (Rolfe, 1969), belotelsonid (Schram, 1981), and pygocephalomorph eumalacostracans are among invertebrate shell crushers. Signor and Brett (1984) excluded Silurian eurypterids such as *Pterygotus* from their tally of durophages because they lacked calcified chelicerae. Nevertheless, Vermeij (1987) conceded that pterygotid eurypterids, with their well-worn, pointed teeth on their massive chelicerae, could have been durophagous. Signor and Brett (1984) also excluded nautiloids as Mid-Paleozoic shell breakers because they possessed a non-calcified jaw, or rhyncholite, that may have been capable of crushing calcareous exoskeletons. However, several investigators attributed scars in strophomenid (Alexander, 1986) or productid (Sarycheva, 1949; Elliott and Brew, 1988) brachiopods to nautiloids, despite the probability that the beaks were chitinous. Horny (1997) likewise attributed repair in the Ordovician mollusc *Sinuitopsis neglecta* to nautiloids. Ammonoids are also possible durophages on Paleozoic gastropods and bivalves, although Signor and Brett (1984) were uncertain of their diet and excluded them as durophages. Bond and Saunders (1989) attributed sublethal damage in Late Mississippian ammonoids to attacks by other ammonoids. Placoderm and ptyctodont fishes and chondrichthyans comprise the bulk of Mid-Paleozoic durophagous taxa (Signor and Brett, 1984; Brett and Walker, in press), although some Late Mississippian sharks are disputed shell crushers. Cladodontids have been implicated as shell piercers (Brunton, 1966; Mapes and Hansen, 1984; Mapes *et al.*, 1989), but Signor and Brett (1984) doubted that these sharks were shell crushers. Equal spacing and geometry of punctures occasionally preserved in Carboniferous brachiopod valves suggest that cladodontids could have broken shelled prey. Chimaeras also have been implicated as shell crushers of Permian bivalves (Brett and Walker, in press). In addition, chondrostean fishes (Doryopteridae) may have been durophagous (Brett and Walker, in press).

The modern (Mesozoic to Recent) molluscan fauna was preyed upon by an even larger diversity of durophages than the Paleozoic fauna (Vermeij, 1977); this interval represents the second major episode of diversification of shell-breaking predators. Shell crushers arguably capable of sublethal, repairable damage on marine bivalves and gastropods, during the Mesozoic, were reviewed by Vermeij (1977, 1987). Triassic shell crushers included placodont reptiles (von Huene, 1956; Walker and Brett, in press). Pycnodontid reef fishes evolved shell-crushing teeth in the Jurassic (Walker and Brett, in press). Dominant shell crushers of the Jurassic and Cretaceous were ptychodid sharks and batoid rays and skates (Walker and Brett, in press). Kauffman (1972) implicated *Ptychodus* predation on Cretaceous *Inoceramus* bivalves. Other vertebrate shell crushers during the Cretaceous include chelonoid turtles (Hirayama, 1997) and globodentine mosasaurs (Massare, 1987). Among invertebrates, the homarid and palinurid lobsters (Harper, 1991; George and Main, 1968) evolved in the Triassic, and brachyuran (Xanthidae; Late Cretaceous) and pagurid crustaceans first appeared in the Jurassic. It is debatable whether these early forms were capable of crushing shells (Vermeij, 1993).

Invertebrate taxa that evolved a durophagous habit in the Cenozoic include the brachyuran crabs (with functionally differentiated claws: e.g., Portunidae and Cancridae) and neogastropods (Fasciolariidae, Buccinidae, Melongenidae, including some Muricidae, e.g., *Muricanthus fulvescens*, Wells, 1958; *Chicoreus ramosus*, Edward *et al.*, 1992). Some octopi reportedly chip edges of bivalve shells with their beaks (Anderson, 1994). Stomatopods (mantis shrimp) have left diagnostic punctures in molluscs (Geary *et al.*, 1991; Bałuk and Radwański, 1996). However, when the mantis shrimp evolved such shell-hammering behavior is unknown (Vermeij, 1987); the shell-hammering gonodactylid group first appeared in the Miocene (Walker and Brett, in press). Extant limulids have gnathobases capable of grinding up small clams, which constitute much of their diet (Botton, 1984). Under laboratory conditions, the seastar *Orthasterias* chipped the shell margin of its clam prey, *Humilaria*, with the suction generated by its tube feet (Mauzey *et al.*, 1968; Carter, 1968). Harper (1994) observed that the flexible margin of *Isognomon legumen* is chipped during attempts by *Coscinasterias acutispina* to pull the valves apart. Although Ordovician seastars have been preserved with their stomachs distended with small gastropod prey (Spencer and Wright, 1966), Vermeij (1977) cautioned that these seastars did not destroy or crush the prey skeletons in the process of ingestion. In addition, some species of sea urchins are believed to prey on mussels (Seed, 1992).

Major groups of vertebrates that became durophagous in the Cenozoic include teleost fishes with specialized shell-crushing dentition (Vermeij, 1993), diving and other coastal marine birds (Olson and Steadman, 1978; Vermeij, 1987), sea otters (Repenning, 1976a), seals (Repenning, 1976b) and walruses (Oliver *et al.*, 1983). For additional comments and perspectives on potential Phanerozoic shell crushers, see reviews of Carter (1968); Signor and Brett (1984); Vermeij (1987); Brett (1990); Harper and Skelton (1993); Dame (1996); Brett and Walker (in press); Walker and Brett (in press).

The evolution and diversification of predatory animals throughout the Phanerozoic had profound effects upon the structure of benthic communities (Bambach, 1993; Bottjer, 1985) and upon the evolution of shelled invertebrates. How many of these durophages would have inflicted preservable sublethal damage on marine gastropods and bivalves? Most likely, selective pressure for evolution of antipredatory morphologies in gastropods and bivalves over time came in response to attacks by various arthropods, namely phyllocarids (Signor and Brett, 1984), and possibly cephalopods (Alexander, 1986), in the Paleozoic. The crustaceans, neogastropods, birds, and fishes probably inflicted most of the sublethal repaired damage on molluscan shells in the Mesozoic and Cenozoic (Papp *et al.*, 1947; Vermeij, 1987). It should be noted that the arthropods, cephalopods, and gastropods that inflicted considerable sublethal shell damage were not the most powerful shell crushers among their contemporaries. Considering that attacks must be survived if the ability to sustain and repair unsuccessful predation attempts is to be inherited, escalation involving molluscan prey has probably been driven primarily by the relatively weaker, more often unsuccessful, durophages rather than by predators that usually pulverized their prey. Regardless, the diversification and specialization of shell-breaking predators is not the only line of evidence to support escalation, as evident in the ensuing sections.

3. Trends in Antipredatory Morphology in Space and Time

The evolutionary history of the incidence and expression of antipredatory shell armor in gastropods and bivalves was used by Vermeij (1987) to support escalation between shell-breaking predators and their prey. Vermeij (1983a) stated that an increase in repair is expected as the incidence and degree of expression of armor increase. Certain architectures with potential antipredatory value change in their percent contribution to marine molluscan communities through geologic time. Among bivalves, skeletal features that may be resistant to shell breakage, and therefore favored by selection, are: (1) crenulated margins, (2) overlapping valve margins, (3) spinosity, (4) cementation, (5) convexity of the valves, (6) valve thickness, including commissural shelves, (7) skeletal mineralogy, (8) valve size, (9) hermetic valve closure or a tight seal between valves, and (10) flexible valve margins. A skeletal feature that facilitates shell breakage is gaping margins (Vermeij, 1987). Among gastropods, features that increase shell resistance to breakage include: (1) toothed and elongate apertures, (2) inflexible opercula, (3) strong external sculpture, including spines, nodes, and varices, (4) thickness, (5) a callus filling the umbilical region, and (6) foot retractility. A polished shell, characteristic of such gastropod families as the Olividae, Cypraeidae, and Marginellidae, is another skeletal feature that decreases the snail's susceptibility to shell crushers (Kohn, 1999), but the feature may not always be distinguished in the fossil record among extinct families. Only 1% of olivids had shell repairs from the early Pleistocene of Fiji (Kohn and Arua, 1999). Features that facilitate breakage in snails include umbilicate, planispiral, hypostrophic, or disjunct coiling and turbiniform shapes (Vermeij, 1987).

Gastropods and bivalves exhibit an array of defensive behavioral adaptations to durophages, which include adoption of cryptic, nestling, clumping, or boring habits. They also include camouflaging the shell with foreign objects. Active defenses include: (1) escape by burrowing, swimming or leaping, (2) autotomization of portions of the body, (3) "mushrooming" responses in gastropods (covering of shell with mantle), (4) violent twisting of the shell, (5) immobilization of potential predators with byssal threads (mussels), (6) foot flailing, (7) valve flapping, (8) movement upwards on vertical surfaces (such as blades of salt marsh grass or out of tidal pools), and (9) squirting of water from the siphons by adduction of the shell valves (*Tridacna maxima*). For more details interested readers are referred to the following reviews: Ansell (1969), Vermeij (1987), and Harper and Skelton (1993).

Among gastropods, morphologic characters that increase over geologic time include: (1) percent of species with narrow apertures (0% to 28%; Ordovician to Recent), and (2) a thick or expanded adult lip (0% to 39%; Ordovician to Recent; Vermeij, 1983a) (Fig. 1). Over the same time interval, certain skeletal features decrease in importance. Umbilicate species decline from 59% in the Ordovician to 8% or less among Recent gastropod communities (Vermeij, 1983a) (Fig. 1). Concurrently, species with planispiral or hypostrophic coiling decreased from 27% to 0%, and turbiniform species declined from 49% to 10% (Vermeij, 1983a). Proportion of species with open or disjunct coiling decreased from 5% to 0% from the Ordovician to the Recent (Vermeij, 1983a). Shells with an umbilicus characterize less than 5% of Recent Tropical Pacific hard bottom gastropods, 5-10% of tropical Atlantic, and more than 10% of temperate latitude gastropod species (Fig. 2). The structural weakness of the umbilicus can be countered by plugging it with a callus such as is found in many naticids. Despite architectural trends through time, correlation between architecture and the incidence of unsuccessful predation is not always evident at the assemblage level. For example, antipredatory gastropod design is not correlated with repair frequency in the Cretaceous Ripley Formation (Vermeij and Dudley, 1982).

The deterrence value of apertural lips in gastropods has been well documented (Vermeij, 1987). For those snails lacking this defensive structure, the shrinking diameter of the apertural whorls into which the pedal mass withdraws may effectively preclude a continuous purchase on the shell by the predator. Thus *Turritella lueteofasciatus*, which lacks a thickened apertural lip, tolerates considerable shell damage as its foot deeply withdraws from the aperture into a passageway too restrictive to be peeled (Cadée et al., 1997). Whorl diameter of this inner sanctuary, not the terminal apertural diameter, may be the effective deterrent to lethal predation by peeling predators.

Whorl expansion rate determines how rapidly the apertural circumference expands with ontogeny and may bear on the snail's ability to survive and repair attacks by lip peelers. Whorl expansion rates have been calculated for a number of gastropod

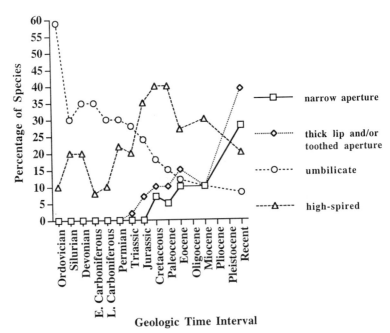

FIGURE 1. Gastropod morphology through geologic time plotted as percentage of species with specific morphologic character. Modified from Vermeij (1987).

geometries (Hickman, 1985). It would be interesting to see if the incidence of unsuccessful predation over time is correlated with whorl expansion rate. Vermeij (1983a) stated that globular shapes generated by a low spire and tight coiling protect many tropical gastropods against shell crushers, whereas a high-spired, turreted shell, which accommodates deep foot withdrawal, appears to be a good defense against peeling calappid crabs. However, Vermeij (1987) and Kohn (1999) both conceded that gastropods with high-spired coiling geometries are especially susceptible to being snapped in half by a strong predator.

Knobs may also deter fish predators (Palmer, 1979). Experimental removal of knobs rendered *Thais kiosikiformis* much more vulnerable to crushing by the puffer fish *Diodon hystrix*. Palmer (1979) stated that knobs reduce the vulnerability of the snail to gape-limited predators such as fish by effectively increasing the diameter of the shell and distributing stress over a larger area of the shell, while localizing stress on the thickest part of the shell. Knobs occur in 20-35% of Indo-Pacific and 10-18% of tropical Atlantic species, but only 2% of gastropod species in molluscan communities from cool temperate latitudes (Fig. 2). A latitudinal cline in height of knobs in *Busycon carica*, increasing from Massachusetts to Georgia, may be related to increased predation pressure southward along the Western Atlantic coast (Edwards, 1988). However, the antipredatory value of knobs may not always manifest itself. Ray and Stoner (1995) noted that mortality rate in experiments of tethered long-spined variants of the queen

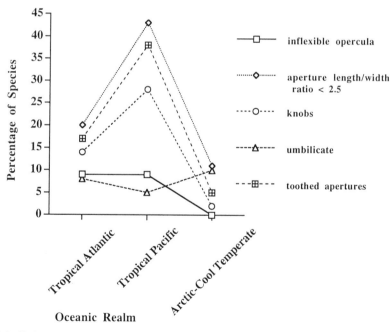

FIGURE 2. Variation in gastropod morphology estimated as percent of species in a molluscan community that have specific morphologic character. Plotted values are midpoints of ranges of values reported from several assemblages in each oceanic realm. Value for Pacific tropical realm is the average between midpoint for Indo-West Pacific, which usually has the highest percent value, and Eastern Pacific realm, which is intermediate between Indo-West Pacific and Atlantic tropical realm. Data from Vermeij (1983, 1987, 1993).

conch *Stombus gigas* was greater than that of short-spined variants (large size and living in aggregations were more important antipredatory defenses). Schmidt (1989) reported that gastropods that were smooth or less ornamented had higher repair frequencies in Bahia la Choya, Gulf of California. Like knobs, varices in gastropods render the shell effectively larger and thereby more difficult for predators to manipulate. A thickened outer lip may become a varix following subsequent accretion at the lip. Varices characterize several tropical families; some genus level examples include *Cassis, Ceratostoma* and *Murex*. Varices have been found in 10-25% of species in tropical assemblages versus 2% in temperate regions (Vermeij, 1993). Varices have been shown to deter lethal predation on *Ceratostoma foliatum* by crabs and seastars (Donovan *et al.*, 1999). When varices were removed experimentally, chelipeds of *Cancer productus* that previously only chipped the apertural margin of unmodified *C. foliatum* were more successful in snapping the gastropods in half (Donovan *et al.*, 1999). Whether these varices bear repairable damage is unreported, although superficial scratches or chips do occur in the thick adult lip of strombids (Randall, 1964), cerithids (Yamaguchi, 1977), and *Ceratostoma foliatum* (Spight and Lyons, 1974).

Some antipredatory ornamentation is oriented toward the shell interior in gastropods. Teeth projecting into the aperture of some gastropods may also reduce the size of the aperture, denying purchase by the predator. Vermeij (1993) observed that toothed apertures occur in 35-40% of Recent tropical Indo-Pacific species, 15%-19% in the tropical Atlantic, and less than 5% of cool temperate species (Fig. 2). An increased amount of whorl overlap will also result in a narrow aperture. Vermeij (1987, 1993) documented that gastropods with an apertural length to apertural width ratio of at least 2.5 comprised 35%-50% of Recent tropical Indo-Pacific species, 15-25% of tropical Atlantic species, and less than 11% of cool-temperate Atlantic species (Fig. 2).

Inflexible opercula, invariably calcareous, are concentrated in the tropics (5-12% of Recent species; Fig. 2). Such opercula are found in Neritidae, Turbinidae, and Naticidae. The feature is found in less than 2% of temperate species. Repairs are found in chitinous opercula of whelks (Fig. 3) and some naticid species, indicating that this aperture-sealing feature is attacked. Calcification of the opercula would render it less prone to breakage.

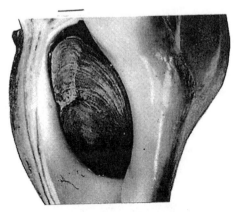

FIGURE 3. Repaired operculum, tracked cleft, in *Busycon sinistrum*. Width of bar = 1 cm.

Thickening of the apertural lip in gastropods may be an effective deterrent to predators (Vermeij, 1993; Kohn, 1999). Accentuated lip thickening occurs in *Cypraea, Drupa, Nassarius, Strombina* and *Strombus* (Kohn, 1999). Growth is determinant in these exemplar genera and lip thickening is the snail's terminal accretionary defense. Another type of localized shell thickening, remodeling, is manifest on the shell interior of the genus *Conus*, and the majority of the genera in the families Neritidae and Ellobiidae. The right side of the gastropod mantle may add two or more successive layers to the apertural margin while concurrently dissolving old shell, thereby maintaining uniform thickness in the last whorl (Vermeij, 1993; Kohn, 1999).

Posteriorly localized spines as in the bivalve *Hysteroconcha* may thwart siphon-nipping attacks (Stanley, 1988), although the character also has been hypothesized to deter gastropod and seastar predation (Carter, 1967). Only two families of Paleozoic bivalves bear spines, the Aviculopectinidae and the Pseudomonotidae. From the Mesozoic to Recent, the percentage of spiny species in the genus *Spondylus* increased from 0% in the Jurassic to almost 90% in the Recent (Harper and Skelton, 1993) (Fig. 4). Furthermore, percentage of extant spiny epifaunal bivalve species increases from 0% in Arctic waters to 26% in the West Indies (Harper and Skelton, 1993). Spines occur in 26-30% of subtropical to tropical epifaunal bivalves in the Eastern Pacific Panamic and Western Atlantic West Indies provinces of North America, respectively. In contrast, only 0-18% of species in Western Atlantic Arctic and the Eastern Pacific Aleutian provinces of North America, respectively, have spines (Harper and Skelton, 1993) (Fig. 5). In addition to thwarting predators by increasing apparent shell size and

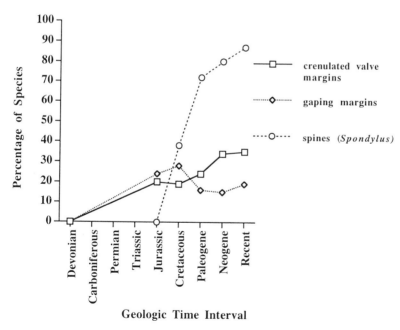

FIGURE 4. Bivalve morphology through geologic time estimated as percent of species in each time interval with specific morphologic character. Data on crenulated and gaping margins from Vermeij (1987). Data on spines in spondylids from Harper and Skelton (1993).

strengthening the shell, spines attract epibionts. Concealment of the valves afforded by encrusters on certain types of spines, such as those on *Spondylus*, may be a more important deterrent to durophages than the spines *per se* (Feifarek, 1987).

Crenulated margins are v-shaped indentations at the commissure of bivalves that function to prevent shearing of the valves in the plane of the commissure and resist compression of the valves. Species show an increase in crenulated valve margins (0% to 47%) from the Devonian to the Recent (Vermeij, 1987) (Fig. 4). Crenulated margins occur in 50-57% of species in the tropical Western Pacific, 35-55% of species in tropical America, and 16-33% in cool temperate species on hard bottoms (Fig. 5). On sandy and muddy bottoms, species with crenulated margins account for 35-50% in tropical communities versus 23-30% in cool temperate communities.

Radial corrugations or concentric folds in the shell wall strengthen the shell of thin bivalve species (Vermeij, 1993; Kohn, 1999). The extent of radial costations increased in the stratigraphic lineage of *Exogyra* in the Late Cretaceous (Dietl *et al.*, 2000). Corrugations are prominent in Recent tropical bivalves. More than 20% of species have radial folds, whereas, in cool water faunas, the incidence of radial folds is less than 15% (Fig. 5).

These folds are prevalent in pectinid scallops and ostreid oysters. Concentric folds also occur in some burrowing clams. Stanley (1988) interpreted these and many other

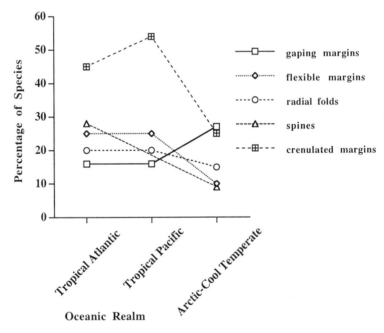

FIGURE 5. Variation in bivalve morphology estimated as percent of species in a molluscan community that have specific morphologic character. Plotted values are midpoints of ranges of values reported from several assemblages in each oceanic realm. Value for Pacific tropical realm is the average between midpoint for Indo-West Pacific, which usually has the highest percentage value, and Eastern Pacific realm, which is intermediate between Indo-West Pacific and Atlantic tropical realm. Data from Vermeij (1983*a*, 1987, 1993).

morphologic characters (e.g., ratchet sculpture, divaricate ribbing, concentric ridges) as primarily aids to burrowing and stabilization of the clams and snails in the sediment. Only secondarily are these features adapted for defense.

Thin, flexible margins, (e.g., Pennidae, Pteriidae, Isognomidae, Ostreidae) are easily damaged by predators that nevertheless do not gain access to tissue of the retracting mantle margin that withdraws deeply toward the dorsal shell margin. Between 16% and 35% of hard-bottom bivalve species found in the tropics today are characterized by flexible margins as compared with 10% or less in temperate latitudes (Fig. 5). Flexible margins may abut dorsally against thick commissural shelves, as is the case in the Exogyrinae. The commissural shelf restricted fractures to the thin, possibly flexible margins of these Cretaceous oysters (Dietl et al., 2000).

Harper (1991) illustrated the chronologic appearance of cementation among bivalve genera and families during the last 300 million years, along with the corresponding appearance of drilling, prying and durophagous molluscivores. More than sixteen families have, or have had, the ability to cement to a hard substratum. She showed experimentally the value of cementation over the byssate habit of mussels in deterrence of manipulation and, ultimately, consumption by crabs.

Hermetically sealed bivalves may prevent leakage of metabolites that chemically cue durophagous predators to the location of prey (Vermeij, 1983b; Harper and Skelton, 1993). Extreme crenulation, or plication (e.g., *Plicatula, Lopho, Tridacna*) increases toward the tropics and may serve to keep the valves tightly sealed and preclude detection of the potential prey by chemical cues. Although gaping margins increased from 0% to 31% from the Ordovician to Recent, (Vermeij, 1987), this feature facilitates breakage, and is concentrated in boreal and high temperate latitudes where durophagous predators are less diverse (Vermeij and Vail, 1978). Gapers constitute on average 16% of Recent species of most tropical molluscan communities, but 22-31% of species in Arctic to temperate latitudes (Fig. 5). Most permanently gaping bivalves are deep burrowers, which mitigates their vulnerability to durophages.

Blundon and Kennedy (1982) and Boulding (1984) showed that curvature of the valves is an important factor in deterrence of lethal shell breakage in bivalved organisms. Boulding (1984) showed that shell-breaking time for the crab *Cancer productus* was greatest when bivalve species with a more inflated shell were attacked. Analogous to convexity of the bivalve shell, globosity in gastropods also protects many species from crushing predators (Vermeij, 1983a). Vermeij and Currey (1980) showed that shell strength increases as shell globosity increases in thaidid gastropods. In addition, Schmidt (1989) observed that squat low-spired snail species had higher repair frequencies. However, Walker (2001) observed that the low-spired species among deep-sea gastropods did not have the predicted high frequencies of repair. Repair frequencies are equally common on robust and slender forms of extant terebrids, prompting Signor (1985) to question whether these scars are a valid index of shell vulnerability to durophagous predators.

With regards to skeletal mineralogy, nacre has been shown experimentally to be the most resistant microstructure in compression, bending, and impact tests, whereas foliated calcite, particularly that of the oyster *Crassostrea virginica,* is extremely weak in these same tests (Taylor and Layman, 1972; Currey, 1980, 1990). Gabriel (1981) showed that microstructures high in organic content, such as nacre, are more resistant to boring, a correlation also noted by Harper and Skelton (1993). Not surprisingly, nacre is characteristic of five epifaunal and semi-infaunal byssate families (Taylor and Layman, 1972) that experience the impact of saltating clasts (Raffaeli, 1978; Shanks and Wright, 1986). This microstructure is also present in four deeply burrowing bivalve families, which may experience compressional forces that crack the valves (Checa, 1993). Nacre would also be useful in dealing with the compressive forces exerted on the shell by durophages. Carter (1980) showed that the combination of external calcitic prisms, oriented with the long axis perpendicular to the shell surface, and the underlying laminated nacre creates a boundary "discontinuity" between the two layers that stops the internal propagation of fractures from the shell exterior in *Pinna*. Pen shells often show huge embayed repair scars. Like pen shells, mussels display a combination of inner nacre buttressing the shell exterior of prismatic calcite. Although prismatic calcite is shown to be weaker than nacre, it is often combined with underlying, buttressing nacre both in bivalve (Taylor and Layman, 1972) and gastropod (Currey, 1980, 1990) architectures. The combination facilitates localization of fractures that can be repaired. The blue mussel *Mytilus edulis* displayed the fourth highest repair frequency (0.19 scars per shell) among New Jersey coastal bivalves (Alexander and Dietl, 2001).

Although crossed-lamellar structures have a lower organic content than nacre, the former skeletal microstructure approaches nacre in strength among gastropods. Possessing less than a few percent organic matter, the crossed lamellar microstructure of *Strombus gigas* is extremely resistant to fracture (Kamat et al., 2000). Scanning electron microscopy reveals that the crossed lamellar microstructure of the queen conch experiences microcracking at low mechanical loads. But crack bridging occurs in the middle layers and renders this shell extremely tough to fracture. Currey and Kohn (1976) also noted that crossed-lamellar microstructure inhibited the propagation of cracks in *Conus*. Kardon (1998) suggested that conchiolin layers, organic-rich microstructural laminae in shells of corbulid bivalves, increase shell strength during durophagous attacks. Compression tests suggested that the layers increase shell strength and toughness by halting propagation of cracks in a similar fashion to nacre and multiple layers of mutually perpendicular crossed-lamellar aragonite.

Most experimental determinations of the strength of shell fabric have utilized previously unbroken, unrepaired portions of shell. But how strong is a resecreted skeletal area? Strength of the repaired area is as great as the original unbroken shell material in *Littorina irrorata* (Blundon and Vermeij, 1983). However, *Busycon* whelk attacks on damaged and repaired *Mercenaria mercenaria* (Fig. 6A) often resulted in propagation of new fractures along healed cracks in the prey (Fig. 6A) (Dietl, pers. observ., 2001). Strength of a particular shell microstructure may be mitigated by fatigue fracture, a process whereby an applied force initially unable to fracture the structure is repeatedly loaded and unloaded against the structure until it fails, as demonstrated in bending test on nacre (Currey and Brear, 1984). Consequently, even strong skeletal microstructures may be breached by this process of repeated loading (Boulding and LaBarbera, 1986). Despite the resistance to breakage by combinations of shell layers with different crystallographies and percentages of organic matter, there seems to be an evolutionary adaptive limit to shell microstructure. Shell microstructures remain remarkably constant in bivalve superfamilies through geologic time. Among bivalves, only the rudists and a few chamids have added external, prismatic shell layers through time (Harper and Skelton, 1993). Even though ratios of calcite to aragonite layers in mussels may vary predictably with temperature (Dodd, 1964), any corresponding changes in strength of mytilids latitudinally may be more dependent on seawater chemistry than selective pressure exerted by durophages. Gastropod orders and superfamilies show even less variability than bivalves in shell microstructure over geologic time (Harper and Skelton, 1993). Although the type of shell microstructure may not have varied considerably within marine gastropod and bivalve clades over geologic time, selection may favor the evolution of shell-crushing resistance in how a given microstructure is secreted. For instance, Lake Tanganyikan gastropods have evolved an increased number of crossed-lamellar layers in their shells to thicken the shell in defense against shell-crushing predators (West and Cohen, 1996).

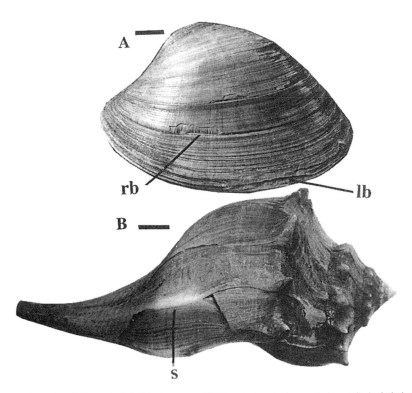

FIGURE 6. Scars in whelks and their bivalve prey. (A) Beveled posterior-ventral margin in lethal whelk predation on *Mercenaria mercenaria*. Note sublethal, repaired striated bevel (rb) due to unsuccessful whelk attack and subsequent lethal bevel (lb). (B) Self-inflicted, repaired, deep scallop (s) (scar depth to length ratio is 0.37) in mid-apertural lip of *Busycon sinistrum*. Note truncated margin of repair against former apertural lip demarcated by growth line intercepting second knob. Width of bar adjacent to capitalized letter = 1 cm.

Experiments indicate that thin-shelled gastropods appear to be more vulnerable to shell-breaking predation than their close relatives that are thick-shelled (Boulding et al., 1999). Similarly, Currey and Hughes (1982) showed that samples of the dog whelk *Nucella lapillus* with thicker apertural lips from wave-protected, rocky intertidal habitats required greater force to fracture than thinner-lipped dog whelk shells from wave-exposed habitats. Although thinner lips may be easier to break, Dietl and Alexander (1998) found that an increase in apertural lip thickness corresponded with increased repair frequency in the Miocene to Recent naticids *Euspira heros* and *Neverita duplicata*. Predator-induced changes in prey shell thickness may be decoupled from changes in repair frequency. Schmidt (1989) found no correlation between shell thickness at the aperture and repair frequency in snails from the northern Gulf of California. Vermeij (1982c) noted no increase in thickness of populations of *Littorina littorea* sampled from Nova Scotia to Cape Cod after invasion of the predatory green crab, *Carcinus maenas*. Paradoxically, although thickness of the periwinkle shell did not increase with increased exposure to the invasive durophagous green crab during the last century, repair frequency in the prey did increase over time (Vermeij, 1982c). Similarly, Raffaelli (1978) found no correlation between thickness, shell breakage, and wave exposure in *Littorina saxatilis* between samples from the wave-exposed, predatory crab-excluded versus the wave-protected, durophagous predator-infested, rocky intertidal. Furthermore, very small differences in gastropod shell thickness between living or fossil populations prone to different predation intensities may be difficult to substantiate statistically. Boulding et al. (1999) conceded as much in a case study of crab predation on littorinid species, *Littorina sitkana* versus *L. subrotundata*. Nevertheless, chemical cues from shell-breakers or their broken prey may stimulate neighboring conspecific gastropods or bivalves to thicken their shells. Gastropod examples include *Littorina obtusata* (Trussell, 1996), *Nucella lapillus* (Palmer, 1990), and *Nucella lamellosa* (Palmer, 1985). The blue mussel *Mytilus edulis* was documented experimentally to thicken its valves when provided chemical cues from predators and damaged conspecific neighbors (Leonard et al., 1999; Smith and Jennings, 2000).

Despite the experimental documentation of predator-induced changes in valve thickness, investigations that show a correlation between repair frequency and valve thickness among extinct bivalves are scarce. Dietl et al. (2000) showed a high correlation (r = 0.90) between repair frequency and mean valve thickness at the muscle scar in eight Cretaceous oysters, but further bivalve studies are needed. Indeed, a high correlation between repair frequency and shell thickness should not necessarily be expected. Leighton (in press) showed in a simulation that decreases in repair frequency with increasing thickness through ontogeny could result in lower correlation (r) values. He argued that a weak correlation between prey armor and repair frequency in time-averaged prey assemblages could be falsely interpreted that predation had little influence on prey evolution. On the contrary, Leighton (in press) reasoned that a sudden decrease in traces of unsuccessful predation during an overall increasing trend through time might be the strongest evidence for predation as an agent of selection. The sudden decrease represents temporal prey adaptation to shell-breakers. A rebound to higher repair frequencies following the decrease may reflect morphological improvements in the crushing elements of the predator induced by the modifications in the prey armor.

Nevertheless, the time-averaged effect may be a low correlation between repair frequencies and thickness as shown in the simulation.

Several studies indicate that, for a given predator size, the success rate of attacks on prey decreases, and handling time prior to shell breakage increases, with increasing prey shell size (Vermeij, 1987; Harper and Skelton, 1993; Kohn, 1999). Size increase has been well documented for many bivalve lineages, the gryphaeid oysters serving as one example (Hallam, 1968). It is possible that one of the factors driving this evolutionary trajectory of increase in size is increasing predation pressure (Harper and Skelton, 1993).

Trends in antipredatory shell form suggest that breakage became more common during the late Mesozoic and Cenozoic than it was in earlier times (Vermeij, 1987). Morphologies that prevent access or strengthen the shell increase in incidence and degree of expression over time, whereas traits associated with mechanical weakness decrease (Vermeij, 1987). These trends, paralleled by the increase in diversity and power of shell-crushing predators, suggest escalation between gastropods and bivalves and their durophagous enemies.

4. Predatory and Non-Predatory Sublethal Shell Breakage

Independent evidence that predation was an increasingly important selective agent for gastropods and bivalves, over the course of the Phanerozoic, is recorded as traces of unsuccessful predation (repair scars) on the shell. Traces of predator-induced damage need to be distinguished from non-predatory damage in order to evaluate the importance of predation as a selective agent. The fossil record of repair frequencies may be overestimated because of non-predatory shell breakage. Abiotically induced breakage during burrowing by clams (Checa, 1993), or by saltating clasts or rolling ice blocks impacting the shells of epifaunal molluscs in turbulent habitats (Shanks and Wright, 1986; Cadée, 1999), may be included, erroneously, in the inferred frequencies of predator-induced sublethal damage to the living or fossilized population.

Burrowing by bivalves may result in internal shell fractures radiating dorsally from the commissure that are subsequently repaired by mantle tissue (Fig. 7A) (Alexander and Dietl, 2001). Fractures are purportedly induced by external sediment pressure against the valves of deep burrowing bivalves, as observed in experiments by Checa (1993). Shallower bivalve burrowers that inhabit muds or sands rich with angular shell hash may repeatedly close their valves with calcareous shards occluded between them during re-burrowing, ultimately causing the shell to crack (Alexander and Dietl, 2001). Saltating clasts may fracture the outer prismatic layers of mussels and limpets living in wave-pounded habitats (Bulkley, 1968; Raffaelli, 1978; Shanks and Wright, 1986). Indirectly, such mechanical breakage may have the same consequence as sublethal predation, namely strengthening the shells of descendent generations (Vermeij, 1989; Alexander and Dietl, 2001); in other words, gastropods or bivalves exposed to shell-breaking agents may adapt to them regardless of whether the agents are biotic or abiotic. Pathological damage to the mantle tissue may be difficult to discriminate from predation-induced damage. Mantle lesions do not necessarily break the shell, although the ensuing aberrant growth deformities mimic predator-induced damage and repair (Boschoff, 1968).

FIGURE 7. Scars in bivalve shells. (A) Posterior ventral-dorsal internal fracture (if) skirting adductor muscle scar in *Mercenaria mercenaria*. (B) Scalloped (s) and cleft (c) valve of *Anadara ovalis*. Note cleft makes right angle deflection into incomplete naticid borehole. (C) Scalloped (s) posterior margin of valve of *Ensis directus*. (D) Anterior divots (d) and ventral scallop (s) in valve of *Mya arenaria*. (E) Multiple posterior divots (d) in valve of *Mercenaria mercenaria*. (F) Cleft (c) in valve of *Crassostrea virginica*. Note shadow in recessed gash and non-conjoined margins of cleft area of valve. (G) Deep scallop (s) succeeded by a tracked cleft (tc) ventrally in *Spisula solidissima*. Width of bar adjacent to capitalized letter = 1 cm.

Predator-induced shell damage can be differentiated from abiotic agents with careful scrutiny. The geometry of bivalve shell repairs has been categorized according to the outline of the scar (Dietl *et al.*, 2000, Alexander and Dietl, 2001). *Scalloped* injuries involve shell breakage that subparallels concentric growth lamellae, often cutting across radial ornamentation, if present. The fracture removes a sliver, or contiguous slivers, of the valve margin (Alexander and Dietl, 2001) (Fig. 7B-D). *Divoted* valves have triangular to chevron shaped depressions in the valve where a piece of the shell was crushed and removed (Fig. 7D-E). The triangular "pits" in the Ordovician *Sinuitopsis neglecta* described by Horny (1997) are probably synonymous with divots. *Cleft* shells display a shear in the valve(s), which extends from the former shell margin radially toward the beak (Fig. 7B, 7F) (Alexander and Dietl, 2001). The opposing valve margins of the gash are not rejoined in the repair process (Alexander and Dietl, 2001), although the repair process may undercoat the gash. Where radially ribbed valves are partially sheared, as in scallops, subsequently accreted ribbing may curve to fill in the gap, leaving an enclosed slit (Alexander and Dietl, 2001, see fig. 2D therein). Although occasionally curved, the shell gash is usually constrained to a straight break between radial ribs on ornamented valves. Cracks in the valves do not always have to be visible externally, but rather may be only expressed and repaired internally (Fig. 7A). More common than true clefts, *Tracked Clefts* project ventrally from a shell breakage, such as a scallop (Fig. 3, 7G) (Alexander and Dietl, 2001). Growth lines show an inverted "V" pattern across the linear depression ventral to the initial shell breakage. These tracked clefts represent trauma to the mantle margin that was not healed immediately during subsequent accretion of growth increments to the shell commissure. The "slashes" crossed by V-ing growth lines described by Horny (1997) are likely synonymous with "tracked clefts." *Embayed* valves have a large piece of the shell removed, outlined by an irregular, jagged fracture (Fig. 9C, D). Repaired areas are often smooth and lack ornamentation that may characterize uninjured parts of the valve. This type of shell fracture resulted in removal of a shell fragment that was ~50% or more as long as the anterior-posterior dimension was wide, a contrast with scalloped margins (Alexander and Dietl, 2001). *Punctured* repairs are internally plastered (Fig. 8A, B). The patch is often recessed and occasionally imbedded with pieces of broken shell in the calcareous repair matrix (Fig. 8B). On the damaged shell interior, the patch is convex like a welt. Bulkley (1968, p. 65) described repair of holes in the limpet *Acmaea scutum* in which "pieces of the shell had been pushed inward, pressing down the viscera. Repair had been achieved by the laying down of new nacreous layers below the pieces of shell, cementing together the crushed fragments of the top of the old shell and partially embedding them to form a solid unit. Any holes left by missing pieces of shell had been covered over on the inside by new shell material." *Beveled* sublethal repair is a narrow swath of the (former) shell margin that was planed, chipped, or filed down, leaving a striated edge (Fig. 6A). The truncated valve margin exposes the mantle edge (Colton, 1908; Magalhaes, 1948; Carriker, 1951). The above categorization has been applied to repaired bivalves (Dietl *et al.*, 2000; Alexander and Dietl, 2001), brachiopods (Alexander, 1986), conularids (Mapes *et al.*, 1989) and could be applied to scars in gastropod apertural lips (Fig. 9).

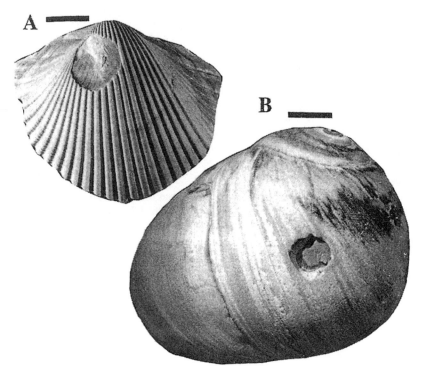

FIGURE 8. Oval fractures and repair in bivalves and gastropod shells. (A) Mantle-mortared shell puncture in *Dinocardium robustum* characteristic of holes made by stomatopods. (B) Mantle-mortared puncture with original broken shell imbedded in repair matrix in *Neverita duplicata*. Width of bar adjacent to capitalized letter = 1cm.

Gastropod shell breakage may show similar outlines to five of the six bivalve categories of repair. Embayed, scalloped, and divoted apertural lips are common (Fig. 9). Less common are repaired punctures (Kohn, 1992; Savazzi, 1991; Bałuk and Radwański, 1996; Bulkley, 1968; Cadée *et al.*, 2000) (Fig. 8A, B) and clefts (Bulkley, 1968). To our knowledge the only reports of cleft-type repair occur in limpets. Bulkley (1968) described severe "cracks" in the shell of the limpet *Acmaea scutum* that extended from the margin to apex. The cleft-like fracture was repaired by new shell material that was laid down over much of the interior of the shell, binding the pieces into a solid unit. Cadée *et al.* (2000) illustrated a similar fracture in the limpet *Patella depressa*.

What predators are incriminated by these various geometries of molluscan shell breakage? Beveled clam margins (Fig. 6A, B), which often resemble exfoliated lamellae when repaired (Fig. 6A), are the consequence of whelk valve chipping or valve wedging based on experiments of clam prey held in captivity with whelks exclusively (Dietl, pers. observ., 2001). Irregular oval fractures (Fig. 8A) are likely the product of stomatopod crustaceans that hammer a hole in their prey's shell (Caldwell and Dingle, 1976; Geary *et al.*, 1991; Kohn, 1992; Bałuk and Radwański, 1996). Shore birds may also punch irregular holes (Fig. 8B) into shells of snails stranded alive on sand bars, beaches, and tidal flats. Holes in oysters also can be attributed to crab predation (Krantz and

FIGURE 9. Scars in gastropod shells. (A) Multiple scalloped (s) and embayment (e) scars in *Busycon carica*. (B) Borderline deep scallop to embayment (s-e), with scar depth to length ratio of 0.47, in body whorl of *Euspira heros*. Example shows transitional, gradational nature between certain categories such as scallop to embayment (C) Embayment (e) on umbilical side of *E. heros*. (D) Embayment (e) on apertural lip of *Littorina littorea*. Width of bar adjacent to capitalized letter = 1 cm.

Chamberlain, 1978; Elner and Lavoie, 1983). Norton (1988) also attributed large punched holes, or a series of small holes, in gastropod prey as traces of fish predation.

Triangular divots localized on the posterior margin of clam valves (Fig. 7E) are attributable to siphon-nipping fish or shore birds. Siphon nipping of *Protothaca staminea* by fish affects 30-92% of individuals (Peterson and Quammen, 1982). Siphon nipping may contribute significantly to minor shell breakage in *Mercenaria mercenaria* (Kraeuter, 2001). Two thirds of shell repairs in *M. mercenaria* are posteriorly concentrated divots that could have been inflicted by siphon nippers (Alexander and Dietl, 2001). Nearly 70% of shells of *Mactra corallina* exhibit posterior shell repair after cropping of the siphons by fish (Cadée, 1999). The razor clam *Ensis*, which often bears posterior scallops (Fig. 7C), is attacked ventrally and posteriorly by birds, which vigorously shake the live clams that partially protrude from their burrows (Cadée, 2001). Cadée (2001, p. 461) illustrated a specimen with "small damage in the middle" of the ventral margin. This magnitude of ventral damage is occasionally repaired (Alexander and Dietl, 2001, see fig. 2L therein).

Embayed or scalloped serrated apertural lips of snails (Fig. 9) and valve margins of clams (Fig.7B, G) are produced by peeling (Shoup, 1968) or crushing crabs (Vermeij, 1978, 1993) and lobsters (Lau, 1987). Seed (1992) illustrated contiguous, minor scallops around the shell margin of the ribbed mussel *Geukensia demissa* killed by blue crabs (*Callinectes sapidus*) in this fashion. Calappid and xanthid crabs are shell peelers (Shoup, 1968; Vermeij, 1982*a*; Kohn and Arua, 1999). Repaired embayed fractures in the naticids *Euspira heros* (Fig. 9) and *Neverita duplicata* (see Dietl and Alexander, 1998), which live where neither calappids nor large xanthids co-occur, are more likely the result of peeling and crushing cancrid or portunid crabs (Lawton and Hughes, 1985). Small scallop-shaped breaks along the shell commissure of *Meretrix meretrix* are the result of chipping with the labial tooth of *Chicoreus ramosus* (Edward *et al.*, 1992); when this damage is viewed in articulated individuals it may resemble a borehole drilled by a gastropod attacking the valve edge. Clefts may also be the result of crabs that attempted to scissor the valves of prey. Hard clams, *M. mercenaria,* that were less than 25 mm in dorsal-ventral length were cleaved in half by *C. sapidus* held in captivity with hard clams (Alexander, pers. observ., 1999). Whelks (*Busycon sinistrum*) also may inflict cleft-like damage to the shell of *M. mercenaria* in their attempt to wedge open prey (occasionally a large triangular piece of one valve is broken off or a single dorsal-ventral crack is formed that breaks one valve in half; Dietl, pers. observ., 2001). Tracked clefts are also found in the operculum of gastropods (Fig. 3), a result of attempts by clawed predators to shear or pry off the protective cover of the foot. Additionally, anteriorly located divots in the valves of deep infaunal bivalves like *Mya arenaria* (Fig.7D) cannot be assigned unequivocally to a specific durophage. Assignment to a specific predator becomes even more difficult in the fossil record. Dietl *et al.* (2000) listed numerous potential vertebrate and invertebrate molluscivores that may have preyed on the Cretaceous oysters of the Mid-Atlantic coast, but appropriately refrained from identifying one culprit as the principal shell breaker.

5. Calculation of Repair Frequencies and Prey Effectiveness

Frequencies of repairs of shell damage inflicted by Phanerozoic durophages on bivalves and gastropods have been the primary means of evaluation of the role of breakage as a component of selection in warm water seas (Vermeij, 1987). But, as noted by Schindel *et al.* (1982), in the analysis of shell repairs in Pennsylvanian gastropods, no single method for quantifying shell breaks can be applied to geometrically and taxonomically diverse gastropods. Schindel *et al.* (1982) stated that highly ornamented forms provide many more indications of repair. Furthermore, high-spired gastropods offer greater surface area between sutures for inspection versus low-spired forms. Planispiral, convolute forms offer only the final whorl, whereas other discoidal forms with low expansion rate leave all penultimate whorls clearly exposed. Schindel *et al.* (1982) resorted to "handicapping" the calculation of repair frequencies among morphologically disparate taxa, counting repairs on only the final two whorls of highly ornamented species, but including those seen over the entire surface of smooth, convolute and rapidly expanding gastropods. Cadée *et al.* (1997) noted this methodological discrepancy in the calculation for repair frequency through time and expressed reservations about escalation inferences based on non-standardized procedures. Another problem in determining repair frequencies in gastropods is that repairs may occasionally accumulate in the calcareous or chitinous opercula (Fig. 3), but not be expressed on the shell, although the likelihood of preservation of a non-calcareous operculum is low. Studies of gastropod opercula on extant species are lacking even though this defensive structure may be readily scraped or chipped in an unsuccessful attack.

Repair frequencies may be used in determining prey effectiveness. Prey effectiveness in gastropods has been defined as the number of repaired breaks in shells divided by the total number of breaks, repaired as well as lethal (Vermeij, 1987). This ratio factors into consideration a measure of predation intensity. However, prey effectiveness has rarely been calculated for any fossilized repaired bivalve or gastropod. Eschewing all the complications affecting repair frequency, including relative abundance of shell-breaking predators, strength of the prey, and the age of the prey, calculation of prey effectiveness requires counts of lethal shell damage that is often unrecognizable or suspect (Walker and Yamada, 1993). A notable exception is the interaction between calappid crabs and their gastropod prey in an early Pleistocene assemblage from Fiji (Kohn and Arua, 1999). Vermeij (1987, 1993) stated that estimates of lethal breakage in gastropods are improved when fractured shells that also contain a drillhole are excluded from the denominator of the ratio of repaired to total number of breaks. If drilled shells were also "lethally" broken (an indication that breakage of the shell occurred after the death of the gastropod), the postmortem artifact would inflate estimates of breakage as a cause of snail death (Vermeij, 1989). Vermeij (1989) reported that the postmortem artifact is small, inflating the estimates of breakage by 10% or less. However, the ecological importance of lethal breakage is underestimated if shells are often completely destroyed by predators. Prey effectiveness in bivalves may yet be determined in special cases. For example, *Busycon* attacks on *Mercenaria* leave repaired and unrepaired (Fig. 6A), beveled margins that cannot be attributed to other agents of breakage (Dietl, 2002).

Bivalves present unique complications in calculations of repair frequencies relative to gastropods. Calculation of repair frequency requires the accurate count of both repaired and unbroken specimens in a sample. Repair frequencies are usually standardized for articulated specimens, not single valves of clams (Vermeij and Dudley, 1985). Disarticulated valves dominate many, if not most, fossil assemblages and determination of the number of individuals present in a sample has spawned different methodologies. Counting the most frequent valve, right or left, has been used to estimate the number of bivalve individuals in a sample. This procedure may be employed when one valve overwhelmingly dominates the sample as it did in the case of Cretaceous oysters where the convex left valve constituted over 90% of the sample (Dietl *et al.*, 2000). Nevertheless, this procedure may significantly underestimate sample size, particularly if the sizes of right and left valves differ.

Alexander and Dietl (2001) used a "matchup" technique with Recent bivalve samples, whereby disarticulated right and left valves of the same length, or width, were paired and treated as an individual, particularly where similarly located scars on each valve indicated that the valves were from the same individual. The remaining right and left valves differed in size and came from separate individuals. Their numbers were added to the denominator, the count of all individuals. This "matchup" technique is reliable *only* where disarticulated valves have not been transported or time-averaged. The technique may still underestimate the actual number of contributing individuals to the sample if "matched" right and left valves belonged to separate individuals, as would be the case in most time-averaged, locally transported fossil assemblages.

In a simulation study, Gilinsky and Bennington (1994) found that the number of individual clams is approximately equal to the number of valves (with < 10% error) when the actual sample of valves is much smaller than the total number of individuals that were preserved in the sampling area (or domain). In other words, it may be reasonable in fossil samples, in which repair scars are often mirrored on left and right valves, to count each disarticulated valve as a single individual for calculations of repair frequency in clams (as is the case for gastropods); in cases where only one scar is produced (e.g., a repaired puncture that results from unsuccessful stomatopod predation), the probability of finding a scarred valve will be two times lower than the probability of finding any valve. Thus a correction factor may have to be applied to cases where the repair scar is limited to a single valve.

Repair frequency may be calculated as the number of individuals with at least one repair divided by the total number of individuals present in the sample—the percent with scars method (Raffaelli, 1978; Bergman *et al.*, 1983; Geller, 1983; Vale and Rex, 1988; Cadée *et al.*, 1997; Walker, 2001). Alternatively, the total number of repairs may be divided by the total number of individuals in a sample—the scars per shell method (Vermeij *et al.*, 1981; Vermeij, 1982a; Vermeij and Dudley, 1982; Dietl and Alexander, 1998; Dietl *et al.*, 2000; Alexander and Dietl, 2001). Other methodologies have analyzed repair scars per shell surface area (Preston *et al.*, 1993). Microscopic techniques utilize acetate peels of sagital sections of clam valves to detect scars (Ramsey *et al.* 2000, 2001).

If no more than one repair occurs on any given damaged shell, then the first two techniques described above produce the same value for repair frequency. However, if the assemblage includes specimens with multiple repairs (e.g. Fig. 7B, 9A), the outcomes may be drastically different depending on which of these techniques is used. Bien *et al.* (1995) noted that as many as nine repairs may accumulate on a large left valve of the Cretaceous oyster *Pycnodonte mutabilis*. Among gastropods, the Recent naticid *Euspira heros* has as many as 12 repairs per shell (Dietl and Alexander, 1998) (Fig. 9B). Robba and Ostinelli (1975) also illustrated multiple repairs in bivalves. Dividing the total number of repairs by the number of individual shells present may give an inflated representation of what percentage of individuals experienced repairable breakage in their life. However, the alternative methodology, counting a repaired individual only once, underestimates the frequency of sublethal attacks experienced by the population.

Kowalewski *et al.* (1997) made the insightful observation that the technique used sometimes depends on the objective of the analysis. They noted that it is important to distinguish between an analysis that targets repairs (scars per shell) and an analysis that targets individual specimens (percent of shells with repairs). A hypothetical example is useful to highlight the advantages and disadvantages of both approaches. If there are 70 repairs among 10 shells in a sample of 100 shells then repair frequency will equal 0.70 (scars per shell). This type of analysis targets number of scars and provides important information about the predator. For instance, the location of the 70 repairs, even though they occur on only 10 shells, has bearing on the issue of where predators attack (stereotypic behavior; see Alexander and Dietl, 2001). On the other hand, in an analysis that targets shells (or individuals), suppose the same 70 repairs were spread over 70 shells in a sample of 100 instead of 10 shells. Repair frequency is still 0.70, but, in this case, 70% of individuals experienced and survived a predatory attack as opposed to only 10% of individuals in the analysis that targets scars. Thus the scars per shell method is not straight forward as to how many individuals in a population experienced sublethal attack; clearly a favorable trait is likely to spread to the entire population faster if more individuals survive and reproduce. We advocate standardization of the way in which data are presented so that both types of calculations can be obtained from data tables.

The metrics, however, both indicate when selection is more likely to favor the evolution of antipredatory traits. In other words, evolution of resistant shell features is more likely when repair frequency is high regardless of which method is used. The frequency of shell repair, calculated in either way, should increase if: (1) the relative abundance of shell-breaking predators, i.e., encounters between predator and prey, increases; (2) the strength of the prey (or probability the predator will be unsuccessful) increases; or (3) the age of prey (or exposure time to potential predators) increases (Vermeij, 1987). A high incidence of repair indicates that many individuals in the population were exposed to unsuccessful predation because predators were either common, prey were resistant, or prey were old. A low frequency of repair indicates that injury-producing attacks were rare because encounters with predators were rare or predators were relatively stronger than prey (Vermeij, 1987). Even if predators are common, the incidence of repair is low if predator success rate is high. Leighton (in press) added an additional scenario to account for an increase in repair frequency that takes into consideration the optimal foraging behavior of the predator. He stated that an increase in repair frequency is likely if there is a shift by the predator in prey preference, perhaps caused by change in encounter rate between predator and prey, or an improvement in the functional attacking ability of the predator. Leighton (in press) stressed the important point that changes in repair frequency can be caused by either a change in number of attacks, in which case predator success rate remains unchanged, or by changes in the likelihood of success of a predator without changing the number of attacks. In other words, increase in repair frequency can be a consequence of either an increase, or a decrease, in predation intensity, which has implications for how trends in repair frequency (and inferred escalation) are analyzed.

Vermeij *et al.* (1981) advocated that it is imperative to calculate repair frequencies standardized for class sizes to avoid ambiguity in the assessment of changes in repair frequencies within communities and stratigraphically sequential samples (see also Leighton, in press). Frequency of shell repair has been normalized for size classes, with repair frequency calculated per successive 10 or 20 mm size intervals of shells (Schindel *et al.*, 1982; Vermeij *et al.*, 1980, 1981; Vermeij, 1982a; Vermeij and Dudley, 1982; Dietl and Alexander, 1998; Dietl *et al.*, 2000). This standardization is required because accumulation of shell repairs is often size dependent (it is important to acknowledge that this standardization assumes shell growth rates are similar between samples, and hence the time of exposure to potential predators was the same between the size classes being compared). Shell repairs accumulate more often in larger (Vermeij *et al.*, 1980;

Vermeij, 1982a; Vermeij and Dudley 1982; Dietl and Alexander, 1998) or intermediate (Schindel et al., 1982; Schmidt, 1989) size classes. However, Cadée et al. (1997) observed higher repair frequencies in juveniles rather than adults of some gastropods species from the Gulf of California.

In addition to the variety of definitions of repair frequency, there is also a lack of consistency regarding multiple repair frequency. Multiple repair frequency has been defined as number of shells with multiple repairs divided by the number of injured individuals (Dietl et al., 2000). An alternative definition of multiple repair frequency is the number of specimens with multiple scars divided by the total number of specimens (Walker, 2001), which yields a lower percentage than the first calculation.

6. Prey Species-, Size-, and Site-Selectivity by Durophages

Studies of durophagous predation on extant molluscs have focused on species-, size- and site-selectivity of the predator. The Strauss Index, L_i, has been applied by Botton (1984) to show species selectivity by durophages in comparison of stomach contents of horseshoe crabs relative to the availability of certain molluscan prey in the ambient habitat. $L_i = r_i - p_i$ where, r_i is the percentage of the species prey in stomach contents among the total shelled individuals and p_i is the percentage of the prey among all available shelled prey (Botton, 1984). Botton (1984) showed that the horseshoe crab, *Limulus polyphemus*, preferred the small clam *Mulinia lateralis* and juveniles of *Ensis directus*, but selected against *Gemma gemma*. The Strauss Index could be applied to fossil molluscan assemblages for which susceptibility to and reparability of breakage of various species, and not predation intensity, is of interest. Percentage contributions of potential bivalve or gastropod prey species (p_i) to the total molluscan assemblage could be determined. Likewise, the percentage of repaired specimens belonging to a particular species among all repaired shells for all species (r_i) could be determined. The procedure would reveal which species show disproportionately low or high susceptibility to repairable damage.

Durophagous predators may show a preference for larger (Zach, 1978), intermediate (Geary et al., 1991), or small (Juanes, 1992; Seed and Hughes, 1995) molluscan prey when given a range of potential prey sizes. Because of this selectivity, the frequency of shell repair is expected to increase as shell size increases because the likelihood that a predator will be successful in its attack declines as shell length increases (Vermeij, 1987).

Size-frequency distributions of the size of the prey at the incidence of shell breakage versus the size of prey at death have been applied to sample populations of bivalves (e.g., Rhoads et al., 1982; Dietl et al., 2000; Alexander and Dietl, 2001). Differences between the size-frequency distribution of the lengths (or widths) of specimens at incidence of injury vs. the size distribution at death are very useful in assessment of shell damage. If shell damage is more sustainable and repairable by young adults, size-frequency distributions of size at incidence of injury will be right or positively skewed. If shell damage accumulates primarily in old age when the shell is thicker, size-frequency distributions at incidence of injury should be left or negatively skewed. If there is no bias regarding when damage occurs and is repaired, scars should be recorded throughout the ontogeny of the shell (Alexander and Dietl, 2001).

Statistical comparison of size-frequency distributions of specimen lengths (or widths) at inception of repair relative to size-frequency distributions of shells at specimen death may facilitate discrimination of any "size refuge" from predation. By this methodology, Alexander and Dietl (2001) statistically confirmed size refuges for the oyster *Crassostrea virginica* and the surfclam *Spisula solidissima* among living bivalve species common to the New Jersey Coast. Similarly, Dietl *et al.* (2000) observed that all exogyrine and pycnodont Cretaceous oysters failed to accumulate repairs beyond a valve length of 100 mm, although several specimens exceeded 150 mm in length. Size refuges from shell breakage may exist because chelate durophagous crabs may minimize the risk of damage to their claws, as well as handling time, by selecting smaller molluscan prey (Seed and Hughes, 1995).

Leighton (2001) discussed a useful application of size-frequency distributions to determine if the success of the predator decreases with time. He argued that repair frequency should be low for the preferred prey size class because attacks are more likely to be successful. Predator success probability decreases with increase in prey size and, consequently, repair frequency will increase. As prey size increases, a size refuge is attained at which the predator will not attack the prey and repair frequency drops to zero. Leighton (2001) postulated that, if a temporal decrease in the size at which a refuge is reached occurs, then the predator's success must have decreased. Such a trend, in concert with data that suggest an increase in prey armor (e.g., ornament or thickness), supports the hypothesis that morphological change was a response to predation (i.e., escalation).

The comparison of size-frequency distributions to determine a size refuge from predation requires one very important caveat. Absence of sublethal repairs in the largest size classes of the shells does not necessarily mean that the largest specimens were spared shell-crushing predation. Such a gap between the largest size at which sublethal breakage occurred and the largest individual sampled only means that predation attempts on large size prey were (1) absent altogether, (2) invariantly successful, killing the fractured individual, or (3) unsuccessful, leaving the shell unmarked.

Standardization of repair frequencies to size classes permits weak inferences to be made about the capacities of predators relative to those of the prey. If predators are small or weak, large prey individuals will show few if any attacks compared to small prey; if predators are relatively powerful, the incidence of repair should increase in successive size classes (Vermeij, 1987)

The location of sublethal damage bears on the issue of site selectivity by a predator that concentrates its attack on certain regions of the prey. These locations on the prey shell may indicate how the predator attempts to minimize energy expenditures while maximizing gain. Lau (1987) noted that chipping or peeling the edge of bivalve or gastropod prey, which is energetically time consuming, is resorted to only if outright crushing of prey is unsuccessful. Beyond the largest size at which edges of specimens can be chipped or peeled, the potential prey has attained a size refuge. However, the margin of the valve that is usually chipped is not often reported. Alexander and Dietl (2001) utilized pie diagrams to show the distribution of bivalve shell repairs according to the percentage of scars located anteriorly (e.g., Fig.7D), posteriorly (e.g., Fig. 7E), or ventrally (e.g., Fig. 7G); repairs in gastropod apertural lips can be apportioned among those that are concentrated apically or umbilically. These variations in position of repair on the shell can be compared with variations in morphology, such as thickness and size, to evaluate the escalation hypothesis.

Lip or valve regions that are commonly attacked and broken should experience, over time, selective pressure for evolution of breakage-resistant morphologic characters (Vermeij, 1987). Concentration of repairs in one area of a clam valve may correlate with exposure of vulnerable tissue parts. Accordingly, posterior divots in *M. mercenaria* (Alexander and Dietl, 2001; Fig 7E) are associated with attacks by siphon-nipping predators (Peterson and Quammen, 1982). Anterior scallops in *Mya arenaria* (Fig.7D) are associated with valve margins around an area where partial exposure of the foot occurs even with the valves closed. Among stratigraphically successive samples, "sudden" appearance or greatly increased accumulation of scars along a valve margin that previously did not bear such scars may favor adaptation by the prey (i.e., escalation; Alexander and Dietl, 2001). Innovations in predatory feeding behavior may impose new selective pressures on prey marginal defenses. For instance, beveled repairs on valves of *Mercenaria* first appeared in the late Miocene with the evolution of valve chipping by busyconine whelks (Dietl and Alexander, 1998; Dietl, 2002).

7. Repair Frequencies by Time, Latitude, and Habitat

Because the presence of repaired damage is a necessary condition for selection in favor of resistance to injury, comparative data on the frequency of repair permit evaluation of temporal and geographical patterns in the importance of predation as a selective agent (Vermeij, 1982b). Repaired molluscan shells have not been detected before the mid-Ordovician, and phenomena such as lip peeling remain rare through the Silurian (Brett, 1990; Lindström and Peel, 1997). Ebbestad and Peel (1997) observed a repair frequency of 7% in Ordovician gastropods from Sweden (see also Ebbestad, 1998). Scars on both Silurian gastropods (Peel, 1984) and bivalves (Liljedahl, 1985) are illustrated, but repair frequencies are not reported. Between 10% and 20% of snail specimens in certain Devonian assemblages suffered peeling (Brett, 1990).

In order to test the hypothesis that shell breakage by predators is a more important agent of natural selection today than it was in the geological past Vermeij *et al.* (1981) calculated repair frequencies in gastropods from the late Paleozoic to the Recent. Comparisons of repair frequencies, by size class of gastropods, from the Pennsylvanian to the Recent support the hypothesis of escalation, with the frequency of shell repair increasing from the Pennsylvanian to Recent (Vermeij *et al.*, 1981)(Fig. 10). Repair frequencies (number of scars divided by total number of shells of all species in a size class) for all species in the Pennsylvanian are low in the largest size classes (0.2). Repair frequencies do not exceed 0.3 in the largest size class by the Triassic (Fig. 10). Frequencies increase to a peak of 0.6 in the Cretaceous (Vermeij *et al.*, 1981). By the Miocene, repair frequency decreased, albeit statistically insignificantly, to 0.4 and remained at that level through the Recent (Vermeij *et al.*, 1981; Vermeij, 1983) (Fig. 10). There have been few studies subsequent to this early analysis, with the result that details of the timing and pace of breakage-related escalation are not well known (see Pan, 1991, for a discussion of the frequency of shell repair among the Late Cretaceous (Turonian) gastropod fauna of the Western Tarim basin, southern Xinjiang, China).

Interpretation of trends in repair frequencies is not always consistent between assemblage-level and taxonomic-level data (Alexander and Dietl, 2001). When repair frequencies through time are determined in a taxonomic context and restricted to the terebrids, Vermeij *et al.* (1980) showed no significant change between the Paleogene and Neogene (Fig. 10). In contrast, Allmon *et al.* (1990) showed that mean repair frequencies (number of scars per shell) in turritellids from the Late Cretaceous to Pliocene peaked in the Late Cretaceous (0.52), dropped in the Paleocene (0.19), rebounded in the Eocene (0.28), fell again in the Oligocene (0.05), and recovered in the Pliocene (0.39) (Fig. 10). This trend is similar to the record of drilling predation on gastropods and bivalves (Kelley and Hansen, 1993). At the generic level, both *Euspira* and *Neverita* show an increase in repair frequency from the Miocene to the Recent (Dietl and Alexander, 1998) (Fig. 10). Vermeij (1987) documented that repair frequencies in *Conus* increased between the Eocene and Miocene of Europe (Fig. 10). Unlike in gastropods, trends in bivalve repair frequencies through geologic time are scarce. Dietl *et al.* (2000) showed that repair frequencies in exogyrine and pycnodont oysters increased, at different rates, from the Late Cretaceous to the early Paleocene of the Northern Atlantic Coastal Plain.

A complication with repair frequencies in some predatory gastropods is that a documented increase over time may reflect damage sustained by the mollusc in its role

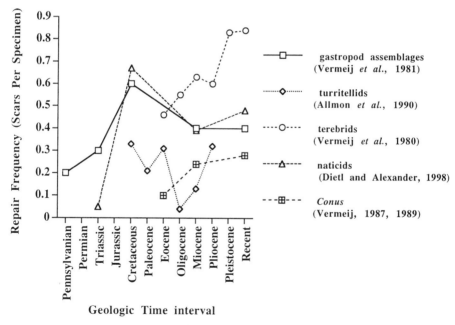

FIGURE 10. Repair frequencies (scars per shell) in gastropods through geologic time. Plotted values for Vermeij *et al.* (1981) and turritellids (Allmon *et al.* 1990) are midpoints of ranges of values reported from several assemblages in each time interval. Data on naticid taxa, tabulated in Dietl and Alexander (1998), also include repair frequencies from Vermeij (1982b), Vermeij and Dudley (1982), and Vermeij *et al.* (1982). Data on terebrids (Vermeij *et al.*, 1980) and *Conus* sp. (Vermeij, 1987) are averages of sampled species in each time interval.

as predator as it developed and employed new modes of breaching prey defenses. Dietl and Alexander (1998) showed that repair frequencies (number of scars per shell) in whelks, *Busycon* spp., and naticids *Euspira heros* and *Neverita duplicata,* increased by a factor of roughly four (1.67 to 7.0), two (0.57 to 0.96) and two (0.8 to 1.4) respectively, from the Miocene to the Recent of the mid-Atlantic Coast. These disparities indicate that whelks, which may have developed valve wedging in the early Miocene, had a much higher rate of increase in repair frequency relative to naticids. Although both are prey to shell-breaking predators, whelks apply their shell lip to wedge or chip open their prey, a process that often leads to self-induced damage to their shell (Fig.6B).

Cadée *et al.* (1997) cautioned that interpretation of trends in shell repair through geologic time should consider variation in shell repair in any one time, given the high interspecific (up to 87%) and interhabitat (up to 31%) variation in repair frequencies among gastropods from the northern Gulf of California. Geller (1983), Schmidt (1989), Reid (1992), and Cadée *et al.* (1997) expressed doubts that escalation can be discerned because variation in repair frequency attributable to microhabitats, time-averaging, and habitat-mixing may be comparable in magnitude to temporal variation in repair frequencies. For example, Schmidt (1989) found that the magnitude of variation across microhabitats (outer flats, inner flats, rocky habitats) of the northern Gulf of California was comparable to the temporal variation Vermeij *et al.* (1981) reported for Pennsylvanian to Recent samples. In the 10-19 mm size class, shell repair frequencies across habitat and through geological time both vary from approximately 0.10 to 0.40 (scars per shell) (see figure 5 in Schmidt, 1989).

Because of this potential bias, similar physical environments must be used in evaluation of patterns of adaptation through time (Vermeij, 1987; Dietl *et al.,* 2000). For example, a temporal trend in repair frequency increase for Cretaceous *Exogyra* oysters is evident if similar offshore environments are compared. But if a nearshore facies is included, the trajectory in repair frequency zigzags, masking the directional trend (Dietl *et al.,* 2000).

Alexander and Dietl (2001) advocated the use of lineage or clade level assessment of repair frequencies to address hypotheses concerned with the role of shell-breaking predation as a driving force in evolution. They argued that, given the complexity of influences on shell breakage and repair, utilization of taphonomic assemblages with admixtures of taxa complicates tests of escalation. Characteristics such as shell size, thickness, ornamentation, inflation, microstructure, presence or absence of a gape between the valves, rate of burrowing (a measure of escape potential), and depth of burial (refuge from predation) have been shown to influence repair frequency in bivalves (Alexander and Dietl, 2001). Lineage or clade level assessment of repair frequencies addresses Alexander and Dietl's (2001, p. 368) question: "do descendent species, which are likely to occupy similar habitats and possess similar habits through time, show an increase or stasis in sublethal breakage?" Comparison of taxa with similar adaptive styles that lived in similar environments standardizes many factors that influence repair frequency (Vermeij, pers. comm., 2001); this problem is difficult to control for in analyses utilizing assemblages of taxa. With detailed data on individual lineages, tests of the hypothesis of escalation become less complicated by the multitude of factors that affect repair frequency.

Latitudinal variations in gastropod repair frequencies show that, for every size class, healed sublethal damage is greatest in the tropics, particularly the Indo-West Pacific (Vermeij et al., 1980). This equatorward increase in repair frequency suggests that predation pressure is greater in the tropics than at more temperate latitudes (Vermeij, 1993). The latitudinal gradient has yet to be documented thoroughly through time. Repair frequencies in terebrids are comparable in warm temperate and tropical latitudes in the Miocene, although repair frequencies are higher in the tropics relative to the subtropics for some Recent species. Allmon et al. (1990), however, showed that repair frequencies for turritellids are significantly higher in low latitudes in the Paleocene. Data on repair frequencies in Recent boreal to Arctic latitudes are scarce. Cadée (1999) reported repair frequencies of 75% in the cold-water limpet *Nacella concinna*. Repair frequencies in thaidids range from 0.09 to 0.63 scars per shell for the tropics versus 0.0 to 0.29 combined for high latitudes in both hemispheres (Vermeij, 1978). Vermeij (1993) reported that repair frequency is less than 5% in shells from the Aleutian Islands, whereas many tropical gastropods have an average of more than one repair per shell. Whether a latitudinal pattern of greater repair frequencies (and hence increased predation pressure) in the tropics persisted throughout the Phanerozoic cannot be determined at this time. Such a latitudinal gradient in gastropod or bivalve repair frequencies in the Paleozoic would be logical based on the prevalence of shell breakage-resistant traits (Leighton, 1999; Dietl and Kelley, 2001) in tropical Paleozoic brachiopods compared to their temperate-latitude contemporaries.

Further gradients in shell repair have been hypothesized to occur among regions, particularly among the tropical shallow-water marine communities of the Indo-West Pacific, Eastern Pacific, and Western Atlantic (Vermeij, 1989). Variation is predicted to have resulted from interoceanic differences in the evolutionary impact of biological interactions (e.g., predation). It is hypothesized that, because the level of antipredatory traits is high in the Indo-West Pacific biota, repair frequencies also will be high. It is also predicted that repair frequencies in the Indo-Pacific will be higher than in the Western Atlantic where traits related to predation are less developed. The Eastern Pacific is intermediate between the other two tropical regions in expression of antipredatory adaptations and is expected to have an intermediate repair frequency among specific groups of potential gastropod prey (Vermeij, 1989). Frequencies of repair in the genus *Nerita* and terebrid, columbellid, and planaxid gastropods support the expected interoceanic pattern, in that the highest frequencies of repair were found in the Indo-West Pacific (Vermeij, 1978, 1989; Vermeij et al., 1980). However, no pattern was found in cerithiid and thaidid gastropods, and a pattern opposite that predicted was found for *Conus*, with highest repair frequencies in the Western Atlantic (Vermeij, 1989).

Regarding bathymetric gradients, repair frequencies in Cretaceous exogyrine and pycnodont oysters in onshore habitats are significantly higher, by a factor of three and two, respectively, than in offshore shelf habitats (Dietl et al., 2000). But this trend should not be extrapolated to the continental slope and deeper habitats. Seventy-five percent of Recent deep-sea hydrothermal vent mussels showed shell repairs (Rhoads et al., 1982). Vale and Rex (1988, 1989) reported repair frequencies (percent of specimens with scars) between 0.8 and 0.48 for prosobranch gastropods from communities below 200 m depths in the Western North Atlantic. These values are comparable for shallow-marine habitats. Even higher repair frequencies (percent of specimens with scars) are reported for the deep-sea bathyal archaeogastropods *Bathybembex* (0.28 to 1.00) and *Gaza* (0.68 to 0.91) (Walker and Voight, 1994). Kropp (1992) did find a bathymetric trend in repair frequency. Repair frequencies are lower in upper slope than inner or outer shelf samples in prosobranch gastropod species from the inner shelf to slope environment off Point Conception, California. Using the percentage of individuals with one or more repairs for Pliocene deep-water gastropods, Walker (2001) found major shell repair (= embayed fractures) occurred on 66% of the 59 species and 25% of individuals. Walker (2001) stated that shell repair frequency in deep-water habitats may be higher than previously thought. Among shallow-water communities, Schmidt (1989) noted that repair frequencies within the same gastropod species are higher in the rocky intertidal habitats than in sandy habitats of Bahia la Choya, Gulf of California. The lack of a persistent bathymetric trend among all of these studies underscores the complexity and importance of habitats from the rocky intertidal to the deep sea.

8. Concluding Remarks

The general trends of (1) diversification of shell-breaking predators, (2) increase in incidence of antipredatory shell armor, and (3) increase in the incidence of sublethal breakage-induced shell repair (independent evidence of the action of the selective agent of predation), observed through time, suggested to Vermeij (1987) that there has been a general increase in hazards to individuals from their enemies over the course of the Phanerozoic. The available evidence suggests that predation is an important component of natural selection for marine bivalves and gastropods; however, much remains to be learned about the tempo and extent of escalation between gastropods and bivalves and their shell-crushing enemies. The timing, directions, and circumstances of escalation are not well understood owing to a lack of detailed data on individual evolutionary lineages in specified environments over intervals of time that span both periods of diversification and events of extinction. We also need to develop a conceptual framework of spatial variation of predation in order to begin to understand the nature, magnitude, history, and causes of geographic differences in ecology of marine gastropods and bivalves.

ACKNOWLEDGMENTS: We thank Royal Mapes and Alan Kohn, who reviewed this manuscript and offered suggestions for its improvement. We also thank editors Patricia Kelley and Michał Kowalewski, who similarly made suggestions that enhanced this contribution.

References

Alexander, R. R., 1986, Resistance to and repair of shell breakage induced by durophages in Late Ordovician brachiopods, *J. Paleontol.* **60**:273-285.
Alexander, R. R., and Dietl, G. P., 2001, Shell repair frequencies in New Jersey bivalves: A Recent baseline for tests of escalation with Tertiary, Mid-Atlantic congeners, *Palaios* **16**:354-371.
Allmon, W. D., Nieh, J. C., and Norris, R. D., 1990, Drilling and peeling of turritelline gastropods since the Late Cretaceous, *Palaeontology* **33**:595-611.
Anderson, R. C., 1994, Octopus bites clam, *Festivus* **26**:58-59.
Ansell, A. D., 1969, Defensive adaptations to predation in the Mollusca, *Proc. Symp. Mollusca* **2**:487-512.

Bałuk, W., and Radwański, A., 1996, Stomatopod predation upon gastropods from the Korytnica Basin and from other classical Miocene localities in Europe, *Acta Geol. Polon.* **46**:279-304.

Bambach, R. 1993, Seafood through time: Changes in biomass, energetics, and productivity in the marine ecosystem, *Paleobiology* **19**:372-397.

Bergman, J., Geller, J. B., and Chow, V., 1983, Morphological divergence and predator-induced shell repair in *Alia carinata* (Gastropoda: Prosobranchia), *Veliger* **26**:116-118.

Bien, W., Wendt, J. M., and Alexander, R. R., 1995, Paleoecology of the Late Cretaceous oysters from New Jersey and Delaware, in: *Contributions to Paleontology of New Jersey* (J. Baker, ed.), Geological Association of New Jersey, Wayne, NJ, pp. 62-71.

Bishop, G. A., 1975, Traces of predation, in: *The Study of Trace Fossils* (R. W. Frey, ed.), Springer-Verlag, New York, pp. 261-281.

Blundon, J. A., and Kennedy, V. S., 1982, Mechanical and behavioral aspects of blue crab, *Callinectes sapidus* (Rathbun), predation on Chesapeake Bay bivalves, *J. Exper. Mar. Biol. Ecol.* **65**:47-65.

Blundon, J. A, and Vermeij, G., 1983, Effects of shell repair on shell strength in the gastropod *Littorina irrorata, Mar. Biol.* **76**:41-45.

Bond, P. N., and Saunders, W. B., 1989, Sublethal injury and shell repair in Upper Mississippian ammonoids, *Paleobiology* **15**:414-428.

Boshoff, P. H., 1968, A preliminary study on conchological physio-pathology, with special reference to Pelecypoda, *Ann. Natal Mus.* **20**:199-216.

Bottjer, D., 1985, Bivalve paleoecology, in: *Mollusks: Notes for a Short Course* (D. J. Bottjer, C. S. Hickman and P. D. Ward, eds.), Univ. of Tennessee, Department of Geological Sciences Studies in Geology 13, pp. 112-137.

Botton, M. L., 1984, Diet and food preferences of the adult horseshoe crab *Limulus polyphemus* in Delaware Bay, New Jersey, USA, *Mar. Biol.* **81**:199-207.

Boulding, E. G., 1984, Crab-resistant features of shells of burrowing bivalves: decreasing vulnerability by increasing handling time, *J. Exper. Mar. Biol. Ecol.* **76**:201-223.

Boulding, E. G., and LaBarbera, M., 1986, Fatigue damage: repeated loading enables crabs to open larger bivalves, *Biol. Bull.* **171**:538-547.

Boulding, E. G., Holst, M., and Pikon, V., 1999, Changes in selection on gastropod shell size and thickness with wave-exposure on Northeastern Pacific shores, *J. Exper. Mar. Biol. Ecol.* **232**: 2317-239.

Brett. C. E., 1990, Predation, in: *Paleobiology: A Synthesis* (D. E. G. Briggs and P. R. Crowther, eds.), Blackwell Scientific, Oxford, pp. 368-372.

Brett, C. E., and Walker, S., in press, Predators and predation in Paleozoic marine environments, in: *The Fossil Record of Predation* (M. Kowalewski and P. H. Kelley, eds.), Paleontological Society Papers, Vol. 8.

Brunton, C. H. C., 1966, Predation and shell damage in a Visean brachiopod fauna, *Palaeontology* **9**:355-359.

Bulkley, P. T., 1968, Shell damage and repair in five members of the genus Acmaea, *Veliger* **11** (Suppl.):64-66.

Cadée, G. C., 1999, Shell damage and shell repair in the Antarctic limpet *Nacella concinna* from King George Island, *J. Sea Res.* **41**:149-161.

Cadée, G. C., 2001, Herring gulls feeding on recent invader in the Wadden Sea, *Ensis directus,* in: *Evolutionary Biology of the Bivalvia* (E. M. Harper, J. D. Taylor, and J. A. Crame, eds.), The Geological Society, London, pp. 459-464.

Cadée, G. C, Cadée-Coenen, J., and Checa, A., 2000, Schelpreparatie bij *Patella depressa* verzameld nabij Cadiz, *Corresp.-blad Ned. Malac. Ver.* **317**:128-130.

Cadée, G. C., Walker, S. E., and Flessa, K. W., 1997, Gastropod shell repair in the intertidal of Bahia la Choya (N. Gulf of California), *Palaeogeogr. Palaeoclim. Palaeoecol.* **136**:67-78.

Caldwell, R. L. and Dingle, H., 1976, Stomatopods, *Sci. Am.* **234**:80-89.

Carriker, M., 1951, Observations on the penetration of tightly closing bivalves by *Busycon* and other predators, *Ecology* **32**:73-83.

Carter, J. G., 1980, Environmental and biological control of bivalve shell mineralogy and microstructure, in: *Skeletal Growth of Aquatic Organisms: Biological Record of Environmental Change* (D. C. Rhoads and R. A. Lutz, eds.), Plenum Press, NewYork, pp. 69-168.

Carter, R. M., 1967, The shell ornament of *Hysteroconcha* and *Hecuba* (Bivalvia): a test case for inferential functional morphology, *Veliger* **10**:59-71.

Carter, R. M., 1968, On the biology and paleontology of some bivalve Mollusca, *Palaeogeogr. Palaeoclim. Palaeoecol.* **4**:29-65.

Checa, A., 1993, Non-predatory shell damage in Recent deep-endobenthic bivalves from Spain, *Palaeogeogr. Palaeoclim. Palaeoecol.* **100**:309-331.

Coen, L. D., and Heck, Jr., K. L, 1991, The interacting effects of siphon nipping and habitat on bivalve (*Mercenaria mercenaria* (L.) growth in a subtidal seagrass (*Halodule wrightii* Aschers) meadow, *J. Exper. Mar. Biol. Ecol.* **145**:1-13.

Colton, H. S., 1908, How *Fulgur* and *Sycotypus* eat oysters, mussels, and clams, *Proc. Acad. Nat. Sci. Phil.* **60**:3-10.

Conway Morris, S., and Whittington, H. B., 1979, The animals of the Burgess Shale, *Sci. Am.* **24**:122-123.

Currey, J. D., 1980, Mechanical properties of the molluscan shell, in: *The Mechanical Properties Of Biological Materials* (J. F. Vincent and J. D. Currey, eds.), Symposia of the Society for Experimental Biology 34, Cambridge University Press, New York, pp. 75-97.

Currey, J. D., 1990, Biomechanics of mineralized skeletons, in: *Skeletal Biomineralization: Patterns, Processes and Evolutionary Trends* (J. G. Carter, ed.), Van Nostrand Reinhold, pp. 11-25.

Currey, J. D., and Brear, K., 1984, Fatigue fracture of mother of pearl and its significance for predatory techniques, *J. Zool. Soc. London* **204**:541-548.

Currey, J. D., and Hughes, R. N., 1982, Strength of the dogwhelk *Nucella lapillus* and the winkle *Littorina littorea* from different habitats, *J. Animal Ecol.* **51**:47-56.

Currey, J. D., and Kohn, A. J., 1976, Fracture in the crossed lamellar structure of *Conus* shells, *J. Materials Res.* **11**:1615-1623.

Dame, R. F., 1996, *Ecology of Marine Bivalves*: An Ecosystem Approach, CRC Press, Inc. Boca Rotan.

Dietl, G. P., 2002, Predators and dangerous prey in the fossil record: evolution of the busyconine whelk-*Mercenaria* predator-prey system. Unpublished Ph.D. Dissertation, North Carolina State University.

Dietl, G. P., and Alexander, R. R., 1998, Shell repair frequencies in whelks and moon snails from Delaware and southern New Jersey, *Malacologia* **39**:151-165.

Dietl, G. P., Alexander, R. R., and Bien, W., 2000, Escalation in Late Cretaceous-early Paleocene oysters (Gryphaeidae) from the Atlantic Coastal plain, *Paleobiology* **26**:215-237.

Dietl, G. P., and Kelley, P. H., 2001, Mid-Paleozoic latitudinal predation gradient: distribution of brachiopod ornamentation reflects shifting Carboniferous climate, *Geology* **29**:111-114.

Dodd, J. R., 1964, Environmentally controlled variation in the shell structure of a pelecypod species, *J. Paleontol.* **38**:1065-1071.

Donovan, D. A., Danko, J. P., and Carefoot, T. H., 1999, Functional significance of shell sculpture in gastropod molluscs: tests of a predator-deterrent hypothesis in *Ceratostoma foliatum* (Gmelin), *J. Exper. Mar. Biol. Ecol.* **236**:235-251.

Ebbestad, J. O. R., 1998, Multiple attempted predation in the Middle Ordovician gastropod *Bucania gracillima*, *Geol. Forenings I Stock. Forhand* **120**:27-33.

Ebbestad, J. O. R., and Peel, J. S., 1997, Attempted predation and shell repair in Middle and Upper Ordovician gastropods, *J. Paleontol.* **71**:1007-1019.

Edward, J. K. P., Ramesh, M. X., and Ayakkannu, K., 1992, Comparative study of holes in bivalves, chipped and bored by the muricid gastropods *Chicoreus ramosus*, *Chicoreus virgineus* and *Murex tribulus*, *Phuket Mar. Biol. Cent. Spec. Publ.* **11**:106-110.

Edwards, A. L. 1988, Latitudinal clines in shell morphologies of *Busycon carica* (Gmelin 1791), *J. Shellfish Res.* **7**:461-466.

Elliott, D. K., and Brew, D. C., 1988, Cephalopod predation on a Desmoinesian brachiopod from the Naco Formation, Central Arizona, *J. Paleontol.* **6**:145-147.

Elner, R. W., and Lavoie, R. E., 1983, Predation on American oysters (*Crassostrea virginica* [Gmelin]) by American lobsters (*Homarus americanus* Milne-Edwards), rock crabs (*Cancer irroratus* Say), and mud crabs (*Neopanope sayi* [Smith]), *J. Shellfish Res.* **3**:129-134.

Feifarek, B. P., 1987, Spines and epibionts as antipredator defenses in the thorny oyster *Spondylus americanus* Hermann, *J. Exper. Mar. Biol. Ecol.* **105**:39-56.

Gabriel, J. M., 1981, Differing resistance of various mollusc shell material to simulated whelk attack, *J. Zool. London*, **194**:363-369.

Geary, D., Allmon, W. D., and Reaka-Kudla, M. L., 1991, Stomatopod predation on fossil gastropods from the Plio-Pleistocene of Florida, *J. Paleontol.* **65**:355-360.

Geller, J. B., 1983, Shell repair frequencies of two intertidal gastropods from northern California: microhabitat differences, *Veliger* **26**:113-115.

George, R. W., and Main, A. R., 1968, The evolution of spiny lobsters (Palinuridae): A study of evolution in the marine environment, *Evolution* **22**:803-820.
Gilinsky, N. L., and Bennington, J. B. 1994, Estimating numbers of whole individuals from collections of body parts: a taphonomic limitation of the paleontological record, *Paleobiology* **20**:245-258.
Hallam, A., 1968, Morphology, palaeoecology, and evolution of the genus *Gryphaea* in the British Lias, *Phil. Trans. R. Soc. London* (B) **254**:91-128.
Harper, E. M., 1991, The role of predation in the evolution of cementation, *Palaeontology* **34**:455-460.
Harper, E. M., 1994, Molluscivory by the asteroid *Coscinasterias acitispina* (Stimpson), in: *The Malacofauna of Hong Kong and Southern China* III (B. Morton, ed.), Hong Kong University Press, Hong Kong, pp. 339-355.
Harper, E. M., and Skelton, P. W., 1993, The Mesozoic marine revolution and epifaunal bivalves, *Scripta Geol.*, Special Issue **2**:127-153.
Hickman, C. S., 1985, Gastropod morphology and function, in: *Mollusks: Notes for A Short Course* (T.W. Broadhead, ed.), Univ. Tennessee, Knoxville, Department of Geological Sciences, Studies in Geology 13, pp. 138-156.
Hirayama, R., 1997, Distribution and diversity of Cretaceous chelonoids, in: *Ancient Marine Reptiles* (J. M. Callaway and E. L. Nicholls, eds.), Academic Press, New York, pp. 235-241
Horny, R. J., 1997, Shell breakage and repair in *Sinuitopsis neglecta* (Mollusca, Tergomya) from the middle Ordovician of Bohemia, *Casopis Narodniho muzea Rada prirodovedna* **166**:137-142.
Huene, F. R., Von, 1956, Palaontologie und Phylogenie der niederen Tetrapoden, *Gustav Fischer*, Jena, Deutschland.
Juanes, F., 1992, Why do decapod crustaceans prefer small-sized molluscan prey?, *Mar. Ecol. Progr. Series* **87**:239-249.
Kardon, G., 1998, Evidence from the fossil record of an antipredatory exaptation: Conchiolin layers in corbulid bivalves, *Evolution* **52**:68-79.
Kamat, S., Su, X., Ballarini, R., and Heuner, A. H., 2000, Structural basis for the fracture toughness of the shell of the conch *Strombus gigas*, *Nature* **405**:1036-1040.
Kauffman, E. G., 1972, *Ptychodus* predation in a Cretaceous *Inoceramus*, *Palaeontology* **15**:439-444.
Kelley, P. H., and Hansen, T. A., 1996, Recovery of the naticid gastropod predator-prey system from the Cretaceous-Tertiary and Eocene-Oligocene extinctions, in: *Biotic Recovery from Mass Extinction Events* (M. B. Hart, ed.), Geological Society Special Publication **102**, London, pp. 373-386.
Kohn, A. J., 1992, *Conus striatus* survives attacks by gonodactyloid!, *Veliger* **35**:398- 401.
Kohn, A. J., 1999, Antipredatory defences of shelled gastropods, in: *Functional Morphology of the Invertebrate Skeleton* (E. Savazzi, ed.), John Wiley and Sons, Ltd, London, pp. 169–181.
Kohn, A. J. and Arua, I., 1999, An Early Pleistocene molluscan assemblage from Fiji: gastropod fanual composition, paleoecology and biogeography, *Palaeogeogr. Palaeoclim. Palaeoecol.* **146**:99-145.
Kowalewski, M., Flessa, K. W., and Marcot, J. D., 1997, Predatory scars in the shells of a Recent lingulid brachiopod: Paleontological and ecological implications, *Acta Palaeontol. Polon.* **42**:497-532.
Krantz, G. E., and Chamberlain, J, V., 1978, Blue crab predation on cultchless oyster spat, *Proc. Nat. Shellfisheries Assoc.* **68**:38-41
Kraueter, J. N., 2001, Predators and Predation, in: *Biology of the Hard Clam* (J. N. Kraeuter and M. Castagna, eds.), Elsevier Science, New York, pp. 441-589.
Kropp, R. K., 1992, Repaired shell damage among soft-bottom mollusks on the continental shelf and upper slope north of Point Conception, California, *Veliger* **35**:36-51.
Lau, C. J., 1987, Feeding behavior of the Hawaiian slipper lobster *Scyllarides squammosus* with a review of decapod crustacean feeding tactics on molluscan prey, *Bull. Mar. Sci.* **41**:378-391.
Lawton, P., and Hughes, R. N., 1985, Foraging behaviour of the crab *Cancer pagurus* feeding on the gastropods *Nucella lapillus* and *Littorina littorea*: comparisons with optimal foraging theory, *Mar. Ecol. Prog. Ser.* **27**:143-154.
Lindström, A., and Peel, J. S., 1997, Failed predation and shell repair in the gastropod *Poleumita* from the Silurian of Gotland, *Vestnik eskeho geologickeho ustavu* **72**:115-126.
Leighton, L. R., 1999, Possible latitudinal predational gradient in middle Paleozoic oceans, *Geology* **27**:47-50.
Leighton, L. R., 2001, New directions in the paleoecology of Paleozoic brachiopods, in: *Brachiopods Ancient and Modern: A Tribute to G. Arthur Cooper* (S. J. Carlson and M. R. Sandy, eds.), Paleontological Society Papers, Vol. 7, pp. 185-205.
Leighton, L. R., In press, Inferring predation intensity in the marine fossil record, *Paleobiology*.

Leonard, G. H., Bertness, M. D., and Yund, P., 1999, Crab predation, waterborne cues, and inducible defenses in the blue mussel, *Mytilus edulis, Ecology* **80**:1-14.

Liljedahl, L., 1985, Ecological aspects of a silicified bivalve fauna from the Silurian of Gotland, *Lethaia* **18**:53-66.

Magalhaes, H., 1948, An ecological study of snails of the genus *Busycon* at Beaufort North Carolina, *Ecol. Mongr.* **18**:379-409.

Mapes, R. H., and Hansen, M. C., 1984, Pennsylvanian shark-cephalopod predation: a case study, *Lethaia* **17**:175-183.

Mapes, R. H., Fahrer, T. R., and Babcock, L. E., 1989. Sublethal and lethal injuries of Pennsylvanian conularids from Oklahoma, *J. Paleontol.* **63**:34-37.

Massare, J. A., 1987, Tooth morphology and prey preference of Mesozoic marine Reptiles, *J. Vert. Paleontol.* **7**:121-137.

Mauzey, K. P., Birkeland, C., and Dayton, P. K., 1968, Feeding behavior of asteroids and escape responses of their prey in Puget Sound region, *Ecology* **49**:603-619.

Nedin, C., 1999, *Anomalocaris* predation on non-mineralized and mineralized trilobites, *Geology* **27**: 987-990.

Norton, S. F., 1988, Role of the gastropod shell and operculum in inhibiting predation by fishes, *Science* **241**:92-94.

Oliver, J. S., Slattery, P. N., E. F. O'Connor, E. F., and Lowry, L. F., 1983, Walrus, *Odobenus rosmarus*, feeding in the Bering Sea benthos, *Fisheries Bull.* **81**:501-512.

Olsen, S. L., and Steadman, D. W., 1978, The fossil record of Glareolidae and Haematopodidae (Aves, Charadriiformes), *Proc. Biol. Soc., Washington* **91**:972-981.

Palmer, A. R., 1979, Fish predation and the evolution of gastropod shell sculpture: Experimental and geographic evidence, *Evolution* **33**:697-713.

Palmer, A. R., 1983, Relative cost of producing skeletal organic matter versus Calcification: evidence from marine gastropods, *Mar. Biol.* **75**:287-292.

Palmer, A., 1985, Quantum changes in gastropod shell morphology need not reflect speciation, *Evolution* **39**:699-705.

Palmer, A., 1990, Effect of crab effluent and scent of damaged conspecifics on feeding, growth, and shell morphology of the Atlantic dogwhelk *Nucella lapillus* (L.), *Hydrobiologia* **193**:155-182.

Papp, A., Zapfe, H., Bachmaer, F., and Tauber, A. F., 1947, Lebbenspurren mariner Krebse, *K. Akad. Wiss. Wien, Math.-Naturewiss. Kl. Sitz.-ber.* **155**:281-317.

Pan, H.-Z., 1991. Lower Turonian gastropod ecology and biotic interation in *Helicaulax* community from western Tarim Basin, southern Xinjiang, China, *Paleoecol. China* **1**:266-280.

Peel, J. S., 1984, Attempted predation and shell repair in *Euomphalopterus* (Gastropoda) from the Silurian of Gotland, *Bull. Geol. Soc. Denmark* **32**:163-168.

Peterson, C. H., and Quammen, M. L., 1982, Siphon nipping: its importance to small fishes and its impact on growth of the bivalve *Prothothaca staminea* (Conrad), *J. Exper. Mar. Biol. Ecol.* **63**:249-268.

Preston, S. J., Roberts, D., and Montgomery, W. I., 1993, Shell scarring in *Calliostoma zizyphinum* (Prosobranchia: Trochidae) from Strangford Lough, Northern Ireland, *J. Moll. Studies* **59**:211-222.

Raffaelli, D. G., 1978, The relationship between shell injuries, shell thickness, and habitat characteristics of the intertidal snail *Littorina rudis* Maton, *J. Moll. Studies* **44**:166-170.

Ramsey, K., Kaiser, M. J., C., Richardson, C. A., Veale, L. O., and Brand, A. R., 2000, Can shell scars on dog cockles (*Glycymeris glycymeris* L.) be used as indicators of fishing disturbance?, *J. Sea Res.* **43**:167-176.

Ramsey, K., Richardson, C.A., and M.J. Kaiser, 2001, Causes of shell scarring in the dog cockles *Glycymeris glycymeris* L., *J. Sea Res.* **45**:131-139.

Randall, J. E., 1964, Contributions to the biology of the queen conch, *Stombus gigas, Bull. Mar. Sci. Gulf Carib.* **14**:246-295.

Ray, M., and Stoner, A. W., 1995, Predation on a tropical spinose gastropod: the role of shell morphology, *J. Exper. Mar. Biol. Ecol.* **187**:207-222.

Reid, D. G., 1992, Predation by crabs on *Littoraria* species (Littorinidae) in Queensland mangrove forest. *Proc. 3rd Int. Symp. Littorinid Biol*, Malacological Society, London, pp. 141-151.

Repenning, C. A., 1976a, *Enhydra* and *Enhydriodon* from the Pacific Coast of North America, *J. Res., U. S. Geol. Surv.* **4**:305-315.

Repenning, C. A., 1976b, Adaptive evolution of sea lions and walruses, *Syst. Zool.* **25**:375-390.

Rhoads, D. C., Lutz, R. A., Cerrato, R. M., and Revelas, E. C., 1982, Growth and predation activity at deep-sea hydrothermal vents along the Galapagos Rift, *J. Mar. Res.* **40**:503-516.
Robba, E., and Ostinelli, F., 1975, Studi Paleoecologici sul Pliocene Ligure. Testimonianze di predazione sui Molluschi Pliocenici di Albenga, *Riv. Ital. Palaeont.* **81**:309-372.
Rolfe, W. D., 1969, Phyllocarida, in: *Treatise on Invertebrate Paleontology, Part. R, Arthropoda. 4* (R. C. Moore and C. Teichert, eds), Geological Society of America, Boulder, Colorado, and University of Kansas, Lawrence, pp. R291-331.
Sarycheva, T. S., 1949, Contribution à l étude des lesions durant la vie des coquille de Productides du Carbonifère, *Trav. Inst. Paleontol. Acad. Sci. U. R. S. S.*, 280-292.
Savazzi, E., 1991, Constructional morphology of strombid gastropods, *Lethaia* **24**:311-324.
Schindel, D. E., Vermeij, G. J., and Zipser, E., 1982, Frequencies of repaired shell fractures among the Pennsylvanian gastropods of north-central Texas, *J.Paleontol.* **56**:729-740.
Schram. F. R., 1981, Late Paleozoic crustacean communities, *J. Paleontol.* **55**:126-137.
Schmidt, N., 1989, Paleobiological implications of shell repair in Recent marine gastropods from the northern Gulf of California, *Hist. Biol.* **3**:127-139.
Seed, R. 1992, Crabs as predators of marine bivalve molluscs, in: *Proceedings International Conference on Marine Biology, Hong Kong and Southern China* (B. Morton, ed.), Hong Kong University Press, Hong Kong. pp. 393-418.
Seed, R., and Hughes, R. N., 1995, Criteria for prey size-selection in molluscivorous crabs with contrasting claw morphologies, *J. Exper. Mar. Biol. Ecol.* **193**:177-195.
Sepkoski, J., 1984, A kinetic model of Phanerozoic taxonomic diveristy. III. Post-Paleozoic families and mass extinction, *Paleobiology* **10**:246-267.
Shanks, A. L., and Wright, W. G., 1986, Adding teeth to wave action: the destructive effects of wave-born rocks on intertidal organisms, *Oecologia* **69**:420-428.
Shoup, J. B., 1968, Shell opening by crabs of the genus *Calappa, Science* **160**:887-888
Signor, P. W. III, 1985, The role of shell geometry as a deterrent to predation in terebrid gastropods, *Veliger* **28**:179-185.
Signor, P. W., III, and, Brett, C. E., 1984, The mid-Paleozoic precursor to the Mesozoic marine revolution, *Paleobiology* **10**:229-245.
Smith, L. D., and Jennings, J. A., 2000, Induced defensive responses by the bivalve *Mytilus edulis* to predators with different attack modes, *Mar. Biol.* **136**:461-469.
Spencer, W. K., And Wright, C. W., 1966, Asterozoans, in: *Treatise on Invertebrate Paleontology, Part U, Echinodermata 3* (R.C. Moore, ed.), Geological Society of America, Boulder, Colorado, and Universisty of Kansas Press, Lawrence, pp.U4-107.
Spight, T. M., and Lyons, A., 1974, Development and functions of the shell sculpture of the marine snail *Ceratostoma foliatum, Mar. Biol.* **24**:77-83.
Stanley, S. M., 1988, Adaptive morphology of the shell of bivalves and gastropods, in: *The Mollusca*, vol. 11, *Form and Function* (E. R. Trueman and M. R. Clarke, eds.), Academic Press, New York, pp. 105-141.
Taylor, J. D., and Layman, M., 1972, The mechanical properties of bivalve (mollusca) shell structures, *Palaeontology* **15**:73-87.
Trussell, G. C., 1996, Phenotypic plasticity in an intertidal snail: the role of a common crab predator, *Evolution* **50**:448-454.
Vale, F. K., and Rex, M. A., 1988, Repaired shell damage in deep sea prosobranch gastropods from the western North Atlantic, *Malacologia* **28**:65-79.
Vale, F. K., and Rex, M. A., 1989, Repaired shell damage in a complex of rissoid gastropods from the upper continental slope south of New England, *Nautilus* **103**:105-108.
Vermeij, G. J., 1977, The Mesozoic marine revolution: evidence from snails, predators and grazers, *Paleobiology* **3**:245-258.
Vermeij, G. J., 1978, *Biogeography and Adaptation: Patterns of Marine Life*, Harvard University Press, Cambridge.
Vermeij, G. J., 1982a, Gastropod shell form, breakage, and repair in relation to predation by the crab *Calappa, Malacologia* **23**:1-12.
Vermeij, G. J., 1982b, Unsuccessful predation and evolution, *Am. Nat.* **120**:701-720.
Vermeij, G. J., 1982c, Environmental change and the evolutionary history of the periwinkle (*Littorina littorea*) in North America, *Evolution* **36**:561-580.
Vermeij, G. J., 1983a, Shell-breaking predation through time, in: *Biotic Interactions in Recent and Fossil Benthic Communities* (M. J. S. Tevesz and P. L. McCall, eds.), Plenum Press, New York. pp. 649-669.

Vermeij, G. J., 1983b, Traces and trends of predation, with special reference to bivalved animals, *Palaeontology* **26**:455-465.

Vermeij, G. J., 1987, *Evolution and Escalation: An Ecological History of Life.* Princeton Univ. Press, Princeton.

Vermeij, G. J., 1989, Interoceanic differences in adaptation: effects of history and productivity, *Mar. Ecol. Progr. Series* **57**:*293-305.*

Vermeij, G. J., 1993, *A Natural History of Shells.* Princeton Univ. Press, Princeton.

Vermeij, G. J., and Currey, J. D., 1980, Geographical variation in the strength of thaidid snail shells, *Biol. Bull.* **158**:383-389.

Vermeij, G. J., and Dudley, E. C., 1982, Shell repair and drilling frequency in some gastropods from the Ripley Formation (Upper Cretaceous) of south-eastern U.S.A., *Cretaceous Res.* **3**:397-403

Vermeij, G. j., and Dudley, E. C., 1985, Distribution of adaptations: A comparison between the functional shell morphology of freshwater and marine pelecypods, in: *The Mollusca, Vol. 10, Evolution* (E. R. Trueman and M. R. Clarke, eds.), Academic Press, Inc., New York. pp. 461-478.

Vermeij, G. J., and J. A. Veil, J. A., 1978, A latitudinal pattern in bivalve shell gaping, *Malacologia* **17**:57-61.

Vermeij, G. J., Zipser, E., and Dudley, E. C., 1980, Predation in time and space: peeling and drilling in terebrid gastropods, *Paleobiology* **6**:352-364.

Vermeij, G. J., Schindel, D. E., and Zipser, E., 1981, Predation through geological time: evidence from gastropod shell repair, *Science* **214**:1024-1026.

Vermeij, G. J., Zipser, E., and Zardini, R., 1982, Breakage-induced shell repair in some gastropods from the Upper Triassic of Italy, *J. Paleontol.* **56**:233-235.

Walker, S. E., 2001, Paleoecology of gastropods preserved in turbidite slope deposits from the Upper Pliocene of Ecuador, *Palaeogr. Palaeoclim. Palaeoecol.* **166**:141-163.

Walker, S., and Brett, C. E., in press, Post-Paleozoic patterns in marine predation: Was there a Mesozoic and Cenozoic marine predatory revolution?, in: *The Fossil Record of Predation* (M. Kowalewski and P. H. Kelley, eds.), Paleontological Society Papers, Vol. 8.

Walker, S. E., and Voight, J. R., 1994, Paleoecologic and taphonomic potential of deepsea gastropods, *Palaios* **9**:58-59.

Walker, S. E., and Yamada, S. B., 1993, Implications for the gastropod fossil record of mistaken crab predation on empty mollusc shells, *Palaeontology* **36**:735-741.

Wells, H. W., 1958, Feeding habits of *Murex fulvescens, Ecology* **39**:556-558.

West, K., and Cohen, A., 1996, Shell microstructure of gastropods from Lake Tanganyika, Africa: Adaptation, convergent evolution, and escalation, *Evolution* **50**:672-681.

Yamaguchi, M., 1977, Shell growth and mortality rates in the coral reef gastropod *Cerithium nodulosum* in Pago Bay. Guam, Mariana Islands, *Mar. Biol.* **44**:249-263.

Zach, R., 1978, Selection and dropping of whelks by north-western crows, *Behaviour* **67**:134-148.

Chapter 7

Predation on Cephalopods
A General Overview with a Case Study From the Upper Carboniferous of Texas

ROYAL H. MAPES and DAVID T. CHAFFIN

1. Introduction 177
2. Background 179
   2.1. Sublethal and Lethal Shell Damage and Abnormalities in Present-day *Nautilus* 180
   2.2. Sublethal Shell Damage and Abnormal Shells in Fossil Cephalopods 182
   2.3. Lethal Damage in Fossil Cephalopods 183
3. Case Study: Lethal Predation on Upper Carboniferous Coiled Nautiloids and Ammonoids ...187
   3.1. Background: Stratigraphic and Paleoenvironmental Considerations 187
   3.2. Methodology and Generic Identities of the Case Study Specimens 188
   3.3. Predation Versus Taphonomy of the Finis Cephalopods 189
   3.4. Analysis of the Coiled Nautiloids 197
   3.5. Analysis of the Ammonoids 201
   3.6. Lethal Damage: Comparisons Between the Ammonoids and Coiled Nautiloids............203
   3.7. A Hypothetical Predation Scenario 203
   3.8. Summary and Conclusions Drawn from the Upper Carboniferous Case Study..............205
4. Studies of Predation and Cephalopods Through Time 206
5. Conclusions and Future Studies 208
   Appendix 209
   References 210

1. Introduction

Predation occurs throughout nature. Predators feed in order to survive long enough to reproduce. Likewise, prey animals attempt to avoid being eaten long enough to reproduce. Avoidance of predation is much more important to the prey because a failed attempt at predation only necessitates that the predator searches elsewhere for a meal.

ROYAL H. MAPES and DAVID T. CHAFFIN • Department of Geological Sciences, Ohio University, Athens, Ohio 45701.

Predator-Prey Interactions in the Fossil Record, edited by Patricia H. Kelley, Michał Kowalewski, and Thor A. Hansen. Kluwer Academic/Plenum Publishers, New York, 2003.

FIGURE 1. Damage created by lethal and sublethal predation on present-day *Nautilus*. (A) Triggerfish (*Balistoides* sp.) attacking *N. belauensis* in Palau after the animal had been released in shallow water; according to Saunders *et al.* (1987), the attack ultimately proved to be lethal despite the rescue of the specimen while the animal was still alive. (B) Damage to *N. pompilius* inflicted by a grouper (*Epinephelus* sp.) after release in shallow water in Manus, Papua New Guinea. The attack produced large jagged, scalloped-shaped breaks at the aperture. L.E. Davis, who collected the damaged shell, witnessed the attack. (C) Collection of damaged *N. macromphalus* collected from a beach on Lefou, Loyalty Islands, by R. A. Davis in 1975. One of these specimens is better illustrated in Fig. 3. Most of this damage probably represents lethal predator damage. (D) Sublethal damage to the aperture of young *N. belauensis*, Palau, in the form of V-shaped embayments. (E) Mandible of *Nautilus* sp. and a V-shaped break in the aperture of *N. belauensis* that fits the shape of the mandible. (Photographs A, B, and D, courtesy of W. B. Saunders, for additional information see Saunders *et al.*,1987, photograph C courtesy of R. A. Davis, and photograph E courtesy of Desmond Collins.)

Successful predation certainly means the prey's demise. Cephalopods can be both predator and prey. Studies of predation on present-day and fossil cephalopods (exclusive of the Coleoidea, which are not considered in this report) are relatively few.

Present-day *Nautilus* is the only living externally shelled cephalopod, and, therefore, virtually all observations on fossil cephalopods eventually come back to a comparison with this living model. Actual photography of a lethal attack involving present-day *Nautilus* is limited to a single event involving a teleost fish (Saunders *et al.*, 1987) (Fig. 1 A, B).

Other reports of lethal and sublethal damage include direct (but not photographed) observations and indirect evidence, such as the stomach contents of predators and the presence of bore holes in *Nautilus* shells that are presumably produced by octopus attacks (Tucker and Mapes, 1978*a*; and see Saunders *et al.*, 1987, for other reports) (Fig.

2 C, D, E). In addition to these lethal and sublethal examples, *Nautilus* inflicts damage on its own kind. Haven (1972) reported new V-shaped shell damage that occurred in a group of newly caught and caged *Nautilus* specimens and speculated that these breaks were bite marks from conspecific fighting and, possibly, attempted cannibalism. Shell damage in the form of these V-shaped bite marks on live-trapped specimens is common in *Nautilus* shells (Haven, 1972; Arnold, 1985; and Saunders *et al.*, 1987) (Fig. 1 D, E).

Additional studies on predation of *Nautilus* are necessary to develop a better understanding of predator-prey relationships. With better understanding of cephalopod predator-prey relationships and variations in shell damage, these organisms will provide a valuable tool for enhanced taphonomic understanding of cephalopod shell accumulations, paleoecological reconstructions, and evolutionary studies of fossil cephalopods through the Phanerozoic.

2. Background

Externally shelled cephalopod predation can be considered from three perspectives. One is the information that we have observed and determined about present-day *Nautilus*; another is specific studies of specific fossil cephalopod collections and the

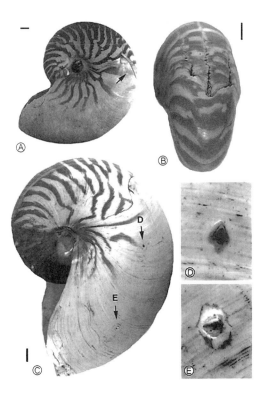

FIGURE 2. (A) *Nautilus* sp. (OUZC 4067) with massive repaired break at arrow (scale bar = 1 cm). (B) Juvenile specimen of *Nautilus pompilius* (OUZC 4068) showing three black-streaked repaired areas on venter that altered shell coloration (scale bar = 1 cm), (C, D, E) *Nautilus* sp.(OUZC 4069) with massive repaired break at posterior end of body chamber (scale bar = 1 cm). Specimen has two octopus borings shown in D, E.

third, which is considered at the end of this chapter, is an overview of the fossil cephalopod record and how it relates to predator evolution.

With both the present-day observations and the record of events preserved on the shells of ancient cephalopods, it is possible to estimate and, in some cases, determine with a high degree of probability whether the shell damage was lethal (in which case, no repairs to shell damage occurred) or sublethal (damage where repair of the shell indicates the animal survived the attack). Shell damage (both lethal and sublethal) can be caused by agencies other than predators, and taphonomic factors may cause post-mortem damage. Indeed, separation of predatory activity from other causes of shell damage is one of the most difficult aspects of investigation of predation in cephalopods. An attempt to distinguish presumed predation events from other causes of shell damage is considered in the case study provided in this overview of cephalopod predation.

In the Finis Shale case study discussed in this report, complete specimens of coiled nautiloids and ammonoids are unusually rare; most specimens were damaged prior to or sometime after burial. Specimen incompleteness may have been caused by storm breakage, shell collapse by lithostatic pressure, crushing and/or dissolution while buried, mechanical damage by weathering and erosion after being exposed, collision with hard substrates while the animal was alive, scavenging, or collision of the empty shell on a hard substrate as it was transported after death, bioturbation of empty shells on the seafloor, or predation. By exploring taphonomic pathways, most of the causes of conch damage on the Finis Shale coiled nautiloids and ammonoids can be partly or completely determined. When the taphonomic causes of damage are eliminated, then it is probable that the damage observed on an ammonoid or a coiled nautiloid is the product of predators.

2.1. Sublethal and Lethal Shell Damage and Abnormalities in Present-day *Nautilus*

Sublethal damage to the soft tissues of present-day *Nautilus* without altering shell growth has been observed and documented on live-trapped specimens (R. A. Davis and W. B. Saunders, pers. comm., 2001). Damage is usually confined to the exposed tentacles and hood, and leaves no record of the event on the shell of the animal. However, if damage to the mantle occurs and the animal survives the attack (Figs. 1 C, D, 2 A, B, C), then shell growth and/or repair at the damaged area will usually reflect this traumatic event.

Most shells of present-day *Nautilus* show signs of repaired shell damage (e.g., Willey, 1903; Arnold, 1985; Bond and Saunders, 1989), suggesting that individuals of this genus are subject to frequent sublethal attacks that damage the mantle tissue. Sometimes the identity of the predator can be determined by the resultant marks on the shell. Tucker and Mapes (1978a) recorded a boring frequency of 28.7% on a purchased set of presumably drift shells of *Nautilus pompilius* from an unknown location in the Philippines (Fig. 2 C, D, E). They concluded that these borings were caused by predatory octopus attacks. Later, Saunders *et al.* (1987, 1991) and Arnold (1985), in live-trapping expeditions in the Indopacific, reported the following bore hole frequencies produced by octopus predation on several different species of *Nautilus*: *N. pompilius* – the ALPHA HELIX expedition in the Philippines had four bored specimens out of 353 live-caught individuals (1.1%); *N. pompilius* – Papua New Guinea had five bored individuals out of 270 live-caught specimens (1.9%); and *N. belauensis* – Palau had 211 bored specimens out of 2720 live-caught specimens (7.5%); Tanabe *et al.* (1988)

reported that in Fiji only one of 41 specimens (0.4%) was bored. Thus, it appears that octopi are capable of sometimes killing *Nautilus* by boring small holes and injecting venom that subdues the prey prior to ingestion. The high percentage of bored shells in the Tucker and Mapes (1978a) study suggests that collections of drift shells may be selectively enhanced in numbers by octopus predation. However, additional study of collections of *Nautilus* drift shells needs to be accomplished before a definitive statement can be made concerning the overall influence of predators on shell damage, geographic distribution, and beach accumulations.

Fish are considered to be major predators of *Nautilus*. Saunders et al. (1987), in their study of *Nautilus* predation, provided the only photographically documented example of fish predation on a living *Nautilus* (Fig. 1 A). In their description of the event, the fish forced the *Nautilus* against a reef and systematically slashed at the aperture and hood of the *Nautilus* and spit out pieces of shell. This attack resulted in scallop-shaped breaks in the apertural edge of the body chamber of the shell (Fig. 1 B). In one case the *Nautilus* was rescued before the fish could eat its prey; however, they reported that the released animal probably died as a result of this attack. Additional evidence of lethal fish predation was provided by Willey (1903), Lehmann (1976), Ward (1984), and others, who indicated that fish, including "conger eels" and "sea perch," prey on *Nautilus*. Some sharks also are known to eat *Nautilus*; both Ward (1984) and Tanabe et al. (1988) reported recovering *Nautilus* mandibles in the stomachs of cat sharks. In a surprising situation, Ward (1998) observed that sea turtles in an aquarium attacked and ate *Nautilus* placed in the same turtle holding tank. Indeed, the turtles were observed to crush the shell in a way that was "like hitting a porcelain plate with a hammer" (Ward, 1998, p. 138), and then the body of the *Nautilus* was eaten with what appeared to be remarkable ease and familiarity.

Sublethal damage in present-day *Nautilus* is relatively common (Fig. 2 A, B, C). Virtually every shell we have examined has repaired damage, ranging from the interruption of a few growth lines to massive removal of large parts of the body chamber. Frequently, in association with the larger shell breaks, there is a secretion of black material at the injury site that is presumably the same as the black layer deposited on the dorsum of the body chamber (Fig. 2 B). When critical parts of the mantle are damaged in predator attacks, shell abnormalities frequently result in *Nautilus* (Arnold, 1985). These abnormal shells have unusual features such as blister pearls, loss of color banding, asymmetry of the section of repaired shell, unusual growth lines, changes from normal umbilical growth, and changes in the rate of coiling.

Under some conditions, *Nautilus* exhibits conspecific shell-damaging behavior (Haven, 1972; Arnold, 1985; Saunders et al., 1989). Such behavior has been observed when numerous *Nautilus* specimens were caged together in a restricted space. Some of these caged specimens that had complete shells, when placed in the cage, developed V-shaped gaps in the apertural margin of their shells that match the shape and size of the mandibles of the other caged *Nautilus* specimens (Fig. 1 D). Because all other large predators were excluded from the cage, other *Nautilus* individuals must have caused the damage. It is not clear whether this V-shaped damage is a byproduct of mating behavior or whether *Nautilus* was exhibiting cannibalistic tendencies. However, it is clear that, if aggressive *Nautilus* individuals had sufficient time to continue the attacks, these sublethal events in the cage would have become lethal events. Indeed, *Nautilus* is known to use scavenging as its major feeding strategy (Tshudy et al. 1989); however, this animal can be an opportunistic predator under certain circumstances. Thus, in a crowded situation in a confining cage, some *Nautilus* individuals are likely to act as opportunistic predators. The differentiation between sublethal *versus* lethal attacks in these artificial situations probably rests on the time span the animals are caged and the availability of supplied food. The fact that these V-shaped divots in the shells of live-trapped specimens are often repaired suggests that sublethal encounters with conspecifics occur with some regularity in natural circumstances.

2.2. Sublethal Shell Damage and Abnormal Shells in Fossil Cephalopods

Shell abnormalities represent one of the largest sources of data on sublethal damage in fossil cephalopods (for examples, see Hengsbach, 1996, and Kröger, 2000, with their extensive bibliographies on this subject). Many of these abnormalities are attributed to parasitic infestations, diseases, or repairs after the body and shell of the cephalopod were attacked and damaged by external forces such as predators. Abnormal shell development also can be produced by artificial environmental conditions, as is exhibited in *Nautilus* shell growth in aquariums (Martin *et al.*, 1978; Arnold, 1985). This latter condition has not yet been documented in the fossil record; however, such shells may exist and must be differentiated from shells malformed as a result of sublethal damage inflicted by predators.

Shapes of breaks in fossil cephalopod shells are often ascribed to different kinds of predators based on present-day observations of the damage produced by predators on other molluscs (mainly gastropods and pelecypods). As previously stated, only one case (Saunders *et al.*, 1987) of shell damage by any specific predator has been documented for present-day *Nautilus*. Thus, the style of damage produced by most types of potential predators on *Nautilus* is inferred, and not available for direct comparison. For example, the repaired crescentic breakage in fossil cephalopods is usually interpreted as damage inflicted by fish or arthropod, including crab, attacks (for examples, see Thiermann, 1964; Roll, 1935; Lehmann, 1976; Keupp and Ilg, 1992). We suspect that, in most fossil cases, shell damage has been attributed correctly. However, some caution should be used in making conclusions about predator-inflicted damage, since studies on present-day *Nautilus* have not been performed, and we do not really know the types of damage that can be produced by different forms of offensive armament (e.g., teeth that are used to cut, as compared to teeth that are used to puncture).

Sublethal damage presumably caused by predation in early and middle Paleozoic cephalopods has not been studied in detail. Abnormalities in shells are known (for example, the orthoconic nautiloid in Barrande, 1869, pl. 299); however, these abnormal shells are simply mentioned or are illustrated with no extensive analysis of the lethal or sublethal damage. In fact, we are aware of no comprehensive studies of predators and of sublethal or lethal damage on any populations of any group of fossil nautiloids; all of the major studies using large sample sizes have concentrated on the Ammonoidea.

The only extensive study of sublethal damage in the Paleozoic is by Bond and Saunders (1989) on an Upper Mississippian ammonoid assemblage from the Imo Formation of Arkansas. Their data set included more than 2000 specimens. They concluded that the Mississippian ammonoids were preyed upon less frequently than is present-day *Nautilus*, with 15% of the ammonoid shells recording sublethal damage, compared to 57% of the *Nautilus* specimens from Palau showing similar repaired injuries. V-shaped breakage and repair on many of the ammonoids described by Bond and Saunders (1989) are consistent with the jaw structure of upper Paleozoic cephalopods (e.g., Tanabe and Mapes, 1995; Doguzhaeva *et al.*, 1998), suggesting that conspecific attacks or attempted cannibalism probably took place.

Relatively few studies of sublethal damage in Mesozoic ammonoids have been published. There are no studies known to us that deal with sublethal damage in Triassic ammonoids. Jurassic ammonoids have received moderate study (see Keupp, 2000; Kröger, 2000 for numerous citations), and in the Upper Cretaceous only the scaphitid heteromorphs have been examined (Landman and Waage, 1986). Abnormal specimens are known to us from all these time periods. However, in the literature on abnormal specimens, reports typically focus on the deformity with little information on the cause of the abnormality, which could be the result of predation damage, or damage caused by parasitic infestations and disease.

Significant sublethal damage to Jurassic ammonoids that was probably in part related to predation was described by Geczy (1965), Guex (1967), Bayer (1970), and Morton (1983). Geczy (1965) reported that ammonoid collections from Hungary contained a variety of abnormal shells including damaged shells that had been repaired. Guex (1967) reported that 2% of the 2000 ammonoid specimens from Aveyron, France, had repaired shell breaks, including 20 of 800 specimens (2.0%) from the *bifrons* zone. Bayer (1970), in his study of Middle Jurassic ammonoids from Germany, noted different proportions of repaired scars in different families (Graphoceratinae with a frequency of 0.3%; Sonniniidae with a frequency of 1.0%; Stephanocerataceae with a frequency of 9.7%, excluding the Sphaeroceratinae which had a frequency of 1.4%). He explained these differences in damage repair frequencies by suggesting that each ammonoid group probably had a different life mode. Morton (1983) analyzed the occurrence of a single ammonoid genus from the Isle of Skye in Scotland. He reported a repair frequency of 0.7%. He attributed the abnormalities as being due to either parasites or disease, but recognized that sublethal predation could also have caused such damage.

In a study of Upper Cretaceous scaphitid ammonites, Landman and Waage (1986) concluded that approximately 10% of the 2000 specimens they analyzed showed repaired external damage. Also, after completion of the Cretaceous study and surveying the limited data available in the literature, they concluded that there was no general increase or decrease in predation from the Mississippian through the Cretaceous. Despite the fact that there was no detectable relationship between predation repair frequencies and morphologic changes among ammonoids, they did discern a gradual increase in the relative numbers of coarsely ornamented ammonoids compared to weakly ornamented ammonoids. They declined to attribute this ornamental change through time to predation as an all-embracing mechanism because they thought that some of the injuries were not caused by predators, and because the correlation between sublethal and lethal injuries could be poor.

Additional studies on sublethal damage in fossil cephalopods are highly desirable. Indeed, this area of paleobiology seems ripe for study, with well-preserved cephalopod faunas lying untouched in fossil repositories around the world. Such studies would, for example, allow the additional testing of Vermeij's (1987) escalation hypothesis on the evolutionary changes in cephalopod shell structure and ornament in the fossil record.

2.3. Lethal Damage in Fossil Cephalopods

Few studies involve shell damage produced at the time the cephalopod was killed and eaten by a predator. Arguably, the most famous report of lethal cephalopod predation was written by Kauffman and Kesling (1960), who analyzed the circular and semicircular perforations in the shell of a Cretaceous *Placenticeras* ammonoid. They suggested that the holes were bite marks from a mosasaur that bit the ammonite sixteen times. Later, Kauffman (1990) expanded the shell perforation model by examining the overall extent of reptilian predation on ammonites during the Cretaceous. Ward and Hollingsworth (1990) reported a Jurassic ammonoid, *Kosmoceras,* that had a shell perforated by an attack by a marine reptile.

The 1960 analysis by Kauffman and Kesling was recently challenged by Kase *et al.* (1998) and Seilacher (1998), who considered the alleged bite marks to be limpet home-scars produced by limpet rasping of the ammonoid shell (Kase *et al.,* 1994), followed by lithostatic crushing of the home scar site, producing circular and semicircular holes. To support their limpet home-scar hypothesis, Kase *et al.* (1998) constructed a mechanical mosasaur jaw apparatus designed to produce circular holes in present-day *Nautilus* shells like those observed on the *Placenticeras.* Their experiment failed to produce circular and semicircular holes in the *Nautilus* shells tested. Based on this experiment, they concluded that all the circular and semicircular holes in *Placenticeras* were caused by the collapse of limpet home-scars. Ward (1998) independently attempted to produce circular perforations in present-day *Nautilus* shells like those seen on *Placenticeras* shells using a mechanical device approximating a mosasaur jaw. He also failed to create circular holes like those seen on the Cretaceous fossils. He concluded that the ammonites bitten by mosasaurs had their shells crushed prior to consumption.

We suggest, as have Davis *et al.* (1999), that the mechanical mosasaur model experiments of Kase *et al.* (1998) and Ward (1998) used to attempt to produce circular holes in present-day *Nautilus* shells are flawed. Arguments against the all-inclusive conclusion of Kase *et al.* (1998) have not yet been developed in detail; however, we are aware of several points that should be considered before their conclusions are accepted. In 1975 R. A. Davis (pers. comm., 2001) collected 25 freshly drifted *Nautilus* shells on the shoreline at Lifou Island, Loyalty Island Group, in the Pacific Ocean (Fig. 3A - D). All of the specimens were massively damaged (Fig. 1C). Some of the damage is in the form of circular holes (Fig. 3A, B) in the phragmocones and the body chambers of the shells. It is arguable that these circular holes, which are symmetrically placed on both sides of the phragmocone and the body chamber in Fig. 3, were caused by predators, but other taphonomic factors, such as empty shells impacting hard substrates by wave activity, must be considered.

However, based on the specimens collected by Davis, there can be no question that circular holes can be made naturally, and it appears that these circular and oval holes were made in these *Nautilus* shells by methods other than boring, rasping, or the collapse of limpet home sites during burial and compaction of the sediment. Additional arguments against the creation of all the holes in *Placenticeras* shells by limpet home-scar crushing are as follows: (1) Mutvei (1967) suggested that *Nautilus* has a different shell structure than ammonoids, which could in part explain why mechanical mosasaurs failed to produce circular holes in *Nautilus*. Ammonoids typically have much thinner shells (although some *Placenticeras* shells are much thicker than that of *Nautilus*), which are constructed with numerous organic membranes between the nacre sheets as compared to the relatively few organic membranes between the aragonite nacre sheets in the shell of *Nautilus.* (2) We are aware of more than 100 *Placenticeras* shells with circular and semicircular holes in the Royal Tyrrell Museum of Palaeontology, Drumheller, Alberta, Canada (R. A. Davis, pers. comm., 2001); most of the external shells of these ammonoids are buried in the concretion matrix, and none of the exposed shell surfaces has limpets attached to the ammonoid shells. In fact, only three ammonite specimens with encrusting limpet colonies are known to us. The disparity between the numbers of holed limpet-bearing ammonoid specimens and specimens with limpets and limpet home-scars that are not crushed inward has not yet been adequately addressed. (3) Hewitt and Westermann (1990) contended that the complex ammonitic septa promote a strong, flexible shell, and this condition explains why the conchs of ammonites could buckle rather than be fragmented under point loads such as those produced by mosasaur teeth. This observation suggests that septal-supported phragmocones should show holes, whereas the body chambers should fragment. This theoretical observation has not yet been fitted into the controversy. Additionally, they rejected implication of a shallow-water life mode for desmoceratacean ammonites suggested by the limpet home scar scenario (Westermann and Hewitt, 1995). (4) The assumption that ammonites were like *Nautilus* with the chambers of the phragmocone being empty of fluid may not be warranted. If the chambers being put under a point stress (such as by mosasaur teeth) were fluid filled, then it is probable that a circular hole could be created. (5) Mapes and Hansen (1984), Hansen and Mapes (1990), Mapes *et al.*

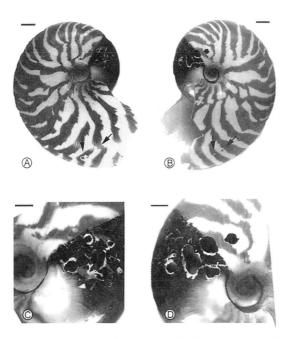

FIGURE 3. *Nautilus macromphalus* (OUZC 4070) collected by R. A. Davis on a beach at Lefou Island, Loyalty Islands. (A, C) Right and left views, respectively, of the specimen showing small diameter circular holes in body chamber at arrows and larger holes on phragmocone in the black layer (scale bar = 1 cm). (B, D) Magnified right and left views, respectively, of holes on phragmocone (scale bar = 5mm). This symmetrical damage was probably caused by a predator with puncture-type teeth. The presence of these circular and oval holes demonstrates that this kind of damage can occur in modern *Nautilus*, contrary to the conclusions of Kase *et al.*, 1998, and Ward, 1998.

(1995), and the case study (herein) report circular holes in both coiled nautiloids and ammonoids of Pennsylvanian age. There is no evidence that limpets or any similar rasping organisms that created circular home sites had evolved by Pennsylvanian time. Thus, they could not have produced the structures necessary to create circular or oval holes in the manner suggested by Kase et al. (1998). The fact that circular holes exist in these Pennsylvanian specimens indicates that there are probably several ways to create circular holes in cephalopod shells.

In the Pennsylvanian case study provided herein, circular and semicircular holes are reported in both coiled nautiloids and ammonoids. Hence an alternative explanation for these features other than the innovative limpet model must be developed. Some of the holes in the case study are similar to the holes described by Mapes and Hansen (1984), who matched the dentition of the Carboniferous shark *Symmorium reniforme* (Zangerl, 1981) to perforations observed on a fragmented Pennsylvanian age coiled nautiloid body chamber. Later, Hansen and Mapes (1990) described and illustrated an orthoconic nautiloid with perforations similar in shape to the teeth of the shark *Petalodontus ohioensis* (Zangerl, 1981). Mapes et al. (1995) studied predation on the ammonoid *Gonioloboceras goniolobum* (Sayre, 1930) from numerous Pennsylvanian localities in Texas, Oklahoma, Kansas, and Missouri. The study included specimens from the same stratigraphic level as the case study provided in this report. They concluded that *Gonioloboceras* shows direct as well as indirect signs of predation from various chondrichthyan predators, including circular and semicircular holes produced by *Symmorium reniforme*.

The controversy regarding the cause of circular and semicircular holes in cephalopod shells is not over. Additional studies on the damage produced by present-day predators on *Nautilus,* as well as examination of cephalopod specimens from throughout the Phanerozoic fossil record, must be done before a conclusive assessment of the kind of damage and the predators (different kinds of fish, crabs, and other cephalopods) that produced the damage can be separated from other kinds of taphonomically produced shell damage such as limpet home site scars and diagenetic crushing.

Evidence of possible lethal cephalopod predation on other cephalopods is the presence of cephalopod mandibles in the crop or stomach area of other cephalopods. In a speculative case, Zangerl et al. (1969), Quinn (1977), and later Dalton and Mapes (1999) and Mapes and Dalton (in press) analyzed the occurrence of large numbers of goniatite ammonoids in carbonate concretion halos around the conchs of large (up to 2 to 3 meters in diameter) actinoceratid nautiloids from a Mississippian shale in Arkansas. All of these researchers concluded that the ammonoids were part of the stomach contents of the nautiloid and that the nautiloid had ingested the ammonoids and then died. Several other workers have observed a similar condition in Mesozoic ammonoids (see summary by Nixon, 1988). All of these cases have only two possible explanations: (1) that one cephalopod attacked and ate another cephalopod, and (2) that a cephalopod scavenged the carcass of a dead cephalopod.

Reeside and Cobban (1960) suggested cephalopods were the prey of predators in the Cretaceous when they speculated that masses of whole and fragmentary shells (often preserved as the nuclei of concretions) were the fecal accumulation of a large unidentified carnivore. Such accumulations of cephalopod debris, including cephalopod mandibles, have been recovered within Carboniferous coprolites preserved as nuclei in phosphate concretions in black shales (Mapes, 1987). Mehl (1978a) reported Late

Jurassic ammonoid aptychi in coprolites and clusters of equal-sized ammonoid shell fragments in the Early Jurassic (Mehl, 1978*b*). This latter case was attributed to teuthoid predators. Westermann (1996) suggested that these teuthoid predators probably were responsible for most of the peristomal mantle injuries in Jurassic and Cretaceous ammonoids, and that when the ammonoids survived attack, the result was an abnormally shaped shell.

3. Case Study: Lethal Predation on Upper Carboniferous Coiled Nautiloids and Ammonoids

3.1. Background: Stratigraphic and Paleoenvironmental Considerations

A diverse collection of cephalopods from the Finis Shale Member of the Graham Formation (Pennsylvanian; lower Virgilian) has been recovered from 13 localities within 80 kilometers of Jacksboro, Young County, Texas (see Chaffin, 2000, for details). The cephalopod fauna encompasses at least twenty-nine genera, which include orthoconic, cyrtoconic, and coiled nautiloids, ammonoids (Miller and Downs, 1950; Boardman *et al.*, 1994), coleoids (Doguzhaeva *et al.*, 1999) and bactritoids (Mapes, 1979). This case study of the coiled nautiloids and ammonoids attempts to determine differences between taphonomic damage and damage caused by lethal predation. After taphonomic damage is eliminated as an explanation, comparisons are made to determine whether shell shape, shell size, and ornamentation influenced predation, and whether there are differences in predation frequency between ammonoids and coiled nautiloids.

The Graham Formation in the Jacksboro, Texas, region is primarily shale with some sandstone and limestone. The formation includes several transgressive-regressive sequences called cyclothems that have received extensive study in north-central Texas by Cleaves (1973), Heckel (1977, 1978, 1980), Boardman *et al.* (1984), Boston (1988) and others. The cephalopods were recovered from part of the Finis Shale Member of the Graham Formation.

In the fossil-bearing interval, there are several different invertebrate communities that could have yielded the study specimens, and, because each community has a different set and number of predators and prey, it was important to identify the specific community and paleoenvironmental conditions that produced the study specimens. Based on community successions identified by Boardman *et al.* (1984), the Finis Shale interval that yielded all of the cephalopods for this study is identified as the "Mature Molluscan Community." This community is characterized by having a fully marine, oxygenated water column from the surface to the water-sediment interface. Ammonoids and other cephalopods, gastropods, pelecypods, rostroconchs, and polyplacophores (Boardman *et al.*, 1984; Hoare and Mapes, 1985; Hoare *et al.*, in press) dominate the Mature Molluscan Community of the Finis Shale. In addition to the molluscs, there are a variety of other fossils, including but not limited to brachiopods, sponges, bryozoans, corals, conulariids, ostracodes, foraminifers, and conodonts (see Boardman *et al.*, 1984) and shark and other fish debris. Thus, the Mature Molluscan Community in the Finis Shale contains a diverse marine fauna that is typical of an oxygenated middle-to-outer-shelf marine environment. Discussions of the Mature Molluscan Community and stratigraphically adjacent communities are found in Boardman *et al.* (1984) and Kammer *et al.* (1986).

The Mature Molluscan Community was probably deposited tens of kilometers from the shoreline (based on basin reconstructions and because the sedimentation rates are modest to low, suggesting an offshore setting relatively far from the active deltas at the shoreline). The water depth, combined with turbidity, was sufficient to deter phylloid algal development that does appear higher in the stratigraphic succession. The rarity of storms is evidenced in part by the lack of reworking of concretions that show a heavier epifaunal growth on one side of the shell. This feature indicates that shell debris was exposed on the bottom prior to burial and was not frequently disturbed by storm events and excessive bioturbation.

Most of the cephalopod specimens collected from the oxic part of the Finis Shale are preserved by infilling the shell with a mud-based carbonate concretion. The quality of the recovered specimens ranges from moderately good to poor, with most specimens retaining some external shell that preserves growth lines. Both the ammonoids and the coiled nautiloids are generally recovered without the body chamber. In most cases, mud apparently entered the empty phragmocones. The early transformation of the mud into concretion material protected the conch from later lithostatic crushing. Where concretionary mud did not fill the phragmocone (i.e., cameral spaces that were probably partly gas filled) and/or the body chamber, later lithostatic pressure crushed the shell, leaving remnants of shell fragments adhering to that part of the three-dimensional conch preserved as a concretion.

3.2. Methodology and Generic Identities of the Case Study Specimens

Coiled nautiloids ($n = 692$) and ammonoids ($n = 193$) from 13 localities were analyzed for this study. Only specimens from the regressive phase of the Finis Cyclothem that fit the preservational and depositional criteria defined for the Mature Molluscan Community by Boardman et al. (1984) were utilized (Figs. 4, 5).

Figured specimens are cataloged with OUZC (Ohio University Zoological Collection) repository numbers. Specimens were measured for height (H), width (W), and diameter (D). Values for incomplete shells were estimated. The specimens were inventoried for presence or absence of the body chamber, encrusting organisms, circular and semicircular holes, irregular holes filled with shell debris, and both lethal and sublethal damage exhibited by missing and repaired shell breakage, respectively. Conch damage was interpreted as potentially lethal if no repair occurred in the damaged area. In the case of sublethal damage, specimens were examined for repaired damage using the criteria developed by Bond and Saunders (1989), which they ranked as minor, moderate, massive, deep-acute, mantle damage, or perforation, depending on the nature and severity of the injury.

The study specimens were identified to the generic level. Because of the great variance in abundance between different genera there is a statistical problem of normalizing the genera so that specimens identified to the generic level can be treated as "populations" that can be evaluated equally. Reasons for the inconsistency in abundance may result from better adaptation by some genera to the paleoecologic conditions, collection bias, or differences in reproduction rates. The problem of normalizing the populations was resolved by using a Z-test for independent proportions. This test is designed for comparisons of large samples for equality of two independent proportions. The test effectively equalizes the numbers, allowing for an equal contribution to the analysis with a 95% confidence interval.

Most of the coiled nautiloids and ammonoids used in the study were recovered loose on the surface of the outcrops. However, some specimens were recovered *in situ*, and these specimens confirm that incomplete shells recovered loose on the surface are not incomplete because of exposure to present-day weathering and erosion. Thus the relative completeness of specimens at the time of burial can be estimated easily by the freshness of any breakage on the specimens recovered from the surface of the shale exposures.

The following coiled nautiloid genera occur in the Mature Molluscan Community paleoenvironment and were utilized in this study: *Domatoceras, Ephippioceras, Liroceras, ?Liroceras, Metacoceras, Neobistrialites, Peripetoceras, ?Peripetoceras,* and *Tainoceras* (see Fig. 4 for some examples). Other genera, such as *Titanoceras* and *Solenocheilus*, were also present in the collection. However, due to their low abundance ($n = 2$ and $n = 5$, respectively) and fragmentary preservation, these latter two taxa were not included in the study. In some cases, the precise generic identity of certain morphologically grouped sets of coiled nautiloids is uncertain because a detailed systematic investigation for the coiled nautiloids for the Finis Shale has not yet been undertaken. This study focuses on the kinds and causes of damage in a population of cephalopods with the goal to establish a rationale to distinguish predatory from non-predatory damage. After that goal was attained, then different aspects of shell diameter, ornamentation, and width (wide and narrow conch forms) were evaluated to see whether these parameters had an impact on the predation frequencies of the different coiled nautiloid genera. In order to provide insight into the features that are significant in the studied taxa, a brief descriptive generic-level overview is provided in the Appendix.

Although 12 ammonoid genera are known from the Finis Shale, only *Gonioloboceras, Glaphyrites, Schistoceras,* and *Neodimorphoceras* are present in the collection in sufficient numbers from the Mature Molluscan Community to use in this predation study. These four taxa are described briefly in the Appendix; each has a body chamber that is about one complete volution.

3.3. Predation Versus Taphonomy of the Finis Cephalopods

The events that will damage or break the shell of a cephalopod can be separated into two distinct phases: (1) events that happen during the lifetime of the animal, including the predatory attack that causes the demise of the animal; and (2) post-mortem events, including those that occur after the dead cephalopod is consumed; this latter phase is called taphonomy.

Taphonomy, generally defined as the forces that acted upon an organism between its death and subsequent discovery, plays an important role in preservation (Canfield and Raiswell, 1991; Maeda and Seilacher, 1996). The taphonomic factors that may interact with a deceased organism include events that happen prior to burial, during burial, and prior to collection, including exhumation and weathering. Prior to burial, shell damage may be caused by the following: (1) breakage due to impact during transportation or reworking, (2) dismemberment by scavengers, (3) mistaken predation on an already dead and empty shell, and (4) utilization of the shell by another organism for shelter. After burial, shells may be broken or destroyed by dissolution, crushing or distortion by compaction, and bioturbation. When shells are exhumed by erosion, they are exposed to the vicissitudes of chemical and mechanical weathering, which will produce shell dissolution and breakage. These phenomena adversely affect the condition of specimens, and, alone or in combination, can complicate the identification of fossils and obscure the damage produced by a predation event.

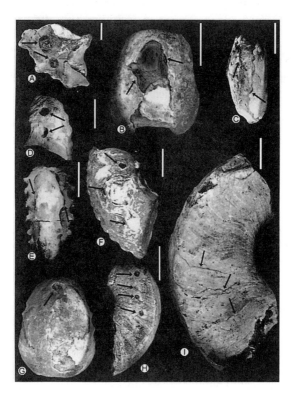

FIGURE 4. Coiled nautiloids from the Finis Shale. For locality information see Chaffin (2000). (Bar scale = 1 cm). (A) *Tainoceras* (OUZC 3774; locality TXV-41) showing circular holes along left lateral side of phragmocone. Based on the size and linear orientation of the holes, these perforations were probably caused by the shark, *Symmorium reniforme*, which co-occurs with these cephalopods in the Finis Shale. (B) *Liroceras* (OUZC 3778; locality TXV-56) showing removal of a massive "U" shaped portion of both the phragmocone and body chamber along the venter. The missing ventral section is interpreted to have been removed by an arthropod. (C) *Domatoceras* (OUZC 3780; locality TXV-200) showing a "U" shaped portion of the phragmocone and body chamber along the venter that was probably removed by an arthropod. (D) *Tainoceras* (OUZC 3775; locality TXV-34) showing circular holes along the left lateral side. These perforations are of approximately the same size, suggesting that an unidentified shark or other large fish may have been the predator. (E) *Metacoceras* (OUZC 3776) taken from the Iola Formation of Oklahoma (Upper Carboniferous) illustrating that the "U" shaped conch damage is not limited to the Finis Shale. (F) *Tainoceras* (OUZC 3779; locality TXV-200) showing a large perforation accompanied by two smaller punctures on the right lateral side. Based on the size of the punctures and their relative orientation, we suggest that *Symmorium reniforme* bit the *Tainoceras* at an angle allowing the maximum penetration of the main cusp and two of the lateral denticles. (G) *Domatoceras* (OUZC 3787; locality TXV-56) with four perforations reasonably aligned along the left lateral side of venter that were probably caused by a predator. (H) *Neobistrialites* (OUZC 3777; locality TXV-34) showing a perforation along the apicad end of the venter. This single perforation suggests predation by a fish. (I) *Domatoceras* (OUZC 3784; locality TXV-56) with repaired damage along right lateral side, indicating that the injuries were sublethal. The sublethal injuries appear to cross several growth lines and are interpreted as "moderate." However, the body chamber is absent, suggesting that lethal predation took place at a later time.

Taphonomic processes often obscure or destroy the evidence of predation, but sometimes these processes enhance the evidence of predation. When concretionary mud fills the empty body chamber and/or phragmocone of a cephalopod, it produces a smooth mold of the internal surface of the shell on the surface of the concretion steinkern. In the case of puncture-type predation, the conch is pierced by the predator's tooth. This hole leaves a void that prevents the formation of a smooth mold on the surface of the steinkern. Conchs with larger irregular pieces of missing shell also possess voids that will not allow the formation of smooth molds. Sometimes the mud that filled the conch body chambers and/or phragmocones became cemented by mineral components that created a solid concretion. This early diagenetic process fortified that portion of the infilled conch and protected it from diagenetic crushing. Based on the Finis specimens, the coiled nautiloid conch phragmocones only rarely became filled with mud, and ultimately, concretions, because the chambers were intact and were probably partially or completely filled with gas. During compaction these gases could not protect the shell from collapse. Thus, in the Finis Shale, there are few complete coiled nautiloid specimens with both body chambers and phragmocones intact. The most common preservational condition of the Finis coiled nautiloids is an incomplete concretion-filled body chamber. The inner whorls of these nautiloid specimens are preserved as fragmented, crushed shell that adheres to the dorsum of the body chamber. These specimens presumably had complete phragmocones, but, because the phragmocones were not concretion-filled during compaction, this part of the conch has been crushed by lithostatic pressure. However, if the chambers in the phragmocone were filled with water, then there is a tendency to preserve these chamber spaces with a crystallized mineral infilling, and these fillings, as described by Maeda and Seilacher (1996), preserved the three-dimensional shape of the conch. In the ammonoids, the body chamber (which was rarely present) or the damaged phragmocone often filled at least partly with sediment that formed a concretionary infilling and with crystallized mineral infillings that preserved the shell from the crushing effects of lithostatic pressure. Where diagenetic crushing occurred, pieces of the shell are usually crushed inward in irregular patterns.

One of the more obvious challenges in isolating damage produced by predators during lethal attacks is determining whether the processes involved with transportation created the shell damage. To evaluate this factor, the distribution of the cephalopod shells within the Mature Molluscan Community must be considered.

Cephalopod specimens are distributed randomly through the Finis Shale (Mapes *et al.*, 1996) suggesting that, at the time of death, the negatively buoyant specimens sank to the bottom and were not transported any appreciable distance. We suggest that many more specimens potentially lived and died in the Finis area, but most shells were positively buoyant at the time of death because of the loss of shell material and the removal of body tissues. These shells drifted away and are no longer part of the fossil record in the Jacksboro region. Thus, shell damage on the recovered specimens could be due to predation, or it could be the result of diagenesis or post-mortem trauma.

Post-mortem scavenging could have produced the same effect. Scavengers may have damaged the conch of moribund nautiloids to obtain any remaining soft tissue within the conch. The degree to which this activity damaged the conch of the dead animal is uncertain; however, larger predators/scavengers were more likely to have been involved with the removal of the tissue of the dead animal and less involved with the

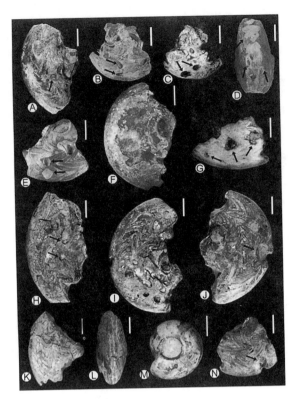

FIGURE 5. Selected ammonoids from the Finis Shale with damage produced by lethal predator attacks. All ammonoids except M were recovered without their body chamber. For locality information see Chaffin (2000). (Bar scale = 1 cm). (A) *Gonioloboceras* (OUZC 3791; locality TXV-34) with a circular perforation along the right lateral side of the phragmocone. Based on the circular puncture (arrow) and the missing body chamber, this specimen is interpreted to have been attacked by an unknown predator with puncture-type teeth. (B, C) Lateral views of *Gonioloboceras* (OUZC 3796; locality TXV-41) phragmocone with multiple circular perforations (arrows) on either side. Based on the size and orientation of the punctures, we suggest that these holes were produced by a symmoriid shark, perhaps *Symmorium reniforme*. (D) Ventral view of *Neodimorphoceras* (OUZC 3818; locality TXV-99) showing U-shaped removal of the shell material along the venter on the phragmocone (arrows). This specimen is missing the body chamber, suggesting that it may have been attacked by a fish or possibly an arthropod. (E) Lateral view of a partial phragmocone of *Gonioloboceras* (OUZC 3795; locality TXV-34). This specimen has two circular perforations on the side of the phragmocone (arrows), and the specimen is missing the body chamber. The orientation of the holes suggests that the specimen was killed by a symmoriid shark. (F) Lateral view of a *Schistoceras* phragmocone (OUZC 3802; locality TXV-200). This specimen lacks perforations but is missing the body chamber and about half of the phragmocone. The lack of perforations suggests that shell crushing was involved. (G) Lateral view of a *Gonioloboceras* phragmocone (OUZC 3793; locality TXV-34). The three perforations occur in a line, suggesting that they were caused by a large fish with puncture-type teeth (arrows). (H) Lateral view of a *Gonioloboceras* phragmocone (OUZC 3794; locality TXV-42). Note the alignment of the two perforations (arrows). The combination of the linear holes and the missing body chamber suggests that this specimen was attacked by a predator with puncture-type teeth. (I) Lateral view of a *Gonioloboceras* phragmocone (OUZC 3798; locality TXV-42). The numerous holes on the conch are interpreted to be caused by dissolution rather than predation. However, the body chamber is missing, suggesting this specimen was attacked by a predator with a crushing-type dentition. (J) Lateral view of a *Gonioloboceras* phragmocone (OUZC 3792; locality TXV-56). This specimen has a semicircular hole located along the venter (see arrow). It is not known whether the hole is related to predation; however, the missing body chamber suggests that predation was the cause. (K,

shelly material that has no nutritive value. Smaller scavengers/predators were probably attracted to smaller tissue volumes, and even the siphuncular cord may have been of interest to the smaller scavengers. Moreover, such small scavengers may not have had the ability to do major shell damage.

Missing body chambers are probably an indirect indicator that lethal predation occurred. Likewise, missing sections of shell along the venter, either on the phragmocone or on the body chamber, suggest that damage was a product of lethal predation (Figures 4, 5). Other evidence of possible predation imprinted on coiled nautiloid and ammonoid conchs may include the following features: (1) Indentations or holes from teeth, which may occur in a straight line with counter marks on the counter side of the shell being reasonably aligned; however, marine predators, such as sharks and other fish, often lose teeth, leaving gaps in such lines. Also, small cephalopod genera and immature specimens may show only one or two tooth marks in the case of larger predators with larger mouths and widely spaced teeth. (2) Perforations caused by teeth may show consistency in shape, size, and spacing on the conch, while post-depositional compaction perforations are irregular in size and position. (3) Theoretically, shell material broken due to puncturing by predators could be driven inside the cameral chamber (Buckowski and Bond, 1989), and internal structures such as septa should show damage consistent with surface damage produced by long conical puncturing teeth (Mapes *et al.*, 1995).

The marine system during regression was within storm wave base because the maximum flooding surface located near the base of the Finis Shale was removed during the regressive phase by a storm (Rothwell *et al.*, 1996). The oxygenated nature of the marine shale above the storm bed suggests some low current energy sources were active in the environment. However, storms and other high-energy events could not have been the main factor in producing the broken cephalopod conchs. If storms played a role in the preservation of the cephalopods, they would be size-sorted and transported into clusters or lenses, and these concentrations would occur as part of lag deposits with preferential orientations to the conchs and other shell debris. The coiled nautiloids and other cephalopods appear to be randomly distributed throughout the exposed shale sections at all the collected localities. Minor accumulations of the shells of invertebrates (brachiopods, bivalves, gastropods, etc.) occur as discrete lensoidal concretions or as pods in the shale that may represent storm or current concentrations; however, the amount of energy was relatively low because larger shells are not part of these concretions. It is also possible that some of these shell concentrations were produced by organisms and would therefore be considered trace fossils. This conclusion is supported by the presence of shell-packed burrow linings throughout the Mature Molluscan Community interval.

L) Phragmocone of *Neodimorphoceras*, lateral and ventral views, respectively (OUZC 3801; locality TXV-56). There are no perforations present on the conch. However, the specimen retains about one-third of the phragmocone and is missing the body chamber, suggesting that it was attacked by a predator with a crushing-type dentition. (M) Lateral view of a *Glaphyrites* (OUZC 3800; locality TXV-56). There are no holes present on the conch, and this *Glaphyrites* retains the body chamber. There is no evidence that this specimen was killed by a predator. (N) Lateral view of a *Goniloboceras* (OUZC 3797; locality TXV-54) phragmocone. The specimen has a circular perforation located on the umbilical shoulder and a missing body chamber, suggesting that it may have been attacked by a predator with puncture type teeth (arrow).

It is possible, though unlikely, that conchs were damaged by impact with other organisms having robust skeletons (e.g., rugose corals) that co-existed on the mud bottom that later formed the Finis Shale. However, organisms such as rugose corals are not abundant, and it is doubtful that the energy level was consistently high enough during regression to cause the observed shell damage on the coiled nautiloids and other cephalopods. Additionally, both the geographic and vertical stratigraphic distribution of the coiled nautiloids and other cephalopods is random. The breakage patterns and the areas of shell loss observed on the conchs are often repetitive, suggesting there may be common causes for the damage. Collisions with hard substrates probably would cause random breakage patterns on the coiled nautiloid conchs and on the body chambers of the ammonoid specimens. We have observed that the other invertebrate faunal elements (bivalves, gastropods, brachiopods, etc.) are not as damaged as are the cephalopods. All of these observations argue against high energy levels causing the damage observed on the cephalopod specimens.

It is conceivable that the shells of cephalopods living in shallow water where wave motion is pronounced and floating nekroplanktic cephalopod specimens concentrated by wave and current patterns on the surface could have collided with one another during storm events. This type of damage has not been analyzed on present-day *Nautilus* nor have studies been done under such conditions. Our presumption is that this kind of damage would be rare, and the collision of shelled cephalopods would cause random breakage patterns that would be confined mostly to the outer whorl and to the body chamber. Also, we think it unlikely that cephalopods would have collided with one another often enough to account for the damage we have observed in the cephalopod data set; one would expect such broken conchs to be accompanied by lag deposits and other signs of a storm event showing that high-energy wave activity had occurred. We cannot totally reject the scenario that shell damage was caused by collision with other hard substrates; however, if the collision of conchs did occur on rare occasions, it is unlikely that it would have produced damage like the circular perforations seen on present-day *Nautilus* (Fig. 3) and the fossil coiled nautiloids and ammonoids described herein (Figs. 4, 5).

There is no evidence that the Finis Shale was subaerially exposed, which eliminates terrestrial weathering and erosional influences as sources of conch damage during the Finis cycle. Immediately after the Finis cycle, and for an unknown number of subsequent subaerial exposures, erosional damage probably did occur. However, it would seem that more recent weathering and erosion has had a greater impact. The specimens recovered *in situ* and as loose specimens on the surface from some localities (see especially locality TXV-200 in the locality register of Chaffin, 2000) have only been uncovered in the past seven years as opposed to the other localities, which may have been exposed for decades. Therefore, freshly exposed outcrop specimens can be compared directly with those from localities that have sustained long-term weathering and erosion. The conch damage produced by subaerial exposure is entirely different from the types of damage produced by diagenetic and taphonomic processes or biologic events. Thus, newly broken surfaces produced by present-day weathering are quite different in appearance from the ancient broken surfaces in terms of color and texture. Many of the TXV-200 specimens have not been oxidized from the original gray to the weathered brown color, and therefore freshly exposed shells retain their gray color that was developed during fossilization. Thus, subaerial exposure cannot be used to explain all the damage to coiled nautiloids and other cephalopods in the Finis Shale. Subaerial exposure damage is characterized by fresher breaks caused by the various mechanical and chemical processes associated with recent weathering and erosion that have acted upon the specimens since the time they were uncovered.

Presumably the Mature Molluscan Community ecosystem of the Finis Shale was typical in that it contained a variety of both predators and prey. Based on observations of present-day *Nautilus*, coiled nautiloids and ammonoids of the Late Pennsylvanian were probably opportunistic predators and scavengers. They also served as a food resource for larger fish and other larger cephalopods. Thus, it seems a reasonable assumption that the Finis coiled nautiloids and ammonoids were attacked and eaten by predators in their natural habitat.

Coiled nautiloid conchs missing the body chamber are frequently recovered, suggesting that the damage was caused by predation. In cephalopods, the body chamber houses the soft tissue of the animal, which would serve as nourishment for the predator. The phragmocones are also often damaged, possibly as a result of previous attempts by the predator to subjugate the coiled nautiloids. Predators could produce multiple punctures on the conchs of the coiled nautiloids that would flood with water, thereby causing them to become negatively buoyant and sink (Chamberlain *et al.*, 1981). We suggest that cephalopods were preyed upon throughout the geographic extent and time span of the Mature Molluscan Community, and that some prey sank within a reasonable distance of where they died. This process would account for the random distribution of conchs in the stratigraphic sequence.

Any remaining soft parts in otherwise empty shells resting on the ocean floor were likely to have been consumed by scavengers. Scavengers may have removed parts of the shell in order to extract the tissue. Burrowing organisms also may have damaged the shells as they moved through the mud. Both scavenging and bioturbation could have damaged the shells, affecting their preservation quality but not their stratigraphically random distribution. Recovered conchs often show damage, which, in many of the coiled nautiloid cases ($n = 328$), we have concluded was caused by predation. In some cases the conchs show the "puncture type" predation style. The least controversial lines of evidence probably are the circular and subcircular holes in the conch. Some of these specimens show larger holes often accompanied by one or two smaller perforations. The main holes are as large as 11 mm in diameter, and we suggest that they were produced by a predator with a dagger-like tooth cusp with smaller cusps on either side. The large holes and smaller lateral perforations support the conclusion that these features were probably caused by a shark, most likely *Symmorium reniforme* (Figures 4A, D; 5B, C, E, G, H, I, J), which is the only known Pennsylvanian age predator to have an offensive armament of this size and morphology (Mapes and Hansen, 1984). In other cases, there is just one perforation in the shell, suggesting that either the penetration was not deep enough for the lateral cusps to contact the shell or that a different predator, such as another type of shark or other large fish, produced the damage. When singular holes are observed, they are in the phragmocone, and the body chamber is missing. Also the phragmocones are typically only partly complete, suggesting that there may have been additional punctures that produced the conch damage.

While shell perforations are interpreted as the best available evidence of predation, they are not the most common form of shell damage in the ammonoid collection, because only 30% of the total specimens exhibit this kind of damage. The possible size difference between predator and prey may account for the lack of multiple punctures on the ammonoid conchs, and it is logical that a predator with a sufficient size advantage and crushing dentition could have completely shattered an ammonoid conch without leaving tooth holes. In all cases, a comparison of the morphology of the holes in the Finis Shale cephalopod data set with modern hole producers has been made. The holes in the fossil cephalopods are circular to subcircular and have slightly crushed, untapered edges. Borings produced by octopi have small irregular slit-like holes (Fig. 2 C, D, E), and those produced by gastropods (described elsewhere in this volume), are smooth, circular, and evenly tapered. There is no evidence that either octopi or gastropods produced the punctures in the Finis Shale cephalopod shells.

Carnivorous organisms such as sharks and other fish lived in the Finis ecosystem, including the Mature Molluscan Community. Coiled nautiloids and ammonoids were likely prey for sharks or other fish that were the dominant predators in the ecosystem at that time. Predation would explain the random distribution of conchs in the stratigraphic section, as well as the recurrent absence of the body chambers that contained the majority of the animals' tissue, the missing parts of phragmocones, and the circular holes in phragmocones. Thus, the damage to the conchs is probably the result of predators attacking and eating the ammonoids. The partially fragmented conchs probably were damaged by predators that were much larger than the ammonoids. Conchs containing multiple punctures, as well as missing pieces of phragmocone or body chamber, usually would have filled with water and lost their positive buoyancy, causing them to sink (Chamberlain *et al.*, 1981). After the predation event, scavengers and burrowing organisms presumably consumed any remaining soft tissue within the otherwise empty shells resting on the ocean floor. Scavengers may have affected the appearance of the ammonoid conchs; however, the presence of encrusting organisms concentrated mostly on one side of the shell suggests that, after the scavengers were finished, the conchs were generally left undisturbed on the seafloor prior to burial. While encrusting organisms and micro-boring organisms also may have affected the preservation of the conchs, it is unlikely that any of this damage was sufficient to alter the overall appearance of the conch significantly.

Other specimens in the analyzed cephalopod collection have damage expressed in the form of small segments or chips of shell that appear to have been removed piecemeal from the venter of the body chambers of coiled nautiloids and ammonoids (Figures 4B, C, E; 5D). Damage of this type has been observed on present-day mollusks, and was caused by crabs and other arthropods (Vermeij, 1987). Although no arthropods with massive crushing claws have been recovered from the Finis Shale, decapods of the genus *Palaeopalaemon*, from the Late Devonian (Schram *et al.*, 1978), possessed the requisite armament for producing the chipping effect. Additionally, phyllocarid arthropods have strong crushing jaws that were capable of producing the shell damage observed on the cephalopods, and these arthropods are known to have been present in the Finis seas. As a result, such arthropods cannot be ruled out as possible predators or scavengers on coiled nautiloids and ammonoids (a more complete treatment of arthropods as predators and scavengers is provided by Babcock, Ch. 3, this volume). Possibly the peeled ammonoid and nautiloid phragmocones represent mistaken predation by a decapod or by an animal creating a "home place" to avoid predation (Walker and Yamada, 1993) (Fig. 5 D). Another possibility for this kind of damage is that other coiled nautiloids or ammonoids, seeking cryptic locations for egg-laying, peeled the moribund conchs themselves and laid their eggs in the vacant shells. While possible, we do not think this scenario is likely.

A significant, and as yet unappreciated, possible evidence of cephalopod predation is the presence of massive damage to, or complete removal of, the body chamber. Because the body chamber contains the nutritive tissue, this part of the conch will most concern the predator. In the Finis Shale collection of cephalopods, the body chamber is seldom found attached to the conch, indicating that it must have been removed. Only 12% of the ammonoids in the Mature Molluscan Community were recovered with the body chamber attached to the conch. It seems to us that it is unlikely that this amount of body chamber removal on so many specimens could have been caused by diagenesis, mechanical weathering, or post-mortem breakage; thus, we suggest that it was caused by predators attacking and eating the cephalopods.

Both coiled nautiloids and ammonoid conchs were recovered from the Mature Molluscan Community that are missing massive portions of phragmocone or body chamber along the venter. This type of damage, which we refer to as "peeling," may have been caused by arthropods attacking or savaging the living or newly dead, respectively, nautiloids and ammonoids when they were on the seafloor (see Babcock, Ch. 3, in this volume, for an extended discussion of arthropod predation and scavenging). Historically arthropods (particularly crabs) have been known to attack cephalopods as well as other ectocochleates (Thiermann, 1964; Roll, 1935; Lehman, 1976; Keupp and Ilg, 1992). Also, Walker and Yamada (1993) reported instances of Recent arthropods mistakenly attacking empty molluscan shells. This kind of post-mortem damage may explain why some ammonoid and coiled nautiloid phragmocones, which contain virtually no tissue except for the siphuncle cord, have been "peeled."

3.4. Analysis of the Coiled Nautiloids

The recovered specimens show the following percentages of potentially lethal predation: *Domatoceras* 54%, *Ephippioceras* 83%, *Liroceras* 33%, *Metacoceras* 39% *Neobistrialites* 33%, *Peripetoceras* 47% and *Tainoceras* 72% (examples of the damage are illustrated in Fig. 4). Genera were recovered in varying abundances, with one genus accounting for more than a third of the total number of recovered specimens (*Liroceras*, $n = 297/692$), whereas other genera recovered, such as *Ephippioceras* and *Neobistrialites*, account for fewer than 10 specimens.

While *Domatoceras* has the largest apparent conch diameter of the genera analyzed (excluding *Titanoceras*), it seldom reached full maturity. The smaller conchs are recovered much more commonly, and show high frequencies of predation (Fig. 6A). The predation frequencies remained constant as the animals grew from post-hatchling to mature specimens, whereas the total numbers of specimens decreased through ontogeny. Although infrequent, larger specimens are always fragmented, which we interpret as due to predation. This suggests that *Domatoceras* was heavily preyed upon throughout ontogeny, and that the decrease in recovered mature specimens is due to attrition rather than collecting bias.

The size-frequency distribution of *Metacoceras* forms a bell-shaped curve both for total numbers of specimens and preyed-upon specimens (Fig. 6B). This suggests that *Metacoceras* could have had different predators at different sizes. Although there appear to be more instances of predation at smaller diameters, one must consider the possibility that fewer specimens survived into maturity to be attacked.

The size-frequency distribution of *Liroceras* forms a bell-shaped curve for total numbers as well as those specimens that were preyed upon (Fig. 6C). The *Liroceras* distribution is skewed to the smaller diameter specimens. We suggest that this distribution is due in part to the presence of two species with small diameters. However, *Liroceras* specimens were heavily preyed upon at smaller diameters, and fewer specimens survived into maturity to be preyed upon.

The size-frequency distribution of *Peripetoceras* seems to form a bimodal distribution in both total numbers of specimens and in terms of numbers of preyed-upon specimens (Fig. 7A). This distribution suggests that the sample represents two distinct taxa. Alternatively, because part of the distribution is represented by only two specimens, this part of the distribution may represent collecting bias.

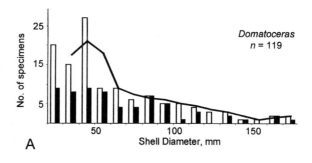

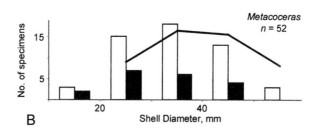

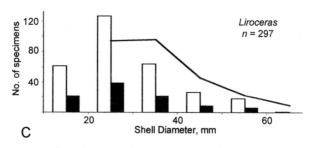

FIGURE 6. Histogram showing the presumed maximum growth based on conch diameter (in gray) at the time of death of *Domatoceras* (A), *Metacoceras* (B), and *Liroceras* (C), as well as predation frequencies (in black). An estimated 1/3 volution has been added to the diameter when appropriate to better evaluate the conch's maturity at death.

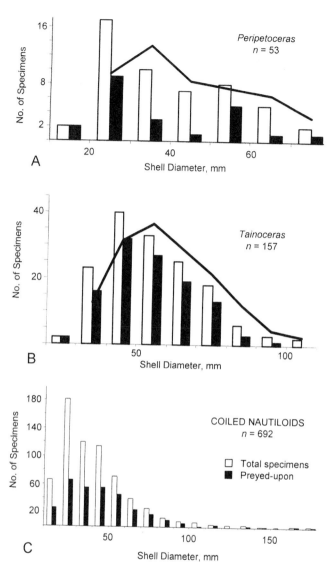

FIGURE 7. Relative abundances of *Peripetoceras* and *Tainoceras* and the overall diameter and predation distribution for all the coiled nautiloid genera. (A) The *Peripetoceras* distribution has a trend that gradually declines as specimens attain larger diameters. Maximum predation is in the 25 mm diameter size class. Based on slight morphological differences, two species may be present. (B) The *Tainoceras* distribution is skewed with mean shell diameter between 40 and 50 mm. Predation is most frequent in this size range. (C) A composite histogram of all the coiled nautiloid specimens for the five genera analyzed for predation from the 10 localities in the Finis Shale. The distribution shows that, while smaller diameter specimens are relatively intensely attacked, they are not attacked as intensely as the larger diameter specimens. Note that *Liroceras* and *Peripetoceras* are smaller diameter at maturity and that *Domatoceras* and *Tainoceras* make up the entire right side of the distribution at >100 mm diameters. These differing maturity sizes shift the overall shape of the distribution but they do not detract from the fact that intense predation occurs in all the taxa.

Although recovered conchs of *Tainoceras* have a size range of 20-110 mm, the size-frequency distribution of preyed-upon specimens is slightly skewed toward the smaller to mid-size specimens (Fig. 7B). One interpretation of this distribution is that there were fewer predators of larger-diameter specimens. Another interpretation is that few specimens of *Tainoceras* lived to reach maturity. This second interpretation is supported by the overall decrease in numbers of recovered specimens with diameters between 80 and 110 mm.

As stated earlier, it was necessary to equalize the genera statistically in order to compare them in terms of predation preference. The statistical analysis determined that, in some cases, genera were preferentially selected. *Tainoceras* appears to have been preferentially selected over all other genera, with the exception of *Ephippioceras*. *Liroceras* was not preferentially selected. *Domatoceras* was preferentially selected over *Liroceras*, but equally selected with all of the other genera with the exception of *Tainoceras*. *Peripetoceras*, *Metacoceras*, and *Neobistrialites* were not preferentially selected. *Ephippioceras* was preferentially selected over *Liroceras*, *Peripetoceras*, and *Metacoceras*. However, the findings on *Ephippioceras* are suspect due to the paucity of specimens.

Based on the fact that both *Domatoceras* and *Tainoceras* have the largest diameters of the genera analyzed and have the highest proportions of predation, there appears to be a relationship between conch diameter and predation. Thus, larger-diameter conchs appear to have attracted predators more readily than smaller specimens.

Ornamentation does not appear to affect predation greatly. The larger coiled nautiloids, such as *Domatoceras*, have a smooth conch, whereas the largest diameter conchs with ribs and nodes belong to *Tainoceras*. These two conch forms (smooth versus nodose) both have frequencies of predation greater than 50%; however, the statistical analysis shows that *Tainoceras* was more frequently selected than *Domatoceras*. When coiled nautiloids with smaller diameter conchs at maturity with nodes (*Metacoceras*) and smooth surfaces (*Liroceras*) are compared, the frequencies of predation are similar at 39% and 33%, respectively; these taxa were also much less likely to have been attacked than were the larger-diameter genera.

Genera with relatively wide body chambers were not preferred over those with relatively slim body chambers. Since the size of the body chamber is directly related to the amount of tissue (= the amount of food value), predators should prey selectively upon the volumetrically larger tissue source if possible. Thus, it would seem that a wide body chamber with proportionately more tissue than a narrow body chamber should be preyed upon more frequently. *Domatoceras* has a relatively narrow body chamber, whereas *Liroceras*, *Peripetoceras*, and *Neobistrialites* have relatively wide body chambers. *Domatoceras* exhibits a 54% predation frequency, whereas the taxa having wide body chambers have frequencies of 33%, 47%, and 33%, respectively. In other words, contrary to expectations, the narrow form has a higher predation frequency than the wider forms. The reason for this is not clear, but may relate to other biological factors such as swimming ability, camouflage, or even the possibility that the wider taxa did not taste as good.

In addition to body chamber width and ornamentation, there are other characteristics that could have affected predation. Both *Tainoceras* and *Domatoceras* have a large diameter-to-width ratio; this may have resulted in a smaller hyponome relative to some of the wider genera such as *Liroceras* and *Peripetoceras*. Present-day *Nautilus* ejects water through the hyponome to move rapidly. Fossil genera that possessed a large hyponome might have had a mobility advantage in avoiding predators. In addition, certain genera may have been better camouflaged, and this may have contributed to successful avoidance of predators.

3.5. Analysis of the Ammonoids

Of the recovered ammonoid shells ($n = 206$), 88% are broken. Predation is the most likely cause of this breakage when taphonomic or other post-mortem phenomena can be ruled out. In most cases ($n = 169/206$), the ammonoid conchs were recovered without body chambers, and many specimens are also missing large segments of the phragmocone (Fig. 5). Because the body chambers of these ammonoids are a complete whorl, the loss of the body chamber significantly reduces the overall volume of the conch.

Gonioloboceras dominates ($n = 155$) the ammonoid fauna. Every specimen in the collection is missing its body chamber, which is about one volution in length. Pieces of the body chamber of this ammonoid are frequently recovered on the outcrop; however, since such pieces cannot always be identified with absolute certainty, they usually are not collected. When the diameters of the specimens are plotted, their size-frequency distribution approximates a bell-shaped curve for total numbers and those damaged by predators (Fig. 8 A). Predation levels are consistently high for all size classes, suggesting that the lack of larger-diameter specimens may be due to attrition, with high numbers of juvenile specimens being killed.

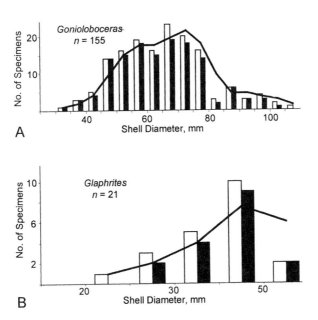

FIGURE 8. Histograms of *Gonioloboceras* (A) and *Glaphyrites* (B) populations from Finis Shale localities near Jacksboro, Jack County, Texas, showing the calculated growth (gray) at the time of death as well as predation frequency (black) at each size level. Note that there are no specimens below the 20 mm range, suggesting that post-hatching and early juvenile specimens lived in a different biofacies, and that when a certain level of growth was attained, the animal migrated into the Mature Molluscan Community.

Glaphyrites (*n* = 21) is the next most abundant ammonoid. Most specimens fall in the 35–40 mm diameter size range. Predation frequencies are consistently high, suggesting that all size classes are equally preyed upon (Fig. 8B). This overall high rate of predation may explain the lack of larger diameter specimens. Of the four ammonoid genera used in this study, this is the only genus with some shells that do not exhibit some form of damage. Some specimens retain the complete body chamber, and, other than some slight crushing around the aperture, there is no appreciable damage to the conch. This does not mean that these undamaged conchs were not killed by predators; rather, it means only that there is not any evidence of a predator attack that damaged the shell.

Specimens of *Neodimorphoceras* and *Schistoceras* constitute a relatively small part of the ammonoid collection (*n* = 12 and *n* = 18, respectively). Both taxa have been subjected to intense predation; virtually all the specimens are missing the body chambers, and most are missing parts of the phragmocone. All of the specimens are more than 20 mm in diameter with no early juveniles being recovered, and there seems to be no evidence of any size selectivity by predators (Fig. 9A, B).

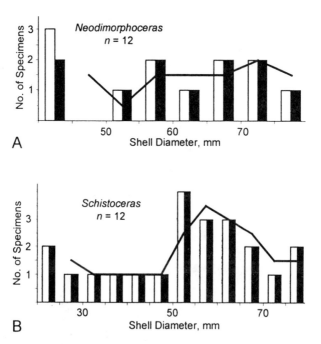

FIGURE 9. Histogram and trend line of *Neodimorphoceras* and *Schistoceras* from Finis Shale localities near Jacksboro, Jack County, Texas, showing the size (gray) at the time of death as well as predation mortality (black) at each size level. Note that there are no specimens <20 mm diameter, suggesting that post-hatching and early juvenile specimens are absent because of different biofacies preferences, as with *Gonioloboceras* and *Glaphyrites*. (A) *Neodimorphoceras* exhibited nearly equal predation frequencies throughout all size ranges. The irregular distribution is due to the number of specimens rather than a lack of predation. The lack of specimens may be due to a collecting bias; however, more probably this genus was not competitive in this environment. *Neodimorphoceras* shows nearly 100% predation, suggesting that there were predators for all sizes of animals. (B) *Schistoceras* reveals no appreciable pattern, probably due to sample size. As with *Neodimorphoceras*, the predation level of 100% suggests these prey are selected without regard to size.

Overall, the Finis Shale ammonoids appear to have been preyed upon heavily, with 88% of the specimens exhibiting one or more kinds of damage that was probably caused by predatory attacks. The specimens were preyed upon with the following frequencies: *Gonioloboceras*, 87% ($n = 155$), *Glaphyrites*, 80% ($n = 21$), *Neodimorphoceras*, 100% ($n = 12$), and *Schistoceras*, 100% ($n = 18$). Results of the Z-test for independent proportions concluded that *Glaphyrites, Gonioloboceras, Schistoceras,* and *Neodimorphoceras* all had equal frequencies of predation despite the differences in percentages of preyed-upon specimens.

In all of the ammonoid collections from the Mature Molluscan Community from the localities that provided the research material for this study, there are no specimens that have diameters of 20 mm or smaller. Indeed, the smaller-diameter specimens are usually recovered from the underlying dark gray shale that is interpreted as having been deposited in a dysaerobic environment (see Boardman et al., 1984 for an extended discussion); some localities have yielded large numbers (thousands to tens of thousands of ammonitella and early juvenile steinkerns up to about 20 mm in diameter) (see Tanabe et al., 1994, for an exceptional case). This size distribution is not an artifact of collecting; rather it is more likely a reflection of a paleobiological condition, in which the smaller growth stages of most ammonoid genera in the Finis Shale preferred a habitat or biofacies that was lower in dissolved oxygen and relatively predator poor, compared to the habitat preferred by adults. Thus, the earliest growth stages are more or less confined to the dysoxic Juvenile Molluscan Community in the Finis Shale (see Boardman et al., 1984 for details). Thus, when growth approached a diameter of about 20 mm (this diameter is somewhat variable, depending on the taxon being considered), the ammonoids appear to have migrated into the Mature Molluscan Community where they became prey for a host of different predators.

3.6. Lethal Damage: Comparisons Between the Ammonoids and Coiled Nautiloids

A much larger proportion of ammonoids were preyed upon (88%) than the coiled nautiloids (47%). This suggests that ammonoids were preferentially selected as prey over coiled nautiloids. *Gonioloboceras, Neodimorphoceras,* and *Schistoceras* show preferential predation over all the coiled nautiloids, with the exception of *Ephippioceras*. *Glaphyrites* shows preferential predation over all of the coiled nautiloids, with the exception of *Ephippioceras* and *Tainoceras*.

Reasons for the difference in predation frequencies between ammonoid and coiled nautiloid genera may be due to collecting bias, in that there are more coiled nautiloids in the collection ($n = 692$) than ammonoids ($n = 193$). However, 193 specimens should be adequate as an indicator of predation selectivity given the Z-test for independent proportions, which is designed to make comparisons between samples of different sizes. Based on percentage differences, the predation frequencies for both ammonoids and coiled nautiloids are substantial, and we consider this a real phenomenon. We speculate that these differences in predation frequencies between Ammonoidea and Nautiloidea may be due to life mode, habitat, mobility, and other biological factors.

3.7. A Hypothetical Predation Scenario

There are several kinds of predation scenarios that can be developed from this analysis. The Carboniferous predators that must be considered are sharks and other fish, arthropods (especially the phyllocarids), and other cephalopods. The fish had a wide variety of dental armament in the Carboniferous, including crushing plates, piercing teeth and cutting teeth (Fig. 10). Predatory effects on the conchs of coiled nautiloids and ammonoids include loss of body chamber and punctures in gas chambers that result in a buoyancy control problem.

It would seem likely that there were different fish predators that employed slightly different methods of subjugation. Based on the analysis by Mapes and Hansen (1984), we conclude that predators may have tried to surprise nautiloids by attacking from the rear. Sharks and other fish with conical teeth may have attempted to bite into the phragmocone to grasp the nautiloids and possibly to reduce the prey's ability to escape due to a loss of buoyancy control. They could then either shake the conch until the shell was fragmented and the animal was dislodged or continue to bite the conch to expose the tissue. Continued biting could cause the conch to fragment, exposing the tissue of the animal without the shell fragments that are of no nutritional value.

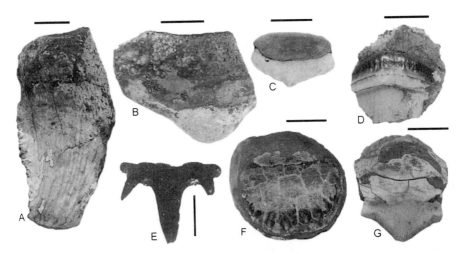

FIGURE 10. Shark teeth recovered from the Finis Shale and other Upper Carboniferous units from Ohio, Oklahoma, and Texas. The teeth were used for a variety of purposes but are generally classified as cutting, puncturing, crushing, or a combination thereof. Though the sizes of the teeth vary, all recovered teeth would have caused significant shell damage to any of the cephalopods analyzed from the Finis Shale. (Bar scales = 1 cm). (A) *Edestus* sp. The tooth is missing the crown. Note the size of this tooth relative to the other teeth as well as the serrated edge (left side). This tooth was used for cutting, but, given the size, could have easily crushed small cephalopods. (B) *Ordus* sp. The tooth is missing the right portion of the crown and root. Note the blunt rounded surface of the tooth. This tooth was used for crushing. It seems likely that this kind of dentition could have decimated the shell of almost all Finis cephalopods. (C) Petalodontid tooth with root. This tooth has a serrated edge and was likely used for cutting. It is also relatively wide and flat, which might have made it effective for crushing as well. (D) Petalodontid tooth and root that is missing the bottom of the crown. This tooth was likely used for cutting. (E) *Symmorium reniforme* tooth. Note the large main cusp and the smaller lateral denticles. This tooth was primarily used for puncturing and grasping prey. The symmoriid sharks were probably one of the more formidable predators of cephalopods during the Late Carboniferous. (F) Petalodont? tooth in a concretion. Judging by the serrations, this tooth was probably used for cutting and crushing. (G) Petalodont tooth with portions of the root intact. This tooth was interpreted to have been used for cutting as well as crushing, given the size, width, and edge.

Sharks and other fish also may have attempted to target the body chamber to access the animal. Predators in this scenario were likely to have attacked from behind and beneath the coiled nautiloid. The teeth would have acted as scissors, effectively cutting the body chamber from the phragmocone. In some cases, the body chamber may have been crushed to free the tissue.

In the case of a coiled nautiloid or ammonoid with a punctured conch or a missing body chamber, the conch probably would have risen toward the surface while the broached chambers filled with water. The weight of the extra water could ultimately cause the conch to become negatively buoyant and sink to the seafloor. Chamberlain *et al.* (1981) explained the ascent/pressure relationships of cephalopod conchs in detail. It is unknown how many conchs retained enough post-mortem buoyancy to float to the water surface, and eventually float away from the site of the attack. We suggest that the specimens utilized in this study represent the specimens that were negatively buoyant at the time of death, and that the collection used in the case study represents only a fraction of the animals that inhabited the Finis Sea in the Jacksboro, Texas, region.

The arthropod predators include a wide range of possibilities during Carboniferous time (see Babcock, Ch. 3, in this volume). Likely candidates include trilobites, crustaceans (e.g. phyllocarids and decapods) and eurypterids. The remains of the latter have not been observed in the Finis Shale. Most of these arthropods are inferred to be nektobenthic, and they probably procured their food resources by predation and scavenging. Thus, an ambush scenario would be as follows: some of these arthropods probably hid among the mud and broken shells on the bottom of the sea. When the cephalopod searched the seafloor for its food, an arthropod attacked, overcame the cephalopod, and then, with jaws or pincers, opened the cephalopod shell like a "can of sardines" (Fig. 4B, C, E). A more likely scenario is that the arthropod was an opportunistic predator, attacking and eating wounded or dying cephalopods, and the best-case scenario is probably that the arthropods scavenged moribund specimens on the bottom for soft tissue. All of these scenarios could be used to explain the damage seen in Fig. 4B, C, and E.

Cephalopods may have preyed upon other cephalopods. Many cephalopods in the Carboniferous and at later times possessed mandibles that would have been capable of removing pieces of shell (and flesh) from other cephalopods. Using present-day *Nautilus* as a model, we conclude that ancient cephalopods may have preyed on juvenile and post-hatchling specimens, as *Nautilus* generally eats small organisms (Nixon, 1988) or the shells or bodies of larger organisms in small-capacity bites. Furthermore, it is doubtful whether coiled nautiloids had the ability to prey upon other cephalopods when those prey animals had grown sufficiently large to be immune from such opportunistic predators. We suspect that there is a somewhat greater probability that hatchlings and young juveniles of cephalopods were prey for larger cephalopods.

3.8. Summary and Conclusions Drawn from the Upper Carboniferous Case Study

Predation is inferred to be largely responsible for the fragmentary condition of the recovered conchs. As at present, ancient marine organisms rarely died a natural death. Thus, it can be reasonably assumed that both ammonoids and coiled nautiloids were attacked and eaten. This could have caused the conchs either to sink to the substrate or to float to the air/water interface. In the Finis Sea, only the negatively buoyant specimens formed the collections used in the case study. The sedimentation rate was sufficiently rapid to cover the empty, negatively buoyant specimens gradually, allowing time for some encrusting organisms to colonize some of the conchs, yet quickly enough to ensure preservation. After the specimens were buried, the sections of the phragmocone not filled by mud-based concretion material or mineral deposits were crushed by compaction.

A lethal predation frequency of 47% (coiled nautiloids and ammonoids, combined) may seem high compared to the findings of Mapes *et al.* (1995), who reported a lethal predation frequency of less than 2% for the specimens of the ammonoid *Gonioloboceras* they studied. However, if Mapes *et al.* (1995) had considered the loss of the body chamber as a sign of predation, their conclusion as to the percent of preyed-upon specimens would have changed dramatically. In the collection of specimens they analyzed, every specimen having a diameter of over 35 mm ($n = 954$) from every locality they analyzed is missing the body chamber, and this would give a predation frequency of 100%. However, we suggest that our numbers in this case study are quite conservative, as many specimens were not included due to poor preservation and excessive fragmentation. These poorly preserved specimens and excessively fragmented conchs probably represent the remains of preyed-upon specimens.

When the results of this case study are compared to the frequencies of sublethal damage and/or abnormal shell growth reported in other studies, there is a remarkable difference in the frequencies reported. For example, Bond and Saunders (1989) reported, in their Carboniferous goniatite ammonoid study, a range from 9 to 38% for different taxa. For different Jurassic ammonoid taxa, Guex (1967) reported a frequency of 2 to 2.5%, Morton (1983) a frequency of 8.1%, and Bayer (1970) a range of 1.4 to 9.7%. Landman and Waage (1986) reported a range of 10% to 40% for Cretaceous scaphitid ammonites. By comparison, the test case has 88% as its lowest frequency in the ammonoids, with all other genera at or near 100%. In the coiled nautiloids the frequency of predation ranges widely in the seven genera, from 33% to 83%, which is remarkably different than the predation frequencies of 88% to 100% seen in the ammonoids.

Based on our study, we suggest that the levels of lethal predation cannot be evaluated accurately by an analysis of sublethal repairs. Also, ammonoids appear to have an overall higher level of predation than coiled nautiloids. As Bond and Saunders (1989) observed, present-day *Nautilus* exhibits a sublethal repair frequency of over 50%, and most specimens that we have observed have more than one repaired area on the shell. Thus, this present-day analogue may not provide an accurate model for the ancient cephalopod record.

4. Studies of Predation and Cephalopods Through Time

Overviews of the relationship between predation and morphologic adaptation of cephalopods in the Phanerozoic are limited to three major summary reports. In 1981, Ward quantified the ornament of ammonoids from the Paleozoic to the Upper Cretaceous and determined that ornament roughness increased through time. He hypothesized that this increase in ammonoid ornamentation served as a protective (defensive) function against the increased numbers of durophagous predators such as brachyuran crabs, lobsters, teleost fish, and rays during the mid-Mesozoic radiation. Ward recognized that the ornament on the ammonoids probably served additional functions, and he provided a list of these alternative functions. Later, Ward (1996) utilized this information to help explain the evolutionary patterns of the ammonites prior to their extinction at the Cretaceous-Tertiary boundary.

Ward (1981), in an insightful analysis, noted that shell changes could be a response to the sublethal damage created by the cephalopod prey on the predator in its efforts to escape. In our opinion, this explanation has considerable merit as it can be used to explain the occurrence of some of the sublethal damage observed in many of the juvenile shells of fossil cephalopods and present-day *Nautilus* specimens that have very thin apertural margins and that could be damaged easily during any kind of major struggle with an uncooperative victim.

Signor and Brett (1984) examined the mid-Paleozoic record of durophagous predators (specifically fish, arthropods, and cephalopods) and applied this information to the morphological changes that could be detected in several invertebrate groups, including the coiled nautiloids (specifically the tarphycerids and barrandeocerids for the pre-Devonian and the nautilids for the post-Silurian). Using sculpture classes similar to those used by Ward (1981), they noted that these sculptural features probably had multiple functions. They concluded that nautiloids did respond to the origin of durophagous predators by a gradual increase in the degree of ornamentation robustness. Their overall conclusion, based on all the invertebrate groups that they analyzed, was that the origin of durophagous predators had a profound effect on the mid-Paleozoic invertebrate biota.

Vermeij (1987) proposed that evolution is in part driven by escalation of evolutionary innovation with enhancement of predator abilities followed by increased efficiency in prey defenses. In this context, he noted that externally shelled cephalopods were much more common in the Paleozoic and Mesozoic with, according to Saunders (1981), only one family (the Nautilidae) surviving to the present. From this fact, Vermeij (1987, p. 271) inferred "...that passive shell armor has not proven successful for cephalopods in the long run." While one may agree with this conclusion, given the current diversity of externally shelled versus internally shelled cephalopods, we disagree with the implication that externally shelled cephalopods were doomed to extinction because of the inherent limitations of an external shell. We see the externally shelled cephalopods as having been successful for more than 400 million years. In addition to achieving considerable diversity over a long time span, they survived four major extinctions. The fact that externally shelled cephalopods are still with us today suggests to us that they are still successful in the race against extinction.

Vermeij (1987) explored two different facets of cephalopod paleobiology to support his escalation hypothesis: (1) the problems of rapid locomotion with the obvious corollary of predator avoidance, and (2) the effect of pressure on the gas-filled chambers with depth. His analysis of cephalopod evolution and body plan limitations and the development of predators in the Phanerozoic seems to make sense, given some of the morphological changes in cephalopods that can be tracked through the fossil record. However, some caution should be exercised before Vermeij's (1987) hypothesis is accepted. We have little evidence that the externally shelled fossil cephalopods were predators in the strict sense as Vermeij (1987) assumed. Present-day *Nautilus* can only be considered an opportunistic predator and, in fact, seems to prefer scavenging as its major method of food gathering (Tshudy *et al.*, 1989). Also, as Vermeij (1987, p. 289) indicated, additional studies of the details of cephalopod history are needed. Analysis of architectural shell features in relation to paleoecological conditions and the life mode of different cephalopod taxa at a number of different times in the Phanerozoic should be integrated into the escalation hypothesis framework. Also, studies of both sublethal and lethal damage on cephalopods need to be undertaken to determine the effects of predation on cephalopod evolution.

5. Conclusions and Future Studies

Vermeij (1987) invoked predation as a driving mechanism of evolution. His challenge to cephalopod workers is to evaluate the fossil record of this molluscan class to see whether there was an evolutionary response in the Cephalopoda to the diversification of predators in the Phanerozoic. Only his work (Vermeij, 1987), the work on some mid-Paleozoic nautiloids by Signor and Brett (1984), and the ammonoid study of Ward (1981) directly address the issue of predator/prey responses in cephalopod evolution. Reasons as to why more studies have not been done may be due in part to the unresolved taphonomic problems in quantifying the degree and cause of damage in cephalopod shells. The quantification problem was resolved in part by studies like those of Bond and Saunders (1989), Kröger (2000), and Landman and Waage (1986), who evaluated sublethal damage and repairs that created abnormal shells in ammonoid populations. The other problem has been the assignment of a causal agent to the damage seen on cephalopod shells. Few studies have separated shell damage due to non-biological agencies (e.g. lithostatic crushing, partial dissolution, shell breakage from impact on hard substrates, breakage due to burrowers, etc.) from different kinds of damage caused by predator attacks (shells altered by piercing, cutting, crushing, etc.). This difficulty has been compounded by the lack of data on damage caused by predators on present-day *Nautilus*. Thus, we recommend that actualistic taphonomic studies be performed on *Nautilus* to gain a reasonable idea of the kinds of damage created on *Nautilus* shells by different kinds of predators with different kinds of offensive armament. Probably equally important is whether the damage is randomly inflicted on the shell or is mostly directed to specific places on the conch. When this kind of information becomes available, it can be applied using taxonomic uniformitarianism to the fossil record of coiled nautiloids and ammonoids to compare the effect of evolution in predators to lethal and sublethal damage in the Cephalopoda.

Evidence of predation on coiled nautiloids, orthoconic nautiloids, and coleoids has not received the same level of attention as evidence on ammonoids. Coiled nautiloids can be treated in the same ways as ammonoids; however, because the sample sizes for nautiloids are usually smaller, it is more difficult to obtain good data sets for this group of cephalopods. Coleoids will continue to be difficult to evaluate because most damage to coleoids involves the flesh-covered exterior; predatory traces on the internal shell may not be present. Orthoconic nautiloids, because of their tendency to break into multiple segments, create difficulties in counting the actual numbers of injured individuals in a collection. The number of orthoconic nautiloid segments in a collection will not necessarily equal the number of individuals, and this will create a degree of unreliability in the conclusions that are drawn from any predation study.

ACKNOWLEDGMENTS: We thank Richard Arnold Davis, Patricia Kelley, David Kidder, and Greg Nadon for their reviews of earlier drafts of this manuscript. We also thank Michael Hansen and Rodney Feldmann for providing information on Paleozoic sharks and arthropods, respectively. We offer a special thanks to Dina Lopez for her assistance with the statistical and computational components of this report. We would also like to thank Mike Dunn and Genaro Castillo-Hernandez for their technical support with electronic images. Richard Arnold Davis allowed us to illustrate some of the specimens of *Nautilus macromphalus* he collected on the beaches of Lifou, the Loyalty Islands, the Pacific. Davis and other colleagues, such as Neil Landman, Kazushige Tanabe, and Haruyoshi Maeda, have provided many interesting discussions on this subject. We are grateful to W. Bruce Saunders at Bryn Mawr College and Desmond Collins at the Royal Ontario Museum for their permission to reproduce the photographs in Fig. 1 A and B, and E, respectively. In addition to the above, we are very grateful to Loren E. Babcock at Ohio State University and an anonymous reviewer for their positive suggestions to improve this report.

Appendix

The following is a brief description of the coiled nautiloid and ammonoid genera utilized in the case study. The coiled nautiloids are as follows:

Tainoceras has a relatively large diameter (up to 130 mm) at maturity. The conch at the late juvenile and mature state is subrectangular in cross section, with a slightly impressed dorsum. *Tainoceras* has an evolute, perforate umbilicus and is heavily ornamented with nodes on the umbilical shoulders, ventro-lateral margins, and on the venter (Miller *et al.*, 1933).

Liroceras has a relatively small diameter (up to 39 mm) conch with the width typically greater than the height. The umbilicus makes up approximately 9% of the total conch diameter. The innermost whorl of the conch is covered with coarse longitudinal lirae around the umbilical region (Tucker and Mapes, 1978*b*). Conchs that contain more than two and one-half volutions are considered mature; the shell is smooth.

Specimens that have similar proportions to mature *Liroceras* but with diameters exceeding 50 mm were grouped under the name *?Liroceras*. *?Liroceras* is relatively wide, although less so than *Liroceras*. *?Liroceras* also lacks longitudinal lirae in the earliest volution on the umbilical region.

Domatoceras has the greatest conch diameter of the genera being evaluated and possesses a relatively narrow, smooth conch. Specimens of *Domatoceras* are evolute, with a deep ventral lobe, and square venter and umbilical shoulders (Tucker and Mapes, 1978*b*). Two species of *Domatoceras* exist within the collection. The differences are based on external features such as conch width *versus* diameter and ornamentation on the ventrolateral shoulders.

Peripetoceras has a wide body chamber and a relatively small (< 40 mm) conch diameter. The shell has a reniform to rounded subtrapezoidal cross section at maturity. The suture has shallow ventral and lateral lobes and a moderately deep dorsal lobe. The umbilicus makes up approximately 14% of the total conch diameter, and, at maturity, has flanges that partly cover the umbilical opening (Tucker and Mapes, 1978*b*).

Specimens with proportions similar to *Peripetoceras* but larger than 40 mm in diameter were assigned to *?Peripetoceras*. These specimens lack the flanges that cover the umbilicus in *Peripetoceras*. *?Peripetoceras* also has a smaller height-width ratio than *Peripetoceras*.

The overall shape of *Metacoceras* is hexagonal in cross section. The umbilical shoulders range in shape from subangular to broadly rounded. Ventrolateral nodes are prominent, and the venter is smooth. *Metacoceras* is somewhat similar in appearance to *Tainoceras*; however, it does not have the ventral nodes, and it never attains the conch diameters seen in *Tainoceras*. It has been suggested that the two genera are closely related (Miller et al., 1933).

Ephippioceras has an involute conch, which expands rapidly and consists of about three whorls. *Ephippioceras* possesses a small umbilicus that makes up approximately 10% of the conch diameter. *Ephippioceras* is also characterized by having a suture with a U-shaped ventral saddle (Tucker and Mapes, 1978b).

The conch of *Neobistrialites* is slightly involute, with a convex venter and a flattened dorsum. The umbilicus is roughly 30 percent of the total conch diameter. The internal mold has nodes on the umbilical shoulders, and a conchal furrow is present (Tucker and Mapes, 1978b).

The following is a general overview of some of the morphological features that separate the ammonoid taxa used in the case study at the generic level; additional morphological descriptive details can be found in Miller and Downs (1950).

Glaphyrites has a subglobular to globular conch with a moderate-sized umbilicus. The conch is moderately evolute, with a goniatitic 8-lobed suture. *Eoasianites* can have a similar conch and suture, with the major difference being that *Eoasianites* has conspicuous umbilical ribs during the early juvenile growth stages. It is possible that some of the specimens identified as *Glaphyrites* in this study will prove to be *Eoasianites* when all the specimens are evaluated as part of a future systematic redescription of this ammonoid fauna.

Gonioloboceras has a discoidal conch and a small umbilicus that is not closed at late juvenile and mature growth stages. The whorls are compressed and somewhat convex laterally while strongly impressed dorsally. The medial portion of the venter is slightly flattened at maturity. *Gonioloboceras* sutures have eight lobes and saddles, with all but the first saddle being rounded.

Neodimorphoceras possesses a discoidal conch with a small umbilicus at maturity. The venter is characterized by a prominent groove. Growth lines form ventral and lateral sinuses and dorsolateral and prominent ventrolateral salients. The suture is similar to that of *Gonioloboceras* except that *Neodimorphoceras* develops an adventitious lobe in the first lateral saddle.

Schistoceras possesses a rounded conch with a moderate to large umbilicus and reticulated ornament. Nodes adorn the umbilical shoulders in early ontogeny. Sutures are goniatitic, with prongs of the ventral and lateral lobes being flared and lancelate. A critical diagnostic feature for this genus is the small umbilical lobe element on the suture that forms early in ontogeny and migrates to an umbilical shoulder position.

References

Arnold, J. M., 1985, Shell growth, trauma, and repair as an indicator of life history for *Nautilus*, *Veliger* **27**:386-396.

Barrande, J., 1869, Systême silurien du centre de la Bohême, Premiére Partie: Rescherches paléontologiques, v. 2, Class des Mollusques, Ordre des Céphalopodes, part 1, 712 pp.

Bayer, U., 1970, Anomalien bei Ammonitin des Aaleniums und ihre Beziehung zur Lebensweise. *N. Jahrb. Geol. Paläont., Abh.* **135**:19-41.

Boardman, D. R. II, Mapes, R. H., Yancey, T. E., and Malinky, J. M., 1984, A New Model for the Depth-Related Allogenic Community Succession within North American Pennsylvanian Cyclothems and Implications on the Black Shale Problem, in: *Limestones of the Mid-Continent* (N. J. Hyne, ed.), Tulsa Geological Society Special Publication No.2, pp.141-182.

Boardman, D. R., II, Work, D. M., Mapes, R. H., and Barrick, J. E., 1994, Biostratigraphy of Middle and Late Pennsylvanian (Desmoinesian-Virgilian) ammonoids, *Kansas Geol. Surv. Bull.* **232**:1-48.

Bond, P. N., and Saunders, W. B., 1989, Sublethal injury and shell repair in upper Mississippian ammonoids, *Paleobiology* **15**:414-428.
Boston, W., 1988, The surficial geology, paleontology, and paleoecology of the Finis Shale (Pennsylvanian, Lower Virgilian) in Jack County, Texas. Unpublished M.S. thesis, Ohio University.
Buckowski, F., and Bond, P., 1989, A predator attacks *Sphenodiscus*, *Delaware Valley Paleont. Soc.*, **IV**:69-74.
Canfield, D. E., and Raiswell, R., 1991, Pyrite formation and fossil preservation, in: *Taphonomy: Releasing the Data Locked in the Fossil Record* (P. A. Allison and D. E. G. Briggs, eds.), Plenum Press, New York, pp. 338-382.
Chamberlain, J. A., Ward, P. D., and Weaver, S., 1981, Post-mortem ascent of *Nautilus* shells: implications for cephalopod paleobiogeography, *Paleobiology* **7**:494-509.
Chaffin, D. T., 2000, Predation of Pennsylvanian cephalopods from the Finis Shale, Texas. Unpublished M. S. Thesis, Ohio University.
Cleaves, A. W., II, 1973, Depositional systems of Upper Strawn Group of north-central Texas, in: *Pennsylvanian Depositional Systems in North-central Texas: A Guide for Interpreting Terrigenous Clastic Facies in a Cratonic Basin* (F. L. Brown, A. W. Cleaves II, and A. W. Erxleben, eds.), pp. 31-42.
Dalton, R. B., and Mapes, R. H. 1999, Scavenging or predation: Mississippian ammonoid accumulations in carbonate concretion halos around *Rayonnoceras* (Actinocerida) body chambers from Arkansas, *V International Symposium, Cephalopods – Present and Past*,Vienna, Abst. Vol., p. 30.
Davis, R. A., Mapes, R. H., and Klofak, S. M., 1999, Epizoa on externally shelled cephalopods, in: *Fossil Cephalopods: Recent Advances in Their Study* (Yu. A. Rozanov, and A. A. Shevyrev, eds.), Russian Acad. Sci. Paleontol. Inst., Moscow, pp. 32-51.
Doguzhaeva, L. A., Mapes, R. H., and Mutvei, H., 1998, Breaks and radulae of early Carboniferous goniatites, *Lethaia* **30**:305-313.
Doguzhaeva, L. A., Mapes, R. H., and Mutvei, H., 1999 (1998), A late Carboniferous spirulid coleoid from the southern Mid-Continent (USA), in: *Advancing Research on Living and Fossil Cephalopods* (F. Oloriz and F. J. Rodriguez-Tovar, eds.), Plenum Press, New York, pp. 47-52.
Geczy, B., 1965, Pathologische jurassische Ammoniten aus dem Bakony-Gebirge, *Annales Univ. Scientiarium Budapestensis, sectio Geol.* **9**:31-37.
Guex, J., 1967, Contribution à l'etude des blessures chez les ammonites, *Bull. Laboratories de Géol. Minér., Géophy. et du Musée Géol. de l'Univ. de Lausanne* **165**:1-16.
Hengsbach, R., 1996, Ammonoid pathology, in: *Ammonoid Paleobiology* (N. H. Landman, K. Tanabe, and R. A. Davis, eds.), Plenum, New York/London, pp.581-605.
Hansen, M. C., and Mapes, R. H., 1990, A predator-prey relationship between sharks and cephalopods in the late Paleozoic, in: *Evolutionary Paleobiology of Behavior and Coevolution* (by A. Boucot), Elsevier, Amsterdam, Netherlands, pp.189-192.
Haven, N., 1972, The ecology and behavior of *Nautilus pompilius* in the Philippines, *Veliger* **15**:75-81.
Heckel, P. H., 1977, Origin of phosphatic black shale facies in Pennsylvanian cyclothems of Mid-Continent North America, *Am. Assoc. Petrol. Geol. Bull.* **61**:1045-1068.
Heckel, P. H., 1978, Field guide to Upper Pennsylvian cyclothemic limestone facies in eastern Kansas, *Kansas Geological Survey Guidebook Series* **2**, pp. 1-69, 76-79.
Heckel, P. H., 1980, Paleogeography of eustatic model for deposition of Mid-Continent Upper Pennsylvanian cyclothems, in: *Paleozoic Paleogeography of West-central United States, West-Central U.S. Paleogeography Symposium 1* (T. D. Fouch and E. R. Magathan, eds.), Rocky Mt. Section SEPM, pp. 197-215.
Hewitt, R. A., and Westermann, E. G., 1990, Mosasaur tooth marks on the ammonite *Placenticeras* from the Upper Cretaceous of Alberta, Canada, *Earth Sci.* **27**:469-472.
Hoare R. D., and Mapes, R. H., 1985, A new species of Pennsylvanian Polyplacophora (Mollusca) from Texas, *J. Paleontol.* **59**:1324-1326.
Hoare, R. D., Mapes, R. H., and Yancey, T. E., in press, Structure, taxonomy and epifauna of Pennsylvanian rostroconchs (Mollusca), *J. Paleontol.*
Kammer, T., Brett, C., Boardman D. R. II, and Mapes, R. H., 1986, Ecologic stability of the dysaerobic biofacies during the late Paleozoic, *Lethaia* **19**:109-121.
Kase, T., Shigeta, Y., and Futakami, M., 1994, Limpet home depressions in Cretaceous ammonites, *Lethaia* **27**:49-58.
Kase, T., Johnson, P. A., Seilacher, A., and Boyce, J. B., 1998, Alleged mosasaur bite marks on Late Cretaceous ammonites are limpet (patellogastropod) home scars, *Geology* **26**:947-950.
Kauffman, E. G. and Kesling R., 1960, An Upper Cretaceous ammonite bitten by a mosasaur, *Univ. Michigan Contrib. Mus. Paleont.* **15**:193-248.

Kauffman, E. G., 1990, Mosasaur predation on ammonites during the Cretaceous – an evolutionary history, in: *Evolutionary Paleobiology of Behavior and Coevolution* (by A. Boucot), Elsevier, Amsterdam, Netherlands, pp.184-199.

Keupp, H., 2000, Ammoniten, *Thorbecke Band 6*, Herausgegeben von Wighart v. Koenigswald, 165 pp.

Keupp, H., and Ilg, A., 1992, Palaopathologische Interpretation der Ammoniten Fauna des Ober-Callovium von Viller sur Mer/Normandie, *Berliner Geowiss. Abh.*, Reihe E, **3**:171-189.

Kröger, B., 2000, Schalenverletzungen an jurassischen Ammoniten – ihre paläobiologisch und paläoökologische Aussangefähigkeit, *Berliner Geowiss. Abh.* Reihe E, Band 33, 97 pp.

Landman, N. H., and Waage, K. M., 1986, Shell abnormalities in scaphitid ammonites, *Lethaia* **19**:211-224.

Lehmann, U.,1976 (trans. 1981), *The Ammonites: Their Life and Their World*, Cambridge University Press, Cambridge, 246 pp.

Maeda, H., and Seilacher, A., 1996, Ammonoid taphonomy, in: *Ammonoid Paleobiology* (N. H. Landman, K. Tanabe and R. A. Davis, eds.), Plenum, New York, pp. 543-578.

Mapes, R. H., 1979, Carboniferous and Permian Bactritoidea (Cephalopoda) in North America, *Univ. Kansas Paleontol. Inst.*, Art. 64, 75 pp.

Mapes, R. H., 1987, Upper Paleozoic cephalopod mandibles: frequency of occurrence, modes of preservation and paleoecological implications, *J. Paleontol.* **61**:521-538.

Mapes, R. H., and Dalton, R. B., in press, Scavenging or predation: Mississippian ammonoid accumulations in carbonate concretion halos around *Rayonnoceras* (Actinocerida) body chambers from Arkansas, *V International Symposium, Cephalopods – Present and Past*.

Mapes, R. H., and Hansen, M. C., 1984, Pennsylvanian shark-cephalopod predation: a case study, *Lethaia* **17**:175-182.

Mapes, G., Rothwell, G. W., and Mapes, R. H., 1996, Intriguing ovulate fructifications from Upper Pennsylvanian dysoxic marine shale (Finis Shale) in mid-continent North America, *Am. J. Botany* **74**:1205-1210.

Mapes, R. H., Sims, M. S., and Boardman, D. R,. II, 1995, Predation on the Pennsylvanian ammonoid *Gonioloboceras* and its implications for allochthonous vs. autochthonous accumulations of goniatites and other ammonoids, *J. Paleontol.* **69**:441-446.

Martin, A., Catala-Stucki, I., and Ward, P., 1978, The growth rate and reproductive behavior of *Nautilus macromphalus*, *N. Jahrb. Geol. Paläont. Abh.* **156**:207-225.

Mehl, J., 1978*a*, Ein Koprolith mit Ammoniten-Aptychin aus dem Solnhofer Plattenkalk, *Jber. Wetterau. Ges. Naturkunde* **129-130**:85-89.

Mehl, J.,1978*b*, Anhaufungen scherbenartiger Fragmente von Ammonitenschalen im suddeutschen Lias und Malm und ihre Deutung als Frassreste, *Ber. Naturforsch Ges. Freib. Breisgau* **68**:75-93.

Miller, A. K., Condra, G. E., and Dunbar, C. O., 1933, The nautiloid cephalopods of the Pennsylvanian system in the Midcontinent region, Nebraska Geol. Surv. Second Ser. 9, 240 pp.

Miller, A. K., and Downs, H. R., 1950, Ammonoids of the Pennsylvanian Finis Shale of Texas, *J. Paleontol.* **24**:185-218.

Morton, N., 1983, Pathologically deformed *Graphoceras* (Ammonitina) from the Jurassic of Skye, Scotland, *Palaeontology* **26**:443-453.

Mutvei, H., 1967, On the microscopic shell structure in some Jurassic ammonoids, *N. Jahrb. Geol. Paläont.* **129**:157-166.

Nixon, M., 1988, The feeding mechanisms and diets of cephalopods-living and fossil, in: *Cephalopods Present and Past*, Second International Cephalopod Symposium (J. Wiedmann and J. Kullman, eds.), E. Schweizerbart'sche Verlagbuchhandlung (Nägele u. Obermiller), Stuttgart, pp. 641-652.

Quinn, J. H., 1977, Sedimentary processes in *Rayonnoceras* burial, *Fieldiana Geol.* **33**:511-519.

Reeside, J. B., Jr., and Cobban, W. A., 1960, Studies of the Mowry Shale (Cretaceous) and contemporary formations in the United States and Canada, U.S. Geol. Survey Professional Paper 355, 126 pp.

Roll, A., 1935, Über Frasspuren an Ammonitenschalen, *Zbl. Miner. Geol. Paläont. Abt. B* **1935**:120-124.

Rothwell, G. W., Mapes, G., and Mapes, R. H., 1996, Anatomically preserved vojnovskyalean seed plants in Upper Pennsylvanian (Stephanian) marine shales of North America, *J. Paleontol.* **70**:1067-1079.

Saunders, W. B., 1981, The species of living *Nautilus* and their distribution, *Veliger* **24**:8-17.

Saunders, W. B., Spinosa, C., and Davis, L. E., 1987, Predation on *Nautilus*, in: *Nautilus* (W. B. Saunders and N. H. Landman, eds.), Plenum, New York, pp. 201-212.

Saunders, W. B., Bond, P. N., Hastie, L. C., and Itano, D., 1989, On the distribution of *Nautilus pompilius* in the Samoas, Fiji and Tonga, *Nautilus* **103**:99-104.

Saunders, W. B., Knight, R. L., and Bond, P. N., 1991, Octopus predation on *Nautilus*: evidence from Papua, New Guinea, *Bull. Mar. Sci.* **49**:280-287.

Sayre, A. N., 1930, The fauna of the Drum Limestone of Kansas and western Missouri, *Univ. Kansas Bull.* **17**:73-160.

Schram, F. R., Feldmann, R. M., and Copeland, M. J., 1978, The Late Devonian Palaeopalaemonidae and the earliest decapod crustaceans, *J. Paleontol.* **52**:1375-1387.
Seilacher, A., 1998, Mosasaurs, limpets or diagenesis: how *Placenticeras* shells got punctured, *Geowissenschaftliche Reihe* **1**:93-102.
Signor, P. W., and Brett, C. E., 1984, The mid-Paleozoic precursor to the Mesozoic marine revolution, *Paleobiology* **10**:229-245.
Tanabe, K., and Mapes, R. H., 1995, Jaws and radula of the Carboniferous ammonoid *Cravenoceras*, *J. Paleont.* **69**:703-707.
Tanabe, K., Shinomiya, A., and Fukuda, Y., 1988, 4. Notes on shell breakage in *Nautilus pompilius* from Fiji, in: *Marine Ecological Study on the Habitat of Nautilus pompilius in Fiji (The Second Operation)*, (S. Hayasaka, ed.), *Kagoshima Univ. Res. Center S. Pac. (KUSP), Occasional Papers* **15**:52-55.
Tanabe, K., Landman, N. H., Mapes, R. H., and Faulkner, C. J., 1994, Analysis of a Carboniferous embryonic ammonoid assemblage – implications for ammonoid embryology, *Lethaia* **6**:215-224.
Thiermann, A., 1964, Über verheilte Verletzungen an zwei kretazischen Ammonitengehäusen, *Fortschr. Geol. Rheinld. Westf.* **7**:27-30.
Tshudy, D. M., Feldmann, R. M., and Ward, P. D., 1989, Cephalopods: biasing agents in the preservation of lobsters, *J. Paleontol.* **63**:621-626.
Tucker, J. K., and Mapes, R. H.,1978a, Possible predation on *Nautilus pompilius*, *Veliger* **21**:95-98.
Tucker, J. K., and Mapes, R. H.,1978b, Coiled nautiloids of the Wolf Mountain Shale (Pennsylvanian) of North Central Texas, *J. Paleontol.* **52**:596-603.
Vermeij, G., 1987, Evolution and escalation: An ecological history of life, Princeton University Press, 527 pp.
Walker, S. E., and Yamada, S. B., 1993, Implications for the gastropod fossil record of mistaken crab predation on empty mollusc shells, *Palaeontology* **36**:735-741.
Ward, D. J. and Hollingsworth, N. T. J., 1990, The first record of a bitten ammonite from the Middle Oxford Clay (Callovian, Middle Jurassic) of Bletchley, Buckinghamshire, *Mesozoic Res.* **2**:153-161.
Ward, P. D., 1981, Shell sculpture as a defensive adaptation in ammonoids, *Paleobiology* **7**:96-100.
Ward, P. D., 1984, Is *Nautilus* a living fossil?, in: *Living Fossils* (N. Eldredge and S. Stanley, eds.), Academic Press, London, pp. 247-56.
Ward, P. D., 1996, Ammonoid extinction, in: *Ammonoid Paleobiology* (N. H. Landman, K. Tanabe, and R. A. Davis, eds.), Plenum, New York, pp. 815-824.
Ward, P. D., 1998, Time machine: scientific explorations in deep time, Copernicus, Springer-Verlag, New York, 241 pp.
Westermann, G. E. G., 1996, Ammonoid life and habitat, in: *Ammonoid Paleobiology* (N. H. Landman, K. Tanabe, and R. A. Davis, eds.), Plenum, New York, pp. 607-707.
Westermann, G. E. G., and Hewitt, R. A., 1995, Do limpet pits indicate that desmoceratacean ammonites lived mainly in surface waters?, *Lethaia* **28**:24.
Willey, A., 1903, Contribution to the natural history of the pearly Nautilus: Zoological Results based on material from new Britain, New Guinea, Loyalty Islands and elsewhere, collected during the years 1895, 1896 and 1897, University Press, Cambridge, England, Part 6, pp. 691-830.
Zangerl, R., 1981, Chondrichthyes I, Paleozoic Elasmobrachii, in: *Handbook of Paleoichthyology* 3A (H. P. Schultze, ed.), Gustav Fisher Verlag, Stuttgart, New York, pp. 1-155.
Zangerl, R., Woodland, B. G., Richardson, Jr., E.S., and Zachry, Jr., D.L., 1969, Early diagenetic phenomena in the Fayetteville black shale (Mississippian) of Arkansas, *Sed. Geol.* **3**:87-119.

Chapter 8

Predation on Brachiopods

LINDSEY R. LEIGHTON

1. Introduction 215
   1.1. Why Would Anything Eat a Brachiopod? 215
   1.2. So What Ate Paleozoic Brachiopods? 216
   1.3. Was Predation Important During the Paleozoic? 220
2. Methods 222
3. Results 224
4. Discussion 225
5. After the Paleozoic 229
6. Final Thoughts and Future Directions 230
   Appendix – Ornament and Population Data 230
   References 234

1. Introduction

If someone were to have written a chapter reviewing brachiopod predation thirty years ago, the paper would have been quite short. Since then, numerous papers have been written on the subject, including many excellent case studies. Nonetheless, despite a growing body of literature on the topic, there are certain questions related to brachiopod predation that I am often asked. So the first part of this review will attempt to answer these questions.

1.1. Why Would Anything Eat a Brachiopod?

I think that the logic behind questioning why anything would eat a brachiopod is based on the idea that there are better things to eat, such as "meatier" pelecypods.

LINDSEY R. LEIGHTON • Department of Geological Sciences, San Diego State University, San Diego, California, 92182.

Predator-Prey Interactions in the Fossil Record, edited by Patricia H. Kelley, Michał Kowalewski, and Thor A. Hansen. Kluwer Academic/Plenum Publishers, New York, 2003.

Brachiopods are mostly mantle cavity and lophophore; the actual body of a Recent brachiopod is proportionally small relative to the shell size (Rudwick, 1970). Eating a brachiopod would involve penetrating the shell for relatively minimal gain. However, for much of the lower and middle Paleozoic, the normal, open-marine communities were dominated by brachiopods, bryozoans, and pelmatozoans. Pelecypods, although present, generally were not as common. Moreover, many of the pelecypods that were present in these communities were small, and often capable of burrowing. Large pelecypods, including epifaunal taxa, existed by the Ordovician, but usually were confined to nearshore and restricted habitats. For many of the predators living in the lower to middle Paleozoic, eating something with more nutritional value than a brachiopod or crinoid might not have been an available option.

By the late Paleozoic, epifaunal pelecypods and brachiopods more often were part of the same communities (Miller, 1988; Olszewski and Patzkowsky, 2001). However, some of the late Paleozoic productides, including members of the Linoproductidae, Productidae, Echinoconchidae, and Aulostegidae, were more than five cm wide. Moreover, based on their muscle scars, some of these taxa had proportionally much larger body cavities than their smaller cousins, large enough to rival some of the Paleozoic bivalves as a menu choice.

Of course, predation also is a matter of scale. A small brachiopod might be the prey of choice for a small predator. When we think of a placoderm, many of us would think of the gigantic *Dunkleosteus*. But most placoderms were small. Ptyctodontid placoderms have been inferred to be durophages (shell-crushers) based on their molariform teeth, which are very similar to those of modern durophagous predators such as rays (Moy-Thomas and Miles, 1971). In addition, ptyctodontids, unlike most other placoderms, have a fixed jaw structure (Moy-Thomas and Miles, 1971), which is typical of shell-crushers. Many ptyctodontids were less than 20 cm long (Moy-Thomas and Miles, 1971; Denison, 1978). Given their fixed jaws, they probably had restricted gapes, and would have had difficulty in taking prey more than 2-3 cm wide – but many brachiopods would have been the appropriate size. Similarly, Alexander (1989) suggested that predation scars on Ordovician brachiopods, which he attributed to nautiloids, indicated that the predator had a gape < 2 cm.

For less active predators, such as drillers, the small size of many brachiopods might have made them desirable prey items. Furthermore, some drillers may have parasitized, rather than killed, their prey (Baumiller *et al.*, 1999); thus, prey size might have been less of a factor in foraging choice.

1.2. So What Ate Paleozoic Brachiopods?

This is a much more difficult question to answer. Some Recent asteroids prey upon brachiopods (Mauzey *et al.*, 1968, Richardson, 1997). There is controversy over whether Paleozoic asteroids were capable of extraoral feeding through extrusion of the stomach (Donovan and Gale, 1990; Blake and Guensburg, 1994). If Paleozoic asteroids were not capable of such an attack, then presumably they would have had to ingest their prey whole, which would have limited the predator to smaller prey. Unfortunately, it is unlikely that evidence for either mode of attack would be preserved.

There is more hope of finding evidence for two other attack methods: drilling and crushing. Paleozoic drilling predation has been controversial (Carriker and Yochelson, 1968; Sohl, 1969; Taylor *et al.*, 1980; Kaplan and Baumiller, 2000). Drill holes (Fig. 1) can be caused by factors other than predation; e.g., some organisms bore into postmortem shell material for domiciles (Richards and Shabica, 1969). However, there have been numerous case studies of Paleozoic drilling predation in which a stereotypic pattern of boring has been demonstrated, either with respect to choice of valve (Brunton, 1966; Buehler, 1969; Watkins, 1974; Rohr, 1976; Ausich and Gurrola, 1979; Baumiller et al., 1999; Leighton, 2001), or position on the valve (Brunton, 1966; Buehler, 1969; Rodriguez and Gutschick, 1970; Watkins, 1974; Rohr, 1976; Sheehan and Lesperance, 1978; Ausich and Gurrola, 1979; Smith et al., 1985; Chatterton and Whitehead, 1987; Baumiller et al., 1999; Leighton, 2001). In addition, particular size classes of prey have more cases of drilling than would be expected due to chance (Sheehan and Lesperance, 1978; Ausich and Gurrola, 1979; Leighton, 2001). All of these examples document borings that also fit Carriker and Yochelson's (1968) criteria for borings made by predators: (a) holes drilled perpendicular to the shell surface and cylindrical in cross-section; (b) holes penetrating only one valve of articulated specimens; (c) there is only one successful (complete) hole per individual prey. This last criterion is conservative; there are some situations in which a single individual might be drilled more than once. For example, octopods may drill more than one hole to inject venom (Bishop, 1975); juvenile predators will sometimes attack a prey item that is being drilled by another juvenile conspecific (Brown and Alexander, 1994), and parasitism may result in more than one hole (Kosuge and Hayashi, 1967; Baumiller, 1993). Of course, care must be taken when examining holes in shells, and the null hypothesis must be that the holes are not predatory, but when all criteria, including stereotypy of drillhole position, or of prey size, are met, then this is strong evidence for predation.

If the borings are predatory, what animal made them? The drillholes themselves may provide some answers. As in the Recent, there are two types of drill holes in Paleozoic brachiopods: Type A – small (< 1.6 mm wide), cylindrical holes [*Oichnus simplex* (Bromley, 1981)]; and Type B – large (> 1.6 mm wide), sometimes beveled, holes [*Oichnus paraboloides* (Bromley, 1981)].

FIGURE 1. Type A (small, cylindrical) predatory borehole (diameter = 0.7 mm) on a specimen of *Pholidostrophia nacrea* from the Devonian (Givetian) Silica Shale of Ohio, USA.

Type A and B holes are morphologically similar to muricid and naticid borings respectively, although these Paleozoic borings were not made by muricid or naticid gastropods as these clades did not evolve until the Mesozoic. It should be emphasized that these are end-members; some small Paleozoic borings are beveled (Smith et al., 1985; Grant, 1988), and some of the large borings are not (personal observation). Ausich and Gurrola (1979) originally suggested that Type A holes were parasitic and Type B holes were predatory, but Baumiller et al. (1999) observed Type B holes in stereotyped position over the fold and sulcus of *Brachythyris* and *Spirifer*, suggesting parasitization of the lophophore. In addition, Type A holes from the Devonian of Iowa included only one instance of a brachiopod that had more than one complete drillhole (Leighton, 2001), suggesting that these Type A holes were made by a predator, not a parasite.

Platyceratid gastropods are present wherever Type B holes have been reported (personal observation). Baumiller (1990, 1993) demonstrated that platyceratids were capable of drilling pelmatozoans, so it is quite plausible that platyceratids were the Type B driller. However, platyceratids are not always associated with Type A holes (Leighton, 2001). Ausich and Gurrola (1979) suggested that the Type A driller might be soft-bodied, although this hypothesis would be extremely difficult to test. One Recent turbellarian flatworm, *Pseudostylachus ostreophagus*, is capable of predatory drilling (Woelke, 1957), but the boring is very small and keyhole shaped (Yonge, 1964).

Preference for one valve of a given species is one common form of stereotyped drilling on brachiopods (Brunton, 1966; Watkins, 1974; Rohr, 1976; Ausich and Gurrola, 1979; Baumiller et al., 1999; Leighton, 2001). The preferred valve usually is the one inferred to be the lowermost (relative to the substrate) valve, regardless of whether that valve is the brachial or pedicle valve (Ausich and Gurrola, 1979; Leighton, 2001), although the example presented by Baumiller et al. (1999) is a notable exception. Ausich and Gurrola (1979) noted that, in cases where the reconstructed life-orientation has been inferred to be nearly upright, there was no valve preference. I can suggest three possible explanations for this mode of stereotyped valve preference:

(a) The predator was capable of overturning the prey. For some taxa, such as productides and umbonate strophomenides, drilling the posterior of the pedicle valve would put the predator closest to the prey's viscera. Attacking this region of these brachiopods would require overturning. However, for biconvex taxa, it is not immediately apparent why overturning would have been beneficial. The driller's grappling approach might also have been stereotyped; for example, a driller that frequently attacked and overturned productides might attack all brachiopods the same way.

(b) Similar to modern predatory naticids, the Paleozoic predator was a burrower, and could attack its brachiopod prey from below.

(c) The predator specialized in attacking prey that had been overturned by other causes, such as currents.

Drilling success also may prove helpful in identifying the predators. There have been two documented cases (Leighton, 1999a, 2001) of brachiopod ornament, such as spines, inhibiting successful predation. In the Upper Devonian of Iowa, the spinose *Devonoproductus* has a statistically lower rate of successful predation than the weakly ornamented *Douvillina* (33% vs. 54%, chi-square, $p < 0.02$; Leighton, 2001). Similarly, in the Lower Mississippian of Indiana, the ornamented athyrides *Cleiothyridina* and *Athryis* experienced less successful predation than did the smooth athyride *Composita*

(52% vs. 80%, chi-square, $p < 0.01$; Leighton, 1999a). In the first case, ornament provided better protection for the brachiopod, even though the more weakly ornamented taxon had thicker shells; in the second example, the taxa had approximately similar shell thickness. It should be stressed that spines do not necessarily inhibit the act of drilling (although see Smith *et al.*, 1985, for a case in which the driller preferred taxa that were not strongly ornamented), but compared to weak ornament, they decrease the likelihood of success.

There are many ways that spines could inhibit predator success after drilling has commenced. For example, spines increase the effective size of the prey, and also may root the prey into the substrate or entangle the prey with conspecifics (Dolmer, 1998), in all cases inhibiting the predator from orienting the prey for optimal drilling. Modern predatory gastropods that are unable to orient the prey optimally are less likely to drill successfully (Kitchell *et al.*, 1981; Anderson, 1992). Spines would generally increase handling time for the predator as well, increasing the likelihood that the predator would be disturbed, possibly by its own predators. All of these defenses would be more effective against a predator that had to grapple the prey; a soft-bodied predator that could slip between the spines would have less problems. Asteroids do not drill, but experiments have demonstrated that the relatively "soft" asteroids are less inhibited by prey ornament than are shelled, predatory gastropods (Stone, 1998). Taking all of the above evidence together, I think that the working hypothesis must be that the Paleozoic predatory drillers are gastropods. But confirming this hypothesis and identifying which gastropod clade or clades are responsible for the borings still remain important tasks.

Durophagous (shell-crushing) predators also were present in the Paleozoic. Crushed brachiopod fragments have been found within the skeletal cavities of two chondrichthyans, *Janassa* and *Fadenia* (Moy-Thomas and Miles, 1971). However, most of our evidence for shell-crushing predators comes from the prey, in the form of crushing scars. Although crushing could occur as a result of processes other than predation, predatory crushing scars tend to be point fractures with distinctive geometries (Alexander, 1986a; Elliot and Bounds, 1987) that would differ from damage inflicted by such factors as post-mortem compaction. The evidence for predation is most compelling when the crushing scar has been repaired (Fig. 2); such repair scars demonstrate that the brachiopod was alive at the time of the damage. However, it should be noted that younger individuals may repair damage more easily; Alexander (1981), using a large data set of Mississippian brachiopods, demonstrated that, for any given taxon, repaired damage generally occurred in smaller size classes than did unrepaired damage.

Alexander (1986a) documented sublethal crushing scars on more than 50 brachiopod species from 13 different assemblages ranging from the Ordovician to the Permian. Other examples of crushing and repair scars have been described by Brunton (1966) and Alexander (1981, 1986b).

As mentioned previously, ptyctodontid placoderms probably preyed upon brachiopods. Placoderms were almost extinct by the end of the Devonian, but the Mississippian saw a major radiation of chondrichthyans (Signor and Brett, 1984; Pough *et al.*, 1989). Signor and Brett (1984), Alexander (1986a), and Pough *et al.* (1989) suggested that petalodontid, cladodontid, or helodontid sharks were durophages, based on their molariform pavement teeth, and could have included brachiopods in their diet. Multiple, regularly spaced punctures on brachiopods fit well with the multi-cusped teeth of cladodontid sharks (Brunton, 1966; Alexander, 1986a), and helodontid teeth fit into

FIGURE 2. Sublethal crushing scar, showing repair (lower right of image). Note distortion of growth lines to repair damage. Specimen is a *Pholidostrophia nacrea* from the Devonian (Givetian) Silica Shale of Ohio, USA. Scale in mm.

tetrahedron-shaped scars on brachiopods (Alexander, 1986a; but see Elliott and Bounds, 1987, for a differing viewpoint).

Other potential durophages include arthropods such as phyllocarids, non-decapod eumalacostracans (Schram, 1981), and eurypterids. Modern phyllocarids are very small, but their Paleozoic ancestors attained large sizes (Copeland and Rolfe, 1978; Mikulic *et al.*, 1985). However, unlike phyllocarids and eumalacostracans, eurypterids lacked calcified appendages (Plotnick, personal communication, in Signor and Brett, 1984), and so probably were not capable of crushing brachiopods, although it is conceivable that some prey were ingested whole.

Whether Paleozoic nautiloids or ammonoids were durophages also is controversial. Bond and Saunders (1989) attributed shell damage on Mississippian ammonoids to attacks by other ammonoids. Elliott and Brew (1988) attributed chevron-shaped scars on a Pennsylvanian *Linoproductus* to a cephalopod attack. Alexander (1986b) described a rhyncholite, embedded in a brachiopod, from the Late Ordovician but this is the only known pre-Mississippian rhyncholite. Nonetheless, something was crushing brachiopods during the Late Ordovician; 9.7% of individual brachiopods in a Late Ordovician assemblage had crushing scars (Alexander, 1986a). Twenty-six percent of *Rafinesquina alternata* and 10% of *Strophomena planumbona* specimens had crushing scars. There was only one large phyllocarid, *Ceratiocaris*, during the Late Ordovician (Signor and Brett, 1984), so nautiloids may have inflicted the damage.

1.3. Was Predation Important During the Paleozoic?

Having established evidence for Paleozoic predation on brachiopods, we can move onto asking questions about the influence of predation on brachiopod evolution and ecology. During the Phanerozoic, there have been two major increases in predator diversity and intensity, the "mid-Mesozoic Marine Revolution" (Vermeij, 1977), and a mid-Paleozoic increase (Signor and Brett, 1984). The former occurred at a time when brachiopods no longer were major players in normal marine communities, whereas much of the evidence for the mid-Paleozoic increase in predation intensity is from brachiopods. Similarly, the vast majority of examples in the literature of predation on brachiopods are Paleozoic.

What major patterns have been observed?

(1) There were major increases (> 75%) in diversity of durophagous predators, primarily fish and arthropods, during the Early Devonian, the Mid-Devonian, the Late Devonian, and the Mississippian (Signor and Brett, 1984). Moreover, Signor and Brett's tabulation was conservative; some clades, such as nautiloids or eurypterids, that might have been capable of taking shelled prey were not included because, as described previously, these predators may have lacked appropriate teeth, jaws, or appendages for crushing shells. The increase in durophage diversity is not simply an increase in the number of predators, but also an increase in predator abilities, and an increase in the different types of predation that prey would have to encounter.

(2) Possible examples of drilling predation on brachiopods are known from as early as the Late Cambrian (Miller and Sundberg, 1984) and Ordovician (Cameron, 1967; Kaplan and Baumiller, 2000). However, such examples are uncommon and controversial (Wilson and Palmer, 2001). There was an increase in the number of cases of drilling predation on brachiopods during the Devonian (Kowalewski *et al.*, 1998). In addition, maximum drilling frequency, within assemblages, also reached a new high during the Devonian (Kowalewski *et al.*, 1998). Subsequent to the Mississippian, examples of drilling predation on brachiopods decreased, as did drilling intensity within assemblages (Kowalewski *et al.*, 1998, 2000; Leighton, 1999a), although it should be stressed that individual genera may have exceptionally high frequencies, such as the Pennsylvanian *Cardiarina* studied by Hoffmeister *et al.* (2001), in which almost 33% of individuals were drilled.

(3) The percentage of "articulated" brachiopod (Subphylum Rhynchonelliformea) genera that are multiplicate, rugate, lamellose, or spinose increased slightly during the Silurian, and then experienced a greater increase during the interval from the Givetian to the Famennian (Middle to Late Devonian) (Alexander, 1990).

(4) The absolute number of articulated brachiopod genera possessing spines steadily increased from the Late Silurian through the Early Mississippian (Signor and Brett, 1984).

(5) There is a latitudinal gradient in brachiopod ornament during the Devonian (Leighton, 1999b) and the Early Carboniferous (Dietl and Kelley, 2001); strongly ornamented taxa were more diverse in the tropics, and they formed a higher proportion of the tropical fauna than of the temperate fauna. As the tropics are inferred to be a zone of higher predation intensity, by analog with the Recent, the results of Leighton (1999b) and Dietl and Kelley (2001) suggest that predation may have influenced community structure and biogeography during the mid-Paleozoic.

Together, these patterns corroborate the hypotheses that (a) some brachiopod ornament was anti-predatory; and (b) the increase in taxa with these types of ornament is the result of an adaptive response to an increase in predation intensity during the Devonian, or possibly an exaptation that decreased the likelihood of extinction compared to unornamented taxa.

However, analyses conducted on global scales may not capture critical processes at smaller scales, and global patterns may not reflect patterns on finer spatial or temporal scales accurately. The increase in spinose genera is partly a function of the radiation of the Order Productida. Did productides evolve spines to resist predators, for example, drillers? This idea seems reasonable, and the papers cited above would corroborate the hypothesis. However, if we examine this problem more closely, two additional salient facts appear: productides first appear in Asia (Lazarev, 1987), but nine of the ten examples of Type A drilling cited by Leighton (2001) were from North America. This pattern may be a function of worker bias (most of the paleontologists working on Paleozoic drilling predation have worked in North America), and it does not constitute a refutation of the hypothesis that productide spines were anti-predatory, or even that spines evolved in response to predation. However, it does highlight the need for more studies on finer spatial and temporal scales, as suggested by Signor and Brett (1984).

Moreover, the translation of pattern from one spatial scale to another may require assumptions that may be questionable. For example, does predator diversity correlate with predation intensity? A community with a single, highly efficient, species of predator may have higher prey mortality than a community with many predatory species. Similarly, global prey diversity (in the sense of taxon richness) may have little to do with the structure of a community. The generic richness of Devonian rhynchonellides is very high; there are more genera of Devonian rhynchonellides than of atrypides and athyrides combined. Clearly, high rhynchonellide diversity would increase the proportion of multiplicate brachiopods. But closer examination reveals that the majority of these rhynchonellides are highly endemic, and many genera are rare. Atrypides, on the other hand, are very abundant, and often dominant, in most Devonian normal-marine assemblages.

These comments emphatically are not criticisms of research at global scales. Seminal papers such as Signor and Brett (1984), Alexander (1990) and Kowalewski *et al.* (1998) have laid most of the groundwork for future study of predation on brachiopods, and predation in the Paleozoic in general. Nevertheless, as pointed out by Alexander (1990), the observed global patterns do not prove a causal link between predation and the evolution of brachiopod ornament. The next part of this chapter will be a preliminary attempt to examine these hypotheses on a finer spatial scale.

2. Methods

The approach used here is to examine Ordovician through Devonian rhynchonelliform (Subphylum Rhynchonelliformea incorporates the old Class Articulata) brachiopod assemblages from a specific region. New York was chosen because (a) the region has a fairly continuous stratigraphic record from the Ordovician through Devonian; (b) it has many richly fossiliferous intervals; (c) some of the intervals have documented cases of predatory activity; and (d) depositional environments of New York units have been well studied in the past. Stratigraphic coverage of this analysis is not complete; if the patterns observed herein are suggestive, similar studies can be conducted using finer temporal resolution. All assemblages are from level-bottom communities interpreted to have lived in shallow shelf and outer platform habitats, above maximum storm wave base, with "normal-marine" circulation, salinity, and oxygen levels. These communities generally would be classified as BA-3 or BA-4 in the benthic assemblage system of Boucot (1975) and Brett *et al.* (1993). New York was part of the Eastern North America (ENA) biogeographic realm of Witzke and Heckel (1989). During the Ordovician through Devonian, New York generally was within the tropics or subtropics (Scotese and McKerrow, 1990). By examining assemblages within a specific region and within a limited range of habitat, the analysis partly will control for geographic and environmental factors.

Four of the units were personally collected or counted by the author; the remaining data are from the literature. Within each stratigraphic unit, the brachiopods were bulk-collected, or field-counted, at several beds from at least two localities (this also is true for units not collected by the author). The taxa were identified, counted, and classified into ornament categories. The ornament categories are: (1) smooth (except for growth lines) or weakly ribbed; (2) strongly costate or lamellose; (3) multiplicate or rugose; (4) spinose. Costae are radial ornament that do not deflect the commissure, whereas plicae do so. Lamellae and rugae are progressively more exaggerated and elevated concentric growth lines. These ornament classes are identical to those used in Leighton (1999b) and Dietl and Kelley (2001).

A weighted mean ornament (WMO) was then calculated for each assemblage by summing the products of the number of individuals in an ornament category by the category number (1 for smooth, 2 for costate or lamellose, etc.), and then dividing the sum by the total number of individuals.

$$WMO = (\Sigma (N_i * category_i)) / (N)$$

For example, if an assemblage consisted of three taxa, with populations = 10, 20, and 30 respectively, and ornament categories of 1, 2, and 3, respectively, then:

$$WMO = ((10*1) + (20*2) + (30*3)) / 60$$
$$= (10 + 40 + 90) / 60 = 140 / 60$$
$$= 2.33$$

Weighted Mean Ornament accounts for the relative abundance of taxa, the evenness and dominance within an assemblage, whereas simple presence and absence data would not. For example, imagine an assemblage consisting of taxa A, B, C, and D. There are 100 individuals of taxon A, which belongs in Ornament Category 4. There is one individual each of taxa B, C, and D, each of which belongs in Ornament Category 1. If mean ornament was calculated based only on presence/absence, then the assemblage would have a mean ornament = $(4+1+1+1) / 4 = 1.75$. However, the WMO = 3.91, which more accurately reflects the range of prey that a predator would have had to encounter and subdue. Note that WMO can be calculated from either absolute or percentage data or both; the result will not change. Indeed, some of the WMO values were calculated from percentage data. Taxa that comprised < 1% of a total assemblage were not included in the analysis.

The WMO within New York was then examined through time to determine if (a) the pattern is similar to that observed globally, and (b) to determine if there is any relationship between WMO and known instances of drilling predation (based on drilling traces) within the region. Eleven assemblages from New York were examined. Although the fossil record generally is excellent in New York, there are some notable gaps and problems with the analysis. For example, stratigraphic coverage in the analysis generally is denser in the Devonian. In addition, the Frasnian-age units primarily preserve fossils as molds and casts; consequently, direct evidence of predation, such as drillholes, is unavailable from these units. Data sources for brachiopod abundance, along with the relative abundance (percentage) data and ornament categories for each taxon, are listed in the Appendix.

3. Results

WMO values for New York through time are plotted on Fig. 3. Four patterns are revealed.

(1) There is a general increase in WMO through time (Pearson's Correlation, $r^2 = 0.50$, $p < 0.02$; Spearman's Rank Correlation, $r = 0.81$, $p < 0.01$).

(2) The general increasing trend occurs in a stepwise manner. There is an increase in WMO after the Silurian. Lower to Middle Devonian (405-375 Ma) assemblages have statistically larger WMO values than do pre-Devonian assemblages (one-tailed t-test, $p < 0.004$; Mann-Whitney U test, $p < 0.04$). Similarly, there is a jump in WMO between the Givetian (Mid-Devonian) and the Frasnian (Upper Devonian, 375-365 Ma) (one-tailed t-test, $p < 0.009$; Mann-Whitney U test, $p < 0.04$).

(3) Although there is no overlap in WMO values of the three intervals (pre-Devonian, Lower to Mid-Devonian, and Upper Devonian), the trend of increasing WMO is not apparent *within* any of the three intervals; indeed, all three end with a *decrease* in WMO.

(4) There is not an obvious relationship between traces of drilling predation and WMO values (Fig. 3). Those Devonian assemblages that experienced drilling do not have statistically different WMOs from those assemblages lacking traces of drilling.

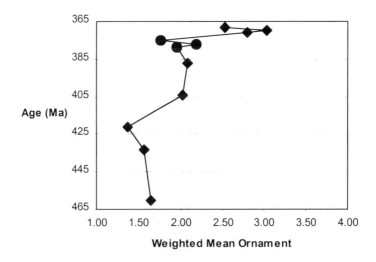

FIGURE 3. Relationship between Weighted Mean Ornament (WMO) and absolute age for eleven assemblages of New York brachiopods. Circles represent assemblages with a high frequency (> 3% of all individuals) of predatory drillholes (all of which are Givetian assemblages). Note stepwise increases in WMO through time.

4. Discussion

The regional, assemblage-level, patterns of a stepwise increase in brachiopod ornament from the Silurian to the Devonian documented in the present study are consistent with the global trend documented by Alexander (1990). However, the fact that the Givetian assemblages, which have the most traces of drilling predation (Buehler, 1969; Smith *et al.*, 1985), do not have the highest WMO values raises questions about the relationship between drilling predation and brachiopod ornament. Admittedly, this result probably is biased by the moldic preservation of the three Upper Devonian (Frasnian) assemblages, as it is extremely difficult to determine if these assemblages, which have the three largest WMO values, experienced drilling. If they did, then assemblages that experienced drilling would have significantly higher WMO values than assemblages without drilling. Frasnian predatory drilling has been documented elsewhere, in Iowa (Leighton, 2001). The WMO for the Frasnian assemblage in Iowa is 2.34, which is lower than any of the Frasnian values from New York, but higher than the Givetian values. Nonetheless, any pattern involving Frasnian data does not explain the pattern within the Early and Middle Devonian, i.e. why is there no increase in WMO during the Givetian?

The earliest reported predatory drilling of brachiopods in New York is from the Upper Ordovician Trenton Group (Cameron, 1967), so it could be argued that the Hamilton fauna are descendants of taxa that already were exposed to drilling and adapted anti-predatory ornament. However, even if the lineages that produced the Hamilton fauna previously were exposed to drilling predation, the Givetian seems to be an interval in which drilling predation is more intense (Smith *et al.*, 1985) and more common. Hamilton-age drilling predation also occurs in the Silica Shale of Ohio and the Traverse Group of Michigan (personal observations). But, despite the increase in drilling in the Givetian, the WMO values for the Hamilton beds are not significantly different from those of older Devonian assemblages in New York.

The Givetian pattern of WMO values is particularly intriguing because the Hamilton driller preferentially avoided strongly ornamented prey (Smith *et al.*, 1985). Predation presumably would have been an agent of natural selection; an adaptive response within weakly ornamented lineages might be expected. Alternatively, weakly ornamented taxa might be more prone to extinction, or marginalized to predation-free refugia, and subsequent communities would consist of strongly ornamented taxa and have a greater WMO. Why is such a response not observed?

Before considering possible answers to this question, another pattern needs to be examined. Up to now, the focus of this study has been on the Devonian increases in both drilling predation and brachiopod ornament. Do these trends continue throughout the Paleozoic? Drilling predation after the Early Mississippian becomes much less frequent within brachiopod assemblages (Kowalewski *et al.*, 1998; Leighton, 1999a) – but by the Pennsylvanian, productides are the dominant brachiopod clade in terms of diversity and abundance. Pennsylvanian assemblages generally would have much higher WMOs because half of the taxa would be spinose. This situation is the opposite from the Givetian – predation pressure is low, but WMO is high. Growth of ornament presumably involves an energetic tradeoff of some sort; for example, taxa that expend energy to grow ornament may produce fewer gametes. So if predation pressure decreased, it potentially could be beneficial for a prey lineage to secondarily lose ornament. But productides did not. Of course, spines have many useful functions other than retarding predation, including rooting the brachiopod in the substrate, inhibiting scour of surrounding sediments (Alexander, 1984; Leighton, 2000), stabilizing the individual on soft substrates (Grant, 1966, 1968), and possibly acting as a sensory system (Rudwick, 1970; Wright and Novak, 1997). Productides may have been so successful precisely because their spines preadapted them for many different conditions.

Nonetheless, there is a paradox. All of the benefits that spines provide would have been useful at any time during the Paleozoic. Yet there are no heavily spined brachiopods prior to the Devonian. The productide radiation occurs at exactly the same time as an increase in drilling predation, and as a radiation in durophages. All of this would seem to suggest that productide spines evolved to deter predators, but then why do we not observe a relationship between predation and ornament at the assemblage level? Several hypotheses to explain this pattern need to be tested.

(1) The first possibility is that the similar global trends in brachiopod ornament and predation are coincidental, or perhaps spines have so many useful functions that no one agent of selection would correlate with spines over time. More studies through space and time examining the relationship between ornament and predation need to be conducted at the level of lineages and assemblages.

(2) There is more than one type of predation, and there is no reason to assume *a priori* that drilling predation and crushing predation always follow similar trends. Drilling is a slow process; thus durophages generally are capable of taking more prey per time than are drillers. Although the greatest source of mortality is not necessarily the strongest agent of selection (Vermeij, 1982, 1987), it is possible that brachiopod ornament primarily was a response to durophagy, rather than drilling. In an analysis of a Recent pelecypod assemblage from Guam, Vermeij (1980) suggested that ornament in these pelecypod lineages primarily was a response to crushing, rather than drilling. Even if brachiopod ornament retarded drilling (Leighton, 1999a, 2001), WMO through space and time might correlate better with crushing predation intensity. If durophagy was intense during the late Paleozoic, then the decline in drilling predation may have had little effect on ornament.

(3) Just as there is more than one type of predation, there is more than one type of defense. WMO does not take any of these alternative defenses into account. Thayer (1983) suggested that cementation would be a defensive strategy. Harper (1991) demonstrated that cemented bivalves were better at resisting crab attacks than were byssate bivalves. However, cementation is not a significant factor in the present study as this mode of attachment did not become common among brachiopods until the Carboniferous. Of the taxa from New York, only the two species of *Schuchertella* may have been capable of cementation.

McGhee (2000) demonstrated a trend towards increasing convexity in biconvex brachiopods. All things being equal, a more spherical object tends to resist point-loading better, and would be less likely to be crushed.

Some Recent terebratulides have biochemical defenses (Curry, 1983; Thayer, 1985). A species that is poisonous does not need heavy armor. It is possible that there is a relationship between punctae and toxins (Curry, 1983). Moreover, punctate brachiopods store much of their organic tissue in the punctae. Most predators would be unable to gain access to this tissue; a predator eating a punctate brachiopod might gain less benefit relative to cost. In this regard, it is interesting to note that there is a strong correlation between the presence of punctae and weak ornament (Fitzgerald and Leighton, in prep).

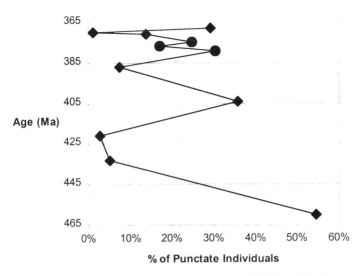

FIGURE 4. Relationship between percentage of punctate individuals and absolute age for eleven assemblages of New York brachiopods. Symbols as in Fig. 3.

However, it is apparent that punctae are not correlated with drilling predation in New York (Fig. 4).

Even if ornament were the only defense, different types of ornament may serve very different purposes. Spines may retard predation, but plicae may not. Moreover, a lineage evolving spines does not have to evolve plicae or rugae beforehand. WMO may be an oversimplification in some cases.

Hypotheses to explain the lack of a relationship between ornament and predation in specific assemblages need to be examined. One approach would be to analyze crushing scar frequency through time and space. The results of such an analysis, however, may be difficult to interpret. The main problem is that repair scars indicate that the predator failed. For a given prey species, low repair frequencies could indicate either that the predator did not frequently attack that prey, or that the predator was extremely successful and often crushed that species completely (Vermeij, 1982, 1987; Alexander, 1986a; Leighton, 2002).

Alexander (1986a) examined 18 brachiopod assemblages through the Phanerozoic for sublethal crushing damage. He demonstrated that strophomenates (with the exception of chonetidines) had significantly higher frequencies of crushing damage than did other brachiopods. Alexander (1989) demonstrated that typical strophomenate features, such as geniculation and rugae, tended to prevent fractures from propagating beyond the mantle cavity. Alexander (1989) also suggested that pseudopunctae might aid in this process. Strophomenates may have been more likely to survive damage than other brachiopods. In contrast, strongly biconvex brachiopods, such as rhynchonellides, atrypides, and some spiriferides, would better resist initial point-loads, but once failure occurred, the entire shell would be shattered.

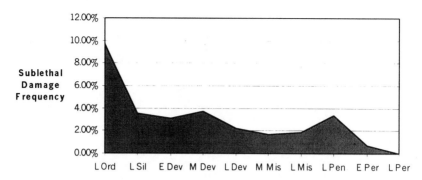

FIGURE 5. Sublethal Damage Frequencies for 14 Paleozoic brachiopod assemblages. Frequencies calculated from data in Alexander (1986b).

Alexander's (1986a) data set showed a general decrease in sublethal crushing frequency through time, from 9.7% in the Late Ordovician to < 1% in all Permian and younger assemblages (Fig. 5). Examining trends in crushing scars through time using this data set, as I do here, should be done with considerable caution, as some of these assemblages were from museum collections, the crushing frequency may not be representative of the original populations. Also, the assemblages are geographically and environmentally widespread. For example, the Early Permian assemblage consisted exclusively of 300 specimens of *Neochonetes*, which obviously was not the only common brachiopod genus alive during that time.

The Late Ordovician assemblage has an exceptionally high repair frequency (9.7%), more than twice that of any other interval. This may be an anomaly because the frequently damaged *Rafinesquina alternata* comprises almost 25% of the assemblage. Removing *R. alternata* would lower the repair frequency to 4.2%, although this still would be high. Similarly, the abundance of *Composita subtilita* (53%) may be artificially lowering the sublethal damage frequency of the Late Pennsylvanian assemblage; if *C. subtilita* is not included, the rest of the assemblage experienced a crushing frequency of 6.0%. Of course, these assemblages might be representative of the original populations, and removal of abundant taxa would be unwarranted, but the exercise also demonstrates that a single taxon might bias the results, and it is possible that there is something unusual about any given taxon that makes it more or less likely to have repair scars. For example, why does *C. subtilita* so rarely (1.5%) have sublethal damage in this assemblage? *Composita* was an extremely successful genus during the late Paleozoic; it was very abundant and had a broad environmental range, despite a smooth shell, and even though it was commonly drilled during the Mississippian. Had the lineage evolved chemical defenses by the Pennsylvanian that made it undesirable prey? Or, because of its biconvex shell, was it prone to being completely shattered by most durophagous attacks, and so left few repair scars?

Alexander (1986a) attributed the overall decrease in sublethal damage through the Paleozoic to an increase in forces generated by durophages over time; late Paleozoic durophagous condrichthyans were significantly more powerful than any early Paleozoic predator. The decreasing trend in sublethal damage to brachiopods would have been accentuated after the Paleozoic as durophages switched to larger pelecypods then available.

If the decrease in sublethal damage to Paleozoic brachiopods is real, and *if* this decrease is caused by an increase in predator efficiency, then trends in brachiopod armor during the Paleozoic would be consistent with those of durophagy, not drilling. One interesting fact pertaining to this issue: examining the Late Pennsylvanian assemblages from Alexander (1986*a*) reveals that the largest (size) taxa also have the highest frequency of sublethal damage (*Echinaria*, 17%; *Derbyia*, 9%; and *Linoproductus*, 29%; the six smaller taxa all have sublethal damage frequencies < 3%). This pattern isn't that surprising; it suggests that larger taxa were either less easily crushed, or that the predators preferred the larger taxa as prey. What is noteworthy is that all three of these taxa had potential defenses to durophagy besides large size; two of them are spinose, and *Derbyia* is a strongly costate cementer.

5. After the Paleozoic

After the Paleozoic, there are few reported cases of predation on brachiopods in the fossil record. Predatory drillholes on brachiopods have been documented in the Jurassic (Kowalewski *et al.*, 1998; Harper and Wharton, 2000) and in the Cretaceous (Surlyk, 1972; Harper and Wharton, 2000). One interesting difference between Harper and Wharton's (2000) results and examples from the Paleozoic is that, although there was some stereotypy of drilling with respect to valve, the preferred valve in the Cretaceous is the uppermost valve for the brachiopod in its inferred life position. These Cretaceous holes may be assigned more confidently to muricid predation (Harper and Wharton, 2000). Other Cretaceous predatory borings in *Cloristothyris* from the Navesink Formation of New Jersey can be attributed to naticids (Alexander, personal communication, 2002), based on the countersunk holes typical of those predators. Predatory drillholes also have been observed on Eocene brachiopods (Baumiller, personal communication, 1999). Predatory drilling on brachiopods has been reported in the Recent (LaBarbera, 1977; Witman and Cooper, 1983; Logan, 1990; Stewart, 1981). However, it is clear that post-Paleozoic drilling on brachiopods is far less common than during the Devonian (Kowalewski *et al.*, 1998). Similarly, sublethal crushing frequencies are < 1% for all of the post-Paleozoic assemblages examined by Alexander (1986*a*).

Subsequent to the Paleozoic, brachiopods may be a poor choice of prey, either because of increased toxicity (Thayer, 1985) or because they provide too little nutritional value compared to other prey. In addition, brachiopods simply are not available to many predators because they no longer dominate, or even live in, many normal marine communities, especially in the tropics.

One argument for the absence of brachiopods from many communities, and for their apparent antitropical distribution, is that those lineages that survived the Permian mass extinction were not as well adapted to deal with predation as were the high-metabolic pelecypods (Vermeij, 1987). As pointed out by Harper and Wharton (2000), there is a certain paradox to this hypothesis – brachiopods may be restricted to refugia because of predation, and yet there are few examples of predation on brachiopods during the Mesozoic. It would be extremely instructive to determine at what time brachiopods became more restricted in their distribution, and whether this event co-occurred with increases in predation intensity. Bambach (1990) pointed out that rhynchonellates

(Class Rhynchonellata includes most of the biconvex brachiopods, including the living "articulate" orders Rhynchonellida and Terebratulida) do not show a latitudinal diversity gradient during the Carboniferous. This raises the possibility that the ancestors of the Recent Rhynchonellides and Terebratulides might have had restricted environmental and geographic distributions by the late Paleozoic, long before the mid-Mesozoic Marine Revolution. Of course, these taxa might have been restricted by the rise of predation in the Mid-Paleozoic!

6. Final Thoughts and Future Directions

Over the past thirty years, we have discovered many more examples of predation on brachiopods in the fossil record. In doing so, we have built a large enough data set that we can start to ask questions about the processes behind the patterns we observe. Nonetheless, I think understanding these processes will require research at several spatial and temporal scales. More studies examining predation *through* time at the level of assemblages are required. Controlling for geography and environment, as I have attempted to do in this preliminary study, provides potentially critical tests of many hypotheses. Similarly, studies in which predation is examined across environmental parameters, such as depth, substrate type, or even latitude, within a single time slice, also hold great promise. These studies will be particularly beneficial if they can be performed on very fine spatial or temporal scales. Moreover, all of these approaches would be enhanced by examining the results within a phylogenetic framework; we need to perform phylogenetic analyses on predators and prey to place the evolution of anti-predatory strategies into an evolutionary framework.

Not surprisingly, my attempt to answer some deceptively simple questions about brachiopod predation has led to many more questions. One thing is clear, however. The relationship between predation and brachiopod evolution provides many interesting possibilities to explore. With their excellent preservation, brachiopods are a good choice for continued studies of predation in the fossil record.

Appendix – Ornament and Population Data

Taxon	Orn.	%	Formation	Age
Paucicrura rogata	2	54%	Trenton	L. Ordovician
Sowerbyella subovalis	1	23%	Trenton	L. Ordovician
Sowerbyella sericea	1	5%	Trenton	L. Ordovician
Rafinesquina spp.	1	9%	Trenton	L. Ordovician
Platystrophia spp.	3	5%	Trenton	L. Ordovician
Strophomena spp.	1	1%	Trenton	L. Ordovician
Dinorthis spp.	2	1%	Trenton	L. Ordovician
Eoplectodonta transversalis	1	39%	Williamson	M. Silurian
Atrypa "reticularis"	2	18%	Williamson	M. Silurian
Leptaena "rhomboidalis"	3	8%	Williamson	M. Silurian
Eospirifer radiatus	1	8%	Williamson	M. Silurian
"Chonetes" cornutus	1	3%	Williamson	M. Silurian

Resserella elegantula	1	3%	Williamson	M. Silurian
Stegerhynchus spp.	3	3%	Williamson	M. Silurian
Skenidioides pyramidalis	2	3%	Williamson	M. Silurian
Cyrtia meta	1	3%	Williamson	M. Silurian
Coolinia subplana	2	2%	Williamson	M. Silurian
Dicoelosia biloba	2	2%	Williamson	M. Silurian
Eocoelia sulcata	3	2%	Williamson	M. Silurian
Plectatrypa nodostriata	3	2%	Williamson	M. Silurian
Strophodonta corrugata	2	2%	Williamson	M. Silurian
Amphistrophia striata	1	62%	Rochester	L. Silurian
Coolinia subplana	2	17%	Rochester	L. Silurian
Eospirifer radiatus	1	7%	Rochester	L. Silurian
Leptaena rhomboidalis	3	7%	Rochester	L. Silurian
Resserella elegantula	1	2%	Rochester	L. Silurian
Stegorhynchus neglectum	3	2%	Rochester	L. Silurian
Eoplectodonta transversalis	1	1%	Rochester	L. Silurian
Atrypa reticularis	2	1%	Rochester	L. Silurian
Dalejina oblate	2	22%	New Scotland	L. Devonian
Macropleura macropleura	3	17%	New Scotland	L. Devonian
Leptaena rhomboidalis	3	15%	New Scotland	L. Devonian
Meristella arcuata	1	15%	New Scotland	L. Devonian
Isorthis perelegans	1	13%	New Scotland	L. Devonian
Mesodouvillina	1	8%	New Scotland	L. Devonian
Kozlowskiellina perlamellosa	3	6%	New Scotland	L. Devonian
Howellella cycloptera	2	2%	New Scotland	L. Devonian
Atrypa "reticularis"	2	31%	Onondaga	Eifelian
Coelospira camilla	2	14%	Onondaga	Eifelian
Nucleospira ventricosa	1	12%	Onondaga	Eifelian
Acrospirifer duodenaria	3	8%	Onondaga	Eifelian
Kozlowskiellina raricosta	3	6%	Onondaga	Eifelian
Pentagonia unisulcata	1	3%	Onondaga	Eifelian
Cupularostrum sp.	3	3%	Onondaga	Eifelian
Schizophoria multistriata	1	3%	Onondaga	Eifelian
Mucrospirifer sp.	3	3%	Onondaga	Eifelian
Dalejina sp.	2	2%	Onondaga	Eifelian
Pentamerella arata	3	2%	Onondaga	Eifelian
Leptaena rhomboidalis	3	2%	Onondaga	Eifelian
Cyrtina hamiltonensis	3	2%	Onondaga	Eifelian
Elita fimbriata	2	1%	Onondaga	Eifelian
Athyris spiriferoides	2	31%	Ludlowville	Givetian
Rhipidomella sp.	1	13%	Ludlowville	Givetian
Tropidoleptus carinatus	2	9%	Ludlowville	Givetian

Cyrtina hamiltonensis	3	8%	Ludlowville	Givetian
Pseudoatrypa devoniana	2	7%	Ludlowville	Givetian
Mediospirifer audaculus	3	6%	Ludlowville	Givetian
Mucrospirifer mucronatus	3	6%	Ludlowville	Givetian
Parazyga hirsute	1	4%	Ludlowville	Givetian
Pholidostrophia nacrea	1	4%	Ludlowville	Givetian
Protodouvillina inequistriata	1	3%	Ludlowville	Givetian
Kozlowskiellina sculptila	3	2%	Ludlowville	Givetian
Nucleospira concinna	1	2%	Ludlowville	Givetian
Spinocyrtia granulosa	3	1%	Ludlowville	Givetian
Megastrophia concava	2	1%	Ludlowville	Givetian
Meristella haskinsi	1	1%	Ludlowville	Givetian
Strophodonta demissa	2	1%	Ludlowville	Givetian
Camarotoechia spp.	3	20%	Moscow	Givetian
Pseudoatrypa devoniana	2	16%	Moscow	Givetian
Nucleospira concinna	1	15%	Moscow	Givetian
Tropidoleptus carinatus	2	12%	Moscow	Givetian
Athyris spiriferoides	2	12%	Moscow	Givetian
Meristella haskinski	1	5%	Moscow	Givetian
Spinatrypa spinosa	4	5%	Moscow	Givetian
Mucrospirifer mucronatus	3	3%	Moscow	Givetian
Mediospirifer audaculus	3	3%	Moscow	Givetian
Rhipidomella spp.	1	3%	Moscow	Givetian
Cyrtina hamiltonensis	3	2%	Moscow	Givetian
Kozlowskiellina sculptila	3	2%	Moscow	Givetian
Spinocyrtia granulosa	3	1%	Moscow	Givetian
Atrypa reticularia	2	28%	Lower Tully	Givetian
Schuchertella arctostriata	2	23%	Lower Tully	Givetian
Emmanuella subumbona	1	18%	Lower Tully	Givetian
Hypothyridina venustula	2	15%	Lower Tully	Givetian
Rhyssochonetes aurora	1	6%	Lower Tully	Givetian
Protoleptostrophia spp.	1	2%	Lower Tully	Givetian
Spinulicosta spinulicosta	4	2%	Lower Tully	Givetian
Schizophoria tulliensis	1	1%	Lower Tully	Givetian
Strophodonta sp.	2	1%	Lower Tully	Givetian
Mucrospirifer mucronatus	3	1%	Lower Tully	Givetian
Praewaagenoconcha	4	20%	Sonyea	Frasnian
Camarotoechia	3	18%	Sonyea	Frasnian
Mucrospirifer	3	11%	Sonyea	Frasnian
Tylothyris	3	8%	Sonyea	Frasnian
Leiorhynchus	3	7%	Sonyea	Frasnian
Pseudoatrypa	2	6%	Sonyea	Frasnian
Tropidoleptus	2	5%	Sonyea	Frasnian

Spinocyrtia	3	5%	Sonyea	Frasnian
Devonochonetes	1	4%	Sonyea	Frasnian
Rhipidomella	1	4%	Sonyea	Frasnian
Schizophoria	1	4%	Sonyea	Frasnian
Ambocoelia	1	2%	Sonyea	Frasnian
Devonochonetes spp.	1	22%	L. West Falls	Frasnian
Nervostrophia nervosa	2	20%	L. West Falls	Frasnian
Orthospirifer mesastrialis	3	11%	L. West Falls	Frasnian
Tylothyris mesacostalis	3	11%	L. West Falls	Frasnian
Praewaagenoconcha speciosa	4	10%	L. West Falls	Frasnian
Spinatrypa sp.	4	9%	L. West Falls	Frasnian
Ambocoelia umbonata	1	8%	L. West Falls	Frasnian
Camarotoechia spp.	3	7%	L. West Falls	Frasnian
Thiemella danbyi	1	1%	L. West Falls	Frasnian
Tylothyris mesacostalis	3	22%	Java	Frasnian
Schizophoria striatula	1	15%	Java	Frasnian
Praewaagenoconcha speciosa	4	12%	Java	Frasnian
Schuchertella chemungensis	2	11%	Java	Frasnian
Cyrtospirifer sp.	2	10%	Java	Frasnian
Leiorhynchus laura	3	10%	Java	Frasnian
Spinatrypa compacta	4	8%	Java	Frasnian
Athyris sp.	1	5%	Java	Frasnian
Cariniferella spp.	3	3%	Java	Frasnian
Ambocoelia sp.	1	2%	Java	Frasnian
Douvillina sp.	2	2%	Java	Frasnian

Data Sources:

Upper Ordovican, Trenton Group	Titus, 1986
Middle Silurian, Williamson and Willowvale Fms.	Eckert and Brett, 1989
Upper Silurian, Rochester Shale	Brett, 1999
Lower Devonian, Kalkberg and New Scotland Fms.	Author's collection
M. Devonian, Eifelian, Onondaga Fms.	Feldman, 1980
M. Devonian, Givetian, Ludlowville Fm., Hamilton Grp.	Smith *et al.*, 1985
M. Devonian, Givetian, Moscow Fm., Hamilton Grp.	Smith *et al.*, 1985
M. Devonian, Givetian, Lower Tully Fm.	Author's collection
U. Devonian, Frasnian, Sonyea Fm.	Author's collection
U. Devonian Frasnian, Lower West Falls Group	McGhee and Sutton, 1985
U. Devonian Frasnian, Java Fm.	Leighton, 2000

ACKNOWLEDGMENTS: The author wishes to thank Patricia Kelley and Michał Kowalewski for inviting this contribution; Richard Alexander and Alan Hoffmeister for their thoughtful reviews, and Peter Roopnarine, Richard Alexander, Geerat Vermeij,

Tomasz Baumiller, Daniel Fisher, Peter Kaplan and Sandra Carlson for numerous interesting discussions of brachiopod predation over the years.

References

Alexander, R. R., 1981, Predation scars preserved in Chesterian brachiopods: Probable culprits and evolutionary consequences for the articulates, *J. Paleontol.* **55**:192-203.

Alexander, R. R., 1984, Comparative hydrodynamic stability of brachiopod shells on current scoured arenaceous substrates, *Lethaia* **17**:17-32.

Alexander, R. R., 1986a, Frequency of sublethal shell-breakage in articulate brachiopod assemblages through geologic time, in: *Les Brachiopods Fossils et Actuels*, First International Brachiopod Congress (P. R. Racheboeuf and C. C. Emig, eds.), *Biostratigraphie du Paleozoique* **4:**159-166.

Alexander, R. R., 1986b, Resistance to and repair of shell breakage induced by durophages in Late Ordovician brachiopods, *J. Paleontol.* **60**:273-285.

Alexander, R. R., 1989, Influence of valve geometry, ornamentation, and microstructure on fractures in Late Ordovician brachiopods, *Lethaia* **22**:133-147.

Alexander, R. R., 1990, Mechanical strength of selected extant articulate brachiopods: Implications for Paleozoic morphologic trends, *Hist. Biol.* **3**:169-188.

Anderson, L. C., 1992, Naticid gastropod predation on corbulid bivalves: effects of physical factors, morphological features, and statistical artifacts, *Palaios* **7**:602-620.

Ausich, W. I., and Gurrola, R. A., 1979, Two boring organisms in a Lower Mississippian community of southern Indiana, *J. Paleontol.* **53**:335-344.

Bambach, R. K., 1990, Late Palaeozoic provinciality in the marine realm, in: *Palaeozoic Palaeogeography and Biogeography*, (W. S. McKerrow and C. R. Scotese, eds.), *Geological Society Memoir* **12**:307-323.

Baumiller, T. K, 1990, Non-predatory drilling of Mississippian crinoids by platyceratid gastropods, *Palaeontology* **33**:743-748.

Baumiller, T. K, 1993, Boreholes in Devonian blastoids and their implications for boring by platyceratids, *Lethaia* **26**:41-47.

Baumiller, T. K, Leighton, L. R., and Thompson, D. L., 1999, Boreholes in Mississippian brachiopods and their implications for Paleozoic gastropod drilling, *Palaeogeogr. Palaeoclim. Palaeoecol.* **147**:283-289.

Bishop, G. A., 1975, Traces of predation, in: *The Study of Trace Fossils* (R. W. Frey, ed.), Springer-Verlag, New York, pp. 261-281.

Blake, D. B., and Guensburg, T. E., 1994, Predation by the Ordovician asteroid *Promopalaeaster* on a pelecypod, *Lethaia* **27**:235-239.

Bond, P. N., and Saunders, W. B., 1989, Sublethal injury and shell repair in Upper Mississippian ammonoids, *Paleobiology* **15**: 414-428.

Boucot, A. J., 1975, *Evolution and Extinction Rate Controls*, Elsevier, Amsterdam, 427 pp.

Brett, C. E., 1999, Wenlockian fossil communities in New York State and adjacent areas, in: *Paleocommunities: A Case Study from the Silurian and Lower Devonian* (A. J. Boucot and J. D. Lawson, eds.), Cambridge University Press, Cambridge, UK, pp. 592-637.

Brett, C. E., Boucot, A. J., and Jones, B., 1993, Absolute depths of Silurian benthic assemblages, *Lethaia* **26**:25-40.

Bromley, R. G., 1981, Concepts in ichnotaxonomy illustrated by small round holes in shells, *Acta Geol. Hisp.* **16**:55-64.

Brown, K. M., and Alexander, J. E., 1994, Group foraging in a marine gastropod predator: benefits and costs to individuals, *Mar. Ecol. Prog. Ser.* **112**:97-105.

Brunton, C. H. C., 1966, Predation and shell damage in a Visean brachiopod fauna, *Palaeontology* **9**:355-359.

Buehler, E. J., 1969, Cylindrical borings in Devonian shells, *J. Paleontol.* **43**:1291.

Cameron, B., 1967, Oldest carnivorous gastropod borings, found in Trentonian (Middle Ordovician) brachiopods, *J. Paleontol.* **41**:147-150.

Carriker, M. R., and Yochelson, E. L., 1968, Recent gastropod boreholes and Ordovician cylindrical borings, *U. S. Geol. Surv. Prof. Paper B* 593-B, 26 pp.

Chatterton, B. D. E., and Whitehead, H. L., 1987, Predatory borings in the inarticulate brachiopod *Artiotreta* from the Silurian of Oklahoma, *Lethaia* **20**:67-74.

Copeland, M. J., and Rolfe, W. D. I., 1978, Occurrence of a large phyllocarid crustacean of Late Devonian – Early Carboniferous age from Yukon Territory, *Paper – Geol. Surv. Can.* **78-1B**:1-5

Curry, G. B., 1983, Microborings in Recent brachiopods and the function of caeca, *Lethaia* **16**:119-127.

Denison, R., 1978, Placodermi, in: *Handbook of Paleoichthyology* (H. P. Schultze, ed.), G. F. Verlag, Stuttgart, 128 pp.

Dietl, G. P., and Kelley, P. H., 2001, Mid-Paleozoic latitudinal predation gradient: Distribution of brachiopod ornamentation reflects shifting Carboniferous climate, *Geology* **29**:111-114.

Dolmer, P., 1998, The interactions between bed structure of *Mytilus edulis* L. and the predator *Asterias rubens* L., *J. Exper. Mar. Biol. Ecol.* **228**:137-150.

Donovan, S. K., and Gale, A. S., 1990, Predatory asteroids and the decline of the articulate brachiopods, *Lethaia* **23**:77-86.

Eckert, B., and Brett, C. E., 1989, Bathymetry and paleoecology of Silurian benthic assemblages, Late Llandoverian, New York State, *Palaeogeogr. Palaeoclim. Palaeoecol.* **74**:297-326.

Elliot, D. K., and Bounds, S. D., 1987, Causes of damage to brachiopods from the Middle Pennsylvanian Naco Formation, central Arizona, *Lethaia* **20**:327-355.

Elliot, D. K., and Brew, D. C. 1988, Cephalopod predation on a Desmoinesian brachiopod from the Naco Formation, central Arizona, *J. Paleontol.* **62**:145-147.

Feldman, H. R., 1980, Level-bottom brachiopod communities in the Middle Devonian of New York, *Lethaia* **13**:27-46.

Grant, R. E., 1966, Spine arrangement and life habits of the productoid brachiopod *Waagenoconcha*, *J. Paleontol.* **40**:1063-1069.

Grant, R. E., 1968, Structural adaptations in two Permian brachiopod genera, Salt Range, West Pakistan, *J. Paleontol.* **42**:1-32.

Grant, R. E., 1988, The family Cardiarinidae (late Paleozoic rhynchonellid Brachiopoda), *Senckenbergiana Lethaea* **69**:121-135.

Harper, E. M., 1991, Role of predation in the evolution of cementation in bivalves, *Palaeontology* **34**:455-460.

Harper, E. M., and Wharton, D. S., 2000, Boring predation and Mesozoic articulate brachiopods, *Palaeogeogr. Palaeoclim. Palaeoecol.* **158**:15-24.

Hoffmeister, A. P., Kowalewski, M., Bambach, R. K., and Baumiller, T. K., 2001, Intense drilling predation on the brachiopod *Cardiarina cordata* (Cooper 1956) from the Pennsylvanian of New Mexico, *Geol. Soc. Am. Abstr. Prog.* **33(6)**: A-248.

Kaplan, P., and Baumiller, T. K., 2000, Taphonomic inferences on boring habit in the Richmondian *Onniella meeki* epibole, *Palaios* **15**:499-510.

Kitchell, J. A., Boggs C. H., Kitchell, J. F., and Rice, J. A., 1981, Prey selection by naticid gastropods: experimental tests and application to the fossil record, *Paleobiology* **7**:533-552.

Kosuge, S., and Hayashi, S., 1967, Notes on the feeding habits of *Capulus dilatatus* A. Adams, 1860 (Gastropoda), *Sci. Rep. Yokosuka City Mus.* **13**:45-54.

Kowalewski, M., Dulai, A., and Fürsich, F. T., 1998, A fossil record full of holes: The Phanerozoic history of drilling predation, *Geology* **26**:1091-1094.

Kowalewski, M., Simões, M. G., Torello, F. F., Mello, L. H. C., and Ghilardi, R. P., 2000, Drill holes in shells of Permian benthic invertebrates, *J. Paleontol.* **74**:532-543.

LaBarbera, M., 1977, Brachiopod orientation to water movement. I. Theory, laboratory behavior and field observation, *Paleobiology* **3**:270-287.

Lazarev, S. S., 1987, The origin and systematic position of the main groups of productids (Brachiopods), *Paleontol. J.* **4**:39-49.

Leighton, L. R., 1999a, Antipredatory function of brachiopod ornament, *Geol. Soc. Am. Abstr. Prog.* **31**(7):A-43.

Leighton, L. R., 1999b, Possible latitudinal predation gradient in middle Paleozoic oceans, *Geology* **27**:47-50.

Leighton, L. R., 2000, Environmental distribution of spinose brachiopods from the Devonian of New York: Test of the soft-substrate hypothesis, *Palaios* **15**:184-193.

Leighton, L. R., 2001, New example of Devonian predatory boreholes and the influence of brachiopod spines on predator success, *Palaeogeogr. Palaeoclim. Palaeoecol.* **165**:53-69.

Leighton, L. R., 2002, Inferring predation intensity in the marine fossil record, *Paleobiology* **28**:328-342.

Logan, A., 1990, Recent Brachiopoda from the Snellius and Luymes expeditions to the Surinam-Guyana Shelf, Bonaire-Curacao, and Saba Bank, Caribbean Sea, 1966 and 1969-1972, *Zoologische Mededelingen* **63**:123-136.

Mauzey, K. P., Birkeland, C., and Dayton, P. K., 1968, Feeding behaviour of asteroids and escape responses of their prey in the Puget Sound region, *Ecology* **49**:603-619.

McGhee, G. R., 2000, The optimum biconvex brachiopod in the theoretical spectrum of shell form, in: *Functional Morphology of the Invertebrate Skeleton* (E. Savazzi, ed.), John Wiley & Sons, New York, pp. 415-420.

McGhee, G. R., and Sutton, R. G., 1985, Late Devonian marine ecosystems of the lower West Falls Group in New York, in: *The Catskill Delta* (D. L. Woodrow and W. D. Sevon, eds.), *Geol. Soc. Am. Spec. Paper* **201**:199-209.

Mikulic, D. G., Briggs, D. E. G., and Kluessendorf, J., 1985, A new exceptionally preserved biota from the Lower Silurian of Wisconsin, USA, *Phil. Trans. Roy. Soc. London, Ser. B: Biol. Sci.* **311**(1148):75-85.

Miller, A. I., 1988, Spatio-temporal transitions in Paleozoic Bivalvia; an analysis of North American fossil assemblages, *Hist. Biol.* **1**:251-273.

Miller, R. H., and Sundberg, F. A., 1984, Boring Late Cambrian organisms, *Lethaia* **17**:185-190.

Moy-Thomas, J. A., and Miles, R. S., 1971, *Palaeozoic Fishes*, Philadelphia, W. B. Saunders, 259 pp.

Olszewski, T. D., and Patzkowsky, M. E., 2001, Measuring recurrence of marine biotic gradients: a case study from the Pennsylvanian-Permian mid-continent, *Palaios* **16**:444-460.

Pough, F. H., Heiser, J. B., and McFarland, W. N., 1989, *Vertebrate Life*, 3rd ed., MacMillan, New York, 904 pp.

Richards, R. P., and Shabica, C. W., 1969, Cylindrical living burrows in Ordovician dalmanellid brachiopod beds, *J. Paleontol.* **43**:838-841.

Richardson, J. R., 1997, Ecology of articulated brachiopods, in: *Brachiopoda* (A. Williams, ed.), Treatise on Invertebrate Paleontology, Part H, revised, GSA, Boulder, and U. Kansas, Lawrence, pp. 441-462.

Rodriguez, R. P., and Gutschick, R. C., 1970, Late Devonian-early Mississippian ichnofossils from western Montana and northern Utah, in: *Trace Fossils* (J. P. Crimes and J. C. Harper, eds.), *Geol. J.* Special Issue, **3**:407-438.

Rohr, D. M., 1976, Silurian predator borings in the brachiopod *Dicaelosia* from the Canadian Arctic, *J. Paleontol.* **50**:1175-1179.

Rudwick, M. J. S., 1970, *Living and Fossil Brachiopods*, Hutchinson, London, 199 pp.

Schram, F. R., 1981, Late Paleozoic crustacean communities, *J. Paleontol.* **55**:126-137.

Scotese, C. R., and McKerrow, W. S., 1990, Revised world maps and introduction, in: *Paleozoic Palaeogeography and Biogeography* (W. S. McKerrow and C. R. Scotese, eds.), Geol. Soc., London, pp. 1-21.

Sheehan, P. M., and Lesperance, P. J., 1978, Effect of predation on population dynamics of a Devonian brachiopod, *J. Paleontol.* **52**:812-817.

Signor, P. W., and Brett, C. E., 1984, The mid-Paleozoic precursor to the Mesozoic marine revolution, *Paleobiology* **10**:229-245.

Smith, S. A., Thayer, C. W., and Brett, C. E., 1985, Predation in the Paleozoic: gastropod-like drillholes in Devonian brachiopods, *Science* **230**:1033-1037.

Sohl, N. F., 1969, The fossil record of shell boring by snails, *Am. Zool.* **9**:725-734.

Stewart, I. R., 1981, Population structure of articulate brachiopod species from soft and hard substrates, *N. Z. J. Zool.* **8**:197-208.

Stone, H. M. I., 1998, On predator deterrence by pronounced shell ornament in epifaunal bivalves, *Palaeontology* **41**:1051-1068.

Surlyk, F., 1972, Morphological adaptations and population structures of the Danish Chalk brachiopods (Maastrichtian, Upper Cretaceous), *Det Kongelige Dansk Videnskabernes Selskab Biol. Skrifter* **19**(2):1-53.

Taylor, J. D., Morris, N. J., and Taylor, C. N., 1980, Food specialization and the evolution of predatory prosobranch gastropods, *Palaeontology* **23**:375-409.

Thayer, C. W., 1983, Sediment-mediated biological disturbance and the evolution of marine benthos, in: *Biotic Interactions in Recent and Fossil Benthic Communities* (M. J. S. Tevesz and P. L. McCall, eds.), Topics in Geobiology 3, Plenum, New York, pp. 479-625

Thayer, C. W., 1985, Brachiopods vs. mussels: Competition, predation, and palatability, *Science* **228**:1527-1528.

Titus, R., 1986, Fossil communities of the upper Trenton Group (Ordovician) of New York State, *J. Paleontol.* **60**:805-824.

Vermeij, G. J., 1977, The Mesozoic marine revolution: Evidence from snails, predators, and grazers, *Paleobiology* **3**:245-258.

Vermeij, G. J., 1980, Drilling predation of bivalves in Guam: some paleoecological implications, *Malacologia* **19**:329-334.

Vermeij, G. J., 1982, Unsuccessful predation and evolution, *Am. Nat.* **120**:701-720.

Vermeij, G. J., 1987, *Evolution and Escalation: an Ecological History of Life*, Princeton Univ. Press, Princeton, NJ, 527 pp.

Watkins, R., 1974, Carboniferous brachiopods from northern California, *J. Paleontol.* **48**:304-325.

Witman, J. D., and Cooper, R. A., 1983, Disturbance and contrasting patterns of population structure in the brachiopod *Terebratulina septentrionalis* (Couthouy) from two subtidal habits, *J. Exper. Mar. Biol. Ecol.* **73**:57-79.

Wilson, M. A., and Palmer, T. J., 2001, Domiciles, not predatory borings: A simpler explanation of the holes in Ordovician shells analyzed by Kaplan and Baumiller, 2000, *Palaios* **16**: 524.

Witzke, B. J., and Heckel, P. H., 1989, Paleoclimatic indicators and inferred Devonian paleolatitudes of Euramerica, in: *Devonian of the World* (N. J. McMillan, A. F. Embry, and D. J. Glass, eds.), *Can. Soc. Petrol. Geol. Mem.* **14(1)**:49-66.

Woelke, C. E., 1957, The flatworm *Pseudostylochus ostreophagus* Hyman, a predator of oysters, *Proc. Nat. Shellfisheries Assoc.* **47**:62-67.

Wright, A. D., and Nolvak, J., 1997, Functional significance of the spines of the Ordovician lingulate brachiopod *Acanthambonia*, *Palaeontology* **40**:113-119.

Yonge, C. M., 1964, Rock borers, *Sea Frontiers* **10**:106-116.

Chapter 9

Predation on Bryozoans and its Reflection in the Fossil Record

FRANK K. MCKINNEY, PAUL D. TAYLOR, and SCOTT LIDGARD

1. Introduction 239
2. Predation on Living Bryozoans 240
   2.1. Colony-level Predation 242
   2.2. Predation on Individual Zooids 244
3. Fossil Record of Predation 245
   3.1. Multizooidal to Whole-Colony Predation...245
   3.2. Predation on Individual Zooids, Brood Chambers, etc. 247
4. Possible Evolutionary Responses to Predation..250
   4.1. Increased Calcification..250
   4.2. Possible Defensive Structures...252
   4.3. Growth Habits 256
5. Summary and Conclusions...256
   References...257

1. Introduction

Bryozoans are present in many benthic marine habitats, where they range from minor to dominant ecological elements. At the present day and apparently throughout their history, bryozoans reached peak levels of taxonomic richness in middle to outer shelf locations (Bottjer and Jablonski, 1988; McKinney and Jackson, 1989; Clarke and Lidgard, 2000). Many shelf-depth carbonate deposits from the Middle Ordovician to the present have been dominated by their skeletal remains (James and Clark, 1997; Taylor and Allison, 1998). Predation intensity can be high in such shallow waters, and living

FRANK K. MCKINNEY • Department of Geology, Appalachian State University, Boone, North Carolina, 28608-2067, and Honorary Research Fellow, Department of Palaeontology, The Natural History Museum, Cromwell Road, London SW7 5 BD, UnitedKingdom.   PAUL D. TAYLOR • Department of Palaeontology, The Natural History Museum, Cromwell Road, London SW7 5 BD, United Kingdom.   SCOTT LIDGARD • Department of Geology, The Field Museum, Roosevelt Road at Lake Shore Drive, Chicago, Illinois 60605.

Predator-Prey Interactions in the Fossil Record, edited by Patricia H. Kelley, Michał Kowalewski, and Thor A. Hansen. Kluwer Academic/Plenum Publishers, New York, 2003.

bryozoans are the targeted or incidental prey of a wide diversity of predators, from fishes to pycnogonids.

Predators of living bryozoans seldom leave unambiguous skeletal indication of their attacks. Where breakage is preserved, it is often difficult or impossible to distinguish a proximate cause. Despite several factual reviews of predators on living bryozoans over the past century (Osburn, 1921; Ryland, 1976; Cook, 1985), our general knowledge remains mostly anecdotal, and ecological apportioning of predation intensities is at best provisional. Moreover, very few papers describe skeletal indications of inferred predation events on fossil bryozoans. Consequently, the history of predation on bryozoans cannot be determined easily from evidence reported thus far. A more or less inductive approach to inferring the history of predation is to extrapolate from examples of predation on modern bryozoans coupled with the appearance of putative anti-predator adaptations in fossil skeletons (cf. Vermeij, 1987). There is considerable scope for future research observing the damaging effects of predators on Recent bryozoan skeletons and seeking similar damage in the fossil record.

In this paper, we first give an overview of what currently is known about predation on living bryozoans, at both zooidal and colonial levels. We then review the evidence suggesting specific predation events on fossil bryozoans at individual zooidal and colonial levels. This is followed by an assessment of the evolution of skeletal attributes that have been or might be inferred to have functioned in predation resistance.

2. Predation on Living Bryozoans

The maximum dimension of individual bryozoan zooids is seldom greater than 1 mm, but bryozoan colonies may contain millions of zooids and exceed 1 m in maximum diameter, though most sexually mature colonies reach maximum diameters of a few mm to a few cm. Consequently, there are many scales at which predators may attack bryozoan colonies, from ingestion zooid-by-zooid, to simultaneous ingestion of multiple zooids, to consumption of entire colonies with one encompassing attack. Our focus excludes predation on bryozoans during other phases of the life cycle that are poorly documented for living taxa and virtually unknowable in the fossil record. These include free-swimming gametes, larvae and post-metamorphosis ancestrulae. As with most marine invertebrates, larval stages of bryozoans are presumed to be subject to extreme levels of mortality due in part to predation. Newly metamorphosed ancestrular zooids and very small colonies are predictably more vulnerable to mortality from predatory attacks than are larger colonies that have achieved an escape in size, spatial dispersion, or regenerative ability (Keough and Downs, 1982; Osman et al., 1990; Turner and Todd, 1991).

Various functional classifications of predators have been proposed, some emphasizing degrees of selectivity, others pursuit versus ambush, still others focussing on a parasite/grazer/predator continuum (Hughes, 1980; Thompson, 1982; Underwood and Fairweather, 1992). Here we will refer to all of these categories broadly as "predators" on bryozoan colonies, dividing them roughly into two groups: those that ingest single zooids, and those that ingest multiple zooids – whether entire colonies or parts of large colonies – in each attack (Table 1). We therefore divide our discussion into predation on colonies (including multizooidal groups) and on individual zooids.

TABLE 1. Taxonomic and functional diversity of some predators on living bryozoan colonies. Literature reports of predation are widely scattered, and often fail to discriminate between implied predator/prey relationships (associations) and direct evidence for a predatory relationship. This list is based on field observations, gut contents, or laboratory manipulations. The list is intended to be denotive rather than comprehensive. Example sources are as follows: 1, Haderlie, 1969; 2, Stekhoven, 1933; 3, Wyer and King, 1973; 4, Coleman, 1989; 5, Buss and Iverson, 1981; 6, Chessa et al., 1990; 7, Lidgard, pers. obs. 1981, S. Ivanenko pers. comm. 1995; 8, Barnawell, 1960; 9, Gordon, 1972; 10, Osman et al., 1990; 11, Cook, 1985; 12, Todd and Havenhand, 1989; 13, Vietti and Balduzzi, 1991; 14, Mauzey et al., 1968; 15, Foster, 1975; 16, Lagaaij, 1963; 17, Vance, 1979; 18, Randall, 1967; 18, Bell et al., 1978; 19, Joubert and Hanekom, 1980; 20, Russ, 1980.

Predator phylum	Predator taxon	Predator size relative to zooid	Zooids per ingestion	Skeletal damage	Example sources
Platyhelminthes	polyclad turbellarian	larger	multiple	none	1
Nematoda	free-living nematode	smaller/same	single	none	2
Arthropoda	pycnogonid	same/larger	single	none	3
Arthropoda	amphipod crustacean	larger	multiple	breakage	4
Arthropoda	isopod crustacean	larger	single	none	5
Arthropoda	galatheid crab	much larger	multiple	breakage	6
Arthropoda	copepod	smaller	single	none	7
Mollusca	chiton	larger	multiple	breakage	8, 9
Mollusca	neogastropod	larger	multiple	breakage	10
Mollusca	nudibranch gastropod	same/larger	single/multiple	none/boring/breakage	11-13
Echinodermata	asteroid	much larger	multiple	none/breakage	9, 14, 15
Echinodermata	ophiuroid	much larger	multiple	breakage	9
Echinodermata	holothurian	much larger	multiple	breakage	16
Echinodermata	echinoid	much larger	multiple	breakage	9, 17
Chordata	euteleost fishes	much larger	multiple	breakage	18-20

2.1. Colony-level Predation

Erect bryozoans ostensibly are far more vulnerable to whole-colony predation than are encrusters (Jackson, 1983; McKinney and Jackson, 1989; Walters and Wethey, 1991) so that they may be reduced or eliminated from environments that otherwise would be suitable. This greater vulnerability exists whether the erect bryozoans are specifically selected (2.1.1) or are grazed by non-selective omnivores (2.1.2). Larvae of some erect bryozoan species exhibit a presumed evolutionary response in seeking small- to intermediate-scale nooks in which to metamorphose, and their survival rate is higher in such nooks than on more exposed surfaces (Walters and Wethey, 1991, 1996; Walters, 1992).

2.1.1. Bryozoan Colonies as Specific Prey

Juveniles of two species of asteroids have been observed to feed selectively on colonies of the cyclostome bryozoan *Tubulipora*, everting their stomachs over the colonies and apparently digesting the bryozoans' soft tissues in situ (Day and Osman, 1981). Feeding of the asteroids on *Tubulipora* left the colonies devoid of soft tissues but with the skeletons intact. There was no apparent taphonomic signature, and any slight dissolution of the calcitic *Tubulipora* skeletons by the asteroids' stomach acids would likely be indistinguishable from other sources of surficial corrosion. The asteroids were not seen to prey upon co-occurring cheilostome colonies. Selection of the cyclostome *Tubulipora* in preference to co-occurring cheilostomes may be due to the thinner exterior cuticle that lacks a mechanism for completely sealing the polypide from the ambient environment even when completely withdrawn into the skeleton. In contrast, cheilostomes have a thicker exterior cuticle as well as a sclerotized cap – the operculum – that tightly seals the orifice and better protects the polypide from chemical or physical attack.

Both adult and juvenile filefish (*Monacanthus setifer* Bennett) have been observed to feed preferentially on the erect cheilostome *Bugula neritina* (Linnaeus) in North Carolina coastal waters, removing all juvenile *B. neritina* colonies that did not begin growth in small cryptic pockets (Walters, 1992; Walters and Wethey, 1996). Fragments of an erect cyclostome constituted the primary gut contents of eight of ten examined specimens of an amphipod from the Antarctic (Coleman, 1989). Colony-level predation on erect bryozoan colonies probably is widespread, whether colonies are specifically targeted or are consumed along with other attached benthos. High predation intensity may be inferred from secondary metabolites such as alkaloids and terpenoids that occur in bryozoans and may have an anti-predatory role.

Several flexible erect gymnolaemate bryozoan species have yielded terpenoid and alkaloid toxins (e.g. *Alcyonidium diaphanum* (Hudson), *A. gelatinosum* (Linnaeus), *Flustra foliacea* (Linnaeus); see Carlé and Christophersen, 1980; Dyrynda 1985*a,b*; Morales-Ríos *et al.*, 2001), as has the loosely encrusting species *Sessibugula translucens* Osburn (Carté and Faulkner, 1986; see section 2.2, below). Extracts from three of four erect bryozoan species examined by Dyrynda (1985*a,b*) contained secondary metabolites with anti-bacterial properties, but only one of the four (*Flustra foliacea*) was acutely toxic to grazing fish. Nevertheless, toxins in erect bryozoans need not be acute to serve anti-predatory roles; lower levels of toxicity or even an offensive taste may suffice to deter or reduce predation on erect bryozoans by grazing vertebrates.

The existing literature on secondary metabolites in bryozoans suggests that they are widespread in erect cheilostome bryozoans and possibly not as common in encrusting bryozoans. However, broadly systematic surveys have not been undertaken, and information on secondary metabolites in cyclostome bryozoans is entirely lacking. A biased view may result from a tendency to report positive results and omit negative results among erect species examined, and also because the search for secondary metabolites may have been focused on the more readily harvested erect species rather than encrusting species.

2.1.2. Incidental Predation

Grazing fish as well as benthic predators and grazers (sea stars, echinoids, gastropods) consume bryozoans along with other benthic organisms such as cyanobacteria, chlorophytes, rhodophytes, sponges, cnidarians, and ascidians attached to hard substrata. On average, bryozoans grow more slowly than many other common attached benthos (e.g. Gordon, 1972; Kay and Keough, 1981; Jackson and Winston, 1982; Russ, 1982), and their abundance, distribution, and the types of colony morphologies present are sometimes predictably affected by predation.

Two sets of fouling panels placed in a reef environment on the north coast of Jamaica resulted in contrasting settlement and community development patterns that appeared to be related to presence of a vigorously grazing male yellowtail damselfish, *Microspathodon chrysurus* (Cuvier), at one site with no similar grazer at the other (Winston and Jackson, 1984). At the ungrazed site, some colonies of an encrusting species of the cheilostome *Steginoporella* lived for three years, reaching large sizes and producing larvae after two years; at the grazed site, this species recruited more slowly, did not produce long-lived or large colonies, and did not reproduce. In contrast, the encrusting cheilostome *Drepanophora tuberculatum* (Osburn) recruited in huge numbers onto the grazed panel where some colonies lived for three years, though never reaching large sizes; on the ungrazed panel, this species recruited much more slowly, grew at a slower rate, and had a shorter average life-span (177 vs 263 days). *Drepanophora tuberculatum* colonies at both sites produced larvae when three to six months old. The contrasting success of *Steginoporella* sp. and *D. tuberculatum* colonies at the two sites appears to have been the result of grazing by the damselfish at the site where *Steginoporella* sp. was unsuccessful and *D. tuberculatum* was highly successful. Part of the reason for the success of *D. tuberculatum* at the grazed site was the damselfish's reduction of space occupation by overgrowth competitors, including *Steginoporella* sp., whereas *D. tuberculatum* apparently survived the grazing relatively unscathed. In a laboratory study of the skeletal strength of encrusting bryozoans from this Jamaican reef, Best and Winston (1984) found the compressive strength of *D. tuberculatum* to be the highest of the nine species studied, and relative to *Steginoporella* sp. to be four times higher along the growing edges and twice as high in inner regions of colonies. Field-collected specimens of *Steginoporella* sp. revealed single "bite-marks" covering groups of zooids several mm^2 in area, with both older parts of colonies and growing margins exhibiting even larger grazed areas; in contrast *D. tuberculatum* showed signs of pruning by grazers along colony margins but only some abrasion of the frontal surface in inner regions of colonies. Apparently the compression generated by feeding damselfish falls between the compressive strengths of these two species of encrusting bryozoans. Interestingly, brood chambers (ovicells) of *D. tuberculatum* become deeply buried by secondary calcification (Winston and Jackson, 1984; McKinney and Jackson, 1989), possibly a response to surface abrasion by grazers; *Steginoporella* does not produce heavy secondary calcification, and its ovicells are more exposed.

Caging experiments have shown that grazing by gastropods significantly reduces the abundance of both encrusting and erect cheilostome bryozoans in the low intertidal zone in New South Wales, Australia (Anderson and Underwood, 1997). Usually, however, field experiments have found that grazing has a selective effect among taxonomic or functional groups of bryozoans, with differential survival of either different clades or different growth habits.

Sea urchins graze virtually indiscriminately on algae, sponges, tunicates, and bryozoans, yet their presence or absence often influences which of these organisms dominate. In shallow subtidal areas off Santa Catalina Island, California, in quadrats grazed by the sea urchin *Centrostephanus coronatus* Verrill, the encrusting bryozoans *Parasmittina* and *Rhynchozoon* were among the community dominants, but where *C. coronatus* was excluded erect bryozoans (*Bugula, Caberea, Crisia, Scrupocellaria*) were dominant (Vance, 1979). A different sea urchin is implicated in reduction of bryozoan abundance in a New Zealand fjord (Witman and Grange, 1998). Two species of nodular to erect bryozoans settle densely in shallow waters off northeastern New Zealand, but these are eliminated except under mesh roofs that prevent access by the leatherjacket fish, *Meuschenia scaber* (Forster), which grazes intensely on the encrusting community (Ayling, 1981).

Grazing by the echinoids commonly leaves distinctive convergent gouges generated by the Aristotle's lantern (e.g. Witman and Grange, 1998), but there may be no comparable signature visible on basal remnants of erect bryozoans fed on by corallaginous fish.

2.2. Predation on Individual Zooids

Diverse small-scale predators feed on bryozoans zooid-by-zooid, including nematodes (Stekhoven, 1933), nudibranchs (e.g. Harvell, 1984, 1986; Yoshioka, 1982; Barnes and Bullough, 1996), annelids (Winston, 1986), isopods (Buss and Iverson, 1981), pycnogonids (Wyer and King, 1973), possibly copepods (Best and Winston, 1984), and other invertebrate taxa with small body sizes roughly within an order of magnitude of zooid size (Table 1).

2.2.1. Zooidal Predation Leaving Little or No Skeletal Evidence

Specimens of three species of encrusting cheilostome bryozoans [*Calpensia nobilis* (Esper), *Reptadeonella violacea* (Johnston), *Schizoporella linearis* (Hassall)] from the northern Adriatic have been observed with small, cryptically colored dorid nudibranchs feeding on them (McKinney, personal observation). Where the nudibranchs had grazed the colony surfaces, the outer cuticles of zooids were gone and orificial opercula were missing or awry (Fig.1A). However, grazed versus ungrazed areas of dried colonies appeared similar under low magnification (Fig. 1B, C), and there was no visible damage to condyles or other parts of zooidal orifices at higher magnification (Fig. 1D, E). More robust nudibranchs than those observed in the northern Adriatic potentially could leave recognizable radular scratches on the zooidal surfaces.

Predation of individual zooids by flatworms, annelids, isopods, and pycnogonids is unlikely to yield skeletal evidence of the predation event. Best and Winston (1984) found that skeletally undamaged zooids lacking an operculum, and in some cases housing a copepod within the zooidal chamber, were relatively common in colonies of *Steginoporella* sp. from northern Jamaican reefs. However, they found no direct evidence that the copepods had killed the zooids whose skeletons they inhabited.

2.2.2. Zooidal Predation Involving Skeletal Damage

Puncture stresses necessary to break through zooidal frontal walls of different cheilostome species vary over two orders of magnitude (Best and Winston, 1984). Anascan cheilostomes, which have lightly calcified frontal walls, are most easily penetrated, and ascophoran cheilostomes, in which there is substantial progressive thickening of frontal wall calcification, are most resistant to penetration. The operculum is generally the easiest point of penetration on the frontal surfaces of cheilostome zooids, giving way along the hinge (Best and Winston, 1984). Exceptions are heavily calcified species in which the operculum consists of a highly thickened plug of cuticle (?calcified) that rests on a strong peripheral ring of skeleton when closed. Best and Winston (1984) found that, if the puncture probe was smaller than the operculum, large stress loads would produce a hole through the operculum, but if the probe was larger than the operculum, the skeletal orifice was shattered. This latter type of injury was not found in any field-collected specimens of the nine species studied. However, a single specimen of *Reptadeonella costulata* (Canu and Bassler) was found to have a puncture through the middle of the frontal wall, centered on the opening to the compensation sac.

There are other, unknown, predators of individual zooids that generate boreholes. Such boreholes have been observed most often in fossil bryozoans (3.2) but are also present in Recent bryozoans, but seldom reported. A specimen of *Stylopoma spongites* (Pallas) from the western Atlantic off South Carolina has the frontal shield of several zooids penetrated by circular borings approximately 50 µm in diameter (Fig. 1F). Many of the borings have healed margins, suggesting unsuccessful predation or successful predation followed by polypide/zooid regeneration. An incomplete boring consists of a 50 µm ring incised into the frontal shield, surrounding a central undamaged area (Fig. 1F, arrow), perhaps the work of a naticid gastropod. Borings described in Recent New Caledonian cheilostomes tend to be slightly larger than the holes in *Stylopoma* (and fossil borings mentioned below), ranging in diameter from 75-220 µm, and are slightly countersunk. Some borings penetrate ovicells, suggesting the predator may have been preying on the yolky embryos contained within (Gordon and d'Hondt, 1991).

3. Fossil Record of Predation

3.1. Multizooidal to Whole-Colony Predation

Shell-crushing vertebrates have evolved many times, first appearing among lungfishes in the Early Devonian, diversifying among the lungfishes and chondrichthyans by the Late Devonian (Vermeij, 1987). The chondrichthyans survived and diversified into the Carboniferous and Permian. The shell-crushing sharks and chimaeroids were joined in the Triassic by shell-crushing reptiles and perhaps as early as Late Triassic by shell-eating actinopterygians, including *Lepidotes* and pycnodonts (Vermeij, 1987). Powerfully bioeroding euteleost fish arose in the Eocene (Labridae - wrasses), Oligocene (Balistidae - triggerfish) and Miocene (Scaridae - parrotfish) (Vermeij, 1987; P. E. Ahlberg, personal communication, September 2001). While

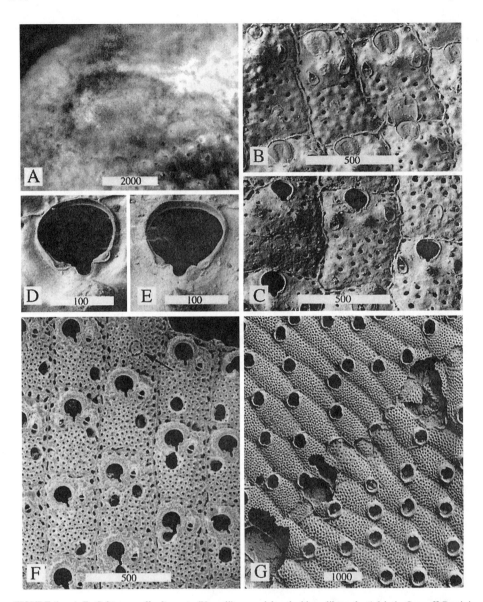

FIGURE 1. A-E, *Schizomavella linearis* (Hassall) grazed by dorid nudibranch, Adriatic Sea off Rovinj, Croatia; NHM 2002.1.16.1. (A) living colony with nudibranch (center) at end of grazing track (upper right). (B) ungrazed portion of colony surface. (C) grazed portion of colony surface, lacking exterior cuticle but with no skeletal damage visible. (D) orifice of ungrazed zooid. (E) orifice of grazed zooid, showing no damage to orifice. (F) *Stylopoma spongites* (Pallas), Atlantic Ocean, shelf off South Carolina, USA; USNM specimen 651385. Note circular boreholes in frontal shield, some of which have healed margins and (arrow) ring-shaped partial boring. (G) Reparative growth along broken border of *Watersipora grandis* (Canu and Bassler), Pinecrest Beds, Florida; NHM D57593. Scale bar lengths are in microns.

bryozoans would seldom have been the primarily targeted food, they often grow on shells and rock surfaces consumed by these skeleton-breaking and -crushing predatory vertebrates.

Fossils can record this type of whole-colony predation directly where predators die and are preserved with residues of their stomach contents, and indirectly if bryozoan-inhabited shell or rock substrata are more frequently preserved as crushed fragments in environments inhabited by the shell crushing and bioeroding vertebrates.

Bryozoans have been found among the stomach contents of the Permian petalodontid shark *Janassa*. An individual described by Schaumberg (1979) contained fragments up to 23 mm long of the erect fenestrate bryozoan *Acanchocladia anceps* (Schlotheim). Preservation of the shark's carcass and stomach contents is due to burial in fine-grained, anoxic sediments away from the oxygenated sea-bed where the shark fed on bryozoans and other epibenthic organisms.

Reparative growth by zooidal budding from fractured edges (Fig. 1G) is often recorded in fossil bryozoans (e.g., Håkansson and Thomsen, 2001), but specific biological causes of breakage are seldom pinpointed. An exception is Baluk and Radwanski's (1977) study of the free-living cheilostome *Cupuladria* spp. from the Miocene Korytnica Clays of Poland. The existence of holothurian spicules in the Korytnica Clays, together with Lagaaij's (1963, p. 187) observation of *Cupuladria* in the stomach of a holothurian from the Gulf of Mexico, led to the supposition that predation by these echinoderms was at least partly responsible for fragmentation of Miocene *Cupuladria* colonies.

Convergent scratches or gouges produced by the Aristotle's lantern of grazing echinoids and known as *Gnathichnus* are common features on Triassic-Recent shell surfaces. A second bioerosional ichnogenus, *Radulichnus*, produced by the radulae of grazing gastropods and chitons, occurs on Jurassic-Recent hard substrates. We know of no reports of *Gnathichnus* or *Radulichnus* traces recording the destruction of fossil bryozoans. However, such traces of grazing over the surfaces of bryozoans can be found in Cenozoic specimens (Fig. 2G), and Bromley and Hanken (1981, fig. 5) have figured an example of incipient *Gnathichnus* on a Holocene bivalve shell associated with destruction of part of a bryozoan colony. In the absence of reparative growth, it would be difficult to distinguish whether the trace maker was grazing on the bryozoan or on algae growing on the skeleton of a dead bryozoan. Other, more localized, points of damage (Fig. 2A) on bryozoan skeletons are enigmatic but could be the result of biting predation (see *Steginoporella*, section 2.1.2 above) or of incidental physical impact.

Large borings penetrate groups of zooids in Eocene bryozoans encrusting echinoids (Gibson and Watson, 1989, p. 319). These borings were interpreted by Gibson and Watson as the traces of gastropods preying on the bryozoans.

3.2. Predation on Individual Zooids, Brood Chambers, etc.

3.2.1. Boring

Most skeletalized Paleozoic bryozoans were free-walled stenolaemates with tubular zooids. Calcified skeleton above the basal lamina comprises interior walls secreted beneath an investment of epithelial, coelomic and cuticular tissues. Typically in free-walled stenolaemates the zooids intersect the colony surface at a high angle and have no apertural constriction at their distal ends. Predation on individual zooids of these

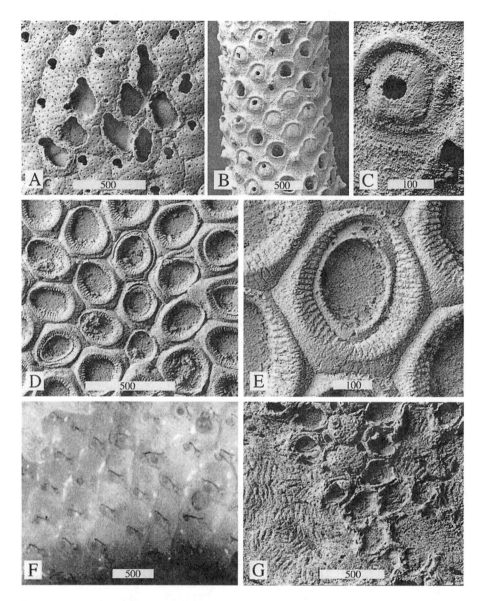

FIGURE 2. (A) Enigmatic damage to *Trypostega venusta* Norman, Waccamaw Formation (Pleistocene), Shallotte, North Carolina, USA; NHM BZ 4885. B, C. Bored *Meliceritites durobrivensis* (Gregory), Chalk (Cretaceous), Luton, Kent, England; NHM BZ3570(b). (B) branch. (C), bored autozooidal operculum (top) and eleozooid (lower right). D, E. Reparative zooidal budding in *Wilbertopora* sp., Prairie Bluff Chalk (Maastrichtian), Livingston, Alabama; NHM BZ 4796. (D) mixture of normal and repaired zooids. (E) single zooid with reparative budding. (F) live colony of *Schizoporella dunkeri* (Heller) with open mandibles of avicularia resulting in curved distal tip of a mandible placed at each zooidal orifice, Adriatic Sea off Rovinj, Croatia. (G) *Radulichnus* grazing traces associated with damaged *Microporella* sp., Waccamaw Formation (Pleistocene), Shallotte, North Carolina, USA; NHM BZ 4886.

stenolaemates is unlikely to leave skeletal indications because the easiest access to zooidal tissues is directly through the skeletally undefended aperture. Some stenolaemates, such as the Paleozoic Fenestrata, have shorter zooids lying parallel with the surface except for a narrowed distal portion that turns abruptly toward the colony surface. The zooidal chambers are protected by a relatively thin sheet of skeleton and conceivably could be the targets of drilling predators, but we know of no holes drilled into zooids of fenestrates or any of the more recent stenolaemates with the same general zooidal arrangement.

Post-Paleozoic stenolaemates of the order Cyclostomata include both free-walled and fixed-walled colonies. In fixed-walled stenolaemates the frontal portion of the zooidal skeletal wall is an exterior wall secreted directly against the outer cuticle, typically where the axis of the zooid parallels the colony surface for some distance. Therefore, in fixed-walled colonies, potential predators can attempt to get access to the zooidal tissues either through the skeletal aperture or by breaching the fixed-walled zooidal skeleton at the colony surface.

Circular and oval borings with diameters of 40-90 µm are known from several fixed-walled cyclostomes from the Middle Jurassic to Paleocene (Voigt and Flor, 1970; Taylor, 1982; Pitt and Taylor, 1990; Taylor and Wilson, 1999). Some of these borings may not have been predatory, especially where multiple borings penetrate individual zooids (e.g., Taylor and Wilson, 1999) or when terminal diaphragms of non-functional autozooids are bored (e.g., Pitt and Taylor, 1990, fig. 86). However, the distribution pattern of others suggests that they were produced by predators. The best evidence for predation is seen in several species of melicerititid cyclostomes ranging from Cenomanian to Campanian (Fig. 2B, C; Taylor, 1982). Uniquely among cyclostomes, melicerititids have calcified opercula closing zooidal orifices. In most species, heterozooids, termed eleozooids, with modified opercula are developed (Taylor, 1985). Borings in one fragment of *Meliceritites durobrivensis* (Gregory) demonstrated a clear pattern (Taylor, 1982): 58% of autozooids that retained their operculum had been bored through the operculum, whereas only one of these had been bored through the frontal wall; 31% of secondary eleozooids (small eleozooids reparatively budded into an autozooidal chamber) had been bored, largely through the frontal wall; and only 16% of the primary eleozooids had been bored, mostly through the frontal wall. Where heterozooids of reduced size occur in living cyclostomes, they possess much smaller polypides than are found in autozooids (Silén and Harmelin, 1974), and the preferential siting of borings in the opercula of Cretaceous melicerititids suggests that predators were focused on the easiest access to the larger, presumably more nutritious, autozooids.

There is also a scattered record of borings in individual cheilostome zooids back to the Late Cretaceous (Voigt, personal communication in Taylor, 1982, for Cretaceous and Paleocene; Cheetham, 2001, personal communication for Neogene). However, the borings have not been studied in detail, and it is impossible to know what proportion were produced by predators. Despite the fact that similar borings are known in Recent cheilostomes (Taylor, 1982, p. 73; Gordon and d'Hondt, 1991; Gordon, 1993), their maker is unknown.

3.2.2. Zooidal Regeneration

Bryozoan zooids characteristically undergo cycles of polypide degeneration and regeneration. A more profound form of zooidal regeneration is less common and involves the formation of an entirely new zooid within the body walls of a dead zooid

(Fig. 2D, E). Levinsen (1907) and Buchner (1918) described this phenomenon as total regeneration. Banta (1969) used the term reparative budding to circumvent confusion with polypide regeneration, and Taylor (1988a) recognized a continuum between intramural reparative budding, where the new zooid is enclosed by the intact skeletal walls of a dead host zooid, and extramural reparative budding where nothing remains of the dead zooid. Intramural reparative buds are extremely common in certain cheilostome taxa, particularly membraniporimorph anascans, and also occur in melicerititid cyclostomes (Taylor, 1994a). They are recognizable by the existence of an additional mural (or apertural) rim located immediately inside the outer rim. Whereas the outer rim belongs to the original zooid, the inner rim defines the aperture of the intramural bud. Sometimes repeated intramural budding leads to a sequence of successively smaller zooids arranged like Russian dolls (e.g., Poluzzi, 1980, pl. 4, fig. 6). The polymorph type of the intramural bud may differ from that of the original zooid, with switches from autozooid to kenozooid or from autozooid to avicularium/eleozooid being particularly common.

Unfortunately, it is not known whether zooid predation is always, sometimes or never the cause of intramural budding. If research on living bryozoans were to show that zooid predation did indeed trigger intramural budding, then quantifying the distribution of intramural buds, which are readily recognizable in fossils, particularly using SEM, could provide a means of studying changes through time in this particular style of zooid-level predation. An important caveat, however, is that the capacity to generate intramural buds may be specific to particular bryozoan taxa, necessitating the use of well-defined taxonomic groups to avoid confounding the taxonomic component.

4. Possible Evolutionary Responses to Predation

Several evolutionary trends within the Bryozoa can be inferred to be responses to predation, at least in part. The trends span the range of bryozoan organization, including attributes of autozooids, development of certain heterozooids, and colonial morphologies. Some possible predator-defensive structures have been reviewed by Taylor (1999, p. 637) but much work remains to be done in establishing their true functional value.

4.1. Increased Calcification

4.1.1. Zooidal Walls

Bryozoans show a broad spectrum of morphologies with regard to the extent to which the frontal surface of the zooids are skeletalized, ranging from taxa with no calcified frontal shields to others in which a calcified frontal shield covers all of the surface apart from a restricted orifice for the passage of the lophophore. Frontal shields can be formed of exterior wall mineralized directly beneath an outer cuticle, interior wall grown beneath an investment of epithelium, coelom/pseudocoel and cuticle, or a combination of the two wall types.

The overwhelming majority of Paleozoic stenolaemate bryozoans lack frontal shields. The only abundant exceptions are fenestrates, in which interior-walled frontal shields are the norm, although the rare Paleozoic cyclostomes all have calcified exterior-walled frontal shields. By contrast, post-Paleozoic stenolaemates comprise dominantly taxa with exterior-walled frontal shields. For example, 67% (33/49) of species in an

Aptian fauna (Pitt and Taylor, 1990), and 83% (30/36) of species in a Recent fauna have exterior-walled frontal shields (Hayward and Ryland, 1985). Among stenolaemates, therefore, there is a clear trend towards an increase in frontal shield skeletalization from the Paleozoic onwards.

A parallel pattern is observed among cheilostomes from the Late Jurassic to the present day. The ancestry of cheilostomes can be traced to a group of soft-bodied ctenostomes which acquired the ability to calcify their zooidal walls (see Todd, 2000 and references therein). The oldest cheilostome fossils date from about the Oxfordian (Taylor, 1994b) but the group remained of very low diversity and abundance until the late Albian/early Cenomanian, some 50 million years later (Taylor, 1988b).

All known early cheilostomes have an extensive, uncalcified region of the frontal wall of the zooid known as an opesium; frontal shield development is restricted to a narrow cryptocyst (an interior wall) bordering the opesium and, in some species, a slender proximal gymnocyst (exterior wall) known as a cauda. Neither the narrow cryptocyst nor the cauda provides appreciable protection to the polypide.

Species with significant frontal shields did not appear until the early Cenomanian. These included the anascan group known as coelostegans (mainly onychocellids), which have frontal shields formed of shelf-like cryptocysts, and cribrimorphs, which have overarching and fused exterior-walled spines that collectively form a "spinocystal" frontal shield.

New grades of frontal shield organization evolved subsequently in the Late Cretaceous and Danian, apparently polyphyletically, defining the "ascophoran" cheilostomes in which a space (termed an ascus in some taxa) beneath the frontal shield is dilated by muscles and filled by sea-water, thereby displacing the lophophore through the orifice and into the feeding position. (Eversion of the lophophore in earlier-evolved cheilostomes was accomplished by depressing the uncalcified portion of the frontal membrane.) The phylogenetic complexities of ascophoran frontal shield evolution have yet to be resolved (see Gordon, 2000, and references therein), but it seems that (1) species with extensive gymnocystal frontal shields had appeared by the Coniacian; (2) exterior-walled shields in which the outer, cuticle-juxtaposed surface forms the shield underside (umbonuloid frontal shields) were present by the Santonian; and (3) shields of cryptocystal skeleton (lepralioid frontal shields) had appeared by the Danian.

Changes through time in the proportion of cheilostome families with different frontal shield organizations have been depicted by Gordon and Voigt (1996, fig. 9; note, however, that this figure does not partition anascans into those with and those without extensive cryptocysts). While the number and proportion of ascophoran families with lepralioid frontal shields increased dramatically through the Paleogene, there was no decline in numbers of anascan families, the group which includes cheilostomes with the least well-developed frontal shields. Therefore, it is not clear from these data whether the trend towards increasing frontal shield development is passive or driven. Should further analysis reveal the existence of a driven trend, investigation would be warranted of which selective forces are responsible. Frontal shield calcification in cheilostomes has been shown to have an important function in mechanically strengthening erect colonies against bending stresses that may result in branch fracturing (Cheetham, 1986). However, it is notable that all of the frontal shield morphologies discussed above can be found commonly in encrusting species attached to hard substrates where bending stresses are irrelevant, and that the earliest examples of species having these morphologies are very often encrusters. Nonetheless, extremely thick frontal shields in cheilostomes are much more common in taxa with erect rather than encrusting growth habits, and gradients of thickening within rigidly erect colonies correlate with the distribution of stress induced by bending (Cheetham and Thomsen, 1981).

4.1.2. Opercula

The orifice through which the lophophore is protruded is arguably the most vulnerable point of a zooid (section 2.2.2). It is the only place where nutrient-rich organs emerge from within the zooidal walls, and it cannot be sealed permanently during the period that a functional polypide is present. The ability to block the orifice when the tentacles are retracted and to strengthen the blocking mechanism by thickening or calcification can be interpreted as a response to zooid-level predation pressure.

In living cyclostome stenolaemates, the orifice is delimited by the thin cuticular orificial and vestibular walls and is closed by contraction of the atrial sphincter muscle (Nielsen and Petersen, 1979; Boardman et al., 1992; Boardman, 1998). In cheilostomes the orifice is usually closed by a proximally hinged cuticular operculum that is a fold of the outer body wall of the zooid. The cheilostome operculum is very heavily sclerotized and in some species it is reinforced by calcification.

There are no clear indications of opercula in Paleozoic stenolaemates, but the Cretaceous-Paleocene melicerititid cyclostomes developed calcified, proximally hinged opercula that presumably had a protective function (Jablonski et al., 1997). The orifice of melicerititids was the focal point of attack for at least some boring predators (section 3.2), and evolution of calcified opercula in the group is likely a manifestation of defense against zooid-level predation.

Calcification of opercula in cheilostomes appears to have evolved repeatedly, as it is present in a small number of taxa apparently widely scattered phylogenetically (Jablonski et al., 1997). The oldest known calcified opercula in cheilostomes are in the anascan *Inversaria flabellula* (von Hagenow) from the late Campanian (Voigt and Williams, 1973), followed a few million year later by independently derived calcified opercula in the anascan *Monoporella exculpta* (Marsson) and cribrimorph *Castanopora lambi* Turner, both from the early Maastrichtian (Voigt, 1989; Turner, 1975). Calcified opercula have evolved independently within other cheilostome clades during the Cenozoic, so that they are present in at least five families.

4.2. Possible Defensive Structures

Possible defensive structures include heterozooids that might deter predators and skeletal elements that project into or above the space occupied by feeding lophophores, making it more difficult for predators to reach exposed lophophores and to penetrate the surface of zooids.

4.2.1. Avicularia and Similar Heterozooids

Avicularia are heterozooids with enlarged, generally distally tapered orifices closed by mandibles, which are enlarged opercula; the avicularium often contains a reduced polypide rudiment (Hyman, 1959; Ryland, 1970). Avicularia may be roughly the same size as autozooids and may occupy the position of autozooids (vicarious avicularia) or occur in spaces between autozooids without substituting for them (interzooidal avicularia), but the majority are generally smaller than autozooids and occur on the exposed surfaces of autozooids and other heterozooids (adventitious avicularia).

Avicularia have been hypothesized to have many different functions (reviewed in Winston, 1984), relatively few of which have any observational support. Some seta-like mandibles of specialized avicularia are used for locomotion (Cook and Chimonides, 1978) and sweeping sand from the colony surface (Cook, 1963). However, mandibles of most avicularia have distally pointed (Fig. 2F), often hooked, or spatulate shapes, and there is some evidence supporting the hypothesis that these defensive-looking structures can function against settling organisms or predators. However, most instances in which defense against predators or foulers has been hypothesized have not held up to scrutiny (reviewed in Winston, 1984).

Large pedunculate avicularia of *Bugula* (Harmer, 1901) and of *Beania* (McKinney, personal observation) have been seen to grasp errant polychaetes. When testing the reactions of avicularia in several shallow-water cheilostomes, Winston (1984) found that sessile adventitious avicularia of a species of *Smittipora* closed vigorously and held on when disturbed. Avicularia of all the other species that she provoked were remarkably passive both to her manipulation and to the nematodes, polychaetes, copepods, and small gastropods that crawled across the colony surfaces. She later (Winston, 1986) found that sessile adventitious avicularia of a species of *Reptadeonella* and a spatulate avicularium of a species of *Celleporaria* captured and held syllid polychaetes that had transgressed on the colonies. The capture of polychaetes by *Bugula*, *Beania*, *Reptadeonella* and *Celleporaria*, and the closure of *Smittipora* mandibles, suggest that at least some avicularia can stop potential predators. However, none of the polychaetes that have been seen to be captured by avicularia are confirmed predators of bryozoans.

Disposition and shape of so many avicularia strongly suggest defense, making one wonder if the relatively few laboratory observations accurately reflect the roles of "normal" avicularia in their natural environment. For example, *Schizotheca serratimargo* (Hincks) has rows of vicarious avicularia along the branch margins and proliferates vicarious avicularia proximally on branches (McKinney, 1989), various phidoloporids have single large interzooidal avicularia located at the base of the fenestrae (Winston, 1984, fig. 11; McKinney, 1998, fig. 2-7), and schizoporellids typically have single or paired adventitious avicularia adjacent to the orifice (e.g. Hayward and Ryland, 1995). Lack of understanding of avicularia and the relationship of function to avicularian morphology is one of the greatest deficiencies in knowledge of the biology of cheilostome bryozoans.

The oldest known avicularia are in *Wilbertopora mutabilis* sensu lato Cheetham from the Albian (Cheetham, 1975). They occur in a high proportion of cheilostome species from the Cenomanian to Recent, constantly present in some but sporadically developed in others.

Avicularia-like structures have evolved in two groups of stenolaemates. The Devonian fenestrate *Fenestrapora* is characterized by abundant rimmed, triangular pits along the frontal keel and across the reverse sides of branches, where they are most common as individual structures at the base of fenestrae, as in the phidoloporid cheilostomes (McKinney, 1998). One angle of the triangular pits is more elevated and more acute than the other two angles, giving the impression that a mandible-like structure hinged on the opposite side from the acute angle and functioned like an avicularian mandible, though no mandibles have been preserved. The Cretaceous - Paleocene melicerititids have mandibulate polymorphs (eleozooids) which bear a general resemblance to the vicarious and interzooidal avicularia of cheilostomes. While some eleozooids are smaller than autozooids and have greatly reduced orifices, others are large and have hypertrophied, elongate orifices over which the calcified mandibles could close (Taylor, 1985; Taylor, 1994a; Jablonski *et al.,* 1997).

4.2.2. Spines and Superstructures

Spines are present in a great many fossil and Recent cheilostome species (Fig. 3A). The most widely distributed are hollow structures believed by some bryozoologists (Silén, 1977) to be polymorphic zooids (spinozooids). Solid spinous projections occur in some cheilostomes, and a third category of spines are the mural spines (or spinules) that grow into the autozooidal chambers of, for example, *Jellyella* (Taylor and Monks, 1997). Basally articulated, hollow spines are first recorded in the Aptian, their existence being indicated by distinctive spine bases encircling the mural rim in the autozooids of *Spinicharixa pitti* Taylor. Basally fused, hollow spines appear to have been derived from articulated spines and had evolved in "myagromorph" and cribrimorph cheilostomes by the early Cenomanian. Hollow spines have considerable importance in cheilostome evolution as modifiable structural units from which frontal shields (Gordon, 2000) and ovicells (Ostrovsky, 1998) can be constructed.

It is becoming apparent that spines may fulfill several functions in cheilostomes, one of which is to ward against predation. Yoshioka (1982) and Harvell (1984, 1986) have shown that nudibranch predators can induce the formation of non-mineralized spines in *Membranipora membranacea* (L.). These spines reduce predation rate on the zooids containing them. Spines may also function in protecting epiphytic bryozoans against wave buffeting (Whitehead *et al.,* 1996; Bayer *et al.,* 1997) and – especially for species with colonies in which long spines are present in marginal zooids but are shed proximal to colony margins – in retarding overgrowth (Stebbing, 1973).

Upper Cretaceous cribrimorphs exhibit a range of elaborate skeletal structures that appear likely to have fulfilled a defensive function against predators. In some taxa (e.g., *Pachydera*, see Voigt, 1993) anastomosing spines form a long peristome which would lengthen the distance to the retracted polypide for any predators attempting to gain access via the orifice. Other cribrimorphs have avicularia closely clustered around autozooidal orifices (e.g., *Ichnopora*). In some genera (e.g., *Phractoporella*) additional calcification above the frontal shield, in the form of tertiary and quaternary frontal walls (Larwood, 1969), forms a superstructure (Fig. 3B).

Post-Paleozoic cyclostomes belonging to the Suborder Rectangulata (lichenoporids and disporellids) typically develop one to several spines around the autozooidal apertures (e.g., Hayward and Ryland, 1985, fig. 46A). These structures are sharply pointed, sometimes bifurcated or bearing subsidiary spinelets, and give the colony surface a formidably thorny appearance to human observers. It seems reasonable to suppose that they have a role in predator defense. Unfortunately, such apertural spines are very seldom preserved in fossils and their geological history has yet to be established.

Many genera of Paleozoic fenestrate bryozoans had long spines extending from their frontal surface, between skeletal apertures through which feeding lophophores extended, and several genera had spine- or keel-based skeletal elaborations that extended as perforated "superstructures" above frontal surfaces (Fig. 3C-F). These structures were especially common and diverse during the Devonian (e.g. McKinney and Kriz, 1986).

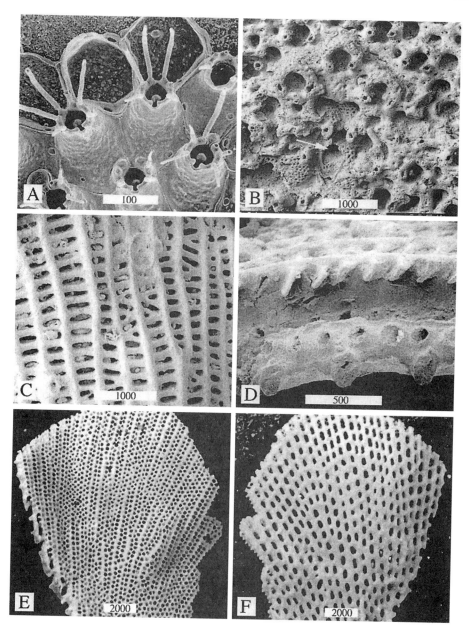

FIGURE 3. (A) *Schizotheca fissa* (Hincks) with elongate orificial spines, Adriatic Sea off Rovinj, Croatia; AMNH bryozoan specimen 942. (B) *Ubaghsia* sp., a cheilostome with ramifying spines coalesced as superstructure above autozooids; arrow indicates opesia of autozooid from which superstructure has been removed. Porosphaera Beds (Maastrichtian), Sidestrand, Norfolk. NHM specimen D55508. C, D. *Unitrypa acaulis* Hall, Upper Helderberg (Devonian), Jeffersonville, Indiana, USA; FMNH (UC) 57428. (C) superstructure formed by high keels (elongate ridges) linked by cross bars. (D) oblique lateral view of branch (lower portion of figure) with zooidal apertures opening into space below superstructure formed by continuous high keels linked along distal edges by thin lateral cross bars (upper portion of figure). E, F. *Hemitrypa* sp., Permian, Glass Mountains, Texas, USA; USNM specimen. (E) view onto surface of fine-meshed superstructure, each opening in superstructure centered over zooidal aperture of underlying branches. (F) reverse side of colony, showing branches much more robust than texture of superstructure.

Superstructures have been hypothesized to have served both for defense against predators (Cumings, 1904; Tavener-Smith, 1975; Bancroft, 1986; McKinney, 1987) and for structural strength (Cumings, 1904; McKinney, 1987). Some superstructures were based on spines and others on keels, and there are several grades of development from simple long spines and high keels to complexly linked, precisely elaborated structures.

High keels and superstructures with outer flanges had I-beam cross sections and undoubtedly did strengthen zoaria. However, some keel-based superstructures were laterally linked by thin bars, which added little or nothing to zoarial strength but subdivided the superstructure into a fine mesh. The superstructure of *Hemitrypa* (Devonian - Permian) was generated by thin lateral extensions from outer ends of delicate spines that met and fused to form an extremely delicate meshwork with each opening centered over a zooecial aperture, above the space into which the lophophore would have extended to feed (Fig. 3E, F). *Hemitrypa* commonly lived in areas of mud accumulation and minimal water motion (e.g. Snyder, 1987). The very delicate construction of the *Hemitrypa* superstructure and the paleoenvironmental distribution of the genus argue against a strengthening role.

Fenestrate bryozoans were typically heavily calcified by lamellar skeleton on reverse sides of branches and less heavily calcified on frontal surfaces of branches. Zooecial living chambers were closer to the frontal surfaces than to reverse surfaces, so drilling or other skeleton-breaching predators would likely have approached the frontal surface, as would orifice-penetrating predators. Interestingly, there is a high statistical correlation in Lower Devonian fenestrates for correspondence of thin frontal walls with occurrence of superstructures and relatively thicker frontal walls in species that do not have superstructures (McKinney, 1987).

The common attribute among the various Paleozoic fenestrates with long spines, high keels, and superstructures is skeletal extension into and beyond the space occupied by feeding lophophores. We interpret them as passive defensive structures against predation.

4.3. Growth Habits

Bryozoan species with erect growth habits increased as a proportion of the total bryozoan fauna throughout the Paleozoic, but species with erect growth habits proportionally decreased from the Early Cretaceous to Holocene (Jackson and McKinney, 1990). In addition, rigidly erect bryozoans have become progressively more restricted to deeper water or cryptic habitats (McKinney and Jackson, 1989). These trends appear to have been driven by the increase in shallow-water gouging, scraping and crushing predators that diversified and became more abundant during the late Mesozoic and early Cenozoic, and which have a greater effect on survival of erect than encrusting bryozoans (Vance, 1979; Walters and Wethey, 1991, 1996; Walters, 1992).

5. Summary and Conclusions

Bryozoans are preyed upon by diverse predators, from zooidal to colonial levels. Many zooid-level predators gain access to soft tissues through the zooidal orifices without leaving skeletal evidence of the attack. Colony-level predators commonly bite or scrape off large portions to entire colonies and, except for echinoid grazing traces, seem to leave little indication that breakage was specifically due to predation. However, erect bryozoans appear to be more vulnerable to colony-level predators than are encrusters, and, among encrusting bryozoans, more heavily calcified colonies appear to survive grazing predators better than do less robustly calcified colonies, at least in some environments.

Most reports of predation on living bryozoans describe local phenomena, and much of the published information is disseminated in more general ecological studies. There is insufficient published information – or it is dispersed and at present un-collated – to determine robust spatial or ecological patterns in distribution and types of predation on living bryozoans.

There is even less literature that addresses possible predation on fossil bryozoans. Although boreholes through the frontal shields may be found in bryozoans from Cretaceous to Recent, there are so few reports that they are in essence anecdotal. The predators that generate the boreholes are at present unknown. *Radulichnus* and *Gnathichnus* damage to bryozoan skeletons, indicating respectively mollusk and echinoid grazing of bryozoans (or of epibionts on dead bryozoan colonies), can be found on Cenozoic and possible Cretaceous bryozoans but to our knowledge have not previously been reported. Note, however, that Bromley *et al.* (1990) have depicted examples of a living chiton seemingly avoiding bryozoans during its grazing activities (their fig. 10) and of incipient *Radulichnus* that was produced both before and after bryozoan encrustation (their fig. 15). Bryozoans in gut contents or feces of grazing or shell-crushing fossil vertebrates are virtually unknown, either because they are rare or because they are under-reported. At present, no temporal or spatial trends in predation of bryozoans can be determined from direct evidence in the fossil record because there are so few published reports.

The best evidence for the predatory history of bryozoans is indirect: increase through time of structures such as avicularia, skeletal thickening, skeletal elements projecting above the level of zooidal orifices, and some changes in proportions of growth habits, can each be interpreted as predation-driven trends. However, much work remains to be done to document these trends more extensively and to relate them to presence, diversity, and abundance of potential predators that might be driving the trends. In addition, some features that can be interpreted as anti-predatory, e.g. skeletal thickening and orificial spines, potentially have other structural or ecological functions as well, and these need to be considered as alternatives that are either entirely or partially responsible for the observed trends.

References

Anderson, M. J., and Underwood, A. J., 1997, Effects of gastropod grazers on recruitment and succession of an estuarine assemblage: a multivariate and univariate approach, *Oecologia (Berl.)* **109**:442-453.

Ayling, A. M., 1981, The role of biological disturbance in temperate subtidal encrusting communities, *Ecology* **62**:830-847.

Baluk, W., and Radwanski, A., 1977, The colony regeneration and life habitat of free-living bryozoans, *Cupuladria canariensis* (Busk) and *C. haidingeri* (Reuss), from the Korytnica Clays (Middle Miocene; Holy Cross Mountains, Poland), *Acta Geol. Pol.* **27**:143-156.

Bancroft, A. J., 1986, The Carboniferous fenestrate bryozoan *Hemitrypa hibernica* M'Coy, *Irish J. Earth Sci.* **7**:111-124.

Banta, W. C., 1969, The body wall of cheilostome Bryozoa. II. Interzooidal communication organs, *J. Morph.* **129**:149-170.

Barnawell, E. B., 1960, The carnivorous habit among the Polyplacophora, *Veliger* **2**:85-88.

Barnes, D. K. A., and Bullough, L. W., 1996, Some observations on the diet and distribution of nudibranchs at Signy Island, Antarctica, *J. Moll. Stud.* **62**:281-287.

Bayer, M. M., Todd, C. D., Hoyle, J. E., and Wilson, J. F. B., 1997, Wave-related abrasion induces formation of extended spines in a marine bryozoan, *Proc. R. Soc. Lond.* **B264**:1605-1611.

Bell, J. D., Burchmore, J. J., and Pollard, D. A., 1978, Feeding ecology of three sympatric species of leatherjackets (Pisces: Monacanthidae) from a *Posidonia* habitat in New South Wales. *Australian J. Mar. Freshw. Res,* **29**:631-643.

Best, B. A., and Winston, J. E., 1984, Skeletal strength of encrusting cheilostome bryozoans, *Biol. Bull.* **167**:390-409.

Boardman, R. S., 1998, Reflections on the morphology, anatomy, evolution and classification of the Class Stenolaemata (Bryozoa), *Smiths. Contrib. Paleobiol.* **86**:1-59.

Boardman, R. S., McKinney, F. K., and Taylor, P. D., 1992, Morphology, anatomy, and systematics of the Cinctiporidae, new family (Bryozoa: Stenolaemata), *Smiths. Contrib. Paleobiol.* **70**:1-81.

Bottjer, D. J., and Jablonski, D., 1988, Paleoenvironmental patterns in the evolution of post-Paleozoic benthic marine invertebrates, *Palaios* **3**:540-560.

Bromley, R. G., and Hanken, N.-M., 1981, Shallow marine bioerosion at Vardø, arctic Norway. *Bull. Geol. Soc. Denmark* **29**:103-109.

Bromley, R. G., Hanken, N.-M., and Asgaard, U., 1990, Shallow marine bioerosion: preliminary results of an experimental study, *Bull. Geol. Soc. Denmark* **38**:85-99.

Buchner, P., 1918, Über totale Regeneration bei chilostomen Bryozoen, *Biol. Zentralbl.* **38**(11):457-461.

Buss, L. W., and Iverson, E. W., 1981, A new genus and species of Sphaeromatidae (Crustacea: Isopoda) with experiments and observations on its reproductive biology, interspecific interactions and color polymorphisms, *Postilla* **184**:1-23.

Carlé, J. S., and Christophersen, C., 1980, Dogger Bank Itch. The allergen is (2-Hydroxyethyl) dimethysulfoxonium ion, *J. Am. Chem. Soc.* **102**:5107.

Carté, B., and Faulkner, D. J., 1986, Role of secondary metabolites in feeding associations between a predatory nudibranch, two grazing nudibranchs, and a bryozoan, *J. Chem. Ecol.* **12**:795-804.

Cheetham, A. H., 1975, Taxonomic significance of autozooid size and shape in some early multiserial cheilostomes from the Gulf Coast of the U.S.A., *Docum. Lab. Géol. Fac. Sci. Lyon – H.S.* **3**(2):547-564.

Cheetham, A. H., 1986, Branching, biomechanics and bryozoan evolution, *Proc. R. Soc. Lond. B* **228**:151-171.

Cheetham, A. H., and Thomsen, E., 1981, Functional morphology of arborescent animals: strength and design of cheilostome bryozoan skeletons, *Paleobiology* **7**:355-383.

Chessa, L. A., Scardi, M., Russu, P. and Fresi, E., 1990, The trophic role of four crustaceans in a *Posidonia oceanica* meadow of Sardinia, Italy, in: *Trophic Relationships in the Marine Environment, Proceedings of the 24th European Marine Biology Symposium* (M. Barnes and R. N. Gibson, eds), Aberdeen University Press, Aberdeen, pp. 347-355.

Clarke, A. and Lidgard, S., 2000, Spatial patterns of diversity in the sea: bryozoan species richness in the North Atlantic, *J. Anim. Ecol.* **69**:799-814.

Coleman, C. O., 1989, *Gnathiphimedia mandibularis* K. H. Barnard 1930, an Antarctic amphipod (Acanthonotozomatidae, Crustacea) feeding on Bryozoa, *Ant. Sci.* **1**:343-344.

Cook, P. L., 1963, Observations on live lunulitiform zoaria of Polyzoa, *Cah. Biol. Mar.* **4**:407-413.

Cook, P. L., 1985. Bryozoa from Ghana - a preliminary survey, *Kon. Mus. Midden-Afrika (Tevuren, België), Zool. Wetensch., Annals* **238**:1-315.

Cook, P. L., and Chimonides, P. J., 1978, Observations on living colonies of *Selenaria* (Bryozoa, Cheilostomata). 1, *Cah. Biol. Mar.* **19**:93-103.

Cumings, E. R., 1904, Development of some Paleozoic Bryozoa, *Am. J. Sci.* **17**:49-78.

Day, R. W., and Osman, R. W., 1981, Predation by *Patiria miniata* (Asteroidea) on bryozoans: prey diversity may depend on the mechanism of succession, *Oecologia (Berl.)* **51**:300-309.

Dyrynda, P. E. J., 1985a, Chemical defences and the structure of subtidal epibenthic communities, in: *Proceedings of the 19th European Marine Biology Symposium* (P. E. Gibbs, ed.), Cambridge University Press, Cambridge, pp. 411-421.

Dyrynda, P. E. J., 1985b, Functional allelochemistry in temperate waters: chemical defences of bryozoans, in: *Bryozoa: Ordovician to Recent* (C. Nielsen and G. P. Larwood, eds.), Olsen & Olsen, Fredensborg, pp. 95-100.

Foster, M. S., 1975, Regulation of algal community development in a *Macrocystis pyrifera* forest, *Mar. Biol.* **32**:331-342.

Gibson, M. A., and Watson, J. B., 1989, Predatory and non-predatory borings in echinoids from the upper Ocala Formation (Eocene), north-central Florida, *Palaeogeogr. Palaeoclimatol. Palaeoecol.* **71**:309-321.
Gordon, D. P., 1972, Biological relationships of an intertidal bryozoan population, *J. Nat. Hist.* **6**:503-514.
Gordon, D. P., 1993, Bryozoa: The ascophorine infraorders Cribriomorpha, Hippothoomorpha and Umbonulomorpha mainly from New Caledonian waters, *Mém. Mus. Natn. Hist. Nat., (A)* **158**:299-347.
Gordon, D. P., 2000, Towards a phylogeny of cheilostomes – morphological models of frontal wall/shield evolution, in: *Proceedings of the 11th International Bryozoology Association Conference* (A. Herrera Cubilla and J. B. C. Jackson, eds.), Smithsonian Tropical Research Institute, Balboa pp. 17-37.
Gordon, D. P., and d'Hondt, J.-L., 1991, Bryozoa: The Miocene to Recent family Petalostegidae. Systematics, affinities, biogeography, *Mém. Mus. Natn. Hist. Nat., (A)*, **151**:91-123.
Gordon, D. P., and Voigt, E., 1996, The kenozooidal origin of the ascophorine hypostegal coelom and associated frontal shield, in: *Bryozoans in Space and Time* (D. P. Gordon, A. M. Smith, and J. A. Grant-Mackie, eds.), NIWA, Wellington, pp.89-107
Haderlie, E. C., 1969, Marine fouling and boring organisms in Monterey harbor II: second year of investigation, *Veliger* **12**: 182-192.
Håkansson, E., and Thomsen, E., 2001, Asexual propagation in cheilostome Bryozoa, in: *Evolutionary Patterns* (J. B. C. Jackson, S. Lidgard, and F. K. McKinney, eds.), University of Chicago Press, Chicago, pp. 326-347.
Harmer, S. F., 1901, Bryozoa in Britain, *Trans. Norfolk Norwich Nat. Soc.*, **7**:115-137.
Harvell, C. D., 1984, Predator-induced defense in a marine bryozoan, *Science* **224**:1357-1359.
Harvell, C. D., 1986, The ecology and evolution of inducible defenses in a marine bryozoan: cues, costs, and consequences, *Am. Nat.* **128**:810-823.
Hayward, P. J., and Ryland, J. S., 1985, Cyclostome bryozoans, *Syn. Br. Fauna (n.s.)* **34**:1-147.
Hayward, P. J., and Ryland, J. S., 1995, The British species of *Schizoporella* (Bryozoa: Cheilostomatida), *J. Zool., Lond.* **237**:37-47.
Hughes, T. P., 1980, Recruitment limitation, mortality, and population regulation in open systems: a case study, *Ecology* **71**:12-20.
Hyman, L. H., 1959, *The Invertebrates: Volume 5, Smaller Coelomate Groups,* McGraw-Hill, New York.
Jablonski, D., Lidgard, S., and Taylor, P. D., 1997, Comparative ecology of bryozoan radiations: origin of novelties in cyclostomes and cheilostomes, *Palaios* **12**:505-523.
Jackson, J. B. C., 1983, Biological determinants of present and past sessile animal distributions, in: *Biotic Interactions in Recent and Fossil Benthic Communities* (M. J. S. Tevesz, and P. L. McCall, eds.), Plenum Publishing Corporation, New York, pp. 39-120.
Jackson, J. B. C., and McKinney, F. K., 1990, Ecological processes and progressive macroevolution of marine clonal benthos, in: *Causes of Evolution* (R. M. Ross, and W. D. Allmon, eds.), University of Chicago Press, Chicago, pp. 173-209.
Jackson, J. B. C., and Winston, J. E., 1982, Ecology of cryptic coral reef communities. I. Distribution and abundance of major groups of encrusting organisms, *J. Exper. Mar. Biol. Ecol.* **57**:135-147.
James, N. P., and Clarke, J. A. D., eds., 1997, *Cool-Water Carbonates, SEPM Spec. Pub.* **56**.
Joubert, C. S. W., and Hanekom, P. B., 1980, A study of feeding in some inshore reef fish of the Natal Coast, South Africa, *South African J. Zool.* **15**:262-274.
Kay, A. M., and Keough, M. J., 1981, Occupation of patches in the epifaunal communities on pier pilings and the bivalve *Pinna bicolor* at Edithburgh, South Australia, *Oecologia* **48**:123-130.
Keough, M. J., and Downs, B. J., 1982, Recruitment of marine invertebrates: the role of active larval choices and early mortality, *Oecologia (Berl.)* 54:348-352.
Lagaaij, R., 1963, *Cupuladria canariensis* (Busk) – portrait of a bryozoan, *Palaeontology* **6**:172-217.
Larwood, G. P., 1969, Frontal calcification and its function in some Cretaceous and Recent cribrimorph and other cheilostome Bryozoa, *Bull. Br. Mus. (Nat. Hist.) Zool.* **18**(5):173-182.
Levinsen, G. M. R., 1907, Sur la régénération totale des Bryozoaires, *Overs. Kgl. Danske vidensk. Selsk. Forhandl.* **1907**(4):151-159.
Mauzey, K. P., Birkeland, C., and Dayton, P.K., 1968, Feeding behavior of asteroids and escape responses of their prey in the Puget Sound region, *Ecology* **49**:603-619.
McKinney, F. K., 1987, Paleobiological interpretation of some skeletal characters of Lower Devonian fenestrate Bryozoa, Prague Basin, Czechoslovakia, in: *Bryozoa: Present and Past* (J. R. P. Ross, ed.), Western Washington University, Bellingham, pp. 161-168.
McKinney, F. K., 1989, Two patterns of colonial water flow in an erect bilaminate bryozoan, the cheilostome *Schizotheca serratimargo* (Hincks, 1886), *Cah. Biol. Mar.* **30**:35-48.
McKinney, F. K., 1998, Avicularia-like structures in a Paleozoic fenestrate bryozoan, *J. Paleontol.* **72**:819-826.
McKinney, F. K., and Jackson, J. B. C., 1989, *Bryozoan Evolution,* Unwin & Hyman, Boston.

McKinney, F. K., and Kriz, J., 1986, Lower Devonian Fenestrata (Bryozoa) of the Prague Basin, Barrandian Area, Bohemia, Czechoslovakia, *Fieldiana (Geol.), n. ser.* **15**:1-90.
Morales-Ríos, M. S., Suárez-Castillo, O. R., Trujillo-Serrato, J. J., and Joseph-Nathan, P., 2001, Total syntheses of five indole alkaloids from the marine bryozoan *Flustra foliacea, J. Org. Chem.* **66**:1186-1192.
Nielsen, C., and Pedersen, K. G., 1979, Cystid structure and protrusion of the polypide in *Crisia* (Bryozoa, Cyclostomata), *Acta Zool. (Stockh.)* **60**:65-88.
Osburn, R. C., 1921, Bryozoa as food for other animals, *Science* **53**:451-453.
Osman, R. W., Whitlatch, R. B., Malatesta, R. J., and Zajac, R. N., 1990, Ontogenetic changes in trophic relationships and their effects on recruitment, in: *Trophic Relationships in the Marine Environment, Proceedings of the 24th European Marine Biology Symposium* (M. Barnes and R. N. Gibson, eds). Aberdeen University Press, Aberdeen, pp. 117-129.
Ostrovsky, A. N., 1998, Comparative studies of ovicell anatomy and reproductive patterns in *Cribrilina annulata* and *Celleporella hyalina* (Bryozoa: Cheilostomatida), *Acta Zool.* **79**:287-318.
Pitt, L. J., and Taylor, P. D., 1990, Cretaceous Bryozoa from the Faringdon Sponge Gravel (Aptian) of Oxfordshire, *Bull. Br. Mus. Nat. Hist. (Geol.)* **46**:61-152.
Poluzzi, A., 1980, I Briozoi membraniporiformi del delta settentrionale del Po, *Atti Soc. Ital. Sci. Nat. Mus. Civ. Stor. Nat. Milano* **121**:101-120.
Randall, J. E., 1967, Food habits of reef fishes of the West Indies, *Stud. Trop. Oceanogr. (Miami)* **5**:665-847.
Russ, G. R., 1980, Effects of predation by fishes: competition and structural complexity of the substrate on the establishment of a marine epifaunal community, *J. Exper. Mar. Biol. Ecol.* **42**:55-70.
Russ, G. R., 1982, Overgrowth in a marine epifaunal community: competitive hierarchies and competitive networks, *Oecologia* **53**:12-19.
Ryland, J. S., 1970, *Bryozoans*, Hutchinson University Press, London.
Ryland, J. S., 1976, Physiology and ecology of marine bryozoans, *Adv. Mar. Biol.,* **14**:285-443.
Schaumberg, G., 1979, Neue Nachweise von Bryozoen und Brachiopoden Nahrung des permischen Holocephalen, *Janass bituminosa* (Schlotheim), *Philippia* **4**: 3-11.
Silén, L., 1977, Polymorphism, in: *Biology of Bryozoans* (R. M. Woolacott and R. L. Zimmer, eds), Academic Press, New York, pp. 183-231.
Silén, L., and Harmelin, J.-G., 1974, Observations on living Diastoporidae (Bryozoa Cyclostomata), with special regard to polymorphism, *Acta Zool. (Stockh.)* **55**:81-96.
Snyder, E., 1987, Bryozoan succession in the Warsaw Formation (Valmeyeran, Mississippian) of the Mississippi Valley, USA, in: *Bryozoa: Present and Past* (J. R. P. Ross, ed.), Western Washington University, Bellingham, pp. 245-252.
Stebbing, A. R. D., 1973, Observations on colony overgrowth and spatial competition, in: *Living and Fossil Bryozoa* (G. P. Larwood, ed.), Academic Press, London, pp.173-183.
Stekhoven, J. H., Jr., 1933, Die Nahrung von *Oncholaimus dujardiniide* Man. *Zool. Anzeiger* **101**:167-168.
Tavener-Smith, R., 1975, The phylogenetic affinities of fenestelloid bryozoans, *Palaeontology* **18**:1-17.
Taylor, P. D., 1982, Probable predatory borings in Late Cretaceous bryozoans, *Lethaia* **15**:67-74.
Taylor, P. D., 1985, Polymorphism in melicerititid cyclostomes, in: *Bryozoa: Ordovician to Recent* (C. Nielsen and G. P. Larwood, eds.), Olsen & Olsen, Fredensborg, pp. 311-318.
Taylor, P. D., 1988a, Colony growth pattern and astogenetic gradients in the Cretaceous cheilostome bryozoan *Herpetopora, Palaeontology* **31**:519-549.
Taylor, P. D., 1988b, Major radiation of cheilostome bryozoans: triggered by the evolution of a new larval type?, *Hist. Biol.* **1**:45-64.
Taylor, P. D., 1994a, Systematics of the melicertitid cyclostome bryozoans; introduction and the genera *Elea, Semielea* and *Reptomultelea, Bull. Nat. Hist. Mus. Lond. (Geol.)* **50**:1-103.
Taylor, P. D. 1994b, An early cheilostome bryozoan from the Upper Jurassic of Yemen, *N. Jahrb. Geol. Paläont. Abh.* **191**:331-344.
Taylor, P. D., 1999, Bryozoan, in: *Functional Morphology of the Invertebrate Skeleton* (Savazzi, E., ed.), Wiley, Chichester, pp. 623-646.
Taylor, P. D., and Allison, P. A., 1998, Bryozoan carbonates through space and time, *Geology* **26**:459-462.
Taylor, P. D., and Monks, N., 1997, A new cheilostome bryozoan genus pseudoplanktonic on molluscs and algae, *Invert. Biol.* **116**:39-51.
Taylor, P. D., and Wilson, M. A., 1999, Middle Jurassic bryozoans from the Carmel Formation of southwestern Utah, *J. Paleontol.* **73**:816-830.
Thompson, J. N., 1982, *Interaction and Coevolution*, Wiley, New York.
Todd, C. D. and Havenhand, J. N., 1989, Nudibranch-bryozoan associations, the quantification of ingestion and some observations on partial predation among Doridoidea, *J. Mollusc. Stud.* **55**:245-259.

Todd, J. A., 2000, The central role of ctenostomes in bryozoan phylogeny, in: *Proceedings of the 11th International Bryozoology Association Conference* (A. Herrera Cubilla, and J. B. C. Jackson, eds.), Smithsonian Tropical Research Institute, Balboa, Republic of Panama, pp. 104-135.

Turner, R. F., 1975, A new Cretaceous cribrimorph from North America with calcareous opercula, *Doc. Lab. Géol. Fac. Sci. Lyon, H. S.* **3**:273-279.

Turner, S. J., and Todd, C. D., 1991, The effects of *Gibbula cineraria* (L.), *Nucella lapillus* (L.) and *Asterias rubens* L. on developing epifaunal assemblages, *J. Exper. Mar. Biol. Ecol.* **154**:191-213.

Underwood, A. J., and Fairweather, P. G., 1992, Marine invertebrates, in: *Natural Enemies: the Population Biology of Predators, Parasites, and Diseases* (J. M. Crawley, ed.), Blackwell Scientific, Oxford, pp 205-224.

Vance, R. R., 1979, Effects of grazing by the sea urchin, *Centrostephanus coronatus*, on prey community composition, *Ecology* **60**:537-546.

Vermeij, G. J., 1987, *Evolution and Escalation: An Ecological History of Life*, Princeton University Press, Princeton.

Vietti, R. C., and Balduzzi, A., 1991, Relationship between radular morphology and food in the Doridina (Mollusca: Nudibranchia), *Malacologia* **32**:211-217.

Voigt, E., 1989, Beitrag zur Bryozoen-Fauna des sächsischen Cenomaniums. Revision von A. E. Reuss' "Die Bryozoen des unteren Quaders" in H. B. Geinitz' "Das Elbthalgebirge in Sachsen" (1872), *Abh. Staatl. Mus. Min. Geol. Dresden* **36**:8-87, 170-183, 189-208.

Voigt, E., 1993, Neue cribrimorphe Bryozoen (Fam. Pelmatoporidae) aus einem Maastrichtium Schrebkreide-Geschiebe von Zweedorf (Holstein), *Mitt. Geol.-Paläont. Inst. Univ. Hamburg* **75**:137-169.

Voigt, E., and Flor, F. D., 1970, Homöomorphien bei fossilen cyclostomen Bryozoen, dargestellt am Beispiel der Gattung *Spiropora* Lamouroux 1821, *Mitt. Geol.-Paläont. Inst. Univ. Hamburg* **39**:7-96.

Voigt, E., and Williams, A., 1973, Revision des genus *Inversaria* v. Hagenow 1851 (Bryoz. Cheil.) und seine Beziehungen zu *Solenonychocella* n. g., *Nachr. Akad. Wiss. Göttingen, II. Math.-Physik. Kl.* **8**:140-178.

Walters, L. J., 1992, Post-settlement success of the arborescent bryozoan *Bugula neritina* (L.): the importance of structural complexity, *J. Exper. Mar. Biol. Ecol.* **164**:55-71.

Walters, L. J., and Wethey, D. S., 1991, Settlement, refuges, and adult body form in colonial marine invertebrates: a field experiment, *Biol. Bull.* **180**:112-118.

Walters, L. J., and Wethey, D. S., 1996, Settlement and early post-settlement survival of sessile marine invertebrates on topographically complex surfaces: the importance of refuge dimensions and adult morphology, *Mar. Ecol. Prog. Ser.* **137**:161-171.

Whitehead, J. D., Seed, R. and Hughes, R. N., 1996, Factors controlling spinosity in the epialgal bryozoan *Flustrellidra hispida* (Fabricius), in: *Bryozoans in Space and Time: Proceedings of the 10th International Bryozoology Conference* (D. P. Gordon, A. M. Smith, and J. A. Grant-Mackie, eds), National Institute of Water and Atmospheric Research, Wellington, pp.367-375.

Winston, J. E., 1984, Why bryozoans have avicularia – a review of the evidence, *Am. Mus. Novitates* **2789**:1-26.

Winston, J. E., 1986, Victims of avicularia, *Mar. Ecol. (Berl.)* **7**:193-199.

Winston, J. E., and Jackson, J. B. C., 1984, Ecology of cryptic coral reef communities. IV. Community development and life histories of encrusting cheilostome Bryozoa, *J. Exp. Mar. Biol. Ecol.* **76**:1-21.

Witman, J. D., and Grange, K. R., 1998, Links between rain, salinity, and predation in a rocky subtidal community, *Ecology* **79**:2429-2447.

Wyer, D. W., and King, P. E., 1973, Relationships between some British littoral and sublittoral bryozoans and pycnogonids, in: *Living and Fossil Bryozoa* (G. P. Larwood, ed.), Academic Press, London, pp. 199-207.

Yoshioka, P. M., 1982, Predator-induced polymorphism in the bryozoan *Membranipora membranacea* (L.), *J. Exper. Mar. Biol. Ecol.* **61**:233-242.

Chapter 10

Predation on Crinoids

TOMASZ K. BAUMILLER and FOREST J. GAHN

1. Introduction 263
2. A Paradigm Shift?..264
3. Neontological Studies of Predation 265
   3.1. Direct Interactions Between Crinoids and Fishes 265
   3.2. Direct Evidence of Predation by Invertebrates.....................................266
   3.3. Indirect Evidence of Crinoids as Prey..267
4. Paleontological Studies of Predation 268
   4.1. Paleoecological Patterns among Crinoids..268
   4.2. Anti-predatory Morphologies in Fossil Crinoids 269
   4.3. Regeneration in Fossil Crinoids..270
   4.4. Regeneration in Paleozoic Crinoids: Quantifying the Patterns 270
5. Conclusion..275
   References 276

1. Introduction

Over the last 25 years a conceptual shift has occurred in paleobiologists' view of the importance of predation as an ecological and evolutionary force. The importance of predation pressure on crinoids and its effects on morphological evolution and on biogeographic patterns of this group serves as a clear illustration of this shift: whereas prior to the 1970's crinoids were generally thought to be immune from predation, since then predation has been invoked as a probable cause of: (a) the success of extant stalkless crinoids, the comatulids, in shallow water settings (Meyer and Macurda, 1977); (b) the retreat of isocrinids, a group of stalked crinoids, into deeper water since the Cretaceous that culminated in their present bathymetric distribution (Meyer and Macurda, 1977; Meyer, 1985; Oji, 1996); (c) trends in arm morphology of isocrinids and comatulids (Oji and Okamoto, 1994); (d) trends in cup plate thickness and spinosity in

---

TOMASZ K. BAUMILLER and FOREST J. GAHN • Museum of Paleontology and Department of Geological Sciences, University of Michigan, Ann Arbor, Michigan 48109-1079.

Predator-Prey Interactions in the Fossil Record, edited by Patricia H. Kelley, Michał Kowalewski, and Thor A. Hansen. Kluwer Academic/Plenum Publishers, New York, 2003.

Devonian and post-Devonian crinoids (Signor and Brett, 1984); (e) evolution of the cladid and disparid anal sac as an anti-predatory device (Lane, 1984); and (f) evolution of morphological and behavioral escape strategies among articulate crinoids (Baumiller *et al.*, 2000), among others. In this chapter we will illustrate this conceptual shift using examples from the literature that span the interval of time during which this shift occurred, review some of the studies that present evidence of predation on crinoids and argue for its importance to their ecology and evolution, and, finally, provide additional data on predation from a couple of Paleozoic examples of regeneration.

2. A Paradigm Shift?

The idea that crinoids may have been easy prey for fishes had occasionally appeared in the paleontological literature prior to the 1970's. For example, Laudon (1957) suggested that the disproportionately high volume of columnals to crowns in Mississippian limestones was a consequence of "cropping" by shell-crushing sharks. Overall, however, there is a striking paucity of references to crinoids being subject to predation pressure and of predation having evolutionary or ecological consequences. Occasionally certain morphological features were hypothesized as having had a protective function, such as the rigidly plated tegmen covering the mouth and ambulacra; tegminal and cup features, such as spines and clefts; and certain types of coiled columns (Ubaghs, 1953; Kesling, 1964, 1965; Springer, 1926), but rarely were these hypotheses linked to any evidence of predation. That predation was not considered important is further illustrated by the fact that Beerbower (1968), in the second edition of his textbook, did not mention it. In that same book, he offered no explanation for the change in depth distribution of stalked crinoids between the present and the pre-Cenozoic, which he noted as one of the dominant paleoecological patterns for the group.

Even as late as 1980, predation was not considered a major factor by echinoderm paleontologists. In the Paleontological Society's Echinoderm Short Course Notes (1980), the word "predation" is not mentioned in chapters on crinoids (Lane and Webster, 1980), blastoids (Macurda, 1980), or Blastozoa (Broadhead, 1980). In that volume only Lewis, in a chapter on taphonomy (1980, p. 29), hints at predation, referring to "[c]rinoid and other benthic invertebrate remains...found in the stomach contents of a juvenile specimen of the Permian ray-like fish *Janassa bituminosa*." But even such remains were not unequivocally ascribed to predation for, as Zangerl and Richardson (1963) warned, they could as easily represent ingested sediment.

A decade later the view of predation on crinoids underwent a dramatic shift as is evidenced in the content of the volume of the 1997 Paleontological Society's Echinoderm Short Course. In a chapter on living comatulids, Messing (1997, p. 21) devoted an entire section to predation, and invoked certain behavioral and morphological features as critical to "their success in shallow-water following the late Mesozoic radiation of durophagous predators." In the same volume, in a chapter on living stalked crinoids, Meyer (1997) also devoted a section to predation. More impressive yet is the fact that in more general textbooks that deal with the history of life, crinoids were used as an example that illustrates the consequences of predation pressure. Thus Stanley (1993, p. 376) in his popular textbook, *Exploring Earth and Life Through Time*, stated that the decline of stalked crinoids, "which were moderately well represented in early Mesozoic seas, can probably be attributed to the diversification of modern predators." Cowen (2000), in his textbook, used essentially the same example.

Thus, sometime during the 1980's a change in perspective occurred. Arguably the seminal paper that contributed most to the change in perspective was Meyer and Macurda's (1977) "Adaptive radiation of comatulid crinoids." This was largely a "perspective" paper in which the authors synthesized crinoid data from in situ observations of behavior and ecology, morphology of modern and extinct forms, patterns of diversity and biogeography, and also published data on the radiation of teleost fishes. They concluded that those patterns, especially the success of comatulids relative to isocrinids in shallow waters, could be explained by invoking increasing predation pressure on crinoids by the diversifying teleosts during the Cretaceous-Cenozoic.

Superficially, what Meyer and Macurda proposed contradicted a long-standing notion that crinoids were distasteful to fishes and thus not subject to much predation pressure. That notion was based on the work of Clark (1915, p. 113-114) who, while studying comatulid crinoids, found that "[f]ishes, often of large size, always swam towards them, but when within a few inches turned and swam away. Either by sight, or some other sense, perhaps stimulated by some exhalation from the comatulids, these animals were recognized as inedible." In fact, despite some extensive SCUBA and submersible work on modern crinoids during the 1970's, fish avoidance of extant crinoids could not be easily refuted. The few examples of supposed fish predation on modern crinoids cited by Meyer and Macurda (1977) were indirect, including presence of crinoid material in fish gut contents or frequencies of arm or cirral injuries on reef-dwelling taxa, patterns that could be explained by invoking other processes. However, Meyer and Macurda provided a way out of the "uniformitarian" assumption by noting that the present immunity of comatulids from fish predation reported by Clark (1915) could simply be a consequence of defense mechanisms (chemical or otherwise) that evolved during earlier times of intense pressure. This was an important step away from strict actualism and one that allowed patterns to be interpreted in terms of processes that may or may no longer be operating today.

While recognizing that present morphological, behavioral, biogeographic and diversity patterns could be a consequence of formerly intense predation, Meyer and Macurda (1977) were nevertheless keen to find some examples of such predation in modern oceans. Observing the "mechanism" would not only strengthen their case but would also provide evidence for the postulated behavioral responses and, most importantly, confirm that predation was the cause of physical damage and frequent examples of regeneration in crinoids.

The Meyer and Macurda paper thus laid the foundation for subsequent studies of crinoid predation by identifying two areas of research into the problem: (1) neontological studies that would address the question of present day predation pressure, either directly or indirectly; and (2) studies of living and fossil crinoids that focused on evolutionary and ecological patterns that may be explained as consequences of predation.

3. Neontological Studies of Predation

3.1. Direct Interactions Between Crinoids and Fishes

As mentioned above, Clark (1915), in his study of living Great Barrier Reef comatulids, noted that they were avoided by fishes. Although this view dominated the thinking of biologists and paleontologists for decades, it was based on rather limited observations: Clark had no access to SCUBA and biotic interactions were not the focus of his research (the "avoidance" observation did not even warrant inclusion in his "summary"). With the advent of SCUBA and submersible research, a more complex picture emerged. Beginning in the 1960's many in situ observations showed that some interactions between fishes and crinoids did in fact occur (Meyer and Ausich [1983] provide an excellent review). By the 1980's instances of attacks by certain fishes, such as triggerfish and pufferfish, on crinoids had been documented (Meyer *et al.*, 1984; Meyer, 1985), as had the presence of crinoid remains in fecal samples and rare instances of fishes with parts of crinoids in their mouths. While those observations suggested that fish predation on crinoids may be occurring, other data seemed to confirm Clark's observations that crinoids were distasteful to fishes. For example, attempts to feed comatulid viscera and other body parts to fishes revealed that fishes find them distasteful and either spit them out after several exploratory nips and bites or outright reject them (Meyer, 1985; Meyer and Baumiller, pers. observations). One explanation of these seemingly contradictory observations is that the interactions between fishes and crinoids represent targeting by fishes of the many commensals that live in the crinoid arms and oral disc, such as shrimp, crabs, and worms (Meyer, 1985). This hypothesis would imply that injuries to comatulids may only represent collateral damage (Meyer, 1985) and that the suggested biochemical properties of comatulids may indeed make them distasteful (Rideout *et al.*, 1979; Bakus, 1981).

The deep-water habitats of living stalked crinoids make direct observations difficult, but photographs and direct submersible observations provide evidence that fishes do interact with these crinoids as well. These photographs reveal that caproid and macrourid fishes may be biting the arms and pinnules of isocrinids (Conan *et al.*, 1981) and that other fishes, such as the boarfish, *Antigonia capros*, often seen in close proximity of isocrinids, may be interacting with them as well (Baumiller *et al.*, 1991; Messing *et al.*, 1988). Again as in the case of comatulids, it is unclear whether the fish-isocrinid interactions involve fish searching for epibionts or feeding on crinoid pinnules and arms. A recent study by McClintock *et al.* (1999) showed that reef fishes offered food pellets containing extracts from isocrinids and control pellets consumed both without preference, suggesting a lack of chemical deterrents to predation in isocrinids. Yet those same fishes, when presented entire isocrinids, either ignored them entirely or showed minimal interest, never exhibiting the aggressive feeding behavior elicited by presenting them with other invertebrates.

3.2. Direct Evidence of Predation by Invertebrates

Although studies of predation on crinoids have typically focused on fishes, field and aquaria observations suggest that certain invertebrates may also be involved in this type of activity. Mladenov (1983), in a study of arm regeneration in the shallow water comatulid, *Florometra serratissima*, observed that when contacted by the predatory sea star, *Pycnopodia helianthoides*, the crinoid escaped by swimming for a distance of 1 to 2 m. When the two were placed together in aquaria, the starfish would capture and eat the comatulid. Mladenov suggested that the crab, *Oregonia gracilis*, may be another predator of *F. serratissima*, as one was observed in the field with a crinoid arm in its claw. He concluded that swimming and arm autotomy may both represent predatory escape strategies.

More recently, an interaction between the sea urchin, *Calocidaris micans*, and the deep water stalked crinoid, *Endoxocrinus parrae*, was captured by a submersible-mounted video camera (Baumiller et al., 2000). This interaction was interpreted as predatory, with the cidaroid biting the distal end of the crinoid stalk. This interpretation is supported by the fact that the guts of several dissected *C. micans* and another cidaroid, *Histocidaris*, which co-occurs with stalked crinoids, contained a large proportion of articulated crinoid stalk and arm elements.

3.3. Indirect Evidence of Crinoids as Prey

Although the most direct evidence of predation on crinoids comes from interactions between crinoids and their enemies observed in the field, there is a wealth of indirect data that support the idea that today's crinoids serve as prey. Most of these data consist of damage/regeneration frequencies observed among stalkless and stalked crinoids (Meyer, 1985; Schneider, 1988; Oji, 1986, 1996). Obviously, only under the assumption that such damage is predation related can these data be viewed as a measure of predation intensity. Such an interpretation is hindered by the fact that active shedding of body parts, or autotomy, and subsequent regeneration has been induced experimentally in crinoids by drastically changing water salinity, temperature, or subjecting individuals to continued physical agitation (Baumiller, in press). However, in those experiments autotomy was induced only by a dramatic change in the ambient environment, to a degree that would be encountered rarely in nature.

Autotomy has also been induced by simulating predatory activity by using scissors to make an incision in the distal part of the crinoid arm (Holland and Grimmer, 1981; Oji and Okamato, 1994). Thus, there is good reason to suspect that most cases of regeneration are more likely to be predation related than caused by changes in the ambient environment. Furthermore, a closer look at regeneration data also supports predation. For example, Meyer (1985) and Schneider (1988) showed that those comatulids that are more cryptic, living within the reef infrastructure, and those that are nocturnal have lower regeneration frequencies than do other comatulids. These patterns are difficult to explain as being causally connected to some dramatic changes in the physical environment, whereas they are consistent with the differing predation intensities on exposed versus cryptic species or diurnal versus nocturnal ones.

Predation is also consistent with patterns observed by Oji (1996), who examined regeneration frequencies in the arms of the deep water isocrinid, *Endoxocrinus parrae*. Because all individuals of *E. parrae* live below the photic zone and, thus, in an environment that is not likely to experience a broad range of temperatures, salinities, energies, or other physical parameters, regenerating arms are evidence of predation.

Instances of regeneration of the stalks of species of the Recent bourgueticrinid *Democrinus* (Donovan and Pawson, 1997) and the arms of cyrtocrinids, *Holopus rangii* (Donovan, 1992) and *Gymnocrinus richeri* (Bourseau et al., 1987), are also potentially a consequence of predation.

Finally, Oji and Okamoto (1994) elegantly demonstrated that arm branching patterns and sites of autotomy in the arms of extant crinoids are organized in such a way as to minimize arm loss from predatory attacks. Because the articulations specialized for autotomy are immovable and, thus, constrain the feeding/movement function of each arm, increasing their number imposes a functional cost on feeding ("harvesting paradigm" of Oji and Okamato, 1994). Thus there is a trade-off between feeding and minimizing arm loss through autotomy. Assuming that a predatory attack is random with respect to the position along each arm, Oji and Okamoto showed that the observed distribution of autotomy articulations for a given branching pattern would minimize arm loss from such attacks. They also compared arm branching patterns and distribution of autotomy articulations between shallow water comatulids and deep water stalked crinoids and found that the design of the former was better for dealing with arm loss, whereas the latter was adapted for feeding; they interpreted this observation as indicating greater predation pressure in shallow water.

4. Paleontological Studies of Predation

Direct evidence from fossils of predation on crinoids does not exist as we neither have examples of "catching the predator in the act" nor the unequivocal signatures of predators in the form of characteristic bite marks, drillholes, or other forms of injuries that would allow us to identify predators. Although drillholes in Mississippian crinoids have been recently reported beneath an infesting platyceratid snail, these were interpreted as representing parasitism rather than predation (Baumiller, 1990) and, therefore, will not be discussed here. Given the absence of direct evidence, can predation be considered a testable proposition in paleontological studies of crinoids? As mentioned above, there are numerous lines of indirect paleontological evidence that point toward predation as being a relevant factor in the ecology and evolutionary history of crinoids.

4.1. Paleoecological Patterns among Crinoids

One of the best documented large-scale patterns that have been interpreted as predation related is the depth distribution of extant crinoids and the timing of the "retreat" of stalked crinoids into deeper water (Bottjer and Jablonski, 1988). Today, only the highly mobile, stalkless crinoids (comatulids) are found on reefs and in shallow waters, whereas the stalked forms are found below 100 m depth and most are restricted to even greater depths (Meyer and Macurda, 1977). Because comatulids differ from stalked crinoids in a variety of behavioral, morphological, and biochemical features that have been interpreted as anti-predatory (e.g., Meyer and Macurda, 1977; Meyer, 1985) and because predation pressure may be higher in shallow water (Oji, 1996), the bathymetric distribution of crinoids has been interpreted as causally linked to predation (e.g., Meyer and Macurda, 1977; Meyer, 1985; Oji and Okamato 1994; Aronson et al., 1997).

Data on the timing of the "retreat" of the most diverse group of stalked crinoids, the isocrinids, from shallow to deeper water has been carefully documented by Bottjer and Jablonski (1988). This taxon shows a broad environmental distribution through the Early Cretaceous, but by the Late Cretaceous these crinoids become exceedingly rare in shallow-water environments. By the Cenozoic, only a few occurrences of these crinoids are associated with shallow-water environments (Weller, 1907; Kutscher, 1980; Eagle and Hayward, 1993; Meyer and Oji, 1993; Baumiller and Gazdzicki, 1996). Even these rare shallow-water occurrences are thought to represent special instances of presumed stalklessness (Meyer and Oji, 1993, but see Baumiller and Gazdzicki, 1996) or low predation levels, as documented by low frequencies of arm regeneration (Aronson et al., 1997). Although data on onshore-offshore patterns are likely influenced by preservation biases that may affect the details (Smith, 1994), the broad correspondence in the timing of retreat of the less well protected stalked crinoids into deep water with the diversification of teleost fishes, neoselachian sharks, and decapod crustaceans (Theis and Reif, 1985; Vermeij, 1987), taxa thought to be potential crinoid enemies, strongly hints at a causal connection between these two patterns.

Ecological patterns in the Paleozoic have also been explored in the context of predation on crinoids. Aronson (1991) compared the frequency of occurrence of dense, shallow-water stalked crinoid assemblages in North America spanning the Ordovician, Silurian, and Mississippian. Assuming an increase in predation on crinoids starting in the Middle Devonian due to the radiation of durophagous predators (Signor and Brett, 1984, also see below), Aronson (1991, p. 123) hypothesized that this increase might have led to a "decline in crinoid dominated communities" across this time interval. His results showed no statistically significant decline from the Ordovician to the Mississippian. Aronson suggested that this result might indicate (1) crinoids keeping up in the "arms race" with their predators by developing defenses; (2) a demise of reefs following the Frasnian-Famennian that would decrease fish habitats and, indirectly, predation pressure on crinoids; or (3) no effect on crinoid abundance by the diversifying predators. The last explanation may well imply, although Aronson does not explicitly state this, that crinoids were not the prey of the newly evolving predators.

4.2. Anti-predatory Morphologies in Fossil Crinoids

Various skeletal features of the tegmen, cup, arms, and stalk in fossil crinoids have been suggested as having a possible anti-predatory function, and Meyer and Ausich (1983) provided a comprehensive summary of the literature on this subject listing the taxa and the features (Figure 1 and Table 1 in Meyer and Ausich, 1983). In nearly all instances, the comatulids being the exception, these claims consist of little more than speculation about the possible anti-predatory function of a particular feature and, thus, represent untested "adaptationist" scenarios. There are, of course, good reasons why features such as spines may serve an anti-predatory function, for example by causing the predator to expend greater energy to overcome the prey, causing damage to the predator, or increasing crinoid size and serving as a deterrent to some predators. Yet, too often, these highly plausible scenarios have been taken as evidence of predation without themselves being tested. In these approaches, an untested functional hypothesis – a given feature as an anti-predatory device – becomes evidence for a process – predation. However, there are several studies where additional evidence has come to bear on the anti-predatory function hypotheses.

Lane (1984) noted that, whereas in extant crinoids specialized proximal pinnules serve as gamete-bearing reproductive structures, such structures have not been recognized among Paleozoic crinoids. He suggested that in Paleozoic crinoids gametes were held inside the calyx and, because they represent the most nutritious part of the animal, they were likely to be the primary aim of predators. Lane proposed that to reduce fatal predatory attacks some cladid and disparid crinoids developed anal sacs, the large tegmenal structures of calyx, and that these served as gamete-bearing devices that would be sacrificed to predators. The well-known regenerative abilities of crinoids, Lane argued, would mean a high chance of recovery from such attacks. Indeed, he cited numerous cases of regenerated anal sacs as evidence of sublethal attacks. This imaginative hypothesis has not been tested, but, if Lane is correct, the high frequency of anal sacs among late Paleozoic inadunate crinoids would suggest strong predation pressure.

Another study that argued for the evolutionary importance of predation on Paleozoic crinoids is Signor and Brett's (1984) often-cited "The Mid-Devonian precursor to the Mesozoic marine revolution." Although these authors acknowledged that evidence for predation on fossil crinoids is "anecdotal," they argued that observed increases in crinoid spinosity and cup robustness across the Mid-Devonian that correlate with increases in anti-predatory morphologies of other taxa represent the evolutionary consequences of intensifying predation driven by the documented concurrent diversification of durophagous predators.

4.3. Regeneration in Fossil Crinoids

The low probability of "catching the predator in the act" and the absence of features that would unequivocally allow us to identify the predator imply that the most direct information about predation available to paleontologists consists of instances of regeneration. However, whereas this problem has been approached rigorously using extant and some Mesozoic and Cenozoic crinoids (Meyer, 1985; Schneider, 1988, Oji, 1986, 1996), generally regeneration has been treated only anecdotally for Paleozoic crinoids. A number of regenerated arms were noted for the Middle Jurassic *Chariocrinus andreae* by Hess and Holenweg (1985), whereas Oji (1985) found no regenerated arms in the Aptian isocrinid *Isocrinus hanaii*. Quantitative measurements of regeneration allowed Aronson *et al.* (1997) to compare frequency of regenerating arms between the extant species of *Metacrinus* and an Eocene species from the Antarctic La Meseta Formation (Meyer and Oji, 1993). They found a statistically significant and a nearly order of magnitude higher frequency of regenerating arms in the extant species (0.04) than in the Eocene species (0.006). They interpreted these differences as reflecting lower predation pressures on the Antarctic Eocene crinoids.

Unfortunately, much of the data on regeneration in Paleozoic crinoids consist of individual examples; we are unaware of quantitative studies of this problem. Regeneration has been described in flexibles (Springer, 1920), camerates (Strimple and Bean, 1966; Ubaghs, 1978; Franzen, 1981), and cladids and disparids (Hall, 1861; Foerste, 1893; Whitfield, 1904; Ulrich, 1924; Springer, 1920; Hattin, 1958; Strimple, 1961; Lane and Webster, 1966; Webster and Lane, 1967; Burke, 1973; Ausich and Baumiller, 1993).

4.4. Regeneration in Paleozoic Crinoids: Quantifying the Patterns

Given the lack of published quantitative data on regeneration in Paleozoic crinoids, we examine in detail two sets of data for frequencies of crinoid arm and spine regeneration.

4.4.1. Arm Regeneration

Crinoids from the Early Mississippian Maynes Creek Formation of Le Grand, Iowa, have long been recognized for their exquisite preservation, but little attention has been directed to understanding the paleoecology of this fauna, despite the availability of thousands of specimens in museum collections worldwide. The Le Grand fauna has typically been interpreted as being deposited in a very tranquil environment. Strimple and Beane (1966, p. 35) described a specimen of *Cribanocrinus watersianus* with regenerated arms and noted that "over a period of 60 years, Beane has collected and studied thousands of crinoids from the limestone beds...near Le Grand, Iowa, but has observed only one specimen with regenerated arms." This prompted them to infer (Strimple and Beane, 1966, p. 35) that "crinoids lived in a favorable environment without serious molestation." A similar conclusion was reached by Laudon and Beane (1937, p. 238) who expressed surprise that "these small areas were spared by the shell feeding sharks...since shark remains are abundant throughout the formation."

A re-examination of the Le Grand fauna from the extensive Beane Collection reposited at Beloit College, Beloit, Wisconsin, suggests that the frequency of damage and regeneration in fossil crinoids at this locality is much greater than previously recognized. Early crinoid workers readily recognized regeneration of complete arms, but apparently failed to observe partial arm loss and regeneration (Fig. 1). This oversight is illustrated by the fact that Strimple and Beane (1966), while describing a single, completely regenerated arm, failed to recognize a partially regenerated arm tip on the same specimen of *Cribanocrinus watersianus* (Strimple and Beane, 1966, Fig. 1, upper left-hand corner, third arm from the top).

We analyzed 328 individuals from eight species as part of a larger paleoecological analysis of the Le Grand fauna. Individuals were chosen based on quality of preservation, and only specimens that were minimally distorted by taphonomic processes were included. Specimen stalk length, cup height, and arm length were measured; each specimen was also examined for evidence of regeneration. Arm regeneration in individual specimens was identified by examining the arms of well-preserved specimens and looking for discontinuities in arm thickness or an abrupt decrease in free brachial size (Fig. 1). Twenty-seven individuals were found with partially regenerated arms, and one specimen of *Holcocrinus longicirrifer* was observed to have a regenerated anal sac (Table 1). In most of the regenerated arms, breakage had occurred near the middle of the arm and, thus, only partial regeneration had occurred.

Among the eight species, *Rhodocrinites kirbyi* displays the greatest frequency (27%) of individuals with regenerated arms. This species is also the most abundant in the fauna

TABLE 1. Summary Table of Beane Collection Specimens from the Lower Mississippian Maynes Creek Formation of Le Grand, Iowa Housed at Beloit College, Wisconsin

Species	# Observed	# Regenerated	% Regenerated
Rhodocrinites kirbyi	70	19	27
Platycrinites symmetricus	50	1	2
Strimplecrinus inornatus	50	1	2
Aorocrinus immaturatus	42	0	0
Cribanocrinus watersianus	38	1	2.6
Holcocrinus longicirrifer	33	3	9.1
Cusacrinus nodobrachiatus	35	0	0
Dichocrinus cinctus	10	2	20

FIGURE 1. Arm regeneration in crinoids from the Lower Mississippian Maynes Creek Formation of Le Grand, Iowa. (A) *Cribanocrinus watersianus*. Note the change in arm width and free brachial size above and below the break of the middle regenerated arm BC-180. Scale is 1 mm. B – D *Rhodocrinites kirbyi*; BC-153. (B) two arms truncated near the arm bases and in initial stages of regeneration. Scale is 1 mm. (C) three regenerated arms; one central arm with a deep proximal truncation, and two adjacent arms with only the most distal arm tips regenerating. Scale is 2 mm. (D) Eight regenerated arms adjacent to one another and displaying medial arm truncation. Scale is 3 mm. Circles indicate sites of regeneration. All specimens are reposited in the Beane Collection, Department of Geology, Beloit College, Wisconsin.

and one characterized by the longest stalk. In the majority of the individuals with regenerated arms, only 1 or 2 arms were involved; however, one individual displays regeneration of eight arms (Fig. 1D; 2). The regenerated arms in this specimen are adjacent to one another and were damaged at nearly the same distance above the calyx. This may represent "cropping" by an unidentified predator followed by regeneration of the damaged arms.

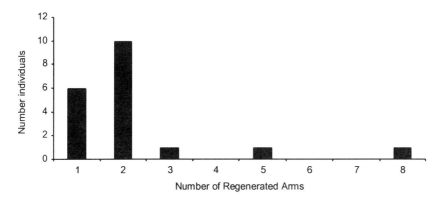

FIGURE 2. Arm regeneration in *Rhodocrinites kirbyi*. Note that most specimens have one or two regenerating arms. 19 of 70 *R. kirbyi* had at least one regenerating arm.

The frequency of regenerated arms in the Mississippian example can be compared to regeneration frequencies in extant crinoids. Mladenov (1983) found that roughly 80 percent of *Florometra serratissima* specimens had at least one regenerating arm. Meyer (1985) observed a range of frequencies (23-77%) of regenerated arms among Great Barrier Reef comatulids, with lower frequencies among cryptic and semicryptic than exposed taxa. Schneider (1988), in his analysis of 34 Pacific and tropical western Atlantic comatulid species, found a similar pattern and a broad range of regeneration frequencies (0-100 percent) with an average of 32% of individuals among all species with at least one regenerating arm. Data on extant isocrinids also show relatively high regeneration frequencies: Oji (1996) found nearly 80% of individuals of the isocrinid species *Endoxocrinus parrae* and *Endoxocrinus prionodes* to have regenerating arms, and Meyer and Oji (1993) found regeneration in 89% of specimens of *Metacrinus rotundus*.

Taken at face value, the higher regeneration frequencies in extant taxa suggest that they may be (1) subject to higher predation intensities; (2) capable of better surviving predatory attacks; or (3) better at regeneration. It must be noted, however, the observed frequencies may not be directly comparable: for fossil individuals typically only half the arms can be examined for damage/regeneration whereas this is not the case for extant crinoids. Ways of assessing regeneration will also need to be standardized; for example, the use of color differences as a sign of regeneration (Mladenov, 1983) cannot be applied to fossils.

4.4.2. Spine regeneration

Morphological features commonly considered as having an anti-predatory function are the spines on the tegmen, cup, and arms of fossil crinoids (e.g., Signor and Brett, 1984, but see Strimple and Moore, 1971). Such spines are found on some camerates, but it is among the cladids that they reach truly large proportions and become widespread (Fig. 3). Unfortunately, in most crinoid-rich localities, especially those of the midcontinental Pennsylvanian and Permian, complete specimens are rare, and one finds specimens with articulated spines only in exceptional circumstances while disarticulated spines are abundant. However, the use of fragmentary crinoid remains has generally been restricted to taxonomic and biostratigraphic problems (Moore, 1939; Moore and Jeffords, 1968), although recently Holterhoff (1997) promoted their use for paleoecological problems. Given the potential anti-predatory function of spines and the rich record of disarticulated crinoid material, including spines, it may prove useful to exploit them as a source of data for studying predation. For example, Burke's (1973, p. 161) suggestion that the regenerated primibrachial spines in the Pennsylvanian *Plaxocrinus* may have been "nipped off by some small fish, like *Peripristis semicircularis*" implies that regenerated spines may be common and may provide us with insights to predation frequencies.

FIGURE 3. Crinoid spine regeneration. (A) *Delocrinus hemisphericus*; well-preserved specimen from the Upper Pennsylvanian Lane Shale of Kansas City, Missouri, demonstrating the placement and extent of spines typical of many Late Paleozoic cladid crinoids, UM6043. Cup diameter is 22 mm. (B) *Delocrinus*? sp.; spine with regenerated tip from the Waldrip Limestone, Coleman County, Texas, Moore and Jeffords Collection UM68705. Length of entire spine is 11 mm.

To test the potential of disarticulated crinoid material for assessing predation intensity, we examined the Moore and Jeffords collections (UMMP) of such material from Pennsylvanian and Permian localities of Kansas, Oklahoma, and Texas. We used material collected at 10 localities and restricted our analysis only to spines of *Dorycrinus*. The percentage of regenerated spines varied from 1 to 18% between localities, with an average of 5% regenerated spines (47 of 904).

5. Conclusion

The last 25 years have seen a substantial increase in studies of predation by crinoid paleontologists, and a process that once was ignored and seemed irrelevant is now invoked as having potentially had a large macroecological and macroevolutionary impact. Fossils have proven critical to this reassessment: for example, the change in bathymetric distribution could not have been established without a fossil record of shallow-water stalked crinoids. The work on crinoid predation, generally carried out by paleontologists, has been two-pronged, combining neontological studies with data from fossils. The neontological results are somewhat puzzling: although regeneration frequencies among crinoids are high, crinoids are not the primary prey of fishes or other animals, and in fact, direct observations of predatory attacks are exceedingly rare. The distaste of crinoids to fishes suggests that damage to crinoids may occur as a byproduct of predators seeking crinoid commensals, though even such interactions have not been observed commonly. It is certain that more field studies are needed to reconcile the high frequencies of injuries and regeneration with the seeming low levels of biotic interactions.

Regardless of how one interprets the neontological observations, the patterns in the post-Paleozoic fossil record are very suggestive of predation having been an important factor in crinoid evolution. The offshore retreat of stalked crinoids and the functional and behavioral attributes of comatulids provide strong indirect evidence for the importance of predation. However, this interpretation could benefit greatly from data on the timing of the diversification of comatulids and paleontological data on predation intensities. To date, only a few studies have attempted to quantify regeneration frequencies and taphonomic problems stymie attempts at solving the diversity question.

With regard to the record and importance of predation on Paleozoic crinoids, hypotheses abound but remain largely untested. However, as the two examples presented above indicate, fossils hold much as yet unutilized information on regeneration and damage that should prove of great value for testing predation hypotheses. It is very difficult to interpret the *Rhodocrinites* data presented here as anything but evidence for predation, and the regeneration frequencies suggest that at least some Mississippian crinoids were subject to predation. The frequencies of regenerating arms in *Rhodocrinites* are higher than those in the only other comparable fossil example, the Eocene *Metacrinus fossilis* (Meyer and Oji, 1993), but fall within the range for those observed among modern crinoids. It will be interesting to see whether crinoids prior to the Mid-Devonian marine revolution (Signor and Brett, 1984) showed regeneration frequencies significantly different from the Mississippian *Rhodocrinites*.

ACKNOWLEDGMENTS: This study was funded in part by grants from the National Science Foundation. We thank the faculty and staff of Beloit College for their generosity and assistance, B. A. Miljour for help with the figures, W. I. Ausich, P.

Kelley, and J. Nebelsick for comments and suggestions, and L. Wingate for editorial help.

References

Aronson, R., Blake, D., and Oji, T., 1997, Retrograde community structure in the late Eocene of Antarctica, *Geology* **25**:903-906.
Aronson, R. B., 1991, Escalating predation on crinoids in the Devonian: negative community-level evidence, *Lethaia* **24**:123-128.
Ausich, W. I., and Baumiller, T. K., 1993, Column regeneration in an Ordovician crinoid (Echinodermata): paleobiologic implications, *J. Paleontol.* **67**:1068-1070.
Bakus, G. J., 1981, Chemical defense mechanisms on the Great Barrier Reef, Australia, *Science* **211**:497-499.
Baumiller, T. K., 1990, Non-predatory drilling of Mississippian crinoids by platyceratid gastropods, *Palaeontology* **33**:743-748.
Baumiller, T. K., in press, Experimental and biostratinomic disarticulation of crinoids: taphonomic implications, in: *Echinoderm Research 2001* (J.-P. Feral and B. David, eds.), A.A. Balkema, Lisse.
Baumiller, T. K., and Gazdzicki, A., 1996, New crinoids from the Eocene La Meseta Formation, Seymour Island, Antarctic Peninsula, *Palaeontol. Polon.* **55**:101-116.
Baumiller, T. K., LaBarbera, M., and Woodley, J. W., 1991, Ecology and functional morphology of the isocrinid *Cenocrinus asterius* (Linnaeus) (Echinodermata: Crinoidea): in situ and laboratory experiments and observations, *Bull. Mar. Sci.* **48**:731-748.
Baumiller, T. K., Mooi, R., and Messing, C. G., 2000, Cidaroid-crinoid interactions as observed from a submersible, in: *Echinoderms 2000* (M. Barker, ed.), A. A. Balkema, Lisse, p. 3.
Beerbower, J. R., 1968, *Search for the Past*, Prentice-Hall, Engelwood Cliffs, N.J.
Bottjer, D. J., and Jablonski, D., 1988, Paleoenvironmental patterns in the evolution of Post-Paleozoic benthic marine invertebrates, *Palaios* **3**:540-560.
Bourseau, J.-P., Cominardi, N. A., and Roux, M., 1987, Un crinoide pedoncule nouveau (Echinodermes), representant actuel de la famille jurassique des Hemicrinidae: *Gymnocrinus richeri* nov. sp. des fonds bathyaux de Nouvelle-Caledonie (S. W. Pacifique), *C. R. Acad. Sci., Paris* **305**:595-599.
Broadhead, T. W., 1980, Blastozoa, in: *Echinoderms: Notes for a Short Course* (T. W. Broadhead and J. A. Waters, eds.), University of Tennessee Department of Geological Sciences Studies in Geology 3, pp. 118-132.
Burke, J. J., 1973, Four new pirasocrinid crinoids from the Ames Limestone, Pennsylvanian, of Brooke County, West Virginia, *Annals Carnegie Mus.* **44**:157-169.
Clark, H. L., 1915, The comatulids of Torres Strait: With special reference to their habits and reactions, *Papers Dept. Mar. Biol., Carnegie Inst., Washington* **8**:97-125.
Conan, G., Roux, M., and Sibuet, M., 1981, A photographic survey of a population of the stalked crinoid *Diplocrinus (Annacrinus) wyvillethomsoni* (Echinodermata) from the bathyal slope of the Bay of Biscay, *Deep-Sea Res.* **28A**:441-453.
Cowen, R., 2000, *History of Life*, Blackwell Science, Inc., Malden, MA.
Donovan, S. K., 1992, Scanning EM study of the living cyrtocrinid *Holopus rangii* (Echinodermata, Crinoidea) and implications for its functional morphology, *J. Paleontol.* **66**:665-675.
Donovan, S. K., and Pawson, D. L., 1997, Proximal growth of the column in Bathycrinid crinoids (Echinodermata) following decapitation, *Bull. Mar. Sci.* **61**:571-579.
Eagle, M. K., and Hayward, B. W., 1993, Oligocene paleontology and paleoecology of Waitete Bay, Northern Coromandel Peninsula, *Rec. Auckland Inst. Mus.* **30**:13-26.
Foerste, A. F., 1893, The reproduction of arms in crinoids, *Am. Geol.* **12**:270-271
Franzen, C., 1981, A Silurian crinoid thanatotope from Gotland, *Geol. Forhandlingar* **103**:469-490.
Hall, J., 1861, Description of new species of Crinoidea from the Carboniferous rocks of the Mississippi Valley, *J. Boston Soc. Nat Hist.* **7**:261-328.
Hattin, D. E., 1958, Regeneration in a Pennsylvanian crinoid spine, *J. Paleontol.* **32**:701.
Hess, H., and Holenweg, H., 1985, Die Begleitfauna auf den Seelilienbanken im mittleren Dogger des Schweitzer Jura, *Tätigkeitsberichte Naturforschende Gesellschaft Baselland* **33**:141-177.
Holland, N. D., and Grimmer, J. C., 1981, Fine structure of syzygial articulations before and after arm autotomy in *Florometra serratissima* (Echinodermata: Crinoidea), *Zoomorphology* **98**:169-183.
Holterhoff, P. F., 1997, Paleocommunity and evolutionary ecology of Paleozoic crinoids, in: *Geobiology of Echinoderms* (J. A. Waters and C. G. Maples, eds.), *Paleontological Society Papers* **3**, pp. 69-106.
Kesling, R. V., 1964, A new species of *Melocrinites* from the Middle Devonian Bell Shale of Michigan, *Contrib. Mus. Paleontol. Univ. Michigan* **19**:87-103.

Kesling, R. V., 1965, Nature and occurrence of *Gennaeocrinus goldringae* Ehlers, *Contrib. Mus. Paleontol. Univ. Michigan* **19**:265-280.
Kutscher, M., 1980, Die Echinodermen des Oberligozäns von Stern berg, *Zeitschrift geol. Wissenschaften, Berlin* **10**:221-239.
Lane, N. G., 1984, Predation and survival among inadunate crinoids, *Paleobiology* **10**:453-458.
Lane, N. G., and Webster, G. D., 1966, New Permian crinoid fauna from southern Nevada, *Univ. California Publ. Geol. Sci.* **63**:1-60.
Lane, N. G., and Webster, G. D., 1980, Crinoidea, in: *Echinoderms: Notes for a Short Course* (T. W. Broadhead and J. A. Waters, eds.), University of Tennessee Department of Geological Sciences Studies in Geology 3, pp. 144-157.
Laudon, L. R., 1957, Crinoids, *Treatise on Marine Ecology and Paleoecology. Vol. 2 Paleoecology* **67**:961-972.
Laudon, L. R., and Beane, B. H., 1937, The crinoid fauna of the Hampton Formation at LeGrand, Iowa, *Univ. Iowa Stud. Nat. Hist.* **17**:226-273.
Lewis, R., 1980, Taphonomy, in: *Echinoderms: Notes for a Short Course* (T. W. Broadhead and J. A. Waters, eds.), University of Tennessee Department of Geological Sciences Studies in Geology 3, pp. 27-39.
Macurda, D. B., Jr., 1980, Blastoidea, in: *Echinoderms: Notes for a Short Course* (T. W. Broadhead and J. A. Waters, eds.), University of Tennessee Department of Geological Sciences Studies in Geology 3, pp. 133-138.
Malzahn, E., 1968, Über neue Funde von *Janassa bituminosa* (Schloth.) im niederrheinischen Zechstein, *Geol. Jahrb.* **85**:67-96.
McClintock, J. B., Baker, B. J., Baumiller, T. K., and Messing, C. G., 1999, Lack of chemical defenses in two species of stalked crinoids: support for the predation hypothesis for Mesozoic bathymetric restriction, *J. Exper. Mar. Biol. Ecol.* **232**:1-7.
Messing, C. G., 1997, Living comatulids, in: *Geobiology of Echinoderms* (J. A. Waters and C. G. Maples, eds.), *Paleontological Society Papers* **3**, pp. 3-30.
Messing, C. G., RoseSmyth, M. C., Mailer, S. R., and Miller, J. E., 1988, Relocation movement in a stalked crinoid (Echinodermata), *Bull. Mar. Sci.* **42**:480-487.
Meyer, D. L., 1985, Evolutionary implications of predation on Recent comatulid crinoids from the Great Barrier Reef, *Paleobiology* **11**:154-164.
Meyer, D. L., 1997, Implications of research on living stalked crinoids for paleobiology, in: *Geobiology of Echinoderms* (J. A. Waters and C. G. Maples, eds.), *Paleontological Society Papers* **3**, pp. 31-43.
Meyer, D. L., and Ausich, W. I., 1983, Biotic interactions among Recent and fossil crinoids, in: *Biotic Interactions in Recent and Fossil Benthic Communities* (M. J. S. Tevesz and P. L. McCall, eds.), Plenum, New York, pp. 377-427.
Meyer, D. L., LaHaye, C. A., Holland, N. D., Arenson, A. C., and Strickler, J. R., 1984, Time-lapse cinematography of feather stars (Echinodermata: Crinoidea) on the Great Barrier Reef, Australia: demonstrations of posture changes, locomotion, spawning and possible predation by fish, *Mar. Biol.* **78**:179-184.
Meyer, D. L., and Macurda, D. B., Jr., 1977, Adaptive radiation of comatulid crinoids, *Paleobiology* **3**:74-82.
Meyer, D. L., and Oji, T., 1993, Eocene crinoids from Seymour Island, Antarctic Peninsula: Paleobiogeographic and paleoecologic implications, *J. Paleontol.* **67**:250-257.
Mladenov, P. V., 1983, Rate of arm regeneration and potential causes of arm loss in the feather star *Florometra serratissima* (Echinodermata: Crinoidea), *Can. J. Zool.* **61**:2873-2879.
Moore, R. C., 1939, The use of fragmentary crinoidal remains in stratigraphic paleontology, *Denison Univ. Bull., Sci. Lab. J.* **33**:165-250.
Moore, R. C., and Jeffords, R. M., 1968, Classification and nomenclature of fossil crinoids based on studies of dissociated parts of their columns, *Univ. Kansas Paleontol. Contrib., Echinodermata, Article* **9**:1-86.
Oji, T., 1985, Early Cretaceous *Isocrinus* from northeast Japan, *Palaeontology* **28**:629-642.
Oji, T., 1986, Skeletal variation related to arm regeneration in *Metacrinus* and *Saracrinus*, recent stalked crinoids, *Lethaia* **19**:355-360.
Oji, T., 1996, Is predation intensity reduced with increasing depth? Evidence from the west Atlantic stalked crinoid *Endoxocrinus parrae* (Gervais) and implications for the Mesozoic marine revolution, *Paleobiology* **22**:339-351.
Oji, T., and Okamoto, T., 1994, Arm autotomy and arm branching pattern as anti-predatory adaptations in stalked and stalkless crinoids, *Paleobiology* **20**:27-39.
Rideout, J. A., Smith, N. B., and Sutherland, M. D., 1979, Chemical defense of crinoids by polyketide sulphates, *Experientia* **35**:1273-1274.

Schneider, J. A., 1988, Frequency of arm regeneration of comatulid crinoids in relation to life habit, in: *Echinoderm Biology* (R. D. Burke, P. V. Mladenov, P. Lambert and R. L. Parsley, eds.), A. A. Balkema, Rotterdam, pp. 531-538.

Signor, P. W., III, and Brett, C. E., 1984, The mid-Paleozoic precursor to the Mesozoic marine revolution, *Paleobiology* **10**:229-245.

Smith, A. B., 1994, *Systematics and the Fossil Record*, Blackwell Scientific Publications, Oxford.

Springer, F., 1920, *The Crinoidea Flexibilia Vol. 1.*, Washington, D.C.

Springer, F., 1926, Unusual forms of fossil crinoids, *Proc. U. S. Nat. Mus. Article* **9**:1-67.

Stanley, S. M., 1993, *Exploring Earth and Life through Time*, W. H. Freeman and Company, New York.

Strimple, H. L., 1961, Late Desmoinesian crinoids, *Oklahoma Geol. Surv. Bull.* **93**:1-189.

Strimple, H. L., and Beane, B. H., 1966, Reproduction of lost arms on a crinoid from LeGrand, Iowa, *Oklahoma Geol. Notes* **26**:35-37.

Strimple, H. L., and Moore, R. C., 1971, Crinoids of the LaSalle Limestone (Pennsylvanian) of Illinois, *Univ. Kansas Paleontol. Contrib., Article 55 (Echinodermata 11)* 1-48.

Thies, D., and Reif, W.-E., 1985, Phylogeny and evolutionary ecology of Mesozoic Neoselachii, *Neues Jahrb. Geol. Paläontologie, Abh.* **169**:625-643.

Ubaghs, G., 1953, Classe des Crinoides, in: *Traite de Paleontologie* (J. Piveteau, ed.), Masson, Paris, pp. 658-773.

Ubaghs, G., 1978, Skeletal morphology of fossil crinoids, in: *Treatise on Invertebrate Paleontology, Part T* (R. C. Moore and C. Teichert, eds.), Geological Society of America and University of Kansas, Lawrence, pp. T58-T216.

Ulrich, E. O., 1924, New classification of the Heterocrinidae, in: *Upper Ordovician Faunas of Ontario and Quebec* (A. F. Foerste, ed.), Quebec Canadian Geological Survey, Memoir, pp. 1-255.

Vermeij, G. J., 1987, *Evolution and Escalation: An Ecological History of Life*, Princeton University Press, Princeton, N. J.

Webster, G. D., and Lane, N. G., 1967, Additional Permian crinoids from southern Nevada, *Univ. Kansas Paleontol. Contrib., Paper* **27**:1-32.

Weller, S., 1907, A report on the Cretaceous of New Jersey, *Geol. Surv. New Jersey, Paleontol. Series* **4**:1-871.

Whitfield, R. P., 1904, Notice of a remarkable case of reproduction of lost parts shown on a fossil crinoid, *Bull. Am. Mus. Nat. Hist.* **20**:471-472.

Zangerl, R., and Richardson, E. S., Jr., 1963, The paleoecological history of the two Pennsylvanian black shales, *Fieldiana: Geol. Memoirs* **4**:1-352.

Chapter 11

Predation on Recent and Fossil Echinoids

MICHAŁ KOWALEWSKI and JAMES H. NEBELSICK

1. Introduction 279
2. Predators of Living Echinoids 280
   2.1. Introduction 280
   2.2. Drilling Gastropods 281
   2.3. Other Invertebrates 284
   2.4. Vertebrates..285
   2.5. Paleontological Implications of Neontological Studies........................288
   2.6. Summary...291
3. The Fossil Record of Predation and Parasitism on Echinoids......................291
   3.1. Introduction 291
   3.2. Drilling Predation and Parasitism 291
   3.3. Durophagous Predation on Echinoids..294
   3.4. The Fossil Record of Parasites...294
4. Future Research Directions..295
   4.1. Actuopaleontological Research..295
   4.2. Paleontological Research 296
5. Summary..297
   References 297

1. Introduction

Predator-prey interactions in marine ecosystems are documented in the fossil record by drill holes, repair scars, tooth marks and other structural damage left by durophagous ("hard-eating") predators on skeletons of their prey. Previous paleoecological research focused primarily on benthic mollusks (e.g., Vermeij, 1977, 1983, 1987; Vermeij *et al.*, 1980, 1981; Kitchell *et al.*, 1981; Kitchell, 1986; Kelley and Hansen, 1993; Kowalewski *et al.*, 1998; Dietl *et al.*, 2000; Hoffmeister and Kowalewski, 2001; and numerous

---

MICHAŁ KOWALEWSKI • Department of Geological Sciences, Virginia Polytechnic Institute and State University, Blacksburg, Virginia 24061. JAMES H. NEBELSICK • Institut und Museum für Geologie und Paläontologie, Universität Tübingen, Sigwartstrasse 10, D-72076 Tübingen, Germany.

Predator-Prey Interactions in the Fossil Record, edited by Patricia H. Kelley, Michał Kowalewski, and Thor A. Hansen. Kluwer Academic/Plenum Publishers, New York, 2003.

references therein). Other groups that received considerable treatment include brachiopods (e.g., Sheehan and Lesperance, 1978; Alexander, 1981, 1986, 1990; Smith et al., 1985; Baumiller et al., 1999; Harper and Wharton, 2000; Kowalewski et al., 2000; Leighton, 2001) and sessile echinoderms (e.g., Baumiller, 1990, 1993, 1996; Donovan, 1991a; Baumiller and Macurda, 1995). However, many important fossil groups of prey have been underrepresented in studies of predator-prey interactions.

In particular, the ecological impact of predator-prey relationships on fossil echinoids rarely has been studied rigorously. Yet, as reviewed below, ecological research of the last several decades has demonstrated clearly that modern benthic ecosystems include a diverse spectrum of predators that can have an extreme impact on local echinoid populations, up to the point of restricting them to cryptic locations (e.g., Levitan and Genovese, 1989; McClanahan, 1999). Moreover, the removal of predators can result in a dramatic increase in echinoid populations and trigger major changes in local ecosystems (e.g., deforestation of kelp beds by grazing sea urchins; Estes et al., 1998). Finally, and perhaps most importantly from the paleontological perspective, predators of echinoids often leave distinctive predatory traces on echinoid tests, and, consequently, echinoids may offer valuable paleontological data on predator-prey interactions. The fossil record of echinoids should be suitable for studying the evolutionary significance of predation in marine benthic ecosystems and may even offer an independent system for testing the hypothesis of escalation (Vermeij, 1987). It should also allow us to evaluate the previously posed hypotheses that link evolutionary changes observed through the history of echinoids (i.e., development of protective armor and infaunalization) to various predation pressures (e.g., Kier, 1965, 1974, 1982; Smith, 1984; Vermeij, 1987). Finally, the fossil record may provide clues regarding the long-term role of predators in controlling intensity of grazing by echinoids and the evolution of kelp ecosystems (Estes and Steinberg, 1988) and reefs (Wood, 1998).

Here, we review the current knowledge of present-day predators of echinoids, with particular focus on those organisms that leave evidence that can be preserved in the fossil record (see also Nebelsick, 1999a). In the second part of this chapter, we examine and analyze the paleontological literature of the fossil record of predation on echinoid prey. The closing section of the chapter outlines future research directions. Our review deals also with sublethal predation, scavenging, and host-parasite systems because such ecological interactions are closely related to predation and difficult to differentiate unambiguously in the fossil record (e.g., Baumiller et al., 1999).

2. Predators of Living Echinoids

2.1. Introduction

Many organisms prey on or parasitize echinoids today. The best known groups are gastropods, crustaceans, echinoderms, fish, turtles, birds, sea otters, and humans (Moore, 1956; Smith, 1984; Nebelsick, 1999a). From the paleontological perspective, drilling snails are particularly important because they produce distinct marks that can be found in the fossil record. Thus, although other groups of predators are discussed in detail, our review pays particular attention to drilling gastropods.

2.2. Drilling Gastropods

Two groups of present-day gastropods are known to drill echinoids: predatory cassid gastropods that typically produce a single hole in the test of their prey and ectoparasitic eulimid gastropods that leave multiple holes in the test of their host and may leave attachment scars or cause spine swelling.

2.2.1. Cassid Gastropods

Drilling predation by present-day cassid (tonnacean) gastropods on both epifaunal regular and infaunal irregular echinoids is well documented (e.g., Moore, 1956; Chesher, 1969; Hendler, 1977; Hughes and Hughes, 1971, 1981; Gladfelter, 1978; Levitan and Genovese, 1989; McClintock and Marion, 1993; Nebelsick and Kowalewski, 1999; McClanahan, 1999). Moore (1956, p. 74) observed an urchin that had escaped an attack by a cassid "trundling along on its secondary spines with a large *Cassis* in hot pursuit." Hughes and Hughes (1971, 1981) and Levitan and Genovese (1989) documented the hunting and killing technique of cassids in detail. Following a slow-motion chase, the snail snares the prey with its extended proboscis, possibly with the aid of toxins (Hughes and Hughes, 1981), and partially or wholly engulfs the test with its foot. Subsequently, the predator cuts a circular groove in the test and punches out the remaining plate material, creating a circular, cylindrical penetration with straight walls (Fig. 1A). Cassids drill with the mechanical use of the radula aided by acid digestion (Hughes and Hughes, 1981). Once the test is penetrated, the snail can extend its long proboscis into the interior of the test.

The diameter of the cassid drill hole is small when compared to the relatively large size of the snail. Nebelsick and Kowalewski (1999) documented drill holes in tests of the minute clypeasteroid species *Echinocyamus crispus* and *Fibularia ovulum* (Fig. 1B) from the Northern Bay of Safaga (Red Sea, Egypt). Holes were cylindrical in shape, with the average diameter less than 1 mm (Fig. 2). Some holes were surrounded by halos, which most likely represented acid-etching traces (Fig. 1C). All holes were complete, though they often show irregular outlines, which, most likely, reflect the plate boundaries and numerous holes (e.g., ambulacral pores, mouth, anus) already present in the test.

Cassids often drill distinct areas of the echinoid test preferentially. In minute clypeasteroids, for example, drill holes occur predominantly on the aboral side of the test (Nebelsick and Kowalewski, 1999). In contrast, McClintock and Marion (1993) observed 100% of drill holes on the oral side of *Leodia sexiesperforata*. The variable site selectivity may correlate with specific hunting techniques, avoidance of defensive spines, internal test morphology, and position of potential food within the echinoid test.

Cassids may be a major source of mortality in both regular (Engstrom, 1982; Levitan and Genovese, 1989; Greenstein, 1995) and irregular echinoids (McClintock and Marion, 1993; Nebelsick and Kowalewski, 1999). In an experimental study, Levitan and Genovese (1989) observed up to 70% mortality of the urchin *Diadema antillarium* from predation by *Cassis tuberosa*. McClintock and Marion (1993) reported that 94.7% of tests of the sand dollar *Leodia sexiesperforata* collected haphazardly from shallow sandy environments at San Salvador (Bahamas) contained a single hole made, most

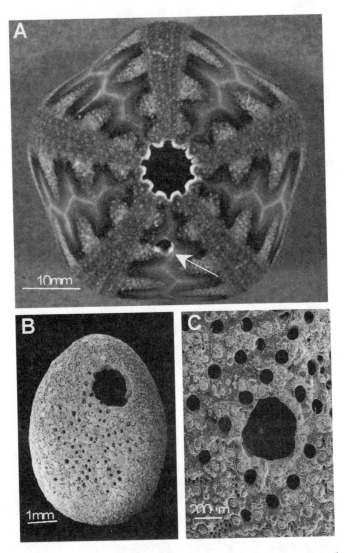

FIGURE 1. Predatory drill holes in tests of modern echinoids attributed to gastropods. All specimens collected in the Northern Bay of Safaga, Red Sea. (A) Hole in the interambulacra of the regular echinoid *Microcyphus rousseaui*. (B) Hole in a test of the minute clypeasteroid *Fibularia ovulum*. (C) SEM close-up of a test of *F. ovulum* showing a drill hole surrounded by a distinct halo interpreted to be the result of acid etching induced during drilling.

likely, by *Cassis tuberosa*. Nebelsick and Kowalewski (1999) reported that up to 83% of tests of the minute clypeasteroids *Fibularia* and *Echinocyamus* were drilled in certain environments of Safaga Bay, Red Sea, with the overall frequency of drilling in the entire bay exceeding 52.6%. However, the mortality rates from drilling predation may vary considerably among taxa and environments (e.g., Nebelsick and Kowalewski, 1999) and cassid predation is not invariably high in Recent environments (e.g., Hendler, 1977).

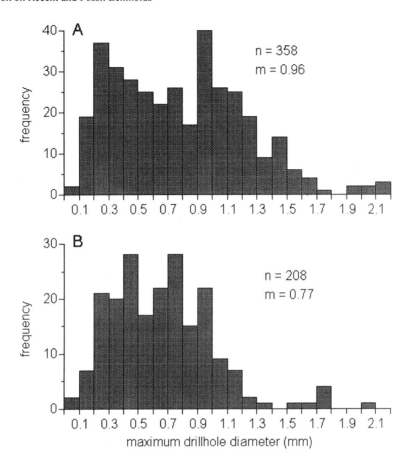

FIGURE 2. Size-frequency distribution of drill holes in tests of (A) *Fibularia ovulum* and (B) *Echinocyamus crispus* from the Northern Bay of Safaga, Red Sea; n = sample size, m = mean drill hole diameter.

Moreover, studies based on frequency of drill holes in empty echinoid tests may seriously overrate the relative importance of cassid predation. Because many other predators of echinoids tend to destroy tests of their prey (see below), the proportion of drilled tests (calculated as percent of preserved specimens) may severely overestimate importance of drilling by ignoring all tests that were destroyed by crushing predators. Nevertheless, controlled experiments suggest that cassids can be, at least in some environments, the primary source of mortality (Levitan and Genovese, 1989).

2.2.2. Eulimid Gastropods

Eulimid gastropods are ectoparasites that live on echinoderm hosts. Whereas eulimids that affect crinoids, ophiuroids, holothurians, and asteroids have been reported (Warén, 1980, 1981), these parasitic snails predominantly infest echinoids (e.g., Warén and Crossland, 1991; Rinaldi, 1994; Warén *et al.*, 1994). Eulimids attach to the host, permanently or temporarily, to gain access to gonad tissue and coelomic fluid (Warén and Crossland, 1991; Crossland *et al.*, 1991). They are known to make minute holes in tests of echinoids with their proboscis. Occasionally, they also leave distinct attachment scars.

Warén et al. (1994) examined the attachment site of a specimen of *Hypermastus mareticola* parasitizing the heart urchin *Maretia planulata* and found a small (~0.8 mm) discolored area with a centrally located hole. A total of 25 holes occurred in 15 specimens. All of the holes were located aborally and some of them had distinct attachment scars. Multiple eulimid parasites are commonly observed on echinoid hosts (e.g., Crossland *et al.*, 1993). Also, they may cause basal swelling of the primary spines by being attached permanently in small groups, as reported for the echinoid *Cidaris cidaris* infested by *Sabinella bonifaciae* (Warén and Misfud, 1990).

The eulimids that parasitize present-day echinoids have a widespread geographic distribution. Parasitic interactions were reported from the Colombian Pacific, Hawaii, coasts of Australia, Mediterranean Sea, Caribbean Sea, and many other regions (Warén *et al.*, 1984; Fujioka, 1985; Cantera and Neira, 1988; Warén and Moolenbeek, 1989; Warén and Misfud, 1990; Crossland *et al.*, 1991, 1993; Oliverio *et al.*, 1994; Rinaldi, 1994).

Eulimids can infest significant proportions of echinoid populations. For example, the sand dollar *Arachnoides placenta* from Queensland, Australia – parasitized by the eulimid *Hypermastus* sp. – displayed an infestation rate ranging from 8 to 15% with a seasonal increase in winter up to 26% (Crossland *et al.*, 1991).

Unfortunately, we cannot provide reliable diagnostic guidelines for differentiating unambiguously drill holes made by eulimids from those made by cassids. First, we know little about the morphological variability and size range of the two types of drill holes (eulimid drill holes are particularly poorly documented). Second, various test heterogeneities (e.g., pores, openings, plate contacts) affect strongly the geometry of drill holes (Nebelsick and Kowalewski, 1999) and thus the morphology of drillings is a questionable tool for identifying the culprit. Consequently, given our current neontological knowledge, the identity of the culprit in the fossil record can be inferred only indirectly using criteria such as presence of attachment scars (eulimid?) or dominance of tests with single holes (cassids?).

2.3. Other Invertebrates

2.3.1. Crustaceans

Crustaceans are diverse predators of modern echinoids. Frequencies and modes of attack depend on the size of both the predator and prey.

Lobsters commonly prey on echinoids and can have a significant effect on local populations of echinoids (e.g., Hagen and Mann, 1992; Cole, 1999). Andrew and MacDiarmid (1991) described how the spiny lobster *Jasus edwardsii* attacks the regular sea urchin *Evechinus chloroticus*. In the case of the smallest echinoids, the predator manipulates the prey using its first walking legs and maxillipeds and crushes the test with its mandibles. In the case of larger sea urchins, the lobster turns the prey over, pierces the peristomal membrane with its first walking legs and expands the oral opening to access the interior of the test. The largest lobsters ingest entire echinoid tests. Similarly, the spiny lobster *Panulirus interuptus* consumes entire tests of small specimens of the echinoids *Strongylocentrotus franciscanus* and *S. purpuratus* (Tegner and Levin, 1983). The consumption process results in a complete fracturing of prey tests. In the case of larger tests, the predator removes the peristomal membrane and produces an asymmetrical hole on the oral surface of the prey by enlarging the oral opening in the test. In the case of the largest tests, the predator removes the peristomal membrane, but both test and spines remain intact. The lobster *Homarus americanus*, on the other hand, cracks the sea urchins in half with its large chelae and ingests the soft tissues using its mouth parts (Scheibling, 1984).

Spider and rock crabs also prey on echinoids. Aquarium observations (Tertschnig, 1988) of the spider crab *Maia crispata* attacking *Paracentrotus lividus* show that the predator punctures the test through the peristomal membrane, enlarges the resulting wound by scissoring away bits of the test, and eats the internal tissues. The rock crab *Cancer irroratus*, feeding on diseased *Strongylocentrotus droebachiensis*, overturns its prey, punctures the peristomal membrane, chips off test material surrounding the peristome, and subsequently picks the soft tissues out of the test (Scheibling, 1984).

In addition to lethal attacks, crustaceans may be responsible for non-lethal predation common among present-day sand dollars (Nebelsick and Kampfer, 1994) and other present-day echinoderms in general (for a review of sub-lethal predation see Lawrence and Vasquez, 1996). In the case of sand dollars, the non-lethal attacks result in irregular subcircular scars along the ambitus of the sea urchin. Specimens with healed scars are common. In extreme cases, echinoids can survive multiple non-lethal attacks and can be recognized by highly irregular outlines of their severely battered tests.

2.3.2. Echinoderms

Mobile echinoderms, especially sea stars, are common predators of echinoids (e.g., Duggins, 1983; Legault and Himmelman, 1993; Tyler *et al.*, 1993). Dayton *et al.* (1977) observed the sea star *Meyenaster* crushing the test of its echinoid prey. Merril and Hobson (1970) described how sea stars totally ingest the sand dollar *Dendraster excentricus*. It is not clear what happens to the test of the sand dollar. However, given that starfish are able to ingest complete tests, it remains feasible that the tests can be regurgitated without damage (though test weakening and crushing is also possible).

Echinoids can attack other echinoids. Quinn (1965) described the attack (in an aquarium) of *Diadema antillarum* on *Tripneustes esculentus*. The predator removed the spines of the prey and produced a ragged hole in the test. Shulman (1990) studied intra- and interspecific aggression in five species of sea urchins from Caribbean coral reefs. Almost all of the antagonistic encounters involved pushing. In addition, biting occurred in 8 to 25% of the trials. In particular, *Diadema antillarum* exhibited predatory/aggressive attacks against *Echinometra* that involved biting behavior in 25% of the trials. Echinoid attacks are often sublethal and can result in repair scars on the test of the victim.

2.4. Vertebrates

2.4.1. Fish

Fish are major predators of sea urchins (e.g., Cole, 1999; McClanahan, 2000), although the distinction between predation and scavenging may be difficult to make (McClanahan, 1995). On Caribbean reefs, for example, at least 34 species of fish eat sea urchins (Hendler et al., 1995). Also, in other regions, diverse fish prey on echinoids (e.g., Keats, 1991; Gillander, 1995).

The hunting strategies and resulting test damage vary greatly depending on the type of fish and the size and type of the echinoid prey. Sala (1997) documented differential test damage of *Paracentrotus lividus* attacked by *Diplodus sargus*. Small urchins are swallowed whole, while larger echinoids are bitten open after the spines have been broken off. In contrast, the Atlantic wolffish *Anarhichas lupus* crushes even large *Strongylocentrotus droebachiensis* without difficulty (Hagen and Mann, 1992). The triggerfish, which attacks both regular and irregular echinoids, can display complex hunting techniques. For example, when attacking *Diadema*, the fish turns its prey over, clips its spines, assaults its peristomal membrane, and then bites the test into small pieces (Fricke, 1974). The grey triggerfish *Balistes capriscus*, when attacking shallowly buried sand dollars (Kurz, 1995), uses jets of water to expose its infaunal prey. Subsequently, the fish grasps its prey between the teeth, lifts it off the substrate and drops it until it lands on its upper side, exposing the lower oral side of the test. The triggerfish thrusts downward with its jaws closed, crushing the center of the prey and creating a hole through which the internal soft tissue of the prey is exposed (Fig. 3). After consuming the exposed tissue, the fish grasps the jagged edge of the wound to access the tissue remaining along the inner edges of the hole. This grasping activity leaves distinct teeth marks.

Predation by triggerfish may thus produce a distinct damage pattern on echinoid tests (see also Faser et al., 1991) that should be recognizable in the fossil record. In the Northern Bay of Safaga (Egypt), such distinct scars can be found on tests of *Clypeaster humilis* (Fig. 3) and *Echinodiscus auritus* (Nebelsick and Kampfer, 1994). The fact that both species have internal supports extending to the ambitus allows the test to be preserved despite the massive predation attack, which kills the animal and removes substantial amounts of the test material. The central wound is characterized by intraplate fracturing and the teeth marks often form parallel patterns that correspond to the teeth arrangement of the predator (Nebelsick and Kampfer, 1994). Triggerfish are among the most important predators of echinoids in some modern ecosystems (e.g., McClanahan, 1995, 1999, 2000; McClanahan et al., 1999).

2.4.2. Tetrapods

Tetrapods known to prey on echinoids include turtles (Chesher, 1969), a variety of birds (mostly shorebirds), marine mammals (primarily sea otters), and arctic foxes (Fell and Pawson, 1966). Except for the birds and sea otters, tetrapod-echinoid interactions have not been studied extensively, and thus the importance of tetrapod predators may be seriously underestimated. Turtles, for example, tend to be non-selective predators preying on all available sessile and slow-moving prey (e.g., Godley et al., 1997). Chesher (1969) observed loggerhead turtles biting and crushing the large irregular echinoids *Meoma ventricosa* and Godley et al. (1997) found echinoderms in the gut content of the loggerhead turtle *Caretta caretta*. However, the role of turtles and many other tetrapods that may prey on echinoids cannot be evaluated from the lamentably scarce data that are available so far.

Avian predators of echinoids are diverse and include gulls, turnstones, eider ducks, oystercatchers, and crows (Moore et al., 1963; Birkeland and Chia, 1971; Hendler, 1977;

FIGURE 3. A test of the echinoid *Clypeaster humilis* from the Northern Bay of Safaga (Red Sea) killed and eaten by a fish (most likely, a triggerfish).

Guillemette *et al.*, 1995; Wootton, 1995, 1997). Predatory strategies vary among birds. Turnstones peck holes in the tests of their echinoid prey (Hendler, 1977). In contrast, herring gulls were observed lifting sea urchins (*Lytechinus variegatus*) into the air and dropping them on the ground to break them open (Moore *et al.*, 1963). Birkeland and Chia (1971) described how the gull *Larus glaucescens* breaks tests, leaving irregular holes on the oral and apical sides. Avian predators can heavily affect echinoid populations. For example, Wootton (1995) estimated that bird predation on Tatoosh Island (Washington State, U.S.A.) reduced the abundance of the sea urchin *Strongylocentrotus purpuratus* by as much as 59%.

Finally, sea otters are among the best known predators of regular echinoids (Kvitek *et al.*, 1993; Heggberget and Moseid, 1994; Estes and Duggins, 1995; Estes *et al.*, 1998; Watt *et al.*, 2000). Echinoids often dominate the diet of sea otters, at least seasonally (Watt *et al.*, 2000). However, softer prey (e.g., fish species with soft integuments) may form a preferred prey in some region (Heggberget and Moseid, 1994). Because sea otters smash completely the test of their prey while feeding, it is highly unlikely that the predation by sea otters can be recognized in the fossil record of echinoids.

2.5. Paleontological Implications of Neontological Studies

2.5.1. Evolutionary and Paleoecological Implications

The above review of the neontological research elucidates the ecological importance of echinoid-predator interactions in modern benthic ecosystems. A great variety of predators hunt echinoids. Moreover, echinoid prey dominate the diet of many of those predators, at least seasonally. Echinoids are also a key host for a variety of parasitic and commensal invertebrates (for numerous examples see Hendler *et al.*, 1995).

Ecological studies also show that, because predators often control local populations of echinoids, the removal of those predators can lead to a substantial increase in population density of sea urchins. Thus, changes in the nature of echinoid-predator interactions can greatly affect marine ecosystems, including their primary productivity, species richness, and species distribution (e.g., Duggins, 1983). Specifically, the decrease in predation pressure often leads to intensification of grazing by echinoids and results in a loss of kelp vegetation and may affect coral cover (e.g., Estes *et al.*, 1998; McClanahan *et al.*, 1999). It is fair to stress that predation does not need to be the primary factor that controls the population dynamics of echinoids (e.g., Rivera, 1979; Sala *et al.*, 1998). For example, in Jobos Bay (Puerto Rico), predation by cassids and wading birds represents only a small source of mortality in echinoids (albeit relatively constant throughout the year), whereas physical mechanisms (spring tides and wind-caused waves) are responsible for mass mortality events that are seasonal in their occurrence (Rivera, 1979).

In summary, ecological studies suggest that fossil echinoids potentially can provide a rich source of information about the long-term evolution of marine ecosystems.

2.5.2. Taphonomic Implications

Although a diverse array of predators hunt echinoids in present-day seas, only a small fraction of those predator-prey interactions is likely to be preserved in the fossil record.

Neontological observations suggest that predatory attacks on echinoids can follow four taphonomic pathways (Table 1). First, in the best case, the predator produces

TABLE 1. Types of Traces Left by Present-Day Predators and Parasites on Echinoid Prey (Modified After Nebelsick, 1999*a*)

Predator (or parasite)	Identifiable damage	Ambiguous damage	Test destruction	No damage
Cassid gastropod	Yes	Yes	Yes	Yes
Eulimid gastropods	Yes	Yes	Unlikely	Yes
Crustaceans	Possible	Yes	Yes	Yes
Asteroids	Unlikely	Yes	Yes	Yes
Echinoids	Unlikely	Yes	Yes	Unlikely
Fish	Possible	Yes	Yes	Unlikely
Turtles	Unlikely	Unlikely	Yes	Unlikely
Birds	Possible	Yes	Yes	Unlikely
Sea otters	Unlikely	Unlikely	Yes	Unlikely

"Identifiable Damage": a distinct, preservable trace in the test that may be attributed to a particular predator (e.g., a gastropod drill hole). Second, the predator leaves "Ambiguous Damage" that may be misidentified as non-predatory damage and cannot be used reliably to identify the predator even at a coarse taxonomic level (e.g., damage made by some fish and crustaceans). Third, the predator's attacks lead to complete "Test Destruction" (as in the case of many crustacean, fish, and tetrapod predators). Finally, the predator leaves "No Damage" on the echinoid test (as possible in the case of cassids, crustaceans, and possibly starfish). Given that almost all groups of predators listed above include cases in which the echinoid test is either completely destroyed or left intact (Table 1), the frequencies of predation that are potentially retrievable from the fossil record are inevitably bound to underestimate the actual intensity of predation in ancient marine ecosystems.

It should be noted that, even in the best case where a predator leaves a distinct trace, the resulting damage may weaken the test. A selective loss of damaged tests due to subsequent taphonomic processes may further bias any estimates of predation retrieved from the fossil record. For example, punctured or scissored tests may be more vulnerable to subsequent disarticulation than intact tests, but we lack experimental data to evaluate this issue rigorously. To further complicate matters, we suspect that this bias may be less severe in the case of irregular echinoids, which tend to have interlocking plates and internal support structures (e.g., Seilacher, 1979; Donovan, 1991b), and thus may offer a better record of predation traces than regular ones. In general, the fossil record of predation on echinoids may have improved following the diversification of irregular echinoids (see also Kier, 1977; Kidwell and Baumiller, 1989; Greenstein, 1991; Martin, 1999). We stress here that all these predictions are not backed up by any rigorous data. Those statements point to some research questions that deserve testing in future studies. Variations in skeletal morphology of prey that may affect the preservational potential of damaged tests occurs at lower taxonomic levels as well (for example, internal support structures that may enhance the chance for preservation of damaged tests occur in only some minute clypeasteroids).

The evaluation of the preservational potential of predation traces may be further complicated by other factors, including (1) size relation of predator and prey (i.e., the same predator may either destroy or only damage the test depending on the size of the prey; see above); and (2) taphonomic pathways that may vary greatly across environments and among taxa.

The highest preservational potential should be associated with predatory attacks that leave a distinct but small trace. This prerequisite is best met by cassid and eulimid gastropods that produce small drill holes that should be readily identifiable in the fossil record. Drill holes are unlikely to seriously weaken the test. Nebelsick and Kowalewski (1999) showed that drilled specimens are as common among tests severely altered by taphonomic processes as among tests that are still pristine; that is, the proportion of drilled tests does not decrease with the increase in taphonomic alteration of the test (Fig. 4). They concluded that drill holes are unlikely to have a serious taphonomic effect even for the small and thin tests of clypeasteroid echinoids used in their study. They noted, however, that the material studied, unlike most of the fossilized tests, was unaffected by compaction, when preferential breakage of drilled tests would be likely to occur (see also Roy et al., 1994). Whereas certain types of fish and crustacean predation are also potentially recognizable, they are more difficult to identify unambiguously than are drill holes. Moreover such durophagous predation on echinoids often causes massive damage to the echinoid test and seriously lowers the chance for preservation in the fossil record.

In addition to taphonomic problems, the interpretation of the fossil record of predation on echinoids may be further complicated by facultative behavior of the predator. Many predators that create marks on the skeleton of their victim are not obligatory in their behavior and occasionally (or even frequently) kill their prey without leaving any distinct marks on its skeleton (e.g., Vermeij, 1980). Cassids, for example, have been observed to be facultative drillers that often access the soft tissue of their prey via the peristomal or periproctal membrane without making a hole in the test of the victim (Hughes and Hughes, 1981). A cassid can also destroy the test during the attack by crushing the test with its muscular foot.

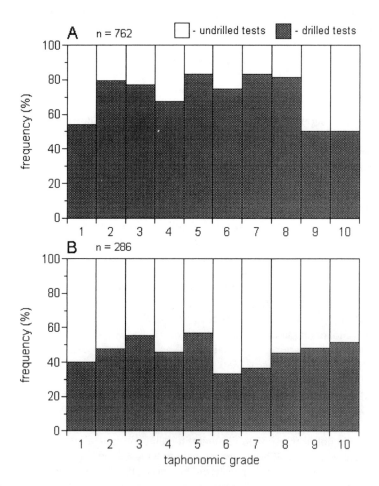

FIGURE 4. Taphonomic comparison of drilled and undrilled tests of clypeasteroid echinoids from the Northern Bay of Safaga, Red Sea. (A) *Fibularia ovulum*. (B) *Echinocyamus crispus*. Symbols: n = sample size.

2.6. Summary

Given the observations gathered from ecological studies, tests of fossil echinoids should be expected to provide a rich and diverse record of their predators and parasites. Moreover, despite all of the caveats and uncertainties, we remain optimistic about the potential value of the fossil record of durophagous predation on echinoids. This is because the historical patterns retrieved from the fossil record are usually analyzed in relative terms and thus represent proxies that provide valuable insights even if taphonomic and behavioral biases exist. The ecological and taphonomic data from neontological studies suggest that drill holes left by cassids and eulimids may represent a particularly rich source of data on the evolutionary interactions of echinoid prey and their predators and thus represent the most attractive target for paleontological studies on predator-prey interactions in the fossil record of echinoids.

3. The Fossil Record of Predation and Parasitism on Echinoids

3.1. Introduction

The review of paleontological literature reveals two facts. First, and perhaps surprisingly, the reports of predation and parasitism on echinoids are sparse in the paleontological literature, although published accounts seem to be increasing in frequency in recent years. Second, as expected, most of the reports focus on drilling predation/parasitism and few deal with other types of damage. Consequently, we focus primarily on the fossil record of drilling predation (or parasitism). Other types of records that may be interpreted as evidence of predation or parasitism on echinoids are discussed in Section 3.3.

3.2. Drilling Predation and Parasitism

Drill holes have been reported for fossil echinoids from the Cretaceous and Cenozoic (Kier, 1981; Sohl, 1969; Beu *et al.*, 1972; Gibson and Watson, 1989; Alekseev and Endelman, 1989; McNamara, 1991, 1994; Rose and Cross, 1993; Cross and Rose, 1994; Abdelhamid, 1999). The fossil occurrences are represented by small round holes (Fig. 5), typically below 1 mm in diameter (e.g., Kier, 1981; Cross and Rose, 1994). These holes may record both parasitic and predatory interactions.

In some cases, the holes can be readily interpreted as parasitic in nature. Kier (1981) described an Early Cretaceous specimen of *Hemiaster elegans* with four holes located in the apical region of the test. One of the holes was located in a deformed petal, suggesting a local cessation of growth following perforation. Kier also noted that these multiple holes were small and at least one of them was maintained during a long period in the life of the echinoid. Drawing on the neontological observations, Kier attributed holes to ectoparasitic, eulimid snails (see Section 2.2.2 above), which are known from Cretaceous rocks (Sohl, 1967). More recently, Alekseev and Endelman (1989) reported possible parasitic holes in Cretaceous echinoids, also suggesting eulimids as possible culprits.

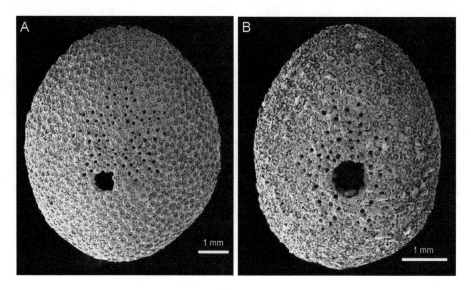

FIGURE 5. Drill holes in tests of Upper Oligocene clypeasteroid echinoids. (A) *Echinocyamus* sp. from Doberg Quarry, Northern Germany. (B) *Fibularia* sp. from Ardlogie Quarry, New Zealand.

In contrast, Sohl (1969; and fig. 3.7 therein) attributed holes in the late Eocene *Haimea alta* to predatory activity of cassids. Similarly, Beu *et al.* (1972) reported predation on spatangoids from upper Oligocene and lower Miocene sediments of New Zealand and noted possible site selectivity for the ambitus and plastron on the oral surface. Gibson and Watson (1989; see also Gibson, 1991) attributed holes in Eocene echinoids from central Florida to gastropod predation. They documented eight different morphological types of traces on echinoid tests but attributed only some of these types to predation. Moreover, Gibson and Watson pointed out that even the holes that may be attributable to predation can record as many as three different types of behavior: (1) a species-specific echinoid-gastropod interaction; (2) opportunistic predation by gastropods; and (3) gastropod predation on organisms encrusting echinoid tests (bryozoans). Cross and Rose (1994; see also Rose and Cross, 1993) attributed holes found in the Cretaceous *Micraster coranguinum* to predation. Similarly to Gibson and Watson (1989), Cross and Rose (1994) and Rose and Cross (1993) recognized that not all traces were made by predators – some may have represented attachment sites for acrothoracic barnacles or other non-predatory organisms. Woodcock and Kelley (2001) reported recently that 5 to 10% of irregular echinoids from the Eocene of North Carolina contained circular drill holes of predatory origin.

McNamara (1994), in a study that has been so far the broadest in temporal scope, examined Paleocene through Miocene echinoids of Australia for records of drilling predation. Drill holes interpreted as evidence of cassid predation were found in many echinoid taxa. Frequencies of drilling varied greatly through time, across environments, and among prey taxa. McNamara noted that the late Oligocene onset of substantial drilling predation coincided with the first appearance of cassid gastropods in the Australian fossil record. The predation intensity was particularly high in the early Miocene (with drilling frequencies exceeding 50% for some echinoids). McNamara

(1994) proposed that the decline of some echinoid groups (e.g., cassiduloids) may have been linked to high predation levels.

Drill holes found in fossil echinoids often provide evidence for site-selectivity (e.g., Cross and Rose, 1994; McNamara, 1994; Abdelhamid, 1999). Interestingly, site-specific patterns may vary across different prey species from the same deposits (McNamara, 1994). Site selectivity provides compelling evidence for the predatory and/or parasitic origin for fossil holes and may provide additional insights into behavior of the drilling organism.

The compilation of the quantitative data reported in the literature, supplemented with new field data, indicates that echinoids have been affected by drilling predation and parasitism since at least the Cretaceous (Fig. 6). Although the frequency of drilling can vary greatly within a given time interval, the data collected so far suggest that the intensity of drilling may have increased through time. When data are grouped into two coarse time-intervals (inset plot on Fig. 6), an average drilling frequency increases notably from the Cretaceous-Oligocene (19%) to the Miocene-Recent (29%). This pattern may thus parallel a similar secular trend in drilling predation observed for mollusk prey (e.g., Kelley and Hansen, 1993; Kowalewski et al., 1998). We stress, however, that the data are insufficient to postulate any far-reaching conclusions. The observed increase may reflect limited sample size (especially for the pre-Oligocene fossil record), geographic or taphonomic biases in the data, changes in predatory behavior (less facultative drilling), or a real increase in predation intensity. More case studies (especially for the Mesozoic and early Cenozoic) are needed to reconstruct the history of drilling predation/parasitism on echinoids.

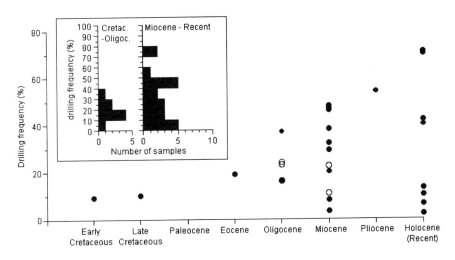

FIGURE 6. Drilling predation and/or parasitism on echinoids through time as estimated by frequency of drilled specimens of fossil and Recent echinoids. Literature data (solid circles) derived from Hendler (1977), Kier (1981), Gibson and Watson (1989), Levitan and Genovese (1989), Rose and Cross (1993), McNamara (1994), Nebelsick and Kowalewski (1999). New field data (open circles) include estimates for Oligocene *Echinocyamus* from Northern Germany (2 localities) for Oligocene *Fibularia* from New Zealand (2 localities) and Miocene *Echinocyamus* from Malta (personal communication, A. Kroh, 2001) and Poland.

The fossil record of drilling predation/parasitism offers great promise for future paleoecological research. Admittedly, as documented by Gibson and Watson (1989) and Cross and Rose (1994), care must be taken to differentiate predatory and parasitic drill holes and to exclude traces of other origin. Nevertheless, fossil data that can be obtained in individual case studies often may be good enough to make statements concerning site selectivity, behavior of predator, defense mechanisms of the prey, or ecological impact of predators/parasites on ancient echinoids. The data also show promise for studying interactions between echinoids and their enemies over the evolutionary time scale (see also McNamara, 1994).

3.3. Durophagous Predation on Echinoids

Structural damage that may represent predation is often observed in fossil echinoid tests from Cretaceous and Cenozoic deposits. Both non-lethal attacks and lethal predatory attacks have been suggested as possible causes for such damage. Non-lethal predation by crustaceans may be common in fossil echinoids. Based on damage pattern -- strikingly similar to that found in Recent sand dollars (see Section 2.3.1) -- non-lethal attacks by crabs on the ambitus of fossil sand dollars have been suggested by Nebelsick (1999*b*; see also Zinsmeister, 1980). Intensity of non-lethal predation events may be substantial: 17% for the lower Miocene *Parascutella* from Austria (Nebelsick, 1999*b*). In contrast, no convincing examples of lethal predation by crustaceans have been reported to our knowledge. Although Rose and Cross (1993) suggested that irregular holes broken through the aboral surface above gonads of the Cretaceous *Micraster* from Great Britain may have represented attacks by crustacean predators, they cautioned about the speculative nature of that interpretation.

Surprisingly few descriptions of potential fish attacks have been recorded. Thies (1985) suggested that pock-marked test surfaces of irregular echinoids from the Cretaceous of Northern Germany may represent evidence of predation by fish. Zinsmeister (1980) reported frequent test margin damage in the Miocene clypeasteroids from Argentina and noted that the damage pattern was similar to that induced today by small fish nipping the thin edges of tests of *Encope* and *Leodia* (Kier and Grant, 1965). Nebelsick (1999*b*) reported rare occurrences of wounds in the lower Miocene fossil sand dollar *Parascutella* from Austria. The gross morphology of the wounds resembles those made by triggerfish preying on modern sand dollars (see Section 2 and Fig. 3 above). Schormann (1987) described possible bite marks made on echinoid tests by other echinoderms.

To our knowledge, no conclusive paleontological evidence for predation by other common present-day predators of echinoids has been reported in the literature.

3.4. The Fossil Record of Parasites

In addition to drill holes attributed to ectoparasitic eulimid gastropods discussed above (Section 3.2), published accounts of fossil parasitism range from distinct traces attributed to specific culprits to general description of wounds and test abnormalities that may have been induced by parasitism. As in the case of other records discussed above, these records date back to the Cretaceous. The Upper Cretaceous and Cenozoic records of ascothoracid cirriped crustaceans are among the most common cases of parasitism documented in the fossil record (Madsen and Wolff, 1965; Cross and Rose, 1994).

Other postulated parasites include copepods infesting the Miocene *Clypeaster* from Turkey (Margara, 1946), the crustacean *Castexia douvillei* infesting Cretaceous echinoids from France (Mercier, 1937), and myzostome parasites infesting the Miocene *Clypeaster* from Morocco (Roman, 1952). Unidentified parasites were suggested for the middle Eocene echinoids from Georgia (Kier, 1981). Abnormal growth and malformation including test bulging, test deepening, and abnormal pores, attributed to the infestation by parasites, were reported for Cretaceous echinoids of France (Saint-Seine, 1950) and Egypt (Abdelhamid, 1999) and for the Miocene *Clypeaster* from Europe (Dimic, 1998; Hiden, 1998).

4. Future Research Directions

The fossil record of echinoids represents an understudied but potentially rich source of paleoecological and evolutionary information on predator-prey and host-parasite interactions. Admittedly, marginal comments and short notes regarding biotic interactions of echinoids are common in the paleontological literature (Section 3). However, few full-scope paleontological research projects (and even fewer actuopaleontological case studies) have been published so far. This makes echinoids a particularly attractive target for paleoecologists. Two key directions are suggested here: (1) actuopaleontological research in modern environments exploring in detail how present-day predators and parasites attack and infest echinoids; and (2) paleontological research documenting events of predation and parasitism recorded by fossilized echinoid tests or their fragments.

4.1. Actuopaleontological Research

The current dearth of paleontologically driven ecological research on present-day echinoids seriously hampers our ability to interpret their fossil record in terms of biotic interactions. More efforts need to be directed at documenting morphology and distribution of wounds inflicted by modern predators and parasites of echinoids. Particularly needed are studies that focus on the most ubiquitous durophagous and drilling organisms (fish, crustaceans, and gastropods). We lack good guidelines regarding differences in morphology of eulimid versus cassid drillholes. Similarly, and perhaps even more importantly, we lack explicit criteria to distinguish traces of parasites from those left by predators.

Taphonomic implications of traces left by predators and parasites on echinoid tests also need to be explored in more detail. Detailed analyses of test fragments (compare Nebelsick, 1992, 1999c) produced by predators should provide valuable guidelines regarding strategies for identifying destructive attacks in the fossil record and differentiating them from mechanically induced fragmentation (this may be especially true for regular echinoid prey from high-energy environments). Experimental studies need to be conducted to evaluate taphonomic effects of drill holes or other structural damage and to assess potential taphonomic biases more rigorously.

Also, we need to know more about the population structures of predators and their echinoid prey. What proportion of predatory events that occur in present-day communities is likely to be preserved in the fossil record? What proportion of mortality is due to drilling predation?

Most importantly, perhaps, the present-day ecosystems are most suitable to evaluate reliably the relative importance of various types of predators/parasites, the overall intensity of predator/parasite-echinoid interactions, and environmental and geographic variation in frequencies of those interactions. Future research targeting those questions should provide an important reference baseline for paleontologists. Finally, present-day communities also represent a time-interval (the Holocene) and should be considered and exploited when studying secular trends in predation patterns.

4.2. Paleontological Research

Given the limited number of detailed paleontological studies, any case study involving fossil echinoids is bound to yield new and valuable information. In particular, however, researchers should attempt to extend our knowledge back beyond the late Mesozoic. In general, the literature points to the potential ubiquity of predatory/parasitic records. It is quite feasible that any careful survey of samples of fossil echinoids (or even cursory review of museum materials) made with respect to predatory/parasitic events is likely to reveal new data on biotic interactions involving echinoids.

At this point, drilling predation and parasitism seem to be the most promising research directions: the fossil record should yield a wealth of credible data on cassid-echinoid and eulimid-echinoid interactions. We suggest three broad research directions: (1) paleoecological studies on echinoid-rich fossil assemblages focusing on morphology and taphonomy of drillings and the behavior of predators (site, size, and species-selectivity); (2) macroevolutionary studies investigating long-term changes in drilling patterns; and (3) spatial studies exploring environmental variability, bathymetric trends, and geographic gradients.

Whereas drilling organisms may likely offer the richest and least ambiguous data, there clearly are other lines of evidence that may be often provided by fossil echinoids and deserve attention from future researchers. The literature suggests that few identifiable records of such ecological interactions have been documented so far. Moreover, those scarce reports are limited primarily to the Cretaceous and Cenozoic.

All in all, the fossil record of predation/parasitism preserved in echinoid tests may provide a great target for future research in evolutionary paleoecology. First, they should allow us to test escalation models and the long-term arms race of prey and predators. So far, this issue was explored rigorously only for mollusks (e.g., Vermeij, 1987; Kelley and Hansen, 1993, 1996; Hansen *et al.,* 1999) and, to some extent, brachiopods (e.g., Leighton, 1999; Dietl and Kelley, 2001; Hoffmeister *et al.*, 2002). Echinoids offer an independent system for evaluating the impact of predators on the evolution of their prey (and *vice versa*). Second, fossilized traces of predation offer direct means for evaluating specific hypotheses (related to the hypothesis of escalation) regarding the role of predation in inducing evolutionary trends through the history of echinoids (e.g., Kier, 1965, 1974, 1982; Smith, 1984; Vermeij, 1987; McNamara, 1990, 1991, 1994; Rose and Cross, 1993; Cross and Rose, 1994). The two main trends that often are linked to predation are (1) evolution of more rigid skeleton and external armor such as spines and (2) infaunalization and reduction of armor (Kier, 1982; Smith, 1984; Vermeij, 1987). Future studies should evaluate if those evolutionary trends match temporal changes in frequency of predatory attacks recorded by traces attributable to predators. Finally, at the ecosystem level, the correlation between secular trends in predation on echinoids and evolutionary success of kelp or reef ecosystems affected by grazing echinoids can be explored (e.g., Estes and Steinberg, 1988; Wood, 1998).

Fossil and recent echinoids should produce substantial data concerning predation intensity, size frequency distributions among populations, site selectivity of attacks, types of wound morphologies, and variation in types and frequencies of predation and parasitism among echinoid taxa and along environmental and latitudinal gradients. Future research may ultimately give us data on predator/parasite-echinoid interactions that will be comparable to those currently available only for mollusks.

5. Summary

Paleontological research on interactions between fossil echinoids and their predators/parasites is limited to few rigorous studies. Yet, observations in present-day marine ecosystems demonstrate that echinoids are affected, often intensely, by a variety of predators and parasites, including gastropods, crustaceans, echinoderms, fish, turtles, birds, and sea otters. Some of the culprits create distinct marks in the tests of their victims and thus leave evidence that can be preserved in the fossil record. Round drill holes left by drilling gastropods (cassids and eulimids) are particularly likely to be preserved and identified in the fossil record. Paleontological literature indicates that drill holes are the most common evidence of predation/parasitism found in fossil echinoids, and few fossilized occurrences of other types of durophagous marks have been suggested so far. The limited quantitative data gathered to date suggest that the intensity of predation/parasitism on echinoids may have increased since the late Mesozoic, although interpretation of this trend remains ambiguous. Future studies will be critical to determine the evolutionary history of predation/parasitism on echinoids and the role of echinoid predators in the Mesozoic and Cenozoic evolution of marine ecosystems.

ACKNOWLEDGMENTS: This study was partly supported by the NSF grant EAR-9909225 (to MK) and the DFG grant NE 537/9-1 (to JN). We thank Antje Teuwsen, Markus Bertling, Klaus-Peter Lanser (all Münster), Ewan Fordyce (Dunedin), Andreas Kroh (Graz) and Michael Strauss (Bünde) for help at various stages of this project. Michael Gibson and one anonymous reviewer provided many useful comments that greatly improved this manuscript.

References

Abdelhamid, M. A. M., 1999, Parasitism, abnormal growth and predation on Cretaceous echinoids from Egypt, *Revue Paleobiol.* **18**:69-83.

Alekseev, A. S., and Endelman, L. G., 1989, Association of ectoparasitic gastropods with Upper Cretaceous echinoid *Galerites,* in: *Fossil and Recent Echinoderm Researches* (Anonymous, ed.), Academy of Sciences of the Estonian SSR, Tallin, pp. 165-174.

Alexander, R. R., 1981, Predation scars preserved in Chesterian brachiopods: Probable culprits and evolutionary consequences for articulates, *J. Paleontol.* **55**:192-203.

Alexander, R. R., 1986, Frequency of sublethal shell-breakage in articulate brachiopod assemblages through geologic time, in *Les Brachiopodes Fossiles et Actuels: First International Brachiopod Congress Proceedings* (P. R. Racheboeuf and C. C. Emig, eds.), *Biostratigraphie du Paleozoique* **4**:159-166.

Alexander, R. R., 1990, Mechanical strength of shells of selected extant articulate brachiopods; implications for Paleozoic morphologic trends, *Hist. Biol.* **3**:169-188.

Andrew, N. L., and MacDiarmid, A. B., 1991, Inter-relations between sea urchins and spiny lobsters in northeastern New Zealand, *Mar. Ecol. Prog. Ser.* **70**:211-222.

Baumiller, T. K., 1990, Non-predatory drilling of Mississippian crinoids by platyceratid gastropods, *Palaeontology* **33**:743-748.

Baumiller, T. K., 1993, Boreholes in Devonian blastoids and their implications for boring by platyceratids, *Lethaia* **26**:41-47.

Baumiller, T. K., 1996, Boreholes in the Middle Devonian blastoid *Heteroschisma* and their implications for gastropod drilling, *Palaeogeogr. Palaeoclim. Palaeoecol.* **123**:343-351.

Baumiller, T. K., and Macurda, D. B., Jr., 1995, Borings in Devonian and Mississippian blastoids (Echinodermata), *J. Paleontol.* **69**:1084-1089.

Baumiller, T. K., Leighton, L. R., and Thompson, D., 1999, Boreholes in brachiopods of the Fort Payne Formation (Lower Mississippian, central USA), *Palaeogeogr. Palaeoclim. Palaeoecol.* **147**:283-289.

Beu, A. G., Henderson, R. A., and Nelson, C. S., 1972, Notes on the taphonomy and paleoecology of New Zealand Tertiary Spatangoida, *N. Z. J. Geol. Geophys.* **15**:275-286.

Birkeland, C., and Chia, F. S., 1971, Recruitment, risk, growth, age and predation in two populations of sand dollars, *Dendraster excentricus* (Eschscholtz), *J. Exper. Mar. Biol. Ecol.* **6**:265-278.

Cantera, K. J. R., and Neira, O. R., 1988, First record of the genus *Echineulima*, new record Lutzen and Nielsen (Gastropoda, Eulimidae), a parasite of sea urchins, at Gorgona Island (Colombian Pacific), *Anal. Instit. Investig. Mar. Punta Betin* **17**:87-94.

Chesher, R. H., 1969, Contribution to the biology of *Meoma ventricosa* (Echinoidea: Spatangoida), *Bull. Mar. Sci. Gulf Caribb.* **19**:72-110.

Cole, R. G., 1999, Trophic relationships between fishes and benthic organisms on northeastern New Zealand reefs, *Vie et Milieu* **49**:201-212.

Cross, N. E., and Rose, E. P. F., 1994, Predation of the Upper Cretaceous spatangoid *Micraster*, in: *Echinoderms Through Time* (B. David, A. Guille, J. P. Féral, and M. Roux, eds.), Balkema, Rotterdam, pp. 607-612.

Crossland, M. R., Alford, R. A., and Collins, J. D., 1991, Population dynamics of an ectoparasitic gastropod, *Hypermastus* sp. (Eulimidae), on the sand dollar, *Arachnoides placenta* (Echinoidea), *Austr. J. Mar. Fresh. Res.* **42**:69-76.

Crossland, M. R., Collins, J. D., and Alford, R. A., 1993, Host selection and distribution of *Hypermastus placentae* (Eulimidae), an ectoparasitic gastropod on the sand dollar *Arachnoides placenta* (Echinoidea), *Austr. J. Mar. Fresh. Res.* **44**:835-844.

Dayton, P. K. R., Rosenthal, R. J., Mahen, L. C., and Antezana, T., 1977, Population structure and foraging biology of the predaceous Chilean asteroid *Meyenaster gelatinosus* and the escape biology of its prey, *Mar. Biol.* **39**:361-370.

Dietl, G. P., and Kelley, P. H., 2001, Mid-Paleozoic latitudinal predation gradient: distribution of brachiopod ornamentation reflects shifting Carboniferous climate, *Geology* **29**:111-114.

Dietl, G. P., Alexander, R. R., and Bien, W. F., 2000, Escalation in Late Cretaceous-early Paleocene oysters (Gryphaeidae) from the Atlantic Coastal Plain, *Paleobiology* **26**:215-237.

Dimic, M., 1998, Tragovi epibioze, predatorstva i parazitizma na skeletu roda *Clypeaster* (Echinoidea) (Traces of epibiosis, predacity and parasitism on Clypeaster (Echinoidea) skeletons), *Geol. Anali Balkan. Poluos.* **62**:213-232.

Donovan, S. K., 1991a, Site selectivity of a Lower Carboniferous boring organism infesting a crinoid, *Geol. J.* **26**:1-5.

Donovan, S. K., 1991b, The taphonomy of echinoderms: calcareous multi-element skeletons on the marine environment, in: *The Processes of Fossilization* (S. K. Donovan, ed.), Belhaven Press, London, pp. 241-269.

Duggins, D. O., 1983, Starfish (*Pycnopodia helianthoides*) predation and the creation of mosaic patterns in a kelp-dominated community, *Ecology* **64**:1610-1619.

Engstrom, N. A., 1982, Immigration as a factor in maintaining populations of the sea urchin *Lytechinus variegatus* (Echinodermata: Echinoidea) in seagrass beds on the south west coast of Puerto Rico, *Stud. Neotrop. Fauna and Environ.* **17**:51-60.

Estes, J. A., and Duggins, D. O., 1995, Sea otters and kelp forests in Alaska: Generality and variation in a community ecological paradigm, *Ecol. Monogr.* **65**:75-100.

Estes, J. A., and Steinberg, P. D., 1988, Predation, herbivory, and kelp evolution, *Paleobiology* **14**:19-36.

Estes, J. A., Tinker, M. T., Williams, T. M., and Doak, D. F., 1998, Killer whale predation on sea otters linking oceanic and nearshore ecosystems, *Science* **282**:473-476.

Faser, T. K., Lindberg, W. J., and Standton, G. R., 1991, Predation on sand dollars by gray triggerfish, *Balistes capriscus*, in the northeastern Gulf of Mexico, *Bull. Mar. Sci.* **48**:159-164.

Fell. H. B., and Pawson, D. L., 1966, General biology of echinoderms, in: *Physiology of Echinodermata* (R. A. Boolootian, ed.), Interscience, New York.

Fricke, H. W., 1974, Möglicher Einfluß von Feinden auf das Verhalten von *Diadema*-Seeigeln, *Mar. Biol.* **27**:59-64.

Fujioka, Y., 1985, Population ecological aspects of the eulimid gastropod *Vitreobalcis temnopleuricola*, *Malacologia* **26**:153-164.

Gibson, M. A., 1991, Evidence of scavenging of an irregular echinoid from the Ocala Formation (Eocene), Florida, *J. Tenness. Acad. Sci.* **66**:62.

Gibson, M. A., and Watson, J. B., 1989, Predatory and non-predatory borings in echinoids from the upper Ocala Formation (Eocene), north-central Florida, U. S. A., *Palaeogeogr. Palaeoclim. Palaeoecol.* **71**:309-321.

Gillander, B. M., 1995, Feeding ecology of the temperate marine fish *Achoeridus viridis* (Labridae): Size, seasonal and site-selective differences, *Mar. Freshwater Res.* **46**:1009-1020.

Gladfelter, W. B., 1978, General ecology of the cassiduloid urchin *Cassidulus caribbearum* [sic], *Mar. Biol.* **47**:149-160.

Godley, B. J., Smith, S. M., Clark, P. F., and Taylor, J. D., 1997, Molluscan and crustacean items in the diet of the loggerhead turtle, *Caretta caretta* (Linnaeus, 1758) [Testudines: Chelonidae] in the eastern Mediterranean, *J. Molluscan Stud.* **63**:474-476.

Greenstein, B. J., 1991, An integrated study of echinoid taphonomy: Predictions for the fossil record of four echinoid families, *Palaios* **6**:519-540.

Greenstein, B. J., 1995, The effects of life habit and test microstructure on the preservation potential of echinoids in Graham's Harbour, San Salvador Island, Bahamas, *Geol. Soc. Am. Spec. Pap.* **300**:177-188.

Guillemette, M., Reed, A., and Himmelman, J. H., 1995, Availability and consumption of food by common eiders wintering in the Gulf of St. Lawrence, evidence of prey depletion, *Can. J. Zool.* **74**:32-38.

Hagen, N. T., and Mann, K. H., 1992, Functional response of the predators American lobster *Homarus americanus* (Milne-Edwards) and Atlantic wolffish *Anarchichas lupus* (L.) to increasing numbers of the green sea urchin *Strongylocentrotus droebachiensis* (Müller), *J. Exper. Mar. Biol. Ecol.* **159**:89-112.

Hansen, T. A., Kelley, P. H., Melland, V. D., and Graham, S. E., 1999, Effect of climate-related mass extinctions on escalation in molluscs, *Geology* **27**:1139-1142.

Harper, E. M., and Wharton, D. S., 2000, Boring predation and Mesozoic articulate brachiopods, *Palaeogeogr. Palaeoclim. Palaeoecol.* **158**:15-24.

Heggberget, T. M., and Moseid, K. E., 1994, Prey selection in coastal Eurasian otters *Lutra lutra*, *Ecography* **17**:331-338.

Hendler, G., 1977, The differential effects of seasonal stress and predation on the stability of reef flat echinoid populations, in *Proceedings, Third International Coral Reef Symposium* (D. L. Taylor, ed.), *Proc. 3rd Int. Coral Reef Symp.* **1**:217-223.

Hendler, G., Miler, J. E., Pawson, D. L. and Kier, P. M., 1995, *Sea Stars, Sea Urchins, and Allies*, Smithsonian Institution, Washington D.C., 390 pp.

Hiden, H. R., 1998, Paläopathologishe Befunden an Echinoideen aus dem Mittelmiozän (Badenium) Ost-Österreichs, *Mitt. Ref. Geol. Paläont. Landesmus. Joanneum* **2**:215-221.

Hoffmeister, A. P., and Kowalewski, M., 2001, Spatial and environmental variation in the fossil record of drilling predation: A case study from the Miocene of Central Europe, *Palaios* **16**:566-579.

Hoffmeister, A. P., Kowalewski, M., Bambach, R. K., and Baumiller, T. K., 2002, A boring history of drilling predation on the Paleozoic brachiopod *Composita*, *Geol. Soc. Am. Abstr. Progr.* **34**(2):A-116.

Hughes, R. N., and Hughes, P. I., 1971, A study on the gastropod *Cassis tuberosa* (L.) preying upon sea urchins, *J. Exper. Mar. Biol. Ecol.* **7**:305-314.

Hughes, R. N., and Hughes, P. I., 1981, Morphological and behavioral aspects of feeding in the Cassidae (Tonnacea, Mesogastropoda), *Malacologia* **20**:385-402.

Kelley, P. H., and Hansen, T. A., 1993, Evolution of the naticid gastropod predator-prey system: an evaluation of the hypothesis of escalation, *Palaios* **8**:358-375.

Kelley, P. H., and Hansen, T. A., 1996, Naticid gastropod prey selectivity through time and the hypothesis of escalation, *Palaios* **11**:437-445.

Keats, D. W., 1991, American plaice, *Hippoglossoides platessoides* (Fabricius), predation on green sea urchins, *Strongylocentrotus droebachiensis* (O. F. Muller), in eastern Newfoundland, *J. Fish Biol.* **38**:67-72.

Kidwell, S. M., and Baumiller, T. K., 1989, Experimental disintegration of regular echinoids: Roles of temperature, oxygen, and decay thresholds, *Paleobiology* **16**:247-271.

Kier, P. M., 1965, Evolutionary trends in Paleozoic echinoids, *J. Paleontol.* **39**:436-465.

Kier, P. M., 1974, Evolutionary trends and their functional significance in the post-Paleozoic echinoids, *Paleontol. Soc. Mem.* **5**:1-96.
Kier, P. M., 1977, The poor fossil record of the regular echinoids, *Paleobiology* **3**:168-174.
Kier, P. M., 1981, A bored Cretaceous echinoid, *J. Paleontol.* **55**:656-659.
Kier, P. M., 1982, Rapid evolution in echinoids, *Palaeontology* **25**:1-9.
Kier P. M., and Grant, R. E., 1965, Echinoid distribution and habits, Key Largo Coral Reef Preserve, Florida, *Smithson. Miscell. Coll.* **149**:1-68.
Kitchell, J. A., 1986, The evolution of predator-prey behavior: naticid gastropods and their molluscan prey, in: *Evolution of Animal Behavior: Paleontological and Field Approaches* (M. H. Nitecki and J. A. Kitchell, eds.), Oxford Press, New York, pp. 88-110.
Kitchell, J. A., Boggs, C. H., Kitchell, J. F., and Rice, J. A., 1981, Prey selection by naticid gastropods: Experimental tests and application to the fossil record, *Paleobiology* **7**:533-552.
Kowalewski, M., Dulai, A., and Fürsich, F. T., 1998, A fossil record full of holes: The Phanerozoic history of drilling predation, *Geology* **26**:1091-1094.
Kowalewski, M., Simões, M. G., Torello, F. F., Mello, L. H. C., and Ghilardi, R. P., 2000, Drill holes in shells of Permian benthic invertebrates, *J. Paleontol.* **74**:532-543.
Kurz, R. C., 1995, Predator-prey interactions between gray triggerfish (*Balistes capriscus* Gmelin) and a guild of sand dollars around artificial reefs in the northwestern Gulf of Mexico, *Bull. Mar. Sci.* **56**:150-160.
Kvitek, R. G., Bowlby, C. E., and Staedler, M., 1993, Diet and foraging behavior of sea otters in southeast Alaska, *Mar. Mammal Sci.* **9**:168-181.
Lawrence, J.M., and Vasquez, J., 1996, The effects of sublethal predation on the biology of echinoderms, *Oceanol. Acta* **19**:431-440.
Legault, C., and Himmelman, J. H., 1993, Relation between escape behaviour of benthic marine invertebrates and the risk of predation, *J. Exper. Mar. Biol. Ecol.* **170**:55-74.
Leighton, L. R., 1999, Possible latitudinal predation gradient in middle Paleozoic oceans, *Geology* **27**:47-50.
Leighton, L. R., 2001, New example of Devonian predatory boreholes and the influence of brachiopod spines on predator success, *Palaeogeogr. Palaeoclim. Palaeoecol.* **165**:53-69.
Levitan, D. R., and Genovese, S. T., 1989, Substratum-dependent predator-prey dynamics: Patch reefs as refuges from gastropod predation, *J. Exper. Mar. Biol. Ecol.* **130**:111-118.
Madsen, F. J., and Wolff, T., 1965, Evidence of the occurrence of Ascothoracica (parasitic cirripeds) in Upper Cretaceous, *Dan. Geol. Foren., Medd.* **15**:556-558.
Margara, J., 1946, Existence de zoothylacies chez des clypeastres (echinodermes) de l'helvetien du Proche-Orient, *Bull. Mus. Nat. d'Hist. Nat.* **18**:423-427.
Martin, R. E., 1999, *Taphonomy, A Process Approach*, Cambridge University Press, Cambridge, 508 pp.
McClanahan, T. R., 1995, Fish predators and scavengers of the sea urchin *Echinometra mathaei* in Kenyan coral-reef marine parks, *Env. Biol. Fishes* **43**:187-193.
McClanahan, T. R., 1999, Predation and the control of the sea urchin *Echinometra viridis* and fleshy algae in the patch reefs of Glovers Reef, Belize, *Ecosystems* **2**:511-523.
McClanahan, T. R., 2000, Recovery of a coral reef keystone predator, *Balistapus undulatus*, in East African marine parks, *Biol. Conserv.* **94**:191-198.
McClanahan, T. R., Muthiga, N. A., Kamukuru, A. T., Machano, H., and Kiambo, R. W., 1999, The effects of marine parks and fishing on coral reefs of northern Tanzania, *Biol. Conserv.* **89**:161-182.
McClintock, J. B., and Marion, K. R., 1993, Predation of the king helmet (*Cassis tuberosa*) on six-holed sand dollars (*Leodia sexiesperforata*) at San Salvador, Bahamas, *Bull. Mar. Sci.* **52**:1013-1017.
McNamara, K. J., 1990, Echinoids, in: *Evolutionary Trends* (McNamara, K. J., ed.), University of Arizona Press, Tucson, pp.205-231.
McNamara, K. J., 1991, Murder and mayhem in the Miocene, *Nat. Hist.* **8/91**:40-45.
McNamara, K. J., 1994, The significance of gastropod predation to patterns of evolution and extinction in Australian Tertiary echinoids, in *Echinoderms Through Time* (B. A. David, A. Guille, J. P. Féral, and M. Roux, eds.), Balkema, Rotterdam, pp. 785-793.
Mercier, J., 1937, Zoothylacies d'echinide fossile provoquees par un crustace *Castexia douvillei* nov. gen., nov. sp., *Bull. Soc. Geol. France* **6**:149-154.
Merril, R. J., and Hobson, E. S., 1970, Field observations of *Dendraster excentricus*, a sand dollar of western North America, *Am. Midland Nat.* **83**:595-624.
Moore, H. B., Jutare, J. C., Bauer, J. C., and Jones, J. A., 1963, The biology of *Lytechinus variegatus, Bull Mar. Sci. Gulf Caribb.* **13**:23-53.
Moore, D. R., 1956, Observations of predation on echinoderms by three species of Cassididae, *Nautilus* **69**:73-76.
Nebelsick, J. H., 1992, The use of fragments in deducing echinoid distribution by fragment identification in Northern Bay of Safaga; Red Sea, Egypt, *Palaios* **7**:316-328.

Nebelsick, J. H., 1999a, Taphonomic legacy of predation on echinoids, in *Echinoderm Research 1998* (M. D. Candia Carnevali and F. Bonasoro, eds.), Balkema, Rotterdam, pp. 347-352.
Nebelsick, J. H., 1999b, Taphonomic comparison between Recent and fossil sand dollars, *Palaeogeogr. Palaeoclim. Palaeoecol.* **149**:349-358.
Nebelsick J. H., 1999c, Taphonomic signatures and taphofacies distribution as recorded by *Clypeaster* fragments from the Red Sea, *Lethaia* **32**:241-252.
Nebelsick, J. H., and Kampfer, S., 1994, Taphonomy of *Clypeaster humilis* and *Echinodiscus auritus* from the Red Sea, in *Echinoderms Through Time* (B. A. David, A. Guille, J. P. Féral, and M. Roux, eds.), Balkema, Rotterdam, pp. 803-808.
Nebelsick, J. H., and Kowalewski, M., 1999, Drilling predation on Recent clypeasteroid echinoids from the Red Sea, *Palaios* **14**:127-144.
Oliverio, M., Buzzurro. G., and Raimondo, A. D., 1994, A new eulimid gastropod from the eastern Mediterranean Sea (Caenogastropoda, Ptenoglossa), *Boll. Malacol.* **30**:211-215.
Quinn, B. G., 1965, Predation in sea urchins, *Bull. Mar. Sci.* **15**:259-264.
Rinaldi, A. C., 1994, Frequency and distribution of *Vitreolina philippi* (De Rayneval and Ponzi, 1854) (Prosobranchia, Eulimidae) on two regular echinoid species found along the southern coast of Sardinia, *Boll. Malacol.* **30**:29-32.
Rivera, J. A., 1979, Echinoid mortality at Jobos Bay, Puerto Rico, *Proc. Assoc. Island Mar. Lab. Caribb.* **1979**:9.
Roman, J., 1952, Quelques anomalies chez *Clypeaster melitensis* Michelin, *Bull. Soc. Geol. France* **2**:3-11.
Rose, E. P. F., and Cross, N. F., 1993, The chalk sea urchin *Micraster:* Microevolution, adaptation and predation, *Geol. Today* **5**:179-186.
Roy, K., Miller, D. J., and LaBarbera, M., 1994, Taphonomic bias in analyses of drilling predation: effects of gastropod drill holes on bivalve shell strength, *Palaios* **9**:413-421.
Sala, E., 1997, Fish predation and scavengers of the sea urchin *Paracentrotus lividus* in protected areas of the north-west Mediterranean Sea, *Mar. Biol.* **129**:531-539.
Sala, E., Ribes, M., Hereu, B., Zabala, M., Alva, V., Coma, R., and Garrabou, J., 1998, Temporal variability in abundance of the sea urchins *Paracentrotus lividus* and *Arbacia lixula* in the northwestern Mediterranean: Comparison between a marine reserve and an unprotected area, *Mar. Ecol. Prog. Ser.* **168**:135-145.
Saint-Seine, R., de, 1950, Lesions et regeneration chez le *Micraster, Bull. Soc. Geol. France* **5**:309-315.
Scheibling, R. E., 1984, Predation by rock crabs (*Cancer irroratus*) on diseased sea urchins (*Strongylocentrotus droebachiensis*) in Nova Scotia, *Can. J. Fish. Aquat. Sci.* **41**:1847-1851.
Schormann, J., 1987, Bissspuren an Seeigeln, *Arbeit, Palaeont. Hannover* **15**:73-75.
Seilacher, A., 1979, Constructional morphology of sand dollars, *Paleobiology* **5**:191-221.
Sheehan, P. M., and Lesperance, P., 1978, Effects of predation on the population dynamics of a Devonian brachiopod, *J. Paleontol.* **52**:812-817.
Shulman, M. J., 1990, Aggression among sea urchins on Caribbean coral reefs, *J. Exper. Mar. Biol. Ecol.* **140**:197-208.
Smith, A. B., 1984, *Echinoid Paleobiology,* Allen & Unwin, London, 190 pp.
Smith, S. A., Thayer, C. W., and Brett, C. E., 1985, Predation in the Paleozoic: gastropod-like drillholes in Devonian brachiopods, *Science* **230**:1033-1035.
Sohl, N. F., 1967, Upper Cretaceous gastropod assemblages of the Western Interior of the United States, in *A Symposium on Paleoenvironments of the Cretaceous Seaway in the Western Interior, Geol. Soc. Am., Rocky Mtn. Sec., 20th Ann. Mtg.,* Colorado School of Mines, Golden, Colorado, pp.1-37.
Sohl, N. F., 1969, The fossil record of shell boring by snails, *Am. Zool.* **9**:725-734.
Tegner, M. J., and Levin, L. A., 1983, Spiny lobsters and sea urchins: analysis of a predator-prey interaction, *J. Exper. Mar. Biol. Ecol.* **73**:125-150.
Tertschnig, W. P., 1988, Predation on the sea urchin *Paracentrotus lividus* by the spider crab *Maia crispata,* in *Actes du VI Séminarie International sur les Èchinodermes* (M. B. Régis, ed.), Fondation océanographique Ricard, Marseille, pp. 95-103.
Thies, D., 1985, Bißspuren an Seeigel-Gehäusen der Gattung *Echinocorys* Leske, 1778 aus dem Maastrichtium von Hemmor (NW-Deutschland), *Mitt. Geol.-Paläont. Inst. Univ. Hamburg* **59**:71-82.
Tyler, P. A., Gage, J. D., Paterson, G. J. L., and Rice, A. L., 1993, Dietary constraints on reproductive periodicity in two sympatric deep-sea astropectinid seastars, *Mar. Biol.* **115**:267-277.
Vermeij, G. J., 1977, The Mesozoic marine revolution: Evidence from snails, predators, and grazers, *Paleobiology* **3**:245-258.
Vermeij, G. J., 1980, Drilling predation of bivalves in Guam: some paleontological implications, *Malacologia* **19**:329-334.
Vermeij, G. J., 1983, Traces and trends of predation, with special reference to bivalved animals, *Palaeontology* **26**:455-465.

Vermeij, G. J., 1987, *Evolution and Escalation: An Ecological History of Life*, Princeton University Press, Princeton.

Vermeij, G. J., Zipser, E., and Dudley, E. C., 1980, Predation in time and space: peeling and drilling in terebrid gastropods, *Paleobiology* **6**:352-364.

Vermeij. G. J., Schindel, D. E., Zipser, E., 1981, Predation through geological time; evidence from gastropod shell repair, *Science* **214**:1024-1026.

Warén, A., 1980, Revision of the genera *Thyca, Stilifer, Scalenostoma, Mucronalia* and *Echineulima* (Mollusca, Prosobranchia, Eulimidae), *Zool. Scripta* **9**:187-210.

Warén, A., 1981, Eulimid gastropods parasitic on echinoderms in the New Zealand region, *N. Z. J. Zool.* **8**:313-324.

Warén, A., and Crossland, M. R., 1991, Revision of *Hypermastus* Pilsbry, 1899 and *Turveria* Berry, 1956 (Gastropoda: Prosobranchia: Eulimidae), two genera parasitic on sand dollars, *Rec. Austr. Mus.* **43**:85-112.

Warén, A., and Mifsud, C., 1990, *Nanobalcis*, a new eulimid genus (Prosobranchia) parasitic on cidaroid sea urchins with two new species, and comments on *Sabinella bonifaciae* (Nordsieck), *Boll. Malacol.* **26**:37-46.

Warén, A., and Moolenbeek, R., 1989, A eulimid gastropod, *Trochostilifer eucidaricola*, new species, parasitic on the pencil urchin *Eucidaris tribuloides* from the southern Caribbean, *Proc. Biol. Soc. Wash.* **102**:169-175.

Warén, A., Burch, B. L., and Burch, T. A., 1984, Description of 5 new species of Hawaiian (USA) Eulimidae, *Veliger* **26**:170-178.

Warén, A., Norris, D. R., and Tempelado, J. T., 1994, Description of four new eulimid gastropods parasitic on irregular sea urchins, *Veliger* **37**:141-154.

Watt, J., Siniff, D. B., and Estes, J. A., 2000, Inter-decadal patterns of population and dietary change in sea otters at Amchitka Island, Alaska, *Oecologia* **124**:289-298.

Wood, R., 1998, The ecological evolution of reefs, *Ann. Rev. Ecol. Syst.* **29**:179-206.

Woodcock, T. C., and Kelley, P. H., 2001, Predation on irregular echinoids (Eocene) from the Castle Hayne Limestone, Atlantic Coastal Plain (Southeastern North Carolina), *Geol. Soc. Am. Abstr. Progr.* **33**(2):A-15.

Wootton, J. T., 1995, Effects of birds on sea urchins and algae: A lower-intertidal trophic cascade, *Ecoscience* **2**:321-328.

Wootton, J. T., 1997, Estimates and tests of per capita interaction strength: Diet, abundance, and impact of intertidally foraging birds, *Ecol. Monogr.* **67**:45-64.

Zinsmeister, W. J., 1980, Observations on the predation of the clypeasteroid echinoid, *Monophoraster darwini* from the upper Miocene Enterrios Formation, Patagonia, Argentina, *J. Paleontol.* **54**:910-912.

Chapter 12

Predation of Fishes in the Fossil Record

JAMES MCALLISTER

1. Introduction 303
2. Origin of Fish Predation 304
   2.1. Introduction 304
   2.2. Origin of Fish Predation: Traditional Model.......................................305
   2.3. Embryological Evidence of the 1980's 308
   2.4. Recently Discovered Fossil Evidence 309
   2.5. Predation and Origin of Fishes: Summary Discussion.......................310
3. Evidence of Fish Predation from the Fossil Record...................................311
   3.1. Introduction 311
   3.2. Direct Evidence of Fish Predation 312
   3.3. Indirect Evidence of Fish Predation...316
   3.4. Diversity of Interpretations of Fossil Fish Predation..........................317
   3.5. Discussion of Evidence for Fossil Fish Predation...............................320
4. Summary..322
   References 323

1. Introduction

This review covers selected highlights of predation of fishes. Understanding this group is important. Approximately one-half of all named vertebrate species are fishes. Fishes are the original bauplan and evolutionary source for the rest of the vertebrates. Our evolutionary history as vertebrates is explainable only with an understanding of fishes, which in turn can only be fully understood in the context of predation. In the first part of this paper, I will review the most influential historical and current theories of early fish predation from the past 100 years. In the second part, I describe the specific evidence for fish predation in the fossil record.

The historical context of fish predation covers the early evolution of vertebrates. It would best be interpreted by observing the historical evidence directly from fossils.

JAMES MCALLISTER • Department of Natural Sciences, Dickinson State University, Dickinson, North Dakota 58601.

Predator-Prey Interactions in the Fossil Record, edited by Patricia H. Kelley, Michał Kowalewski, and Thor A. Hansen. Kluwer Academic/Plenum Publishers, New York, 2003.

However, the direct evidence (fossils) has been extremely limited until recently. The lack of fossils forced the first theories of early vertebrate evolution to be based on comparisons of modern organisms. I will emphasize the Garstang paedomorphic theory as the traditional model (section 2.2). Later, developmental biology would provide major insights and cause a shift in systematics and lifestyle interpretations (section 2.3). Today we have a variety of data, from molecular studies to new fossil finds (section 2.4), which are currently invigorating the field. This is reflected in the abundance of recent overviews; for example: Shimeld and Holland (2000), Zimmer (2000), Purnell (2002). A common theme through the literature is the switch from filter feeding to active predation. The interpretations of the timing and the morphological characteristics necessary for the feeding transition have radically changed over the last two decades.

The second part of this review is a description of the evidence of fish predation from the fossil record. There is both direct evidence (section 3.2) of predation (coprolites, traumatized body fossils) and indirect evidence (section 3.3), which is inferred from modern analogs. The diversity of interpretations based on fossil evidence (section 3.4) is surprisingly broad, including physical characteristics of the predator and interpretations of paleoecology. Some limitations and biases are then discussed (section 3.5).

2. Origin of Fish Predation

2.1. Introduction

The term predator conjures images of dangerous pursuing animals such as the great white shark portrayed in the movie *Jaws*. This anthropomorphic vision of terror is at the extremes of the characteristics that make a predator: selective unrelenting pursuit, large body size, ability to violently and quickly kill, and ability to devour organisms larger than its mouth. However, the full range of predators can include many organisms and body styles beyond these colloquial understandings. Mackenzie *et al.* (1998, p. 107) define predation as the consumption of all or part of another individual. The breadth of this definition can thus include herbivores, omnivores, parasites, unicellular or multicellular organisms. Therefore it is important to define clearly how terms are used.

I will modify slightly the scheme Mallatt (1985) used to separate the behaviors of fish predation. He first differentiated the major food acquisition behaviors. Suspension feeders take in large quantities of materials suspended in the water with limited ability to select items preferentially. Raptorial feeders take in individual particles and are highly selective. Deposit feeders ingest sediment to extract the necessary nutrients. The size of the food particle can be taken into account with the terms microphagous (ingesting particles less than 100 micrometers in diameter) and macrophagous (ingesting particles larger than 100 micrometers). Combining elements from these two sets provides a good simple starting point for description of the lifestyles of fishes. In addition to acquisition method and prey size, the size of the predator relative to the prey is also important. Some predators are limited to food items smaller than their mouth while others have the ability to reduce large prey to manageable size. The ability to eat food items larger than one's mouth is a notable achievement in the evolution of feeding strategies.

2.2. Origin of Fish Predation: Traditional Model

The traditional view of vertebrate origins was proposed by Garstang in 1928. The fossil evidence was very scanty and most speculation on origins of animals was conceived with information gathered from modern representatives that have at least half a billion years of independent evolution from the splitting of the phyla. Understanding vertebrate origins was quite a formidable task. Garstang was a marine biologist who was interested in larval forms. Based on his observations, he envisioned a low-metabolism sessile microphagous suspension feeder with a tunicate morphology as the precursor of fishes. Although the familiar sac-like tunicate adult is not similar in morphology to fishes, the mobile larva has a tadpole-like body suitable for dispersal. It was reasonable to hypothesize an invertebrate origin in the Precambrian by retention of important larval features (paedomorphosis of myomeres, notochord, pharyngeal gill slits) in the adult protovertebrate. This hypothesis has been propounded as the most likely of several competing hypotheses since that time.

2.2.1. Evidence for the Traditional Model

The traditional evidence to support Garstang's view of vertebrate origins includes embryological and morphological similarities. Development of vertebrates places us in the deuterostome group (Animalia whose second body opening forms the mouth and which have radial cleavage; Phyla Echinodermata, Hemichordata, Urochordata, and Cephalochordata; Fig. 1). The adults of these non-vertebrate groups appear too morphologically dissimilar to be closely related to vertebrates but the connecting clue is the larvae. The Echinodermata and Hemichordata have larvae with advanced

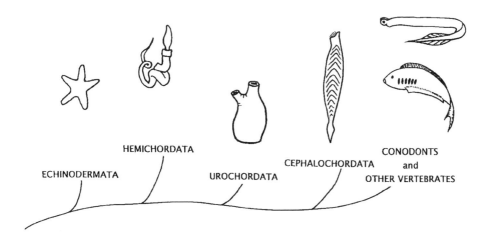

FIGURE 1. Phylogeny of vertebrate relatives.

invertebrate features such as bilateral symmetry, three germ layers, cilia around the oral entrance (the gut-tube being ventrally located and one-way in direction), cilia along the back for locomotion, and presence of an endostyle for mucus production. They are mobile suspension feeders. An interpretation can be created that follows a progressive transition from one group to another by retention of larval characters into the adult stage of the next group (paedomorphically retaining and combining those modifications which we now recognize as classic vertebrate characteristics).

2.2.2. Interpretation of Traditional Model Evidence

Conversion of a vertebrate from an invertebrate larval form is easy to envision as a series of hypothetical transitions based on the body forms of modern organisms. As the duration of the larval stages increased, the size of the organism in the paedomorphic morphology would also increase. If the body increased in size and elongated (while retaining larval characters), it is reasonable to expect that locomotory and feeding changes would compensate for increased demands. The cilia, with associated nerve network, would move further to the dorsum of the organism (prescient of dorsal nerve cord movement in chordates) to increase the effectiveness of the beating motion. The cilia would eventually need to be replaced in function by a segmented musculature along its length (myomeres) and a stiffened rod to resist shortening anteriorly-posteriorly. This would allow lateral undulating (fish-like) swimming. The suspension-feeding capabilities also would be increased (expanding the anterior gut-tube into a pharyngeal gill basket). The result would appear morphologically similar to the modern urochordate larvae. This hypothetical urochordate-like organism would in turn look similar to the modern cephalochordate if it also extended its larval stage and continued the size, locomotory and filter-feeding trends. The modern cephalochordate (lancet or amphioxus) adult is an elongate filter feeder with a notochord extending to the nose, has myomeres along its long axis, and a well-developed gill basket contained within an atrium (the water flows in the mouth and gill basket where the organic matter is strained, and the water passes into the atrium and then out the body via a separate opening, the atriopore). The cephalochordate larvae lack the atrium surrounding the pharynx (metapleural folds come down as a sheet to encase the atrium of the adult form). The cephalochordate larval pharynx morphology is very similar to that of the modern larval lamprey (the larval forms of these agnathan vertebrates are commonly referred to as ammocetes). The modern adult amphioxus and the ammocete have a similar lifestyle, buried in sediment, head regions projecting into the water column to filter feed. Modern representatives of these disparate groups provide a means of visualizing the transition through increasingly complex morphologies (of selected larval conditions) to a fish-like vertebrate.

Comparing characteristics of fishes with those of close non-vertebrate relatives is important in assessing lifestyle strategies. The non-vertebrates are generally small, possessing less segmentation, brain specialization, sensory organs, and possess a large pharyngeal gill structure utilizing cilia to move water. These characteristics describe typical sedentary filter feeders. The characteristics that distinguish fishes from the non-vertebrates begin with nervous system and locomotory innovations. The nervous system innovations (neural crest, described later) provide the sensory input and coordination necessary to orient an active organism in the environment and to become aware of and locate prey. The muscular pharynx and gill structures provide more effective processing of food in the water column. The segmented muscles and stiff notochord provide the ability to pursue prey actively. These abilities would likely provide more food to those taxa that possess them and allow increases in body size over time. Increases in body size would require better metabolism, tissue and organ differentiation, and circulation (kidney, respiration). This interpretation emphasizes interdependence of the new characters and development of vertebrate organs as the organisms evolved into larger active pursuers of food. The transition from microphagous to macrophagous feeding is not as clear in this interpretation (see Mallatt,1985) as the sedentary to active transition.

2.2.3. Discussion of Traditional Model

A hypothesis of origins that seems so clear and easily understood just cannot be without troubles. Gee (1996) provided a well-documented summary of Garstang's theory (as well as alternative theories) in historical context. The philosophical foundations (the adult form of the modern tunicate as ancestral body plan is an example of how body forms could change, not a specific indication of an ancestor) and biases (Garstang emphasized natural selective forces in larval evolution over recapitulation) during Garstang's time were described and then Gee added his considerations to the questions of origins. Gee (1996) cited recent molecular and genetic work indicating tunicates may be primitively mobile and segmented (the sessile form is derived and unusual for urochordates), and also concluded that paedomorphosis may be overemphasized in the transition to the vertebrate morphology. The traditional hypothesis is an innovative concept; however, the evidential foundation it was constructed on does not provide easily tested deductions concerning the transitions. The lack of direct evidence and reliance on modern representatives to create a hypothesis of events that occurred over one-half billion years ago will leave some specifics open to speculation and scenario building, no matter how insightful the main idea. If the hypothesis (deduction) cannot be easily tested, an alternative is to indirectly test the observations and inductions it is founded upon with new observations. New observations and inductions, relevant to the hypothesis, allow the creation of new testable deductions. They are useful if they can be shown to be both necessary but potentially contradictory to the original hypothesis. New information concerning the foundation of evidence, such as reinterpretations of primitive urochordate lifestyles, appears to be the manner in which the traditional model is shown to be weak.

The consequence of emphasizing modern sessile filter feeders in the traditional origin hypothesis is that the protovertebrate (transitional forms) and early vertebrates were regarded as less active and more similar to suspension feeders than may be warranted. In addition, the early fossilized fishes are dominated by flattened heavy-bodied specimens (osteostracans). They have a jawless ventral oral opening, dorsal sensory organs, and large gill pouches. These specimens had been described as bottom dwellers surviving as suspension-feeders, scavengers or detritus feeders. Other early fishes (anaspids, heterostracans) include fusiform bodies with jawless terminal mouths. They are generally interpreted as suspension feeders cruising through the water column. Stahl (1974, p. 41) provided the general perspective of protovertebrate lifestyle and the limitations of interpretation from that time:

> The common ancestor of amphioxus and the vertebrates, if it existed as
> this theory suggests, may have inhabited shallow places, filter feeding
> in quiet water, and swimming away from invertebrates that chanced to

disturb it. Beyond this supposition theorists cannot go. Although they have based their reasoning carefully upon the evidence of comparative anatomy and relied upon the phenomena of variation and natural selection as agents of evolutionary change, they cannot offer the fossil evidence that would substantiate their ideas.

Carroll (1988, p. 30) stated the traditional perspective of early vertebrate lifestyles (specifically the cyathaspid agnathans) and provided a glimpse of the changing attitudes of that time:

> Paleontologists have long presumed that these animals have fed by drawing in water and mud and filtering out the food particles in a manner comparable to amphioxus. The pharynx as a whole could not be expanded, and muscles associated with the gills or a velum must have pumped the water and mud. Northcutt and Gans (1983) doubt these assumptions, pointing out that these and other ostracoderms might have been capable of feeding on soft-bodied invertebrates. They question whether enough food could have been gained by filter feeding to support the metabolic processes of a relatively large and potentially active vertebrate.

2.3. Embryological Evidence of the 1980's

During the process of vertebrate development the embryo forms three germ layers. Each layer develops into predictable parts of the adult's organ systems (e.g., mesoderm forms muscles and bones). Northcutt and Gans (1983) were able to differentiate a fourth germ layer (mesectoderm or neural crest, a wandering tissue formed from the interaction of mesoderm and ectoderm). These cells travel throughout the developing embryo and become the peripheral nervous system; they also invade the lateral plate, which develops into the muscular pharynx and gill arches (and jaws), and they likely become the sensory placodes (lateral line, otic, olfactory, optic, and cranial nerves). Some of the most important characteristics of vertebrates seem to be dependent on the existence of neural crest tissue. These characters provided a basis for an alternate interpretation by Northcutt and Gans regarding the lifestyle of the early vertebrates. They considered predation to be important in the evolution of the early vertebrates. The effects of neural crest are most apparent on the head end of the organism and the new interpretations can be referred to as the new head model.

2.3.1. Discussion of Embryological Evidence

There is plenty of discussion, dissension, and alternate interpretations regarding the morphology and lifestyles leading to the vertebrate transition. The following discussion minimizes the variety and details to include only two alternatives from Northcutt and Gans (1983). Mallatt (1985) summarized differences between his interpretations of lifestyle and those of Northcutt and Gans. He envisioned protovertebrate larvae as raptorial, based on similarity of feeding mode to that of larvae of modern fishes (lamprey larvae and deuterostome non-vertebrate larvae are suspension feeders) or possibly they used a combination of raptorial and suspension feeding. Northcutt and Gans (1983) considered the protovertebrate larvae as mobile suspension feeders and the adults to be benthic predators retaining the features characteristic of fishes: segmented body muscles, elaborated brain, and special sense organs. Mallatt (1985, p. 66) considered both alternatives to have weaknesses. The suspension-feeder model requires an explanation for the need of vertebrate sense organs (later he indicated their usefulness to find food patches and avoid predators) whereas the predator model lacks the backing of the fossil record. Mallatt (1985) considered the earliest preserved vertebrates known at that time to be relatively inactive, small-eyed, and benthic; the bias of a poor fossil record obscures easy interpretation, but Mallatt chose suspension feeding.

Mallatt is also an important voice in the question of the origin of jaws. The jaws of gnathostomes, according to Mallatt (1997), are derived from the gill arches supporting the pharynx as expected, but the driving force of change was to increase the pumping of water through the gills. The origin of jaws was in response not to feeding pressures to grasp and bite, but rather efficiency of ventilation. This can be referred to as the new mouth or ventilatory model.

Jollie (1982) presented a notable twist in the prevertebrate interpretations. He proposed that the ancestral condition of the Urochordates, Cephalochordates, and vertebrates began with a larval form similar to that of echinoderms and hemichordates. The acquisition of chordate features led to an active, cephalized, large-mouthed organism with a tendency to predatory behavior. The vertebrate trends continued to macrophagous raptorial vertebrates or, in the case of urochordates and cephalochordates, a reversion to suspension feeding.

2.4. Recently Discovered Fossil Evidence

Since the work of Northcutt and Gans (1983), many significant fossils have been discovered that help the reinterpretation of the lifestyles of early fishes and vertebrate relatives. Conodonts (previously known only from assemblages of phosphatic tooth-like structures) were systematically problematic and were suspected to be suspension feeders. Beginning with the discovery of the "conodont animal" fossils (Briggs *et al.,* 1983), reevaluation of systematic positions and lifestyles could occur. The tooth elements have been shown to have wear patterns consistent with macrophagy—grasping, slicing, and crushing prey (Purnell and Donoghue, 1997). Evidence of paired eyes, otic capsules, possible branchial structures (Aldridge and Purnell, 1996), and elongate bodies with segmented muscles all indicate well-developed sensory and locomotory abilities in these small animals. Today conodonts are considered to be vertebrates and predators.

Wilson and Caldwell (1993, 1998) described a new order of fork-tailed fossil agnathans from the Silurian and Devonian, the Furcacaudiformes. The compressed bodies, large tails, and large eyes fit the expectation of an active predator. There is even sediment in the gut-tube, suggesting the presence of a stomach (a likely expectation for a predator). However, the sediment infill was interpreted by Wilson and Caldwell to be indicative of either suspension or detritus feeding. Donoghue and Smith (2001) also supported an interpretation of detritus feeding. There is little direct evidence of gut contents of the earliest fossil fishes beyond these examples.

Discoveries of other early vertebrates and relatives (*Pikaia, Cathaymyrus,* and *Hailouella* — cephalochordate relatives; *Myllokumingia* — chordate relative; *Haikouichthys* — Cambrian agnathan) provide new perspectives similar to those for the conodonts. The characteristics of fusiform bodies, myomeres, branchial gill development, neural development, and emphasis on sensory organs all help reinforce the interpretation of early fishes and relatives as active, aware (relative to other organisms of the time), and likely candidates to be predators.

The early fossil vertebrates are different enough from modern fishes that lifestyles are controversial. Their body forms do not force the conclusion that they must have been the top predators of their day. There is little direct evidence of their feeding habits. Purnell (2002) examined fossil heterostracans (early jawless fishes), noting evidence to support interpretations of feeding behaviors. He focused on microscopic wear patterns of plates and oral denticles that line the mouth. Although the function of the denticles cannot be determined, they are interpreted as incompatible with macrophagy due to their forward orientation and delicateness. The lack of expected wear patterns on the oral plates also provided negative evidence for functions in deposit feeding. Purnell concluded these primitive vertebrates were microphagous suspension feeders.

2.4.1. Discussion of Fossil Evidence

The literature reveals a variety of lifestyles advocated for transitional forms as well as the known fossil vertebrates. Generally the expectation is a transition from sessile filter feeding to predation, based especially on the expected invertebrate origins and developmental biology. However, the direct evidence from the fossil jawless fishes appears to contradict the expectation that sensory-rich active vertebrates should reflect the epitome of predation. Purnell (2002, pg. 87) has pondered this dilemma:

> Specifically, the evidence from extant jawless vertebrates and conodonts supports the hypothesis that the most primitive vertebrates were predatory and that a shift occurred at the origin of the clade ...
> but the heterostracans, a more derived group [of fossil jawless vertebrates] ... were not predatory.

Purnell interpreted this evidence to support the hypothesis that predation and microphagous suspension feeding are plesiomorphic to herbivory (herbivores are specialized and evolved from the other two more primitive groups). In the case of vertebrates, he suggested that predation is also plesiomorphic to suspension feeding, resolving the dilemma.

2.5. Predation and Origin of Fishes: Summary Discussion

The origin of vertebrates and the role of predation are currently under intense scrutiny and a clear consensus has not been reached. The invigoration comes from many new lines of evidence in brain development and embryology (neural crest), studies on origin of bone, molecular and genetic studies with implications for segmentation and brain development, studies of Recent vertebrate relatives, and new fossil finds (Zimmer, 2000; Shimeld and Holland, 2000). The traditional model provided the general placement of vertebrates within the diversity of life by employing data primarily from classic embryology and modern analogs. More data from developmental biology redefined the morphological characters necessary to be a vertebrate. The data also allowed refinement of the interpretations of the transitional lifestyles for the organism with a new type of sensory-rich head and segmented muscular body. Recent fossil finds provide new materials (including a new group for the vertebrates, the conodonts), which generally reinforce interpretations of active, sensory aware organisms that likely had predatory lifestyles.

There are many details and questions waiting for more data, study, and hopefully answers. How will correlation of the genetic data fit in with the morphological clues (When can we assume a particular homeobox gene became functional based on fossil evidence?) Will the molecular and genetic data provide a better definition of the chordate, craniate, or vertebrate groups than morphological or developmental characters (such as notochords or neural crest)? The early evolution of the hypothetical protovertebrate (and hence the entire vertebrate group) seems to have been determined by modifications preadapted for, if not specifically evolved for, a predatory lifestyle, but what is the operative definition for predation? Are active pursuers of plankton predators, and what is the size limit of raptorial predator to prey (consider the life history transition of paddlefish — young *Polyodon* pursue individual zooplankton, while adults cruise the water column selectively suspension feeding)? What are the constraints of size to predation and do jawless vertebrates have size limits for predation? Possibly the protovertebrates and the fossil jawless fishes were effective predators at small sizes but not at larger sizes.

The current data require time to assess and integrate. The lack of fossils and other suitable data leaves weaknesses in the interpretations. For example, the traditional model of vertebrate origins provides answers regarding invertebrate relations, mode of life, and process of evolution. This coverage is tremendous, considering the amount of available data. It is desirable to extend the interpretation of the data to provide as complete a hypothesis as possible. However, this extension leads to major controversies concerning details, such as the type and timing of predation in the transition to fishes. The new data of the past 20 years have upset many details of the traditional model and more data will be necessary before a clear consensus can be adopted for the next round of testing. However, when the new consensus is reached, the role of predation will certainly still be central in the interpretations.

3. Evidence of Fish Predation from the Fossil Record

3.1. Introduction

Predation is a behavior of an organism and is not normally known with certainty for extinct organisms. The behavior of fossil vertebrates is not easy to recognize or describe in comparison to interpretations based on most organic remains (body fossils such as bones). Knowledge of fossil behaviors requires more interpretation. (Is this a decayed, scavenged or digested fish? What fish created this coprolite? Does the morphology of the head relate to feeding, respiration, sensory requirements, or all of these factors?) To examine the evidence, interpretations, and problems of fossil predation requires a classification. The fossil evidence for fish predation is divisible into direct and indirect categories (actual vs. interpretative evidence, after Boucot, 1990). Indirect evidence involves comparison of a predator or prey to a modern analog (if one exists) to deduce an expected behavior. An organism with morphological feature X is assumed to exhibit behavior Y (e.g., durophagous teeth indicate crushing behavior). Direct evidence of a particular activity allows behavior to be interpreted from physical remains that have changed due to the behavior (bite marks and coprolites represent transformed prey items).

Coprolitic material is the most common direct evidence. The historical review of coprolite studies parallels the development of most scientific studies. The early literature is rare, scattered and embedded as asides in larger works, and is generally descriptive. As the field has developed, the literature on coprolites has become more common, in-depth, and comprehensive. Buckland (1829) was the first to use the term coprolite to describe fossilized fecal material from the Lias of Lyme Regis of England. He used the similarity of morphology to modern analogs and the presence of included remains to determine the identity as coprolites. Since that time other workers noted coprolitic material and comprehensive lists of these references were compiled by Amstutz (1958) and Hantzschel et al. (1968). Most references were descriptive; however, Amstutz provided criteria for the discrimination of coprolites from structures of non-organic origin (extrusion form, flattening on depositional side, correspondence to gut shape, etc.). This was a noteworthy endeavor with minor new criteria subsequently added (Wilson, 1987; McAllister, 1989; Hunt et al., 1994). After Amstutz there has been an increasing emphasis on the recognition, biological processes, and interpretation of behaviors represented by coprolitic material, rather than pure morphological description. Recent work can also be characterized by increased sample sizes and comparisons covering a diversity of selected stratigraphic units or localities.

Research on fish coprolitic matter that provides important recognition criteria or predator-prey insights includes the following studies: Zangerl and Richardson (1963) described and interpreted different types of coprolitic material from Pennsylvanian black shales of Indiana; Williams (1972) described spiral coprolites and provided a comprehensive literature review; Zidek (1980) described coprolites from the Pennsylvanian Heath Shale of Montana and estimated prey size from coprolite inclusions; Elder (1985) expanded the work on the Indiana black shale coprolites and emphasized detailed recognition criteria; Wilson (1987) described coprolitic material from an Eocene lake deposit from Canada and provided additional insight into recognition criteria; McAllister (1989) described material from the Pennsylvanian aged Hamilton quarry in Kansas with commentary on food preferences and predator-prey interactions; Williams (1990) described Devonian Cleveland Shale material that provided important insight into predator-prey interactions; Viohl (1990) tabulated and discussed numerous fossil examples of fish predation on other fishes; and McAllister (1996) described material from the Devonian Escuminac Formation and discussed predator-prey interactions. Poplin (1986) discussed many of these terms with French translations. Additional references to coprolitic matter of fishes are found within all these papers.

3.2. Direct Evidence of Fish Predation

Direct evidence can be classified as predator or prey dominated. For predator-dominated fossils, the evidence is changed by the life processes of the predator (prey morphology is modified by mechanical and chemical digestive processes of the predator, making identification of the prey difficult — e.g., a coprolite). For prey-dominated fossils, the identifying morphological characteristics of the living prey item are retained (e.g., a swallowed fish intact within the predator fish).

The dichotomy between predator- and prey-dominated direct evidence is gradational but can be generally separated at the pyloric sphincter. In fishes the prey is typically swallowed whole or in large pieces. The maximum change from recognizable prey to a wet chyme (paste) occurs in the stomach. The pyloric valve (sphincter) serves as a gateway into the intestine. The prey cannot pass into the intestine if it has not been reduced to a chyme (watery paste) that the churning stomach can squirt through the pyloric valve. Once past the valve, large prey will no longer be morphologically recognizable as a whole organism. The morphology of the prey is now clearly predator dominated.

The fossil record contains examples of gradations from whole non-predated fishes to coprolites. A basic classification includes generally unmodified specimens; ejecta: bite marked, mouthed (repeatedly manipulated orally) and scavenged evidence; regurgitations (taken into the stomach and expelled via the oral cavity); cololites (fossilized materials contained within the digestive tract), intact swallowed specimens, digested masses; and coprolites (the expelled non-digestibles, fossilized feces).

The degree of change produced by these processes on the fossil remains can be illustrated on a ternary diagram (Fig. 2). There is a continuum between prey-predator dominated but most recognized evidence is reasonably easy to place toward either extreme. Coprolites, cololites, and regurgitates are highly processed while most bitten and "fish in fish" examples are not processed very much. The third angle of the diagram is decay processes. The processes of decay can overprint the initially deposited evidence and confound the interpretation.

3.2.1. Terms and Definitions Applied to Direct Evidence

Defining the terms used in any discussion of predation is particularly important as different groups of workers have their traditional terms and usage. For example, invertebrate paleontologists and ornithologists use the term pellet for small coprolites and relatively large regurgitates respectively, but the term is rarely used in

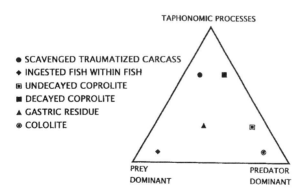

FIGURE 2. Ternary diagram illustrating the relationship of taphonomic processes, predator-dominated, and prey-dominated conditions for coprolitic materials.

paleoichthyology. Commonly used terms describing fossil fish remains and coprolitic material follows (see also Fig. 3).

(1) Scavenged. A dead, dying or disabled organism that cannot effectively take action against predation (e.g., a carcass scavenged by a hagfish or crabs).

(2) Traumatized. A prey item that received a fatal bite, slash or gouge but whose preserved remains are not ingested; the carcass is essentially recognizable and largely intact (e.g., bite-marked prey; tail bitten off by predator but the rest of the fish is intact).

(3) Regurgitations. Prey that was captured and expelled via the mouth by the predator. Includes (a) Ejecta. Material expelled from the mouth relatively quickly with mechanical breakage during manipulation but no real chemical processing or reduction of parts into unrecognizable paste. (b) Gastric residue. Material expelled from the stomach. Both mechanical and chemical digestion have occurred. There is chyme present.

(4) Ingested. Intact *in situ* contents of the esophagus and stomach; prey-dominated morphology (recognizable; e.g., "fish within fish").

(5) Cololite. In situ fossilized gut contents, endocasts. Predator-dominated morphology; normally past the pyloric valve.

(6) Coprolite. Fecal material. Expelled via the anus and fossilized.

3.2.2. Recognition of Direct Evidence

Paleoecologic and taphonomic studies have provided important cautions and clues in the interpretation of the direct evidence of predation. Especially important is distinguishing prey items (within the continuum of processing) and non-predated decayed organisms. For example, a fish with a missing head may have been bitten, scavenged, or may have decayed with the head disarticulating and separating from the body.

Elder (1985) studied the taphonomy of fishes. She described the most important principles that govern the processes that occur between death and burial of fishes and

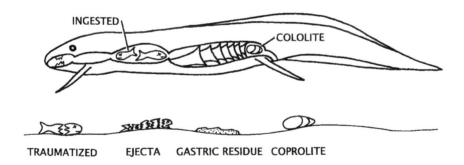

FIGURE 3. Sketch of *Protopterus* (Dipnoi) after a feeding frenzy to illustrate the sites and general conditions of material with potential to be fossilized.

conducted numerous experiments to establish the effects of exposure time, temperature, pressure, buoyancy, currents, bacterial decomposition and scavengers. Modern fishes were shown to decay, explode, and disarticulate in particular patterns and sequences depending on environmental conditions. For example, a floating carcass will lose the lower jaws and fin lepidotrichia first. Then bones from the skull to pectoral girdle will fall off with scales; and the body will break along weak areas (between head and vertebral column, at the caudal peduncle). The middle of the body carcass will be the last part floating, having dropped isolated body parts and bones in sequence along the floor of the sea. Under cold or high pressure conditions the carcass may not float and the bottom currents will disassociate the elements based on hydrodynamic principles (density, surface area, projection of flanges into the water column). Under these conditions bones will group together, with heavy, low profile bones remaining close to the resting place of the carcass. Light, high surface area bones will scatter farther away in the direction of the current.

The importance of taphonomic information in this context is to prevent misidentification or corroborate a determination. A fish specimen without a head or tail may not be due to predation but rather could be the result of normal decay. If a fossil has been truncated in the midline with no head bones missing and no anterior scales disturbed, then it does not show the expected process of decay. Predator induced trauma may be the best interpretation. Elder's (1985) experiments with scavengers also provides guidance to distinguish other taphonomic processes from scavenging. Scavenger activity does not follow hydrodynamic principles or decay patterns. Scavengers disturb elements in proportion to their original proximity to soft tissues of high nutritive value. The distance of scattering of these elements will depend on the size and feeding mode of the scavenger but the directionality of scatter will be randomly away from the body. While it is expected that two paired bones will disassociate from the carcass and be carried in the same direction by currents, a scavenged carcass could likely have those two elements at different distances and directions from the carcass. Schwimmer *et al.* (1997) noted scavenged carcasses with toothmarks and teeth embedded in the skeletal material. The marks were unhealed and the environmental conditions and presence of teeth concentrated in and around the carcasses indicated shark scavenging.

To distinguish a regurgitate from decayed carcasses and coprolites can be difficult. Wilson (1987) distinguished coprolitic material by noting the presence of digestive effects (dissolution of bone, broken bones, and the concentration and arrangement of bones — closely packed but disarticulated, consistent with passage through the gut). He also indicated the difficulty in distinguishing loose fish fecal material from ejecta of waterfowl. McAllister (1996) suggested distinguishing loose materials from compact fecal masses by their relationships to the sediment. Loose material settling to the bottom will spread over and merge with the substrate. A solid mass will create an impact depression. Even if the mass decayed before complete burial, the lower surface of the mass will still be disturbed to a greater degree. Once it is established that the specimen is unlikely to be decayed or scavenged material, the criteria to distinguish regurgitates could include: presence of gastric chyme, little to minor digestive etching, bones broken and disarticulated in a manner inconsistent with decay processes but expected with feeding, presence of indigestible materials and inclusions (spines) of a size unexpected to travel through the pyloric sphincter, and large inclusions extending beyond the border of the specimen. Hattin (1996) identified a probable regurgitate, which had irregular morphology, mixed mineralogy, some articulated cirriped plates and numerous coccoliths with little digestive etching.

Ingested fish specimens are actually fairly common in the fossil record. It is not difficult to recognize the condition, although care must be taken to be certain the prey is actually contained within the predator and does not underlie the predator. These specimens are generally prey-dominant in morphology, as the digestive processes have not fully had their effects. They are often readily identifiable, have less etching, and less disarticulation than cololites. The most famous example is the "fish within a fish" of George H. Sternberg. In 1952 he excavated from Gove County, Kansas, a 14-foot *Xiphactinus audax* containing a *Gillicus arcuatus* which extends from the pectoral fin almost to the anal fin. It now is displayed at the Sternberg Memorial Museum at Fort Hays, Kansas. Williams (1990) tabulated and described many specimens from the Devonian aged Cleveland Shale. Viohl (1990) compiled tables of references to fishes containing prey worldwide. Forty-nine examples came from the Solnhofen lithographic limestone of Germany and were the emphasis of his research. Of special interest is the cause of death. While there are modern analogs and fossil examples of fishes which died from their gluttony (suffocation by blocked gills, piercing of body by prey spines), Viohl (1990) proposed that the Solnhofen Limestone fishes likely were victims of sudden environmentally induced death (rapid changes in salinity or oxygen). If they had not died they would have completed the digestion of the prey.

Cololites are also easy to recognize, as they are contained within the body of the predator. They are predator dominated in morphology. The bones and scales are disarticulated and contained in a ground mass which often follows the contours of the intestines. Seilacher *et al.* (2001) studied the Miocene aged sideritic "coprolites" of Washington, and demonstrated the cololitic nature of these specimens. The criteria cited to establish a cololitic origin included both incompatibilities with excrements ("it cannot be coprolitic because ...") and concordance with cololitic processes ("it is cololitic because ...."). The longitudinal grooves are unlike anal squeeze striations but correspond to *taeniae coli* muscle contractions and other ridges and grooves of the internal intestinal surface. The coiling is unlike excretory twists but does reflect the meanders of the intestinal turns. The pinching of coprolites typically occurs at the end last extruded but the cololites show pinching at both ends consistent with peristaltic contractions along the intestine. Finally, well-preserved specimens can be found which have pinching and seem to be continuous in series with potential connections. These are interpreted as intestinal fillings which had relaxed and contracted areas. They have also identified one specimen which has the form of a cecum.

3.3. Indirect Evidence of Fish Predation

Indirect fossil evidence provides additional clues to possible predatory behaviors. The two main categories of predatory inference are derived from morphology of the locomotory adaptations and morphology of feeding adaptations. Comparison to the most similar modern analogs will increase the reliability of the interpretations; reliability decreases with increasing novelty and unfamiliarity of the morphology (reliability of interpretations is especially low in extinct Paleozoic Era fishes with no modern equivalents).

Modern ichthyologists can distinguish and classify basic fish body shapes and predict their general life habits. Of course, as with most classification schemes, there are transitional forms. Also, since fish are rather opportunistic, they will display unexpected feeding behaviors occasionally; thus some chaos is expected. Moyle and Cech (1982) summarized six broad categories of fish morphology and related them to life characteristics. For example, Rover-predators have extremely streamlined bodies, pointed heads, narrow caudal peduncles, forked tails and terminal mouths (e.g., *Xiphias gladius*, swordfish). This morphology is well adapted for fast endurance cruising in pursuit of prey. Lie-in-wait predators are torpedo-shaped with dorsal and anal fins posteriorly positioned to add thrust to the caudal fin. Their heads are often flat with wide terminal mouths (e.g., *Esox lucius*, pike). These fishes remain in cover and with a burst of speed lunge out to capture passing fish prey.

Likewise the morphology of the teeth, location of the mouth, mobility of the muscles, bones and ligaments will all help predict the possible feeding behaviors of the fish. A fish which can only open a slightly flexible mouth at the hinge joint (*Moythomasia*, an extinct heavy scaled palaeoniscoid) would ram feed (push itself through the water with dropped jaws and engulf all slower objects ahead of itself), while a flexible mouth that can be extended outward from the face while the opercula flare would create negative pressure in the oral cavity for faster more efficient suction feeding (*Perca*, perch). Teeth can occur in a wide range of morphologies. Each shape provides a particular feeding adaptation. For example, sharp elongated teeth (*Lepisosteus*, gar) pierce prey; flat heavy-bodied durophagous teeth (*Neoceratodus*, lungfish) crush hard-shelled organisms; and flat, wide, sharp-edged teeth with serrations (*Carcharodon*, white shark; *Isistius*, cookie-cutter shark) tear chunks off carcasses. The ability to dismember prey was important for the radiation of fishes into new niches. If the predator is required to swallow the entire prey, then the size of the prey is limited to the maximum gape of the predator. Slicing teeth allowed fishes to exploit prey larger than themselves with new behaviors. Based on tooth morphology, Williams (1990) described the predatory behavior in Devonian aged Cleveland Shale sharks, including the biting and head shaking expected with slicing teeth.

3.4. Diversity of Interpretations of Fossil Fish Predation

The available evidence allows a great diversity of interpretations. The interpretations could be categorized in many ways and one example could cover many categories. I will consider a few interpretations found in the literature and place them somewhat arbitrarily under the headings of physical and ecological interpretations, for convenience.

3.4.1. Interpretations of Physical Characteristics of Predator and Prey

Physical characterizations of the predator and prey have been deduced from a variety of direct evidence. The fish-in-fish evidence provides data that can be directly measured, while more obtuse means can be employed for less obvious evidence. Everhart and Everhart (1998) measured the spacing of tooth marks and a tooth (attributed to the Cretaceous aged shark *Cretoxyrhina mantelli*) embedded in a mosasaur to estimate predator size (5.5 meters). McAllister (1996) found a correlation between width and length of coprolites from the Devonian aged Escuminac Formation of Canada. The width of the coprolite was interpreted to vary with the size of the anal sphincter and predator (with presumably large confidence limits). Although sample sizes were small, the acanthodian prey occurred in the smaller coprolites, paleoniscoids occurred in the middle range, and conchostracans occurred at both ends of the range. This could suggest the predators were of different species, or sizes with differing feeding preferences (resource partitioning).

Cololites provide direct evidence of gut morphology of the predator. McAllister (1987) documented the distribution of the intestinal morphologies and referenced many examples of cololites from the literature. One interesting example is the sediment-packed intestines of *Bothriolepis*, first studied by Denison (1941). The large intestines of primitive fishes have many elaborate morphologies, a common type resembling a tightly wound spiral staircase encased in an intestinal tube. The *Bothriolepis* intestines consist of six or seven highly angled spirals that fill the abdomen of the fish. Spiral coprolites also provide morphological information of the predator. Wilson and Caldwell (1993) documented cololitic infill (endocasts) within furcaudiform agnathans and distinguished the form of a stomach and short intestine. McAllister (1985) documented a modern analog for the creation of spiral coprolites. The modern catshark, *Scyliorhinus canicula*, apparently produces spiral fecal masses by continuing the spiralling rotation of the fecal ribbon after the ribbon has passed through the valvular intestine into a widened open area at the posterior end of the intestine. The spiral fecal mass can then be expelled intact from the body. The presence of spiral coprolites provides evidence of spiral valves from primitive fishes. Most spiral coprolites are found in Paleozoic and Mesozoic aged strata, reflecting the abundance of primitive fishes prior to the domination of the modern fishes (Actinopterygii).

3.4.2. Ecological Interpretations

Connections made between predator and prey indicate relative trophic levels. Identifications can be made with teeth or tooth marks left by predators or scavengers on prey remains, cololite, and coprolite inclusions. The most compelling associations are "fish within fish" where the observer interprets a frozen behavior (and inevitably reflects upon progressively philosophical meanings: How can a fish swallow prey half its own body length? Does this represent the rewards of gluttony? Is this a naturalistic representation of the rat race?) At the ecological level these specimens provide trophic links. However, as fishes tend to be opportunistic in their feeding, multiple specimens of the same species provide more confidence in suggesting meaningful associations. Correlation with specific morphological adaptations also solidifies the link as a common predation event (possession of elongate gill rakers would corroborate the association of zooplankton in a single example of a cololite specimen). Of particular interest are two cases of multiple trophic association in fishes. Viohl (1990) cited in a footnote an example of an Eocene *Knightia* within a *Diplomystus* within a *Lepisosteus*. Williams (1990) describes the Devonian aged specimen CMNH 5870 as a conodont eaten by a small shark eaten by a large shark. The size of the sharks as well as the occurrence of conodonts in four other small cladodont sharks was suggestive of resource partitioning of small soft prey based on the small size of the shark. Williams (1990, p. 287) later summed up one basic trophic observation that may be the limit of interpretation for many situations: "When it comes to the predators of the Cleveland Shale, it seems the big fish ate the little ones."

Direct evidence of predator/prey associations can be combined with the indirect morphological features to hypothesize the behaviors and trophic levels of fossil fishes from a stratigraphic unit. The most famous site interpreted this way is the Eocene Green River Formation (Grande, 1980). The following quote provides an incomplete example of his evidence and interpretations of the life of the most commonly excavated fish fossil (*Knightia*) in the world (Grande 1980:85):

> *Diplomystus* has the body form, size, and upturned mouth typical of a surface feeder, and fed on smaller surface-dwelling fish such as *Knightia*. Several specimens of *Diplomystus* have been found with *Knightia* fossilized in their stomachs and mouths.... The more slender-bodied *Knightia* seems to have been a primary to secondary consumer, probably feeding on algae, diatoms, ostracods, insects, and, rarely smaller fish. A low position on the food chain is probably one reason for its abundance in the Eocene fish population.... *Knightia* has been found in the mouths or stomachs of *Diplomystus*, *Lepisosteus*, *Amphiplaga*, *Mioplosus*, *Phareodus*, *Amia*, and *Astephus*.

Grande (1980) was able to construct a table of 17 fish genera and classify them by trophic adaptation. The trophic categories included major predator, predator of small fish, predator of insects, predator of decapods, molluscivore, zooplanktonivore and herbivore. Richter and Baszio (2001) used coprolitic material to trace the trophic connections in the Eocene Messel locality in Germany. The faunal list and fish coprolites provided information that allowed five levels of consumers to be identified. They emphasized two points. Unlike a theoretical foodweb constructed from a faunal list, coprolitic material provides direct evidence of actual predation. Also, because there will always be taphonomic loss, the food web can never be completely known.

From stomach contents, tooth and body morphology, and environmental reconstructions, Viohl (1990) was able to hypothesize behavior and lifestyle of several groups of fishes. For example, *Tharsis dubius* is a small planktonivore which is a common prey item for the larger fishes of the fauna. Possibly they schooled because several specimens occur in a single predator. Williams (1990) also suggested this possibility for three paleoniscoid fish that were all swallowed tail first by a Cleveland Shale shark (*Cladoselache*, CMNH 8114).

These studies examined direct and indirect evidence from both trace and body fossils. The ecological relationships of the time and place studied are more comprehensive than for previous studies. If enough faunal studies are completed then geographic and historical changes can be determined. Olson (1952) compared Permian sites from Oklahoma. He constructed trophic webs for permanent and seasonally ephemeral aquatic localities. The data, taphonomy and paleoecology of that time did not allow for more than the basic links to be established, but the differences among the localities were apparent. Taken to the extreme, large-scale studies of numerous localities covering ecological niches, extinctions, and species diversity across vast time scales (Sepkoski, 1981) provide a picture of changing faunal dominance. For example: the nektonic predatory niche was dominated in the Cambrian by nautiloids and ammonoids; then ancient jawed fishes entered, diversified and radiated throughout that niche in the Silurian. The later evolution of the actinopterygians increased the feeding and locomotory efficiencies with the subsequent domination of the nektonic predatory realm by the modern ray-finned fishes. During the Mesozoic and Cenozoic, reptiles and mammals, respectively, radiated into the aquatic predatory niche, usually at the larger size limits of vertebrate life. The big picture will continue to develop and become more finely focused with the increase in the type and quality of data upon which it is founded (Alroy *et al.*, 2001).

3.5. Discussion of Evidence for Fossil Fish Predation

Predation is a behavior of an organism. To discover behaviors in the fossil record at first seems a hopeless endeavor but trace fossils and morphological interpretations using modern analogs provide clues to behavior of extinct organisms. There is a surprisingly wide variety of evidence to interpret predation behaviors in fishes. However, the evidence has limitations and needs to be carefully interpreted.

3.5.1. Recognition and Classification of Evidence

The identification of predation items in the fossil record is difficult. It is important to articulate the reasoning process, including both the rationale for the identification as well as reasons for eliminating other alternatives. For example, Elder (1985) reinterpreted examples of regurgitations identified by Zangerl and Richardson (1963) as decayed and scavenged carcasses and coprolites. McAllister (1996) was able to document and refine the criteria he used to recognize regurgitates from the specific critiques of Elder. The criteria must be articulated for others to reproduce the results of the study.

Seilaicher *et al.* (2001) questioned if cololites should be considered trace fossils. Coprolites are considered trace fossil but are extruded from the body. Cololites are within the body and could be considered more like body fossils. In my opinion, cololites should not be considered body fossils. They are a result of the behavior and physiological processes of an organism (fitting a trace fossil expectation). Additionally they do not exist as a part of the predator but rather as material that passes through the body lumen. From oral to anal opening, the gut tube has a continuous lumen, which is technically outside the organism although surrounded by it. Material does not become part of the predator until it crosses the intestinal wall during absorption. Cololites are predator-dominated materials which preserve an internal cast of the predator; they are in this manner very similar to tracks. Another important distinction, provided by Seilacher *et al.* (2001), is that cololites are not static and distinct like body fossils. They are ephemeral within the organism, changing shape, and are discharged in relatively short time periods. I conclude that cololites are trace fossils dependent on process and behavior, reflecting morphology of a predator but not actually a part of the predator.

The terms describing the evidence of behavior, what to classify, and how to classify the evidence and behaviors are all still under discussion in the scientific community. Different observers would disagree when classifying borderline specimens. The classification of major trace fossil categories can be based on morphology, ichnofacies (grouping based on ecological associations) or behaviors. The purpose of the user determines the choice of classification scheme. Pickerill (1994) provides a brief history of invertebrate ichnotaxa, including criteria for differentiation of ethologic and non-ethologic (trace fossils and pseudofossils), and ichnotaxobases (what constitutes valid morphological features to designate ichnotaxa). This use of generic and species names is fraught with problems despite the tradition for invertebrates. Even co-authors may not agree philosophically on the utility of classifications (Hunt *et al.*, 1994, is an example of co-authors differing on the use of binomial nomenclature for trace fossils). Before a consensus of the taxonomy/classification of coprolitic or predation behavioral evidence can be stable, there will need to be significant progress in identification of purpose, appropriateness of recognition criteria, and limits of variability for categories (can a behavior have a definition based on a natural limit, similar to the reproductive definition of the biological species concept?) The classification scheme followed here is not perfect for all purposes and specimens. The classification of predatory behavior/coprolitic evidence could emphasize taxa, morphological characteristics, physiological processes, behavior or varying combinations of them all. Some of the categories merge without distinct boundaries (regurgitates, ejecta).

Boucot (1990) discussed many aspects of the interpretation of fossil behavior. He produced a reliability scale in which he classified the range of interpretations from 1 (directly observed behavior) to 7 (speculation). An example of category 1 is the frozen behavior represented by ants preserved *in copula*. The fish-within-fish fossils would be incontrovertible examples of predation behavior classified in category 1. The presence of durophagous dentition would imply crushing ability and an expected behavior of predation on hard shelled prey. However, Boucot notes that it is easy to over-interpret the evidence. The availability and similarity of modern analogs for comparison of both morphology and phylogenetic relationship is important. The level of reliability varies from 2B, for a behavioral interpretation based on strong morphological evidence in a fossil with a close living relative with very similar dentition, to 7 if the fossil is an extinct group with no modern analog and the interpretation is overly specific (predation on a particular shelled invertebrate to the exclusion of other organisms) with no other corroborating evidence. A researcher must know his/her limitations and Boucot (1990) has provided a scale (albeit somewhat subjective) to help assess the limitations.

3.5.2. Limitations of Physical Evidence

One taphonomic bias in the fossil record is for the preservation of hard parts. The preservation of soft invertebrate prey is unexpected. Even preservation of coprolitic material is rare. Most fecal material will be processed into a soft paste, which will have a variable amount of liquidity. The best preserved and most recognizable coprolites will be from fecal matter that was relatively dense, moisture poor, compact in morphology, and quickly buried to ensure protection from the elements and bioturbation. The indigestible parts of the prey that pass through the gut tube are mechanically and chemically macerated. Only the most durable parts remain identifiable and avoid becoming a paste. This type of evidence will never be totally complete, but will depend on numerous factors, including predator digestive processes, prey element durability, and the knowledge of the observer.

Knowledge of Paleozoic coprolites attributable to fishes is also limited. This reflects a lack of information from fossils as well as modern analogs. The confidence in the comparison of modern to fossil fishes is contingent on understanding the processes affecting the modern analog and determining how similar the analogs are to the fossils being studied. Modern fishes are varied and respond to similar conditions in a variety of ways. Elder (1985) noted many differences between modern fishes (actinopterygians);

for example, the buoyancy of carcasses differs for fishes with and without gas bladders. She also noted differences in decay between ganoid-covered fishes (fishes similar to our Paleozoic fossils) and actinopterygians.

3.5.3. Ecological Evidence

Relationships, correlations, and connections cannot be made without a foundation of data. Today there are increased efforts to understand the data limitations, provide clear criteria, understand processes, and specify terminology and classifications. These are important steps beyond pure data gathering; for example, the clear definition of the evidence helps prevent over-interpretation and provides later researchers opportunities to confirm the findings.

Increases in the basic data allow more interpretation, especially in comparisons through time. The change is reflected in the literature. The older literature of any scientific endeavor begins with basic descriptions. Initial studies of coprolitic material, and trace fossils in general, provided simple descriptions. The focus of later studies tends to emphasize behavioral or ecological information as part of faunal analyses. More comprehensive information of predation is derived from examination of whole fossil faunas using all available direct and indirect evidence to make interpretations (Grande, 1980; Wilson, 1987; Williams, 1990; Viohl, 1990). Understanding trace fossils, their behavior and as products of living organisms, is the primary goal. The direction of future research is indicated by the titles of such works as *Evolutionary Paleobiology of Behavior and Coevolution*, by Boucot (1990), and *Fossil Fishes as Living Animals*, edited by Mark-Kurik (1992).

Most workers would agree that large-scale studies covering vast time and sweeping trends will increase in frequency. Prothero (1998) summarized changes in the direction of paleoecological studies over the last 50 years. Early studies were concerned with local organisms and community descriptions and later studies included examples of evolutionary paleoecology. Evolutionary paleoecology differs from basic paleoecological studies in that the scope of study transcends the data set of any one time or place. Prominent examples would be studies of tiering (exploitation of new ecological niches above and below the sediment surface of the ocean through time), escalation (essentially arms races between predators and prey defenses through time), and diversification of organisms. However, although it is important and necessary to understand any discipline at the grand scale, it is also important to acknowledge the small-scale studies as the foundation. The more complete and rigorous the basic data, the less speculative and flawed will be the evolutionary paleobiological interpretations.

4. Summary

The interpretation of the origins of fish predation and the use of direct and indirect evidence to deduce predation has become increasingly sophisticated. Studies of the origins of fishes were initially based on relationships of modern phyla with the emphasis on filter-feeding lifestyles. Important breakthroughs in developmental biology, molecular studies, and new fossils have changed the interpretations of the origin of predation in fishes. For example, today the characters used to describe the entire vertebrate group have an increased emphasis on developmental knowledge of the head (neural crest, sensory placodes), and new fossils provide better interpretations for the acquisition and loss of predatory habits.

The interpretation of evidence of fish predation has also undergone significant change. Individual behaviors, physical characteristics of predator and prey, and ecological relationships have all been interpreted from direct and indirect evidence. The classification, criteria of recognition, confusion with taphonomic processes, and especially the limitations of the data are important to recognize (the interpretations are only as good as the data base). Today the range of research includes descriptions, faunal analysis, and studies that span the Phanerozoic (evolutionary paleobiology).

ACKNOWLEDGMENTS: I wish to sincerely thank the reviewers, Mark Wilson and J. D. Stewart, for their help in dramatically improving the manuscript. Lillian Crook of Dickinson State University facilitated the interlibrary loan process (often with expedited requests), without which I would have been unable to complete the manuscript. I also wish to thank John Kirby of Mansfield University for his help tracking down sources with exceptional bravery in the face of great physical pain.

References

Aldridge, R. J., and Purnell, M. A., 1996, The conodont controversies, *Trends Ecol. Evol.* **11**:463-468.
Alroy, J., Marshall, C. R., Bambach, R. K., Bezusko, K., Foote, M., Fursich F. T., Hansen, T. A., Holland, S. M., Ivany, L.C., Jablonski, D., Jacobs, D. K., Jones, D.C., Kasnik, M. A., Lidgard, S., Low, S., Miller, A. I., Novack-Gottshall, P. M., Olszewski, T. D., Patzkowsky, M. E., Raup, D. M., Roy, K., Sepkoski, J. J., Jr., Sommers, M. G., Wagner, P. J., Webber, A., 2001, Effects of sampling standardization on estimates of Phanerozoic marine diversification, *Proc. Natl. Acad. Sci.* **98**(11):6261.
Amstutz, G. C., 1958, Coprolites — a review of the literature and a study of specimens from southern Washington, *J. Sed. Petrol.* **28**:498-508.
Boucot, A. J., 1990, *Evolutionary Paleobiology of Behavior and Coevolution*, Elsevier, Amsterdam.
Briggs, D. E. G., Clarkson, E. N. K., and Aldridge, R. J., 1983, The conodont animal, *Lethaia* **16**:1-14.
Buckland, W., 1829. On the discovery of coprolites, or fossil feces, in the Lias at Lyme Regis, and in other formations, *Trans. Geol. Soc. London, ser.2*, **3**:223-236.
Carroll, R. L., 1988, *Vertebrate Paleontology and Evolution*, W. H. Freeman and Co., New York.
Denison, R. H., 1941, The soft anatomy of *Bothriolepis*, *J. Paleontol.* **15**:553-561.
Donoghue, P. C. J., and Smith, M. P., 2001, The anatomy of *Turina pagei* (Powrie), and the phylogenetic status of the Thelodonti, *Trans. R. Soc. Edinburgh, Earth Sci.* **92**(1):15-37.
Elder, R. L., 1985, Principles of aquatic taphonomy with examples from the fossil record. Unpublished Ph.D. Dissertation, University of Michigan.
Everhart, M. J., and Everhart, P. A., 1998, New data regarding the feeding habits of the extinct lamniform shark, *Cretoxyrhina mantelli*, from the Smoky Hill Chalk (upper Cretaceous) of western Kansas, *Kansas Acad. Sci. Trans.* **17**(Abstracts):33.
Garstang, W., 1928, The Morphology of the Tunicata, and its bearing on the phylogeny of the Chordata, *Quart. J. Microscopical Sci.* **72**:51-187.
Gee, H., 1996, *Before the Backbone: Views on the Origin of the Vvertebrates*, Chapman and Hall, London.
Grande, L., 1980, Paleontology of the Green River Formation, with a review of the fish fauna, *Bull. Geol. Surv. Wyoming* **63**:1-333.
Hantzschel, W., El-Baz, F., and Amstutz, G. C., 1968, Coprolites: An Annotated Bibliography, Geol. Soc. Am., Mem. 108, Boulder, Colorado.
Hattin, D. E., 1996, Fossilized regurgitate from Smoky Hill Member of Niobrara Chalk (Upper Cretaceous) of Kansas, USA, *Cretaceous Res.* **17**:443-450.
Hunt, A.P., Chin, K., and Lockley, M. G., 1994, The palaeobiology of vertebrate coprolites, in: *The Palaeobiology of Trace Fossils* (S. K. Donovan, ed.), The John Hopkins University Press, Baltimore, Maryland, pp. 221-240.
Jollie, M., 1982, What are the "Calcichordata"? and the larger question of the origin of the chordates, *Zool. J. Linn. Soc.* **75**:167-188.

McAllister, J. A., 1985, Reevaluation of the formation of spiral coprolites, *Univ. Kansas Paleontol. Contrib. Pap.* **89**:1-12.

McAllister, J. A., 1987, Phylogenetic distribution and morphological reassessment of the intestines of fossil and modern fishes, *Zool. Jahrb. Anat.* **115**:281-294.

McAllister, J. A., 1989, Preliminary description of the coprolitic remains from Hamilton Quarry, Kansas, in: *Regional Geology and Paleontology of Upper Paleozoic Hamilton Quarry Area in Southeastern Kansas* (G. Mapes and R. Mapes, eds.), Kansas Geol. Surv., Guidebook Ser. 6 [for 1988], Lawrence, Kansas, pp. 195-202.

McAllister, J. A., 1996, Coprolitic remains from the Devonian Escuminac Formation, in: *Devonian Fishes and Plants of Miguasha, Quebec, Canada* (H.-P. Schultze and R. Cloutier, eds.), Verlag Dr. Friedrich Pfeil, München, Germany, pp. 328-347.

Mackenzie, A., Ball, A., and Virdee, S. R., 1998, *Instant Notes in Ecology*, Springer-Verlag, New York.

Mallatt, J., 1985, Reconstructing the life cycle and the feeding of ancestral vertebrates, in: *Evolutionary Biology of Primitive Fishes* (R. E. Foreman, A Gorbman, J. M. Dodd, and R. Olsson, eds.), NATO Advances Science Institutes Series A, Vol. 103, Plenum Press, New York, pp. 59-68.

Mallatt, J., 1997, Crossing a major morphological boundary: The origin of jaws in vertebrates, *Zool. Jena.* **100**(3):128-140.

Mark-Kurik, E. (ed.), 1992, *Fossil Fishes as Living Animals*, Second International Colloquium on the Middle Palaeozoic Fishes, Academia 1, Academy of Sciences of Estonia, Tallinn, Estonia.

Moyle, P. B., and Cech, J. J., Jr., 1982, *Fishes: An Introduction to Ichthyology*, Prentice-Hall, Englewood Cliffs, New Jersey.

Northcutt, R. G., and Gans, C., 1983, The genesis of neural crest and epidermal placodes: a reinterpretation of vertebrate origins, *Quart. Rev. Biol.* **58**:1-28.

Olson, E. C., 1952, The evolution of a Permian vertebrate chronofauna, *Evolution* **6**:181-196.

Pickerill, R. K., 1994, Nomenclature and taxonomy of invertebrate trace fossils, in: *The Palaeobiology of Trace Fossils* (S. K. Donovan, ed.), The John Hopkins University Press. Baltimore, Maryland, pp. 3-42.

Poplin, C., 1986, Taphocœnoses et restes alimentaires de Vertébrés carnivores, *Bull. Mus. natn. Hist. natl., Paris,* 4th sér., 8, section C, no. 2:257-267.

Prothero, D., 1998, *Bringing Fossils to Life: An Introduction to Paleobiology*, McGraw-Hill Publishing Co., New York.

Purnell, M. A., 2002, Feeding in extinct jawless heterostracan fishes and testing scenarios of early vertebrate evolution, *Proc. R. Soc. London B* **269**:83-88

Purnell, M. A., and Donoghue, P. C., 1997, Architecture and functional morphology of the skeletal apparatus of ozarkodinid conodonts, *Phil. Trans. R. Soc. London B* **352**:1545-1564.

Richter, G., and Baszio, S., 2001, Traces of a limnic food web in the Eocene Lake Messel — a preliminary report based on fish coprolite analyses, *Palaeogeogr. Palaeoclim. Palaeoecol.* **166**:345-368.

Schwimmer, D. R., Stewart, J. D., and Williams, G. D., 1997, Scavenging by sharks of the genus *Squalicorax* in the Late Cretaceous of North America, *Palaios* **12**:71-83.

Seilacher, A., Marshall, C., Skinner, H. C. W., and Tsuihiji, T., 2001, A fresh look at sideritic "coprolites," *Paleobiology* **27**:7-13.

Sepkoski, J. J., Jr., 1981, A factor analytic description of the Phanerozoic marine fossil record, *Paleobiology* **7**:36–53.

Shimeld, S. M, and Holland, P. W. H., 2000, Vertebrate innovations, *Proc. Natl. Acad. Sci.* **97**(9):4449-4452.

Stahl, B. J., 1974, *Vertebrate History: Problems in Evolution*, McGraw-Hill, New York.

Viohl, G., 1990, Piscivorous fishes of the Solnhofen Lithographic Limestone, in: *Evolutionary Paleobiology of Behavior and Coevolution* (by A. J. Boucot), Elsevier, Amsterdam, pp. 287-303.

Williams, M. E., 1972, The origin of "spiral coprolites," *Univ. Kansas Paleont. Contrib., Paper* **59**:1-19.

Williams, M. E., 1990, Feeding behavior in Cleveland Shale fishes In: *Evolutionary Paleobiology of Behavior and Coevolution* (by A. J. Boucot), Elsevier, Amsterdam, pp. 273-287.

Wilson, M. V. H., and Caldwell, M. W., 1993, New Silurian and Devonian fork-tailed 'thelodonts' are jawless vertebrates with stomachs and deep bodies, *Nature* **361**:442-444.

Wilson, M. V. H., 1998, The Furcacaudiformes: a new order of jawless vertebrates with thelodont scales, based on articulated Silurian and Devonian fossils from Northern Canada, *J. Vert. Paleontol.* **18**:10-29.

Wilson, M. V. H., 1987, Predation as a source of fish fossils in Eocene lake sediments, *Palaios* **2**:497-504.

Zangerl, R., and Richardson, E. S., Jr., 1963, The paleoecological history of two Pennsylvanian black shales, *Fieldiana, Geol. Mem.* **4**:1-352.

Zidek, J., 1980, *Acanthodes lundi*, new species (Acanthodii) and associated coprolites from uppermost Mississippian Heath Formation of central Montana, *Ann. Carnegie Mus.* **49**:49-78.

Zimmer, C., 2000, In search of vertebrate origins: beyond brain and bone, *Science* **287**:1576-1579.

Chapter 13

Dinosaur Predation
Evidence and Ecomorphology

THOMAS R. HOLTZ, JR.

1. Theropods as Predators... 325
   1.1. Introduction... 325
   1.2. Theropods in Phylogeny and Ecology.. 326
2. Fossil Evidence of Theropod-Prey Interaction... 327
   2.1. Scavenging versus Predation 327
   2.2. Taphonomic Associations... 328
3. Theropod Ecomorphology 330
   3.1. The Theropod Body Plan 331
   3.2. Variations on Theropod Predatory Ecomorphology.. 332
4. Prey Defenses 337
5. Conclusions 337
   References.. 337

1. Theropods as Predators

1.1. Introduction

Carnivorous dinosaurs (Theropoda) such as *Tyrannosaurus rex*, *Velociraptor mongoliensis*, and *Spinosaurus aegyptiacus* are among the most popularly known fossil species and (perhaps together with the felid *Smilodon* and the synapsid *Dimetrodon*) represent the public's primary vision of extinct predators. Numerous restorations of theropods engaged in mortal combat with each other or with one of the many clades of herbivorous dinosaurs are among the most common illustrations of life in the ancient past.

What is not as apparent is the fact that direct evidence of these hypothesized interactions is difficult to find in the fossil record. Several factors contribute to the difficulty in establishing predator-prey interactions for dinosaurs (and large bodied

---

THOMAS R. HOLTZ • Department of Geology, University of Maryland, College Park, Maryland 20742.

Predator-Prey Interactions in the Fossil Record, edited by Patricia H. Kelley, Michał Kowalewski, and Thor A. Hansen. Kluwer Academic/Plenum Publishers, New York, 2003.

vertebrates in general) compared to similar associations in the marine invertebrate record: the differences between sedimentary styles and modes in terrestrial rather than marine realm; the larger body size (and hence greater difficulty in preservation) of dinosaurs relative to marine invertebrates; the much greater number of hard parts in the vertebrate versus invertebrate skeleton; the vastly greater abundance of invertebrate fossils than vertebrate fossils; and the difficulty in distinguishing predation versus scavenging events. Nevertheless, there are some cases where direct theropod-prey interactions can be established.

Details of the predatory apparatus of theropods (essentially skull and teeth, forelimbs, and hindlimbs) and their differences among the various theropod clades allow for some speculation as to the diversity of predatory modes of varying types of carnivorous dinosaurs.

1.1.1. Institutional Abbreviations

AMNH, American Museum of Natural History, New York City; DNMN, Denver Museum of Natural History, Denver; GI, Geological Institute, Ulaan Baatar; MNN, Musée National du Niger, Niamey; USNM, United States National Museum of Natural History, Smithsonian Insititution, Washington, D.C.

1.2. Theropods in Phylogeny and Ecology

Dinosauria is nested phylogenetically among the archosauriform reptiles (Benton, 1999), a clade comprised of primarily carnivorous taxa. Some archosauriforms (such as *Proterosuchus*, proterochampsids, and parasuchians) possessed elongate snouts and conical dentition, morphologies associated in the modern world with piscivory (as, for example, in *Gavialis* or various small cetaceans). However, many other basal archosauriforms (erythrosuchids, *Euparkeria*, ornithosuchids, rauisuchians, basal crocodylomorphs, etc.) possessed teeth that were blade-like in cross-section and serrated along the carinae (termed the ziphodont condition) and skulls that were relatively tall dorsoventrally and narrow mediolaterally (the "oreinirostral" condition of Busbey, 1995).

Dinosaurs and their closest kin (small archosaurs such as *Lagerpeton* and *Lagosuchus*) are distinguished from other archosauriforms by the fully upright stance of their hindlimbs. Furthermore, Dinosauria proper are ancestrally characterized by a suite of features including obligate bipedality and a grasping hand with reduced fourth and fifth digits (Sereno, 1997, 1999; Holtz, 2000*a*).

In several aspects theropods retain the primitive condition compared to the remaining clades of dinosaurs. The beaked ornithischians (including armored forms such as stegosaurs and ankylosaurs, the horned and frilled ceratopsians, and the duckbilled hadrosaurids) and the long-necked sauropodomorphs (including the Sauropoda proper, the largest terrestrial animals known) are characterized by derived leaf-shaped teeth with large denticles rather than fine serrations; these morphologies are inferred to indicate a herbivorous diet (Hotton *et al.*, 1997; Holtz *et al.*, 2000; Reisz and Sues, 2000). Theropods, however, retain ziphodont dentitions and oreinirostral skulls. Furthermore, all known theropods were obligate bipeds, while many clades of ornithischian and sauropodomorph dinosaurs convergently evolved a quadrupedal stance.

Theropods first appeared in the Late Triassic, and radiated throughout the Jurassic and Cretaceous into a great diversity of forms (Sereno, 1999; Holtz, 2000b). Dinosaurs ancestrally were animals of approximately 1 m length, but many clades of theropods were larger, and some substantially so. The largest theropods achieved lengths of perhaps 14 m and masses of 8-9 tonnes (Coria and Salgado, 1994; Calvo and Coria, 2000). Although many theropods retained the ancestral conditions of ziphodonty and grasping hands, these morphologies were modified within various subclades. Perhaps the most interesting aspect of theropod history is the fact that modern birds represent the direct descendants of Mesozoic theropods (see Gauthier and Gall (2001) for a substantial review). Many anatomical features unique to birds among living animals (such as furculae (wishbones) and air sacs) are widely distributed among extinct theropods. Indeed, recent discoveries from the Lower Cretaceous deposits of Liaoning, China, demonstrate that many clades of coelurosaurian theropods (including dromaeosaurids, therizinosauroids, oviraptorosaurs, and compsognathids) possessed broad feathers and/or simple filamentous integument over much of the body.

When theropods first appeared in the Late Triassic they represented mid-sized carnivores, considerably smaller than various other archosaurian predators that shared their communities (Heckert and Lucas, 2000). However, most of the theropods' competitors disappeared by the end of the Triassic (Sereno, 1997, 1999), and throughout the Jurassic and Cretaceous predatory dinosaurs comprised the sole group of large-bodied terrestrial predators. (It should be noted, however, that potential theropod-killers existed in the form of aquatic crocodyliforms from the Cretaceous of northern Africa (Sereno *et al.*, 2001) and North America (Schwimmer, in press) of sizes rivalling the largest carnivorous dinosaurs).

2. Fossil Evidence of Theropod-Prey Interaction

Establishing direct evidence between theropods and prey items (or between terrestrial predators and prey in general) is confounded by numerous factors. The first set has to do with the demonstration that a particular fossil represents a predation event (that is, the carnivore doing the feeding was also the agent of death for the food item) rather than a scavenging event (that is, the carnivore was consuming the remains of a food item that died by other causes). The second set of problems has to do with the nature of the remains themselves. Vertebrates in general are many times larger than typical invertebrates, and dinosaurian predators include the largest fully terrestrial carnivores in the history of the Earth. As such, depositional environments are unlikely to record actual dinosaur predation events. Nevertheless, some rare finds document such occurrences, as discussed below.

2.1. Scavenging versus Predation

Since the initial description of *Megalosaurus* (Buckland, 1824) theropods have been recognized as carnivorous reptiles. However, some workers have suggested that at least some large theropods (in particular, tyrannosaurids) were primarily or exclusively scavengers on both theoretical ecological (Colinvaux, 1978) and ecomorphological (Horner and Lessem, 1993) grounds. Critical analyses of these arguments are presented elsewhere (Farlow, 1993; Farlow and Holtz, in press; Holtz, in prep.).

However, these arguments do raise the more general issue: how does a paleontologist distinguish a predation event from a scavenging event from fossil remains? This difficulty is due to the fact that scavenging events generally result in the same physical remains as successful predation events: that is, the food item is a set of dead bones and soft tissues at the end of feeding.

There are a few possible ways that particular case studies could be examined to test if they were scavenging events. These include examining the theropod tooth marks to see if one set (from one taxon) were always cross-cut by a second set (which would have therefore fed after the initial feeder) and examining the bones of the food item for signs of non-predator generated fatal pathologies (such as trampling marks and breaks) which suggest the food item had suffered lethal injuries not produced by a carnivore.

2.2. Taphonomic Associations

Dinosaurs, being fully terrestrial animals, are typically preserved in environments with intermittent sedimentation: fluvial overbank deposits or channel conglomerates, eolian beds, and the like. Only rarely are complete specimens of dinosaurs found in death position. However, various partial remains allow some insight into theropod predation (or at least feeding).

2.2.1. Trace Fossils and Gut Contents

The most abundant form of trace fossil documenting theropod feeding is tooth-marked bone. Tooth marks generated by theropod teeth on the bones of ornithischians, sauropodomorphs, other theropods, and pterosaurs are found from the Late Triassic until the latest Cretaceous (Hunt *et al.*, 1994; Currie and Jacobsen, 1995; Erickson and Olson, 1996; Erickson *et al.*, 1996; Jacobsen, 1997, 1998, 2001; Chure *et al.*, 2000). However, while these tooth marks indicate that theropod dinosaurs fed on the animal in question, they do not establish the theropod as the agent of the food item's death. Similarly, the recently described coprolite of a large theropod, almost certainly *Tyrannosaurus rex*, from the Frenchman Formation (Upper Cretaceous, Saskatchewan) indicates that a meat-eating dinosaur consumed a medium-sized ornithischian dinosaur (Chin *et al.*, 1998), but does not indicate that the carnivore killed the plant-eater.

Theropod gut contents have been described for several different taxa. A tyrannosaurid has been described with the acid-etched remains of a juvenile hadrosaurid within its body cavity (Varricchio, 2001), while a spinosaurid has similarly been discovered associated with the partially digested remains of both fish and a juvenile *Iguanodon* (Charig and Milner, 1997). Among smaller theropods, articulated specimens of meter-long compsognathids have been described with the remains of small lizards (Ostrom, 1978) and mammals (Currie and Chen, 2001) as gut contents. Some specimens of the primitive Late Triassic theropod *Coelophysis bauri* seem to have remains of smaller specimens of that species within their gut cavities, strongly suggestive of cannibalism in this form (Colbert, 1989); however, given the jumbled nature of the deposit in question, some of these apparent associations may be due to superimposition of an adult skeleton over an underlying juvenile.

While most theropod trackways simply document locomotion, one particular trace fossil seems to record an episode of attempted predation. A series of footprints of a large theropod (possibly *Acrocanthosaurus*: Farlow (2001)) and a large sauropod have been found along the same bedding plane in the Lower Cretaceous Glen Rose Limestone near the Paluxy River in central Texas (Bird, 1939; Farlow *et al.*, 1989). For about a dozen paces the footprints of the theropod and the sauropod are synchronous; at the end of this synchrony these trackways intersect. At the intersection point the stride length of the herbivore changes and an expected left footprint of the carnivore is not present. Immediately thereafter the trackways of both the carnivore and the herbivore veer leftward (Thomas and Farlow, 2000). This trackway has been interpreted as an attempt by the carnivore to attack (perhaps latch onto) the larger herbivore. As the trackways continue it is apparent that the possible attack (if correctly interpreted) was not immediately fatal. Unfortunately the animals continued travelling beyond the preserved section of this bedding plane, so the result of this apparent interaction will never be known.

2.2.2. The "Fighting Dinosaurs"

A remarkable taphonomic association between probable would-be theropod predator and prey is documented in a pair of fossils commonly referred to as the "fighting dinosaurs" (Fig. 1). Discovered in the Upper Cretaceous Djadokhta Formation of Mongolia in 1971 and initially reported by Kielan-Jaworowska and Barsbold (1972), this fossil set has been the subject of considerable speculation (see Carpenter, 2000, for a recent review). These fossils represent the skeletons of the 130 cm long herbivorous ornithischian *Protoceratops andrewsi* locked in its final moments with the 200 cm long dromaeosaurid theropod *Velociraptor mongoliensis*. These individuals were apparently killed by a sand dune collapse while in the midst of a struggle (however, see Osmólska, 1993, for an alternative interpretation).

These remains document the probable prey dispatch technique of a particular clade of theropod dinosaurs. Dromaeosaurids, and some of their closest relatives, are characterized by a greatly enlarged sickle-shaped claw on the retractable second digit of the foot. Ostrom (1969) speculated that this implement was used as the primary killing weapon of this group of dinosaur. The "fighting dinosaurs" specimens show that, in at least this instance, the sickle-claw of the left foot was used to pierce (and presumably

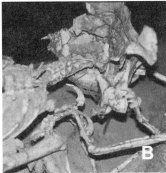

FIGURE 1. The "fighting dinosaurs" specimen, Djadokhta Formation (Upper Cretaceous, Mongolia). (A) *Protoceratops andrewsi* (GI 100/512), preserved standing, left and *Velociraptor mongoliensis* (GI 100/25), preserved lying on its right side, right. (B) Close up of sickle-shaped pedal claw of *Velociraptor* near the cervical vertebrae of *Protoceratops*. Photographs by the author.

tear out) the soft-tissues of the throat (see Fig. 1B), as this fossil was discovered with the bony portion of the claw only a few cms from the ventral surface of the cervical vertebrae of the plant-eater (and thus was deep within the neck of the living animal). The left hand of the *Velociraptor* specimen is preserved grasping the frill of the back of the skull of the *Protoceratops* individual, while the right hand is trapped by the beak of the herbivore. Given the powerful construction of the beak and shearing teeth of ceratopsians, it is quite likely this particular *Velociraptor* would have lost its right arm during this combat, had both dinosaurs not been buried.

It should be noted, however, that while this specimen is generally interpreted as a predation event trapped in time, other possibilities for this association exist. For example, the *Velociraptor* may have been defending itself against the *Protoceratops* individual, rather than trying to kill the herbivore for food. Nevertheless, this specimen does seem to preserve the means by which one particular group of theropods would dispatch its prey.

2.2.3. Failed Predation

As noted above, a successful predation event and a scavenging event would be difficult to distinguish from the fossil record, as in both cases the food item is dead by the end of the feeding event, and hence its immune system cannot respond to the injuries. On the other hand, evidence of failed predation attempts provide direct indication that at least some theropod dinosaurs engaged in attacks on living animals.

Such evidence has been described by Carpenter (2000). DMNH 1943, a specimen of a 7 m long duckbilled hadrosaurid *Edmontosaurus annectens* from the Hell Creek Formation (Upper Cretaceous, Montana), is preserved with a pathology consisting of damage to caudal neural spines 13-17. These have been punctured, and in the case of neural spine 15 a substantial section of bone has been removed. The size and shape of the trauma is consistent with the anterior end of the snout of the only large theropod known to occur in the Hell Creek Formation, the gigantic coelurosaur *Tyrannosaurus rex*.

Significantly, Carpenter (2000) notes that this pathology shows substantial evidence of post-traumatic healing. Thus, the wound was inflicted on a living individual that survived the bite long enough for some regrowth of bone. If this is indeed a theropod wound, it indicates that at least one extremely large carnivorous dinosaur (over 2.9 meters tall, as this is the height of the wound in the mounted skeleton of DMNH 1943) delivered an attack on a living individual of another dinosaur.

3. Theropod Ecomorphology

Although the direct physical evidence of predation and predatory modes in theropod fossils is very limited, some speculation as to these may be made based on the morphology of theropods and (in particular) of their primary implements of prey acquisition and dispatch: skulls and teeth, forelimbs, and hindlimbs. Below is a brief examination of the generalized body plan of carnivorous dinosaurs, as well as an examination of some variations from this basic pattern found in various subclades of Theropoda.

3.1. The Theropod Body Plan

From the appearance of basal carnivorous dinosaurs such as *Herrerasaurus* in the Late Triassic until the Cretaceous-Tertiary extinction events, terrestrial theropod carnivores shared a generalized body plan (Fig. 2).

Theropods retained the ancestral dinosaurian condition of obligate bipedality. As such, unlike most other groups of terrestrial carnivores, their forelimbs were freed from a dual use as organs of locomotion and implements of prey acquisition, and could specialize for the latter mode. Most theropods, therefore, had grasping forelimbs with trenchant claws.

Carnivorous dinosaurs demonstrate fully upright hindlimbs with a digitigrade stance, indicating that these animals were striding cursors (Farlow *et al.*, 2000). A few clades, such as tyrannosaurids, ornithomimids, and troodontids, exhibit elongated distal limb elements and specialized shock-absorbing structures in the metatarsus, consistent with greater cursorial abilities in these groups than in their more generalized kin (Holtz, 1995).

Theropods ancestrally possessed an oreinirostral skull lacking a solid bony palate and with ziphodont dentition (Fig. 2E). As discussed above, this condition is present in many groups of archosauriforms, including the ancestral members of the crocodylomorph lineage (Russell and Wu, 1997). A recent biomechanical analysis of the skull of the carnosaur *Allosaurus fragilis* (see Fig. 2B, E) by Rayfield *et al.* (2001) suggests that this rather open skull design was capable of only weak muscular force, but was extremely strong in bilateral vertical compression, particularly if the lower jaw were not employed in the initial bite. Those authors suggested that *Allosaurus* may have attacked its victims with a high-impact strike by the upper jaw, followed by a bite from the lower jaw to tear out flesh from the victim (Bakker (2000), using different lines of evidence, suggested a similar mode of attack). At present only *Allosaurus* has been subjected to rigorous digital biomechanical finite-element analysis, but it seems likely that (given the relatively unspecialized nature of the cranium of this dinosaur) many other theropods (and perhaps other carnivorous archosauriforms) would have had similar bites. It should be noted that, while strong in compression, this ancestral skull form would be expected to be relatively weak in torsional regimes as it lacks a solid bony palate to transmit forces and possesses teeth which are much thinner mediolaterally than anteroposteriorly (Busbey, 1995; Russell and Wu, 1997).

A distinctive feature of the theropod jaw apparatus is the intramandibular joint. The tooth-bearing dentary is hinged with the postdentary bones, allowing for some flexion in several planes. This particular aspect of theropod anatomy has yet to be subjected to serious biomechanical study, but appears to have served as a shock-absorber to dampen the forces generated by the acquisition, manipulation, and/or consumption of large prey. A similar function has been postulated for the similar intramandibular joint of mosasaurs and some other squamates (Lee *et al.*, 1999).

From these basic observations some general trends among possible theropod predatory behavior can be made. The cursorial nature of the theropod limbs suggests that they were rather swift animals for their body size, and likely used this speed (either in pursuit or in ambush). Prey was likely acquired by a combination of forelimbs and the skull, as both would serve as organs of prehension. However, it should be noted that

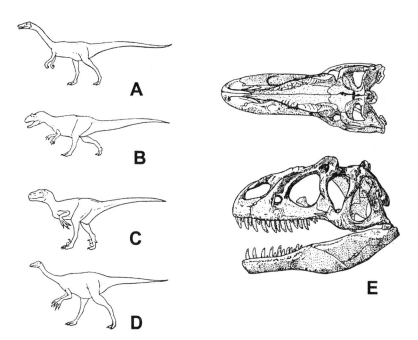

FIGURE 2. A sample of Mesozoic theropod diversity. (A) The Late Triassic coelophysoid *Coelophysis bauri* (approximate length 3 m); (B) The Late Jurassic carnosaur *Allosaurus fragilis* (approximate length 8 m); (C) The Early Cretaceous dromaeosaurid *Deinonychus antirrhopus* (approximate length 3 m); (D) The non-predatory ornithomimosaur *Struthiomimus altus* (approximate length 3.5 m); (E) Skull of *Allosaurus fragilis* (USNM 5434), in dorsal (upper) and left lateral (lower) view (skull length 642 mm). A-D by James Whitcraft, after Farlow *et al.* (2000); E by Tracy Ford.

in theropods ancestrally the skull would have been less resistant to the additional torsional forces associated with struggling prey, and so (unlike living carnivoran mammals or crocodylians) most theropods probably did not bite and hold throughout the kill. Instead, as suggested for *Allosaurus* by Rayfield *et al.* (2001), the theropod predator may have more typically used a "strike-and-tear" bite to generate fatal wounds. The grasping forelimbs with trenchant claws may have served to maintain a purchase on the victim. After the victim was dispatched by a combination of bites and talon-wounds, feeding on the soft tissues could begin; given the ziphodont teeth of most theropods, it is unlikely that they chewed the bones of their prey (Chure *et al.*, 2000).

3.2. Variations on Theropod Predatory Ecomorphology

Theropoda represents a great diversity of forms over their history, and some taxa possessed specializations indicating a deviation from the stereotyped predatory mode suggested above. Below is a brief survey of some of the derived forms, their adaptations, and speculations as to their particular variation on feeding habits (see also Paul, 1988).

3.2.1. Coelophysoids

The coelophysoids are, in general, a rather primitive group of theropods (Sereno, 1997, 1999; Holtz, 2000a,b). These forms represent one of the first major radiations of theropod predators (Heckert and Lucas, 2000), present thoughout much of the world in the Late Triassic and Early Jurassic. Coelophysoids are distinguished by a subnarial kink: a space between the premaxillary and maxillary tooth row. This kink forms part of a "double wave" curvature of the margin of the upper jaw, which differs from the relatively straight or ventrally convex curvature of most theropod jaw lines. However, a similar "double wave" is present in the skulls of many crocodylomorphs (Russell and Wu, 1997), where it serves to enhance the ability to grasp and manipulate prey items with the jaws.

Within Coelophysoidea, the Coelophysidae proper (*Coelophysis* and *Syntarsus*) are characterized by small skull size. While most carnivorous theropods have skull lengths that are subequal with neck length (including herrerasaurids, the coelophysoid *Dilophosaurus*, ceratosaurs, carnosaurs, compsognathids and other basal coelurosaurs, primitive tyrannosaurids, dromaeosaurids, and others), coelophysids have elongated necks and smaller skulls (Fig. 2A). This suggests that they may have typically preyed on smaller sized animals than other theropods of the same body size.

3.2.2. Ceratosaurs

Coelophysoids traditionally have been grouped phylogenetically with *Ceratosaurus* and the Cretaceous abelisauroids in the Ceratosauria (Sereno, 1997, 1999; Holtz, 2000b), but recent analyses suggest that coelophysoids lie outside Ceratosauria proper, having diverged from other theropods basal to the ceratosaur-tetanurine split (Holtz, 2000a; Sampson *et al.*, 2001). True ceratosaurs have forelimbs with greatly reduced manual phalanges (Gilmore, 1920; Bonaparte *et al.*, 1990). In the abelisauroid *Carnotaurus* the forelimb is further reduced in overall size and the forearm elements (ulna and radius) are shortened until they are no longer than the palm of the hand (Bonaparte *et al.*, 1990). These adaptations strongly suggest that the ceratosaurs relied more on the skull and less (if at all) on the forelimb for prey acquisition, dispatch, and manipulation than did typical theropods. The recently discovered bizarre small abelisauroid *Masiakasaurus*, with its procumbent anteriormost dentary teeth, hints at the diversity of predatory modes in this clade (Sampson *et al.*, 2001).

3.2.3. Tyrannosaurids

The coelurosaurian clade Tyrannosauridae (Fig. 3A) is among the most highly derived and highly studied group of Mesozoic theropod (see Holtz, 2001, for a recent review). Some workers suggest that these taxa (or at least the type species *Tyrannosaurus rex*) was an obligate scavenger (Horner and Lessem, 1993); these suggestions are dealt with in greater detail elsewhere (Farlow and Holtz, in press; Holtz, in prep.).

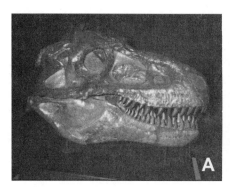 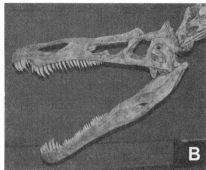

FIGURE 3. Skulls of two large derived theropods. (A), The tyrannosaurid *Gorgosaurus libratus* (AMNH 5336), Dinosaur Park Formation (Upper Cretaceous, Canada), skull length 990 mm; (B) the spinosaurid *Suchomimus tenerensis* (restored cast of MNN GDF501), Elrhaz Formation (Lower Cretaceous, Niger), estimated skull length 1000 mm. Photographs by the author.

Tyrannosaurids show numerous adaptations away from the generalized theropod condition. The skulls of tyrannosaurids are more solidly constructed than those of more basal forms such as *Allosaurus* (Molnar, 2000), with extensive bony palates formed by medial extensions of the premaxillae and maxillae (Molnar, 1991, 2000; Holtz, 2000b). The premaxillary teeth of tyrannosaurids are reduced in size and are incisiform, and were more likely used in scraping meat from bone than in prey acquisition; in contrast, the lateral teeth of tyrannosaurids are incrassate (thickened mediolaterally) rather than ziphodont (Fig. 4). Tooth marks associated with tyrannosaurids suggest that these dinosaurs were capable of generating tooth-on-bone forces that equal or exceed those of any living carnivore (Erickson *et al.*, 1996), and both coprolites (Chin *et al.*, 1998) and bite marks on living dinosaurs (Carpenter, 2000) attributable to *Tyrannosaurus* show that these theropods were capable of crushing bones, unlike typical theropods. The most derived tyrannosaurids (*Daspletosaurus*, *Tarbosaurus*, and *Tyrannosaurus*) are characterized by shortened necks relative to more basal taxa (*Albertosaurus* and *Gorgosaurus*) of the same skull size.

In contrast to their skulls, the forelimbs of tyrannosaurids are reduced: they are proportionately shorter than in most non-ceratosaur theropod clades, and have lost all but two digits. Nevertheless, the arms of tyrannosaurids still were capable of generating substantial forces (Carpenter and Smith, 2001). The hindlimbs of tyrannosaurids are elongate relative to other similar sized carnivorous dinosaurs and to herbivorous dinosaurs and possessed the shock-absorbing arctometatarsalian condition (Holtz, 1995; Farlow *et al.*, 2000), suggesting that tyrant dinosaurs were capable of greater speeds than other large dinosaurs, even if the largest individuals may have been incapable of engaging in more than a fast walk due to mechanical and biophysical constraints (Farlow *et al.*, 1995; Hutchinson and Garcia, 2002).

The morphology of tyrannosaurids suggests that they differed from typical large-bodied theropods (such as *Allosaurus*) in that the skull served as the primary (if not only) organ of prey acquisition and dispatch. The reinforced skull and thickened teeth would be more capable of withstanding the torsional forces associated with holding onto a struggling victim, as well as absorbing the forces generated by tooth-on-bone contact, than the conditions in more basal forms. The forelimbs, if used during predation

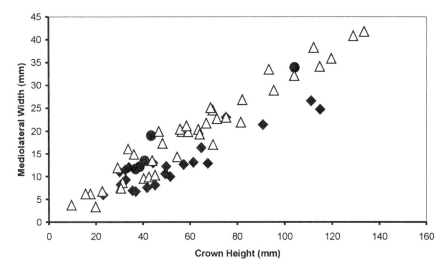

FIGURE 4. Plot of tooth mediolateral width against tooth crown height for typical ziphodont (solid diamond), tyrannosaurid (open triangle), and spinosaurid (solid circle) theropods.

attempts at all, may have been used to brace the body of the victim while the skull was used to kill and dismember the prey.

Busbey (1995) and Russell and Wu (1997) note that the development of substantial secondary palates in crocodylomorph evolution coincides with the replacement of ziphodont with conical teeth, and the probable change of diet from an ancestral archosauriform predatory mode to the modern torsional-based method.

3.2.4. Spinosaurids

The spinosaurids of the Cretaceous of Europe and Gondwana possessed the most crocodyliform-like skulls of all theropods (Fig. 3B). These theropods, which include genera that equal or exceed the largest tyrannosaurids in size, possessed elongate snouts with solid palates (Sereno et al., 1998). Unlike the condition in tyrannosaurids, the premaxillary portion of the solid palates is not produced primarily from medial extensions from these bones, but rather by the extensive direct contact of the main bodies of the premaxillae along the midline throughout their length. The teeth of spinosaurids are conical rather than ziphodont, and in fact might easily be confused with those of large crocodyliforms. Like the incrassate teeth of tyrannosaurids, the conical teeth of spinosaurids are thicker mediolaterally than those of ziphodont teeth of the same height (Fig. 4), and thus were stronger in torsional stresses than more primitive teeth. In contrast to tyrannosaurids, the premaxillary teeth of spinosaurids are their largest rather than their smallest, and the enlarged teeth of the premaxillary and anterior end of the dentary form an expanded "rosette."

The gut contents of at least one spinosaurid contained both the bones of a terrestrial herbivorous dinosaur and the partially digested scales of the large fish *Lepisosteus* (Charig and Milner, 1997). Various other authors have also suggested that spinosaurids were at least partially piscivorous (Sereno et al., 1998; Holtz, 1998), noting that spinosaurids are routinely found in the same sediments that also contain the remains of

1-3 m long fish. Holtz (1998) noted that spinosaurids would be able to access prey that may have been difficult for other large-bodied theropods to obtain, while being capable of travelling from water source to water source with more ease than contemporaneous comparable-sized crocodyiliforms. In this model, spinosaurids would have been heron-like stalkers of fish, striking down vertically and grasping the large fish with their pincher-like terminal rosettes. Note that these authors acknowledge that spinosaurids also fed upon terrestrial dinosaurs, just as many large crocodylians feed on both fish and terrestrial mammals.

3.2.5. Dromaeosaurids

Among the highly specialized maniraptorans of the Mesozoic, dromaeosaurids (Fig. 1, 2C) are the sole group for which a carnivorous (and indeed predatory) habit has never been questioned. The spectacular "fighting dinosaurs" specimen discussed above seems to document at least one possible predatory technique by a member of this clade. Dromaeosaurids and some related maniraptorans have greatly elongate forearms compared to other theropods (Middleton and Gatesy, 2000), which would greatly increase their range as organs of prey acquisition. Biomechanical analysis by Carpenter and Smith (2001) suggests that the forelimbs of dromaeosaurids maximized the speed of deployment over the forcefulness of grasp. Ostrom (1969) speculated that the sickle-shaped claw of the hyperextensible pedal digit II was a weapon of prey dispatch, as seemingly documented by the "fighting dinosaurs" specimen (Fig. 1B). Finally, dromaeosaurids are characterized by bony extensions of the caudal prezygapophyses and chevrons, resulting in a tail that may have served as dynamic stabilizer for rapid turns in locomotion or for balance while striking with its pedal claw (Ostrom, 1969).

The morphologies in which dromaeosaurids differ from the generalized theropod condition suggest a more felid-like predatory mode than in typical carnivorous dinosaurs: an ambush, in which the predator was capable of rapid turns; a quick strike of the forelimbs to acquire the prey, followed by a combination of strike-and-tear bites and strikes from the pedal sickle claw to dispatch it. Some taphonomic evidence suggests that at least the Lower Cretaceous dromaeosaurid *Deinonychus antirrhopus* was a group-hunter (Maxwell and Ostrom, 1995); however, extrapolation of this behavior in absence of other fossil evidence to all members of Dromaeosauridae is at present as unwarranted as extrapolation of the pride-based hunting technique of lions to closely related forms such as tigers, leopards, and jaguars.

3.2.6. Possibly Non-predatory Coelurosaurs

It should be noted that, although theropods are popularly referred to as "carnivorous dinosaurs," there are many clades within Theropoda that show evidence for specialization to non-predatory (and indeed non-carnivorous) diets. Ornithomimosaurs (Fig. 2D) may have had a significant part of their diets composed of aquatic plants and/or small invertebrates (Kobayashi *et al.*, 1999; Norell *et al.*, 2001). Among the feathered maniraptoran theropods, therizinosauroids have been interpreted variously as herbivores (Russell and Dong, 1993) or as piscivores (Barsbold and Perle, 1980), while oviraptorosaurs (Currie *et al.*, 1993) and troodontids (Holtz *et al.*, 2000; Ryan *et al.*, 2000) may have been omnivorous. Mesozoic birds include apparent piscivores, insectivores, and possible herbivores (Chiappe and Witmer, in press), but Mesozoic "birds of prey" seem to be lacking, with one notable exception. The primitive bird *Rahonavis*, a magpie-sized late-surviving form of *Archaeopteryx*-grade from the Late Cretaceous of Madagascar, possesses a foot morphology nearly identical to that of dromaeosaurids coupled with the wings of a fully volant bird (Forster *et al.*, 1998); this suggests a small dinosaur capable of flying down on a prey item and delivering dromaeosaurid-like attacks on the victim.

4. Prey Defenses

Potential prey evolved a number of defenses against theropods throughout the Cretaceous. These include passive defenses such as osteoderm scutes in thyreophoran ornithischians and at least some titanosaurian sauropods; active defenses such as the horns of ceratopsids and tail clubs of ankylosaurids and stegosaurs; herding behavior in sauropods and some ornithischians; and enhanced cursorial ability in some non-predatory theropods such as ornithomimosaurs and troodontids (Holtz, 1995).

5. Conclusions

Theropod dinosaurs first appeared in the Late Triassic and survive in the Cenozoic as birds; during the Jurassic and Cretaceous, however, these reptiles were the dominant group of terrestrial predators. Although direct physical evidence of predation events is rare (or at least difficult to distinguish from scavenging events) in the fossil record, a number of trace fossils and taphonomic associations yield some indications of modes of dinosaur predation. A generalized morphology exists among many non-avian theropods, and (by a combination of biomechanical studies and analogies with modern vertebrate predators) some predictions of the predatory habits of unspecialized carnivorous dinosaurs can be made. Various taxa (including coelophysoids, ceratosaurs, tyrannosaurids, spinosaurids, and dromaeosaurids) possess suites of adaptations suggesting predatory behaviors different from those of the generalized morphs. With the extinction of the non-avian dinosaurs at the Cretaceous-Tertiary boundary, however, the generalized theropod ecomorphotype has vanished. Although the Cenozoic has seen large predatory flightless birds, some terrestrial crocodyliforms, and many clades of carnivorous mammals (including the Neogene appearance of bipedal tool-using hunting primates), striding cursorial bipedal carnivores with grasping hands and serrated teeth have never again evolved in Earth's history. *Sic semper* Tyrannosaurus.

ACKNOWLEDGMENTS: The author would like to acknowledge P. J. Currie and J. O. Farlow for their insightful reviews, and would further like to thank those workers, M. T. Carrano, R. E. Molnar, G. S. Paul, and S. D. Sampson, and the participants of the Internet Dinosaur Mailing List (dinosaur@usc.edu), for sharing their thoughts and comments on the predatory modes of theropod dinosaurs.

References

Bakker, R. T., 2000, Brontosaur killers: Late Jurassic allosaurids as sabre-tooth cat analogues, *Gaia* **15**:145-158.
Barsbold, R., and Perle, A., 1980, Segnosauria, a new infraorder of carnivorous dinosaurs, *Acta Palaeontol. Polon.* **25**:185-195.

Benton, M. J., 1999, *Scleromochlus taylori* and the origin of dinosaurs and pterosaurs, *Phil. Trans. R. Soc. Lond.* B **354**:1423-1446.

Bird, R. T., 1939, Thunder in his footsteps, *Nat. Hist.* **43**:254-261, 302.

Bonaparte, J. F., Novas, F. E., and Coria, R. A., 1990, *Carnotaurus sastrei* Bonaparte, the horned, lightly built carnosaur from the Middle Cretaceous of Patagonia, *Nat. Hist. Mus. Los Angeles County, Contrib. Sci.* **416**:1-41.

Buckland, W., 1824, Notice on *Megalosaurus*, or great fossil lizard of Stonesfield, *Trans. Geol. Soc. Lond.* **21**:390-397.

Busbey, A. B., 1995, The structural consequences of skull flattening in crocodilians, in: *Functional Morphology in Vertebrate Paleontology* (J. J. Thomason, ed.), Cambridge University Press, pp. 173-192.

Calvo, J. O., and Coria, R., 2000, New specimen of *Giganotosaurus carolinii* (Coria & Salgado, 1995) supports it as the largest theropod ever found, *Gaia* **15**:117-122.

Carpenter, K., 2000, Evidence of predatory behavior by carnivorous dinosaurs, *Gaia* **15**:135-144.

Carpenter, K., and Smith, M., 2001, Forelimb osteology and biomechanics of *Tyrannosaurus rex*, in: *Mesozoic Vertebrate Life: New Research Inspired by the Paleontology of Philip J. Currie* (D. H. Tanke and K. Carpenter, eds.), Indiana University Pres, Bloomington, pp. 90-116.

Charig, A. J., and Milner, A. C. 1997, *Baryonyx walkeri*, a fish-eating dinosaur from the Wealden of Surrey, *Bull. Nat. Hist. Mus. Lond.* (Geol.) **53**:11-70.

Chiappe, L. M., and Witmer, L. (eds.), in press, *Mesozoic Birds: Above the Heads of Dinosaurs*, University of California Press, Berkeley.

Chin, K., Tokaryk, T. T., Erickson, G. M., and Calk, L. C., 1998, A king-sized theropod coprolite, *Nature* **393**:680-682.

Chure, D. J., Fiorillo, A. R., and Jacobsen, A., 2000, Prey bone utilization by predatory dinosaurs in the Late Jurassic of North America, with comments on prey bone use by dinosaurs throughout the Mesozoic, *Gaia* **15**:227-232.

Colbert, E. H., 1989, The Triassic dinosaur *Coelophysis*, *Bull. Mus. N. Arizona* **57**:1-160.

Colinvaux, P., 1978, *Why Big Fierce Animals are Rare: An Ecologist's Perspective*, Princeton University Press, Princeton.

Coria, R. A., and Salgado, L., 1994, A new giant carnivorous dinosaur from the Cretaceous of Patagonia, *Nature* **227**:224-226.

Currie, P. J., and Chen, P., 2001, Anatomy of *Sinosauropteryx prima* from Liaoning, northeastern China, *Can. J. Earth Sci.* **38**:1705-1727.

Currie, P. J., and Jacobsen, A. R., 1995, An azhdarchid pterosaur eaten by a velociraptorine theropod, *Can. J. Earth Sci.* **32**:922-925.

Currie, P. J., Godfrey, S. J., and Nessov, L., 1993, New caenagnathid (Dinosauria: Theropoda) specimens from the Upper Cretaceous of North America and Asia, *Can. J. Earth Sci.* **30**:2255-2272.

Erickson, G. M., and Olson, K. H., 1996, Bite marks attributable to *Tyrannosaurus rex*: preliminary description and implications, *J. Vert. Paleontol.* **16**:175-178.

Erickson, G.M., Van Kirk, S. D., Su, J.-T., Levenston, M. E., Caler, W. E., and Carter, D. R., 1996, Bite force estimation for *Tyrannosaurus rex* from tooth-marked bone, *Nature* **382**:706-708.

Farlow, J. O., 1993, On the rareness of big, fierce animals: speculations about the body sizes, population densities, and geographic ranges of predatory mammals and large carnivorous dinosaurs, *Am. J. Sci.* **239-A**:167-199.

Farlow, J. O., 2001, *Acrocanthosaurus* and the maker of Comanchean large-theropod footprings, in: *Mesozoic Vertebrate Life: New Research Inspired by the Paleontology of Philip J. Currie* (D. H. Tanke and K. Carpenter, eds.), Indiana University Press, Bloomington, pp. 408-427.

Farlow, J. O., and Holtz, T. R., Jr., in press, The fossil record of predation in dinosaurs, in: *The Fossil Record of Predation* (M. Kowalewski and P. H. Kelley, eds.), Paleontological Society Papers 8.

Farlow, J. O., Gatesy, S. M., Holtz, T. R., Jr., Hutchinson, J. R., and Robinson, J. M., 2000, Theropod locomotion, *Am. Zool.* **40**:640-663.

Farlow, J. O., Pittman, J. G., and Hawthorne, J. M., 1989, *Brontopodus birdi*, Lower Cretaceous sauropod footprints from the U.S. Gulf coastal plain, in: *Dinosaur Tracks and Traces* (D. D. Gillette and M. G. Lockley, eds.), Cambridge University Press, Cambridge, pp. 371-394.

Farlow, J. O., Smith, M. B., and Robinson, J. M., 1995, Body mass, bone "strength indicator," and cursorial potential of *Tyrannosaurus rex*, *J. Vert. Paleontol.* **15**:713-725.

Forster, C. A., Sampson, S. D., Chiappe, L. M., and Krause, D. W., 1998, The theropod ancestry of birds: new evidence from the Late Cretaceous of Madagascar, *Science* **279**:1915-1919.

Gauthier, J. A., and Gall, L. F. (eds.), 2001, *New Perspectives on the Origin and Early Evolution of Birds: Proceedings of the International Symposium in Honor of John H. Ostrom*, Yale University Press, New Haven.

Gilmore, C. W., 1920, Osteology of the carnivorous Dinosauria in the United States National Museum, with special reference to the genera *Antrodemus* (*Allosaurus*) and *Ceratosaurus*, *Bull. U.S. Nat. Mus.* **110**:1-154.
Heckert, A. B., and Lucas, S. G., 2000, Global correlation of the Triassic theropod record, *Gaia* **15**:63-74.
Holtz, T. R., Jr., 1995, The arctometatarsalian pes, an unusual structure of the metatarsus of Cretaceous Theropoda (Dinosauria: Saurischia), *J. Vert. Paleontol.* **14**:480-519.
Holtz, T. R., Jr., 1998, Spinosaurids as crocodile mimics, *Science* **282**:1276-1277.
Holtz, T. R., Jr., 2000a, Classification and evolution of the dinosaur groups, in: *The Scientific American Book of Dinosaurs* (G. S. Paul, ed.), St. Martin's Press, New York, pp. 140-168.
Holtz, T. R., Jr., 2000b, A new phylogeny of the carnivorous dinosaurs, *Gaia* **15**:5-61.
Holtz, T. R., Jr., 2001, The phylogeny and taxonomy of the Tyrannosauridae, in: *Mesozoic Vertebrate Life: Recent Research Inspired by the Paleontology of Philip J. Currie* (D. H. Tanke and K. Carpenter, eds.), Indiana University Press, Bloomington, pp. 64-83.
Holtz, T. R., Jr., Brinkman, D. L., and Chandler, C. L., 2000, Denticle morphometrics and a possibly omnivorous feeding habit for the theropod dinosaur *Troodon*, *Gaia* **15**:159-166.
Horner, J. R., and Lessem, D., 1993, *The Complete* T. rex*: How Stunning New Discoveries Are Changing Our Understanding of the World's Most Famous Dinosaur*, Simon and Schuster, New York.
Hotton, N., III, Olson, E. C., and Beerbower, R., 1997, Amniote origins and the discovery of herbivory, in: *Amniote Origins: Completing the Transition to Land* (S. S. Sumida and K. L. M. Martin, eds.), Academic Press, San Diego, pp. 207-264.
Hunt, A. P., Meyer, C. A., Lockley, M. G., and Lucas, S. G., 1994, Archaeology, toothmarks and sauropod dinosaur taphonomy, *Gaia* **10**:225-231.
Hutchinson, J. R., and Garcia, M., 2002, *Tyrannosaurus* was not a fast runner, *Nature* **415**:1018-1021.
Jacobsen, A. R., 1997, Tooth marks, in: *Encyclopedia of Dinosaurs* (P. J. Currie and K. Padian, eds.), Academic Press, San Diego, pp. 738-739.
Jacobsen, A. R., 1998, Feeding behavior of carnivorous dinosaurs as determined by tooth marks on dinosaur bones, *Hist. Biol.* **13**:17-26.
Jacobsen, A. R., 2001, Tooth-marked small theropod bone: an extremely rare trace, in: *Mesozoic Vertebrate Life: New Research Inspired by the Paleontology of Philip J. Currie* (D. H. Tanke and K. Carpenter, eds.), Indiana University Press, Bloomington, pp. 58-63.
Kielan-Jaworowska, Z., and Barsbold, R., 1972, Narrative of the Polish-Mongolian Palaeontological Expeditions 1967-1971, *Palaeontol. Polon.* **27**:5-13.
Kobayashi, Y., Lu, J.-C., Dong, Z.-M., Barsbold, R., Azuma, Y., and Tomida, Y., 1999, Herbivorous diet in an ornithomimid dinosaur, *Nature* **402**:480-481.
Lee, M. S. Y., Bell, G. L., Jr., and Caldwell, M. W., 1999, The origin of snake feeding, *Nature* **400**:655-659.
Maxwell, W. D., and Ostrom, J. H., 1995, Taphonomy and paleobiological implications of *Tenontosaurus-Deinonychus* associations, *J. Vert. Paleontol.* **15**:707-712.
Middleton, K. M. and Gatesy, S. M., 2000, Theropod forelimb design and evolution, *Zool. J. Linn. Soc.* **128**:149-187.
Molnar, R. E., 1991, The cranial morphology of *Tyrannosaurus rex*, *Palaeontograph. A* **217**:137-176.
Molnar, R. E., 2000, Mechanical factors in the design of the skull of *Tyrannosaurus rex* (Osborn, 1905), *Gaia* **15**:193-218.
Norell, M. A., Makovicky, P. J., and Currie, P. J., 2001, The beaks of ostrich dinosaurs, *Nature* **412**:873-874.
Osmólska, H., 1993, Were the Mongolian "fighting dinosaurs" really fighting? *Rev. Paléobiol. Spec. Vol.* **7**:161-162.
Ostrom, J. H., 1969, Osteology of *Deinonychus antirrhopus*, an unusual theropod from the Lower Cretaceous of Montana, *Peabody Mus. Nat. Hist. Bull.* **30**:1-165.
Ostrom, J. H., 1978, The osteology of *Compsognathus longipes* Wagner, *Zitteliana* **4**:73-118.
Paul, G. S., 1988, *Predatory Dinosaurs of the World: A Complete Illustrated Guide*, Simon and Schuster, New York.
Rayfield, E. J., Norman, D. B., Horner, C. C., Horner, J. R., Smith, P. M., Thomason, J. J., and Upchurch, P., 2001, Cranial design and function in a large theropod dinosaur, *Nature* **409**:1033-1037.
Reisz, R. R., and Sues, H.-D., 2000, Herbivory in late Paleozoic and Triassic terrestrial vertebrates, in: *Evolution of Herbivory in Terrestrial Vertebrates: Perspectives from the Fossil Record* (H.-D. Sues, ed.), Cambridge University Press, Cambridge, pp. 9-41.
Russell, A. P., and Wu, X.-C., 1997, The Crocodylomorpha at and between geological boundaries: the Baden-Powell approach to change, *Zoology* **100**:164-182.
Russell, D. A., and Dong, Z.-M., 1993, The affinities of a new theropod dinosaur from the Alxa Desert, Inner Mongolia, People's Republic of China, *Can. J. Earth Sci.* **30**:2107-2127.

Ryan, M. J., Currie, P. J., Gardner, J. D., Vickaryous, M. K., and Lavinge, J. M., 2000, Baby hadrosaurid material associated with an unusually high abundance of *Troodon* teeth from the Horseshoe Canyon Formation, Upper Cretaceous, Alberta, Canada, *Gaia* **15**:123-133.

Sampson, S. D., Carrano, M. T., and Forster, C. A., 2001, A bizarre predatory dinosaur from the Late Cretaceous of Madagascar, *Nature* **409**:504-506.

Schwimmer, D. R., in press, *King of the Crocodylians: The Paleobiology of* Deinosuchus, Indiana University Press, Bloomington.

Sereno, P. C., 1997, The origin and evolution of dinosaurs, *Ann. Rev. Earth Planet. Sci.* **25**:435-489.

Sereno, P. C., 1999, The evolution of dinosaurs, *Science* **284**:2137-2147.

Sereno, P. C., Beck, A. L., Dutheil, D. B., Gado, B., Larsson, H. C. E., Lyon, G. H., Marcot, J. D., Rauhut, O. W. M., Sadlier, R. W., Sidor, C. A., Varricchio, D. J., Wilson, G. P., and Wilson, J. A., 1998, A long-snouted predatory dinosaur from Africa and the evolution of spinosaurids, *Science* **282**:1298-1302.

Sereno, P. C., Larsson, H. C. E., Sidor, C. A., and Gado, B., 2001, The giant crocodyliform *Sarcosuchus* from the Cretaceous of Africa, *Science* **294**:1516-1519.

Thomas, D. A., and Farlow, J. O., 2000, Tracking a dinosaur attack, in: *The Scientific American Book of Dinosaurs* (G. S. Paul, ed.), St. Martin's Press, New York, pp. 242-248.

Varricchio, D. J., 2001, Gut contents from a Cretaceous tyrannosaurid: implications for theropod dinosaur digestive tracts, *J. Paleontol.* **75**:401-406.

Chapter 14

Bones of Comprehension
The Analysis of Small Mammal Predator–Prey Interactions

J. P. WILLIAMS

1. Introduction 341
2. Identifying the Predator 342
   2.1. Fossilized Coprolites 343
   2.2. Taphonomic Analysis 343
3. Fossil Evidence 350
   3.1. Case Studies of Predatory Origin 350
   3.2. Problems with Early Fossil Evidence 353
   3.3. Predator Bias? 353
   3.4. Population Fluctuations 353
   3.5. Prey Selection and Small Mammal Evolution 355
4. Summary 356
   References 356

1. Introduction

Prey species abundances in fossil mammal assemblages rarely mirror those in the original community from which they were drawn. This disharmony may result from a number of factors, one of which is the initial prey selection. As a result, the species present in a fossil assemblage may be more representative of the size and habits of the predator than of the ecology of the surrounding area. This is a particularly acute problem in the analysis of small mammal fossil deposits, especially when the overall goal of the analysis is to gain further insight into the environment at the time of

J. P. WILLIAMS • English Heritage, 44 Derngate, Northampton, NN1 1UH, United Kingdom.

Predator-Prey Interactions in the Fossil Record, edited by Patricia H. Kelley, Michał Kowalewski, and Thor A. Hansen. Kluwer Academic/Plenum Publishers, New York, 2003.

deposition. It is therefore necessary to identify the mode of accumulation, and, where appropriate, the predator(s) responsible. By comparison with the present-day behavior of these predator species, it is possible to recognize and account for any bias in the species representation.

Very few fossil sites will contain both the predator and the prey (Andrews, 1990), and it would be unusual for both to succumb at the same time. Equally, the scarcity of predators in the fossil record (Behrensmeyer, 1975, p. 545) reflects a real difference in abundance when compared to the numbers of their prey (Elton, 1966). Mammalian or avian predators (at the top of their food chain) do not usually constitute a large part of another predator's diet, so death is usually a result of old age, infirmity, scarcity of food or a combination of these factors. In autochthonous deposits, such as caves, there is the increased possibility that prey remains may be deposited in a location where the predator may die at a later date. However, allochthonous deposits, such as fluvially derived ones, are less likely to contain the remains of both the predator and the prey, as bones may become scattered. Clearly then there is a need to be able to recognize the origin of fossil deposits where predation has been the cause of death from accumulations of non-predated bones.

The study of fossil mammalian predator-prey interactions dates back to the work of Rev. William Buckland. His work at Kirkdale Cave (Yorkshire, England) demonstrated that the fossils accumulated there had been brought in by predators (hyenas) rather than dying *in situ*, or having been washed in by some catastrophic event. Buckland's observations were not only limited to large mammals. He writes of the origin of bones found in Pontnewydd Cave, Wales (Buckland 1823, p. 63):

> In this earth I found the bones of various birds, of moles, water rats, mice and fish, and a few land snails... and their presence in this most inaccessible spot can only be explained by referring the bones [of birds, moles, rats and mice] to the agency of hawks, and the fish bones to that of sea gulls.

The following review deals principally with small mammals, which are by far the most numerous mammal class. For example, in Britain, the pre-breeding population of field voles (*Microtus agrestis*) is almost twice the size of the human population (data from Harris *et al.*, 1995; Yalden, 1999). In all instances in this review, it is the small mammals that are the prey items of larger mammalian predators, birds, and in some cases reptiles.

2. Identifying the Predator

This chapter will begin by reviewing some of the methods used in the analysis of small mammal predator-prey relationships in the fossil record. These are the analysis of fossil dietary waste, the study of bone modification, and analogy with the behavior of modern comparative species, all of which fall within the study of taphonomy. These will be discussed further, with reference to some more recent archeological and paleontological examples using prey examples drawn mainly from the study of fossil small mammals.

2.1. Fossilized Coprolites

Following consumption of small mammal prey, any undigested solids, such as fur and bones, are returned either in scats or regurgitated pellets. During the burial and fossilization process, the hair and fur in these pellets usually break down, leaving only the bones behind. Only rarely do fossil pellets or scats survive in the fossil record. Eocene predator accumulations, including four probably "mammalian carnivore" scats, have been recorded from Gnat-Out-of-Hell, Utah, USA (Thornton and Rasmussen, 2001). Carnivore scats have also been found in the Plio-Pleistocene deposits of East Turkana, Kenya (Jacobs, 1985) and hyena coprolites have been recovered from a variety of Paleolithic and Holocene archaeological sites in Israel (Horwitz and Goldberg 1989). Walton (1990) reports the discovery of owl pellets from the Upper Oligocene location of Flagstaff Rim, Wyoming, USA, and Kowalski (1990) lists a small number of late Miocene and late Pleistocene examples of owl pellets from Germany. Through the analysis of modifications to these bones (either within or without the pellet or scat matrix), it is possible to recover information about their predatory origin.

2.2. Taphonomic Analysis

Taphonomic analysis is the method used to reconstruct the history of a fossil, including the manner in which it died. For small mammals, when predation was responsible for their mortality, recognizable modifications of bones or teeth are usually visible, the analysis of which can often lead to the identification of the predator. Thus most studies of small mammal taphonomy aim to ascertain the predatory origin of the fossil material and consider how this action may have affected the depositional fidelity of an accumulation.

The studies reviewed below are concerned with actualistic studies of modern day predation. Utilizing either laboratory or wild predator species, bone modifications associated with various modern predators have been measured and used as a comparison with fossil material. In all studies the uniformitarian assumption that the predatory behavior of modern predators has undergone little change has predominated, and is not challenged within the following section. Throughout the last thirty years, these approaches have been improved and new analyses and methodologies have increased the certainty with which predators can be identified. When used together, the methods described below complement and support each other. Criteria most often used in the analysis of small mammal assemblages are the relative abundance and breakage of bones, and the digestion of tooth enamel. Most of these data have come from the analysis of dietary waste, either scat or pellets, of mammalian or avian predators. However, bones are sometimes marked or damaged during the process of prey capture prior to consumption, and studies involving these are considered first.

2.2.1. Predator Marks

Mammalian carnivore deposits can be identified by the analysis of gnawing and puncture marks made by the teeth of these predators. Unfortunately, characteristic predator tooth marks are less common on small mammal bones (Andrews, 1990, p. 42-43), because such bones normally break rather than retain evidence of tooth marks, and many of the prey are small enough to be swallowed whole without the bones ever coming into contact with the predators' teeth. Where present, these marks occur mainly on prey items attributed to foxes and coyotes (Andrews and Evans, 1983). Indentations that fit the size of fox teeth have been discovered on small mammal bones from Penal, Spain (Fernandez-Jalvo, 1995).

Gnawing marks are more common on the bones of large mammal prey (Haynes, 1980). For example, characteristic gnawing marks on bones from Kirkdale Cave led Buckland to suggest that the bones had been accumulated by hyenas, although the action of secondary bone modifiers, such as the porcupine (Brain, 1981) should also be considered. The difference in frequencies of proximal and distal skeletal elements was utilized by Behrensmeyer to identify the activity of carnivores in the large vertebrate assemblage from Koobi Fora, Kenya. She suggested that the loss of proximal elements was due to predator gnawing and bone crushing of the "marrow filled diaphyses" (Behrensmeyer, 1975, p. 540). Puncture marks produced by raptor talons have been recorded on recently collected bones of primates, hyrax and birds from South Africa (Berger and Clarke, 1995), and on small primate skulls from the Eocene fossil locality of Grizzly Buttes, Wyoming, USA (Alexander and Burger, 2001). Cranial damage has also been recorded on a juvenile *Australopithecus* cranium (SK 54) from Swartkrans (South Africa), which matches exactly the lower canines of a leopard (*Panthera pardus*) (SK349) from the same deposit (Brain, 1981).

When assessing predator-prey relations from tooth marks on large mammal prey, it is important to consider that prey items may have interactions with more than one predator or scavenger species, and that marks left on the bones may not be those of the individual responsible for the kill. This problem is less likely to be encountered in the analysis of small mammal accumulations. However, in autochthonous deposits, the possibility that more than one predator has been involved in the accumulation of either small or large vertebrate remains must be considered. The absence of tooth marks should not be used as evidence that an assemblage is not the product of mammalian carnivores. The presence of tooth marks is governed by a large number of variables, including prey vulnerability and predator feeding group size (Haynes, 1983), as well as the relative size of the predator to the prey (Andrews, 1990). Finally, although many species of mammalian predator may leave characteristic marks on large mammal bones, weathering and trampling may then obscure this information (Haynes, 1980).

Another mammalian predator to leave characteristic marks on bones are humans. Tooth marks are rarely recorded, and bone modification is usually confined to cut-marks sustained during the processing of the food (Schmitt and Juell, 1994). Other recognizable cultural modifications occur during cooking (Simonetti and Cornejo, 1991), which in some cases affects only part of the skeleton, enabling a complete reconstruction of the cooking practice to be undertaken (Henshilwood, 1997). Hominid predator-prey relations are the subject of the next chapter, so only data on small mammals as food have been included here.

2.2.2. Relative Abundance and Bone Breakage

Relative abundance and bone breakage criteria were first applied to the study of small mammals by Mellett (1974) in an analysis of material recovered from Mesozoic and Tertiary deposits. He compared the bones with modern carnivore scat material, and found them to be "identical" in appearance (Mellett, 1974). He suggested that all fluvial microvertebrate accumulations were due to deposition by carnivores, "mainly mammalian, but also including predacious fish, reptiles and birds" (Mellett, 1974, p. 349). An analysis of the similarity of disarticulation and fragmentation patterns on bones from fossil deposits with those recovered from modern predators was also carried out by Mayhew (1977) in an analysis of Cromerian deposits at West Runton, England. These early analyses contained little quantitative examination of the bone breakage, and it is not possible to compare these results with those of later studies.

The first truly quantitative analysis of predator diet material was undertaken by Dodson and Wexlar (1979). They analyzed pellets from three species of captive owls: barn owl (*Tyto alba*), great horned owl (*Bubo virginianus*) and Eastern screech owl (*Otus asio*). They recorded the amount of bone returned within the owl pellets and compared it with the expected number of bones based in the minimum number of individuals (MNI) for each pellet. They were therefore able to indicate that the screech owl and the great horned owl were responsible for greater rates of bone element loss than the barn owl. Bone breakage was also recorded in detail, with long bone breakage recorded in categories of either intact, proximal, distal or shaft, with results similar to those for bone loss (Dodson and Wexlar, 1979, p. 279). Analysis of bone loss was carried out by Hoffman (1988) based on the number of prey fed to captive owls.

Quantitative analysis of predator-derived bone breakage was also carried out using similar methods by Korth (1979), using pellets from barn owl (*Tyto alba*) and great horned owl (*Bubo virginianus*) obtained from the wild. His results indicated lower frequencies of breakage and bone loss than had been recorded in earlier studies. The cause of the higher frequencies in captive owls was possibly associated with increased stress through captivity, and uncharacteristic behavior, such as eating dead prey, rather than capturing it. The remainder of the studies outlined below also used only pellet or scat remains collected from wild predators.

TABLE 1. Species of Owl, Diurnal Raptor and Mammalian Carnivore Studied by Andrews (1983), Andrews and Evans (1983), and Andrews (1990)

Owls	Diurnal raptors	Mammalian carnivores
barn owl	common buzzard	red fox
(*Tyto alba*)	(*Buteo buteo*)	(*Vulpes vulpes*)
snowy owl	kestrel	coyote
(*Nyctea scandiaca*)	(*Falco tinnunculus*)	(*Canis latrans*)
long-eared owl	hen harrier	bat-eared fox
(*Asio otus*)	(*Circus cyaneus*)	(*Otocyon megalotis*)
short-eared owl	peregrine	white-tailed mongoose
(*Asio flammeus*)	(*Falco peregrinus*)	(*Ichneumia albicauda*)
great gray owl	red kite	small-spotted genet
(*Strix nebulosa*)	(*Milvus milvus*)	(*Genetta genetta*)
spotted eagle owl		badger
(*Bubo africanus*)		(*Meles meles*)
Verreaux eagle owl		American mink
(*Bubo lacteus*)		(*Mustela vison*)
tawny owl		otter
(*Strix aluco*)		(*Lutra lutra*)
little owl		Arctic fox
(*Athene noctua*)		(*Alopex lagopus*)
		pine martin
		(*Martes martes*)

The previous studies concentrated on only a limited number of predators. Andrews by comparison dramatically increased the number of predator species for which bone breakage data were collected from pellets and scats (Andrews, 1983; Andrews and Evans, 1983; Andrews, 1990). These are shown in Table 1. In these studies both cranial and post-cranial bone breakage were recorded. The categories for post-cranial breakage followed those employed by Dodson and Wexlar (1979). Cranial breakage was defined as damage to the maxilla (complete, broken skull with zygomatic intact, or fragmented with the loss of the frontal and incisors) or the mandible (complete, ascending ramus broken, ascending ramus missing, and inferior border broken) (Andrews, 1990). Relative abundance was further defined by measuring the proportions of cranial/post-cranial elements, and also the proportion of distal elements. This second measure generally indicates that raptors and mammalian carnivores are responsible for higher rates of distal element loss than owls (Andrews, 1990, p. 50). A statistically significant difference between the relative abundance of bones comparing owls and mammalian carnivores was also recorded by Kusmer (1990).

2.2.3. Discussion of Bone Breakage Data

Evidence from these comparative studies suggests that bone breakage can be used to recognize predator groups. Breakage is lowest in owls, and highest in mammalian carnivores, and is most likely related to feeding styles; owls swallow their prey whole, diurnal raptors pull the flesh from the bones and carnivores chew and crunch their prey. Analyses of intra-species variation in rates of bone breakage suggest that detailed analysis of bone breakage at the species level should be treated with caution (Saavedra and Simonetti, 1998). Furthermore, analysis of relative abundance and bone breakage are beset by the problem that many of these species-specific attributes can be altered by post-depositional diagenetic activity. For example, transport and fluvial sorting will affect the relative abundance and survival of particular skeletal elements (Voorhies, 1969; Dodson and Wexlar, 1979; Korth, 1979), and trampling and sediment compaction may lead to further bone breakage and therefore obscure predator-specific breakage patterns (Andrews, 1990; Fernandez-Jalvo, 1996). In extreme cases this breakage and transport may also lead to the loss of evidence for specific species from the assemblage and bias the ecological information (Dauphin et al., 1994).

2.2.4. Bone Digestion

As many small mammal assemblages will have undergone some post-depositional modification, identification of the predator species based solely on bone breakage and abundance data should be viewed cautiously (Fernandez-Jalvo, 1996). However, the analysis of bone digestion is considered to be a more reliable indicator (Andrews, 1990, p. 64):

> It has been found ... that the corrosive effects of digestion on the bones and teeth in the predator's stomach are not duplicated by any other alteration process and so may be used to identify bone assemblages derived from predators.

All the predators mentioned produce a visible record of digestion on the bones and teeth of their prey. The frequency and extent of these modifications can be used to differentiate predator species. In general, the results of analysis of digestion mirror those of other bone modification studies, with the least modification by digestion recorded in owls, and the greatest recorded in diurnal raptors and mammalian carnivores. The reasons for this are varied. Firstly, digestion in the volant predators takes place only in the stomach, through the action of stomach acids and enzymes. The pH level in the stomach (Duke *et al.,* 1975) and the period of time that the bones are subjected to this acid and enzymatic attack (Denys *et al.,* 1995) are suggested as the main mechanisms regulating the extent of digestion witnessed on fossil microvertebrate bones. Furthermore, pre- and post-prandial changes in pH level may also affect the extent and number of bones digested (Grimm and Whitehouse, 1963), and, where food supply is limited, this may be the cause of increased digestion (Andrews, 1990, p. 33). Greater frequency of digestion has also been recorded in younger owls when compared with their parents (Raczynski and Ruprecht, 1974; Andrews, 1990), making it possible to differentiate between nest and roost sites in the fossil record (Williams, 2001).

Similar mechanisms are responsible for bone digestion in mammalian carnivores, except that digestion takes place in the intestines as well as the stomach, thereby increasing the amount of digestion. Bone breakage may also be responsible for higher levels of digestion; when mastication occurs (as opposed to occasions where the prey is swallowed whole), an increased surface area of bone is exposed (Andrews and Evans, 1983). This is pertinent to the case of mammalian carnivores which exhibit relatively high levels of bone breakage, but may also relate to diurnal raptors and owls that break their prey into smaller pieces before consumption. Human digestive effects on small mammals have also been investigated and found to be similar in appearance to that of other mammalian carnivores (Crandell and Stahl, 1995).

2.2.5. Recognizing Digestion

The effects of digestion on small mammal bones and teeth were first recognized by Mayhew (1977), who analyzed pellets from raptors and owls and found a significant difference between the digestion and rounding of the bones from the raptor pellets as opposed to the owls. He suggested that, on the basis of comparison with modern pellet remains, small mammal teeth excavated from the Cromerian freshwater deposits at West Runton, England, were deposited by a diurnal raptor, most likely the kestrel (*Falco tinnunculus*) (Mayhew, 1977, p. 31).

In this study, Mayhew (1977, p. 25) commented on the lack of digestion in the owl species used in his analysis, and a similar conclusion was drawn by Korth (1979, p. 240). Kusmer (1990) also found a low incidence of digestion in her samples. However, these authors were using pellets from species of owls that have since been shown to digest bone infrequently (Andrews, 1990). Corrosion (or digestion) of the bone was also used as a category of bone alteration by Andrews and Evans (1983) in their analysis of mammalian carnivore deposits, although these data were not quantified. Digestion and decalcification of small mammal teeth and bones were witnessed in experimental analysis of crocodile scat, often leading to complete bone destruction, although in some cases teeth with no enamel were preserved (Fisher, 1981).

In his analysis of the Omo microfauna, Wesselman (1984) considered that both diurnal and nocturnal bird species were responsible for the accumulation, as evidenced by the differential degree of enamel etching seen between the diurnal and nocturnal prey species (Wesselman, 1984, p. 185). Bone digestion data were also collected by Geering (1990) from modern pellets of the Tasmanian masked owl (*Tyto novaehollandiae castanops*). The data indicated that digestion varied according to the specific skeletal element, with a higher proportion of tibia digested than mandibles. The study also investigated the link between digestion and age of prey and found that digestion was significantly more pronounced on unfused juvenile and sub-adult bones than fused adult bones (Geering, 1990, p. 139).

2.2.6. Categorizing Digestion

Although the presence of corroded bones had been recognized and commented on by a number of authors (Mayhew, 1977; Dodson and Wexlar, 1979; Fisher, 1981; Andrews, 1983; Andrews and Evans, 1983; Geering, 1990), the first quantification was again carried out by Andrews (1990). Evidence of digestion was collected for most of the species listed previously (see Table 1). Data were recorded for molar and incisor teeth as well as the proximal femur and ulna, and distal humerus. The results were used to produce five groups for molar and incisor digestion, shown in Table 2, with the least amount of digestion recorded for category 1 and the greatest for category 5. In most cases owls are placed in categories 1-3 while diurnal raptors and mammalian carnivores mainly comprise categories 4 and 5 (the mammalian carnivores are not shown below as the number of bones recovered from these species was low). The results indicate a similarity in the categories for both molar and incisor digestion, with species falling into roughly the same categories. However, in all cases the frequency of digestion is higher on the incisors.

These categories have been assigned on the basis of the number of bones digested (frequency) for both molars and incisors. The location of the digestion, e.g. tips only for category 2 incisor digestion, was recorded but not quantified. In the initial analysis Andrews did not test the significance of these digestion categories. However, the frequency of digestion data have recently been tested using ANOVA (Williams, 2001);

TABLE 2 Summary of Molar and Incisor Digestion, Indicating Categories of Avian Predator Assigned on the Basis of Analysis of the Percentage of Molar and Incisor Digestion, from Andrews (1990)

Digestion category	Molar digestion	Incisor digestion
1. digestion absent or minimal digestion molars 0-3% incisors 8-13%	barn owl (*Tyto alba*), long-eared owl (*Asio otus*), short-eared owl (*Asio flammeus*), Verreaux eagle owl (*Bubo lacteus*).	barn owl (*Tyto alba*), short-eared owl (*Asio flammeus*), snowy owl (*Nyctea scandiaca*).
2. moderate digestion molars 4-6% incisors 20-30% (tips only)	snowy owl (*Nyctea scandiaca*), spotted eagle owl (*Bubo africanus*), great gray owl (*Strix nebulosa*).	long-eared owl (*Asio otus*), Verreaux eagle owl (*Bubo lacteus*), great gray owl (*Strix nebulosa*).
3. heavy digestion molars 18-22% incisors 50-70%	European eagle owl (*Bubo bubo*), tawny owl (*Strix aluco*).	European eagle owl (*Bubo bubo*), spotted eagle owl (*Bubo africanus*), tawny owl (*Strix aluco*), little owl (*Athene noctua*).
4. extreme digestion molars 50-70% incisors 60-80%	little owl (*Athene noctua*), kestrel (*Falco tinnunculus*), peregrine (*Falco peregrinus*).	kestrel (*Falco tinnunculus*), peregrine (*Falco peregrinus*).
5. extreme digestion molars 50-100% incisors 100% (dentine corroded)	hen harrier (*Circus cyaneus*), buzzard (*Buteo buteo*), red kite (*Milvus milvus*).	hen harrier (*Circus cyaneus*), buzzard (*Buteo buteo*).

see Table 3. The results for number of molars and incisors digested for categories 1-4 predators indicate that these groups (1-4) are significantly different from each other, and therefore represent discrete categories of predator modification.

The categories of digestion initially described by Andrews (1990) have been further refined by Fernandez-Jalvo and Andrews (1992) with the introduction of a methodology to record the location of digestion, and to categorize its extent. The new categories of digestion are based on the molars and incisors, and the principal difference between this research and that of Andrews (1990) is that digestion is recorded to sub-categories of light, moderate, heavy or extreme digestion. In the case of the incisors, the location of the digestion is recorded as either superficial (along the entire incisor) or tip (restricted to the incisor tips) (Fernandez-Jalvo and Andrews, 1992, p. 412-417).

2.2.7. Digestion and Bone Chemistry

It is also possible to analyze the chemical composition of skeletal elements that have passed through the digestive system of various predators to differentiate owls, raptors and mammalian carnivores. This tool is particularly useful when dealing with sites where more than one predator may have been responsible for the accumulation of microvertebrate remains (Dauphin *et al.*, 1997). It should be noted, however, that digestion can alter other chemical signatures, for example those used in isotopic analysis (Denys *et al.*, 1992), which could lead to erroneous paleoecological interpretations based on isotope analysis of digested small mammal bones.

2.2.8. Digestion and Prey Species Variability

While it has been shown that different predators produce a varied range of digestion, the type of prey caught will also affect the amount of digestion recorded. For example, vole molars are more easily digested than mouse, gerbil or hamster molars, as a result of differences in shape and enamel thickness. Vole molars are commonly digested on the corners of the salient angles of the teeth, leading to a rounding of the occlusal surface, whereas mouse molars are already more rounded in shape, and digestion appears to be less common, and more difficult to detect (Andrews, 1990). This difference is most visible in cases of low digestion (category 1 and 2 predators) and, as a result, could lead to bias in samples containing high proportions of a particular species.

Evidence from pellet analysis in both Britain and southern Africa indicates that it is typical for the dominant prey species of a particular habitat to be captured more commonly than other species, and for the proportions of the other species to rise or fall in response to the availability of the main prey species (Hanney, 1963; Taylor, 1994). This then has an impact upon the taphonomic analysis carried out on small mammal molar digestion. For example, the proportion of mice to voles within a sample will have an effect upon the amount of molar digestion recorded, as mouse molars are

TABLE 3. Results of One-Way ANOVA (Analysis of Variance) Comparing Total Molar and Total Incisor Digestion for the Predator Categories 1-4 (Andrews, 1990)

Digestion variable	"F"	"P"
Total molar digestion	142.699	.000
Total incisor digestion	41.14	.000

less frequently digested than vole molars. A high proportion of the diet of the barn owl (*Tyto alba*) in Britain is made up of voles (Glue, 1974; Andrews, 1990, p. 96; Love *et al.,* 2000), whereas the African barn owl (*Tyto alba affinis*), hunting in South Africa, could be expected to capture a higher proportion of mice, gerbils or hamsters (Coetzee, 1972; Rautenbach, 1978; Brain, 1981). Thus, it is likely that more digestion will be recorded in a sample dominated by vole species than one containing mostly mice. While this has implications for paleoecological reconstructions, it could also lead to erroneous identification of the predator species.

It is suggested that the number of specific species within an assemblage of small mammals plays an important role in the taphonomy of that deposit. Seasonal variations in proportions of prey taken, or the numbers of the most dominant prey type, could then have a dramatic effect on the amount of digestion in part of the sample. In order to recognize this difference, it is essential that all evidence of molar digestion is recorded for each species represented. This level of recording was carried out in the course of research on British Holocene small mammal assemblages (Williams, 2001). The results of all of these analyses were then divided into three classes, mice, voles and large voles. Digestion for each of these classes was tallied, and from this the proportion of digestion commonly associated with each class was recorded. This figure was used as the population, against which individual site data could be standardized. When this calculation was applied to the data, it led to a reduction in the amount of recorded digestion at some sites, where high proportions of digestion had occurred due to high numbers of individual species. At one site, Fox Hole Cave, England, standardization of the result of molar digestion of three samples from the cave reduced the variability of these samples, suggesting a single predatory origin (Williams, 2001).

3. Fossil Evidence

One of the principal reasons for trying to interpret the predatory origin of fossil faunas is to understand how the selection of prey represents the ecology of the local environment at the time of capture. The many ecological analyses of predator species diet and behavior produce a fairly clear picture of how modern day predators sample their prey, and how representative this prey is of the overall prey community. The same behavior is assumed for these predators, or their ancestors, in the past.

3.1. Case Studies of Predatory Origin

Using the techniques described in the sections above, a number of studies of the predatory origin of fossil microvertebrate assemblages have been carried out. Some of these, such as the early studies by Mellett (1974), Mayhew (1977), and Wesselman (1984), have already been covered. A number of other studies of predator assemblages are reviewed below, which concentrate on the identification of the predator, and also the potential bias in the sampling of the small mammal species.

3.1.1. Olduvai FLKN, Tanzania

In an analysis of bone breakage, and also by reference to high levels of digestive attack, Andrews identified the predator responsible for the accumulation of the Olduvai FLKN microfauna as a small mammalian predator, most likely a genet (Andrews, 1983).

Further analysis of the fish and birds, as well as the small mammals from this site, prompted Stewart et al. (1999) to suggest that at least the birds and fish may have been accumulated by fish eagles (genus *Haliaeetus*). They recognized that these predators are not responsible for the entire accumulation but suggested that the patterning of parts of the deposit is suggestive of a fish eagle accumulation, and that the deposit is therefore likely to have multiple predatory origins.

3.1.2. Westbury-sub-Mendip, England

By far the most significant study is the analysis of the Westbury-sub-Mendip small mammals by Andrews (1990). Taphonomic analysis was carried out on 18 different samples from this cave, and the predatory agent of accumulation was identified in most cases. This information was then used to assess the potential bias in the collection of these faunas. For example, in the analysis of unit 11, Andrews identified two predators, a European eagle owl (*Bubo bubo*) that produced sub-unit 11/4, and a barn owl (*Tyto alba*) that accumulated sub-unit 11/1. The different hunting strategies of these two owls are reflected in the prey spectrum of the two assemblages. The eagle owl deposit (11/4) has a much higher species richness (number of species), which results from more opportunistic hunting practices. Unit 11/1, on the other hand, is dominated by voles, a common component in the diet of the barn owl, which tends to hunt the most common species in open grassland environments – in northern latitudes, this tends to be voles.

The contribution that an accurate taphonomic analysis plays in a paleoecological assessment should not be underestimated. The Taxonomic Habitat Index (THI) will provide a more accurate result if it can be shown that all of the potential components of the fossil fauna are represented within the sample. In the example above, the THI is not affected by this difference, and the faunas of the two sub-units indicate an environment dominated by deciduous woodland, with abundant ground vegetation (Andrews, 1990, p. 169-170). The problems associated with different predator hunting strategies are also apparent in the results from the analysis of the main chamber units 18 and 19. The predator responsible for the deposition of unit 18 is identified as the stoat (*Mustela erminea*). This small mammalian carnivore does preferentially select its prey, so individual abundance data may be biased. However, as the stoat is also an opportunistic predator, it is likely that presence / absence data for a particular prey species are valid. The THI for these species has high tundra and boreal forest affinities, indicating cold conditions. The small mammals from unit 19 were accumulated by a barn owl (*Tyto alba*), and, as was the case with sub-unit 11/1 above, this sample was dominated by vole species. The THI for this unit indicates much warmer conditions than in the underlying unit 18. Andrews suggested that, given the known bias towards voles, and therefore potential loss of other species, evidence of even warmer conditions may have been selected against by the action of the predator (Andrews, 1990, p. 176).

3.1.3. Gran Dolina, Spain

The accurate assessment of the predatory origin of the deposits at Gran Dolina, Spain, also allowed Fernandez-Jalvo and Andrews (1992) to indicate how the paleoecological reconstruction of the site was biased by predator selectivity. Opportunistic predators were recognized in units TD3 (European eagle owl, *Bubo bubo*), TD5 and TD6 (tawny owl, *Strix aluco*) and the resultant small mammal accumulations are therefore likely to be an accurate representation of the prey community. The environmental reconstruction suggests a predominantly woodland environment with some open areas. Conversely, a specialist predator, the long-eared owl (*Asio otus*), was recognized in unit TD11. It is suggested that, as this owl feeds mainly on voles, the species richness of the deposit is low, and therefore less representative of the habitat. It is likely that the long-eared owl only hunted in open areas, and this accounts for the lower species richness in unit TD11.

Other lines of evidence (pollen and the larger mammal fauna) imply that the environment of TD11 was similar to that of the lower units, corroborating the suggestion that TD11 was taphonomically biased (Fernandez-Jalvo and Andrews, 1992, p. 426). The fact that the long-eared owl lives in woodland and hunts predominantly over grassland should also be considered, as pellets containing grassland species are likely to be deposited in a woodland environment (Fernandez-Jalvo, 1996, p. 33). Taphonomic analysis of the adjacent site of Penal has also indicated that most of the predators were opportunistic predators, and therefore a representative sample of small mammal species will result from their hunting behavior (Fernandez-Jalvo, 1995).

At both of these sites, as well as at Westbury-sub-Mendip, these opportunistic predators are described as taking all of the prey species within their habitats. While this is usually the case, because most owls and mammalian carnivores are nocturnal, the vast majority of their prey will be nocturnal. The converse is true for diurnal raptors and their prey.

3.1.4. Olduvai Bed 1, Tanzania

Both avian and mammalian predators have been identified in the Olduvai bed 1 deposits by Fernandez-Jalvo *et al.* (1998). At this site the authors used the Gerbillinae/Murinae ratio in each unit to indicate changing habitats, and via a thorough taphonomic analysis were able to identify in which deposits ratio data were a product of predator hunting practices, and those in which they reflected actual proportions. Species richness (number of species) was also examined, and positively correlated with sample size in all but three deposits. These three deposits were accumulated by eagle owls, which usually select a wide range of prey. Because species richness increases in more complex habitats, it is likely that any reduction in species richness was a reflection of a reduction in ecological diversity, and therefore environmental change.

The deposit with the greatest species richness was shown to have been accumulated by a number of predators. While this will have an effect on studies of habitat diversity, from a paleoecological perspective the greater the number of predators, the less the problem of prey selection bias. More species are likely to be captured, and abundance data may be more meaningful. Where studies relate changes in species abundance to environmental change without taking account of predator bias, the possibility that such interpretations may themselves be biased must be considered. However, abundance data from multiple predator sites, where there has been no other selection bias (or where it has been accounted for), could then be used to evaluate changes in environment (Hadly, 1999).

3.2. Problems with Early Fossil Evidence

One aspect that has not yet been explored is the extent to which the behavior of modern predators may have changed over the last few million years. The discovery of small primate fossils from the *Omomys* Quarry (middle Eocene, Bridger Formation, Wyoming, USA) may be a case in point (Murphey *et al.*, 2001). Taphonomic analysis indicated that the bones from this site were fairly fragmented (approximately 70%) and that most of this breakage was pre-depositional and predator derived. However, there was no evidence for digestion at the end of the long bones or on the teeth. This pattern is not consistent with any modern avian or mammalian predators, and it was suggested that, given the level of digestion, these primates were accumulated by an owl (Murphey *et al.*, 2001). As this period was during the early evolutionary stage for owls (with most modern species emerging in the Miocene (Grossman and Hamlet, 1964)), such differences may be expected. These issues are also raised by Thornton and Rasmussen (2001) in relation to the identification of mammalian predators of the small mammals and primates at the Eocene locality of Gnat-Out-of-Hell, Utah, USA. It is likely that occasions where the exact predatory origin may be more difficult to define will increase in proportion to the antiquity of the deposit and the number of extinct predator or prey species represented.

3.3. Predator Bias?

It should be recognized that the fossil record does not truly reflect predator density in any given region. From an analysis of the fossil data, we would conclude that owls are the most prolific predators of small mammals. Mammalian carnivore scat accumulations are, as pointed out by Andrews and Evans (1983), rare in the fossil record, despite the fact that considerable numbers of scats can be accumulated at specific latrine or territory marker sites. This is particularly true for Viverridae and Canidae species, such as the civet (personal observation), genet, mongoose, foxes (Andrews, 1990) and coyotes (Schmitt and Juell, 1994). Recently published data on predation on British field voles (*Microtus agrestis*) indicates that, although voles may constitute a considerable percentage of owl diets, this figure is significantly overshadowed, in terms of predation rates, by mammalian carnivores (Dyczkowski and Yalden, 1998). In this study, weasel (*Mustela nivalis*), kestrel (*Falco tinnunculus*), red fox (*Vulpes vulpes*), and feral cat (*Felis catus*) were shown to consume 85% of total numbers of field voles, with the five British owl species only accounting for an additional 1% each. Even if the recent decline in numbers of barn owls is taken into account, it is clear that a significant part of a prey population does not enter the fossil record. This is certainly a biasing factor in any analysis of predator-prey interactions based purely on a single species of predator, where the effect of prey interactions with other possible predators (such as reduced foraging through predation risk) cannot be adequately identified.

3.4. Population Fluctuations

The ecological record contains evidence for cyclic population fluctuations of small mammals, and the effect that these cycles have on their predators, and *vice versa* (Elton, 1966; Taylor, 1994; Chitty, 1996). There is also evidence of the occurrence of these cycles in the historic record, where downward fluctuations in the lemming (*Lemmus*) population in Canada are mirrored by dramatic decreases in fox pelts sold by the Hudson's Bay Company, recorded for the years 1870 - 1920 (Chitty, 1996, p. 11-14).

However, little evidence of these cycles is recorded in the fossil record. Why has recognizing this phenomenon been so difficult? If, for example, a predator specializes in one species of vole, and hunts it to the exclusion of all other prey when it is available, the probable taphonomic bias in fossil deposits would be towards large numbers of voles, perhaps even one species of vole. This would give an unbalanced view of the local environment. If, however, this predator had to switch prey, as a result of a crash or disappearance of its main prey species, then the resulting prey spectrum would be more wide ranging. However, the prey range would no longer reflect that which we imagine for that predator (a good reason for not using prey species abundance data to identify predators) and may make our understanding of those data more difficult.

3.4.1. Tadcaster, England

An example from research on the British Holocene archaeological site of Tadcaster highlights a few questions worth exploring in more detail (Williams, 2001). A microvertebrate deposit from an abandoned building was shown to have been deposited by a barn owl; taphonomic analysis indicated that both adult birds and their chicks had contributed to the accumulation. The deposit is approximately 300 years old and, because of its autochthonous location, provides an excellent opportunity to test these questions on an undisturbed deposit. The small mammals would have been hunted within a limited radius of the building, no more than 1-2 km. With the exception of modern removal of field hedgerows and boundaries, the local environment should have differed little from what is seen today. This type of environment should support large populations of field voles, and these normally dominate British barn owl diets. However, the numbers of field voles (*Microtus agrestis*) are low at this site when compared to that recorded for Britain as a whole, which may indicate evidence of prey switching. While further research in the area would enable a better understanding of the environment, data presented in Table 4 clearly indicate a difference between the species present at this site and the average result for the rest of the country. Analysis of the data indicates that there is a significant difference between the two data sets, $x^2 = 71.39$, df = 6, $p = .000$.

3.4.2. Marabib Rock Shelter, Namibia

Further evidence of prey switching is given by Brain (1981), who recorded changes in the proportions of desert gerbils (*Gerbillurus vallinus* & *Gerbillurus paeba*) and

TABLE 4. Species Abundance from the Old Vicarage, Tadcaster, Compared with Survey Data from Glue (1974). Species Represented by Letters. A= *Apodemus sylvaticus*, B= *Mus domesticus*, C= *Arvicola terrestris*, D= *Clethrionomys glareolus*, E= *Microtus agrestis*, F= *Sorex araneus*, G= *Sorex minutus*.

SITE / SPECIES	A	B	C	D	E	F	G
Tadcaster	31%	6%	2%	3%	25%	31%	2%
Glue 1974	13%	1%	0%	4%	49%	27%	6%

geckos (*Pachydactylus bibroni*) from the Mirabib rock shelter in Namibia. Over a six thousand year period, the number of each prey species fluctuated widely, with individual species often comprising much of the owl's diet. While it is clear in this case that a barn owl was responsible for the latest accumulations, a lack of detailed taphonomic analysis means that some of this difference in prey type might be the result of different predators contributing to the accumulation.

3.5. Prey Selection and Small Mammal Evolution

In ecological terms, the action of prey selection is likely to mean that some species are not recovered from the fossil record. This will then give a distorted view of the environment, particularly if those species are sensitive ecological indicators. This is certainly true for volant hunters such as the barn owl, which hunt over a particular type of environment – grassland – and therefore catch mainly grassland species. What, however, is the evolutionary effect of prey selection, in terms of our study and understanding of rodent taxonomy? The problem with sampling only part of the potential prey population is that species inhabiting different, perhaps more closed, environments may be evolving and changing without ever entering the fossil record. Similarly, in sites with multiple predators, the presence of a species within one deposit and its disappearance in a later phase of the site should not necessarily be interpreted as the loss of a species, and the possibility of predator bias should be considered.

Changes in rodent behavior and any subsequent evolutionary change are likely to be related to feeding; most of the differences recognized between fossil (and modern day) rodents are made on the basis of differing dental morphology. These changes result from selective pressure on dental structure in order to take advantage of particular food resources within a given habitat. Competition will also encourage individuals to exploit different resources, which may necessitate morphological changes. Part of these adaptations may include the development of anti-predatory measures, for example changes in pelt color or pattern, overall size, or morphological adaptations of specific limbs.

For example, size variation in modern desert rodents (Heteromyidae) appears to be linked to foraging ability in dense brush. Smaller rodents forage more successfully under these conditions than larger species in the same environment. This dense brush provides cover from predation and in the case of the Arizona pocket mouse (*Perognathus amplus*) also represents a foraging habitat for this species. On the other hand, the larger Merriam's Kangaroo rat (*Dipodomys merriami*) spends more time foraging in open environments. It has also evolved mechanisms to exploit this area successfully. An increase in the size of auditory bullae ensures better aural detection of predation risk, while enlarged hind-limbs provide the mechanism for a fast bipedal escape (Brown et al., 1988). Similar results are recorded for desert gerbils (*Gerbillius pyramidum* and *Gerbillius allenbyi*) by Kotler et al. (1991). Fossorial behavior, and the concomitant physical manifestations of this (for example the extent of pro-odonty in voles (Montgomery, 1975), and limb evolution in moles, e.g. the humerus of *Talpa europaea*), may also reflect an attempt to avoid predation.

While it is possible to detect these changes in modern species, it becomes increasingly harder when analyzing rodent anti-predatory behavior in the fossil record. First, any camouflage is unlikely to be recovered, as fur does not usually survive consumption and burial environments. Secondly, in the examples used above, it may be difficult to recognize changes in limb length or bullae size unless there is a relatively complete fossil sequence with which to make the comparison. This may be further hampered by capture, consumption or post-depositional bone breakage, particularly of the fragile cranial bones.

4. Summary

This chapter has focused on small mammal predator-prey interactions, particularly on the identification of predators, using small mammal bones recovered from the fossil record. It has been shown that the analysis of bone breakage and digestion enables the identification of many small mammal predators. The lowest levels of bone modification are found in owl accumulations and the highest levels of damage and digestion are associated with the prey remains of diurnal raptors and mammalian carnivores. These data have been used to demonstrate that, through the identification of the predatory origin of fossil deposits, accurate reconstructions of past environments can be made. General trends can be drawn, suggesting that, where deposits have been accumulated by predators with opportunistic hunting and foraging habits, a more accurate picture of past environments may be produced. This is because these predators usually take a wider range of species from the local area than do specialist predators that may only exploit certain sections of the available environment.

Examples of the occurrence of prey switching by predators in response to low prey numbers have been given. However, further identification of these phenomena will be recorded only if researchers are able to study deposits where they know evidence of environmental change is unlikely, and even then, all other possible bone and deposit modifications must also be investigated. This review has highlighted that through detailed taphonomic studies it is possible to recognize the action of different predators in the fossil record, and that if the correct questions are asked, significant information regarding the possible predator-prey interactions of the time can be investigated.

ACKNOWLEDGMENTS: I would like to thank Peter Andrews, who was kind enough to suggest that I would be well positioned to contribute this chapter, and the editors for believing him. I am grateful to the reviewers for their constructive comments and for directing me to some of the recent North American literature. I would also like to thank Robin Dennell, Andrew Chamberlain, Mike Toms, and Erika Petersen for their constructive comments, support and guidance.

References

Alexander, J. P., and Burger, B. J., 2001, Stratigraphy and taphonomy of Grizzly Buttes, Bridger Formation, and the middle Eocene of Wyoming, in: *Eocene Biodiversity: Unusual Occurrences and Rarely Sampled Habitats* (G. F. Gunnell, ed.), Kluwer Academic/Plenum Publishers, New York, pp. 165-196.

Andrews, P., 1983, Small mammal diversity at Olduvai Gorge, in: *Animals and Archaeology Vol. 1 Hunters and their Prey* (J. C. Brook and C. Grigson, eds.), British Archaeological Reports, International Series, Oxford, pp. 77-85.

Andrews, P., 1990, *Owls, Caves and Fossils*, Chicago University Press, Chicago.

Andrews, P., and Evans, E. M. N., 1983, Small mammal bone accumulations produced by mammalian carnivores, *Paleobiology* **9**:289-307.

Behrensmeyer, A. K., 1975, The taphonomy and paleoecology of Plio-Pleistocene vertebrate assemblages east of Lake Rudolf, Kenya, *Bull. Mus. Compar. Zool.* **146**:473-578.

Berger, L. R., and Clarke, R. J., 1995, Eagle involvement in accumulation of the Taung child fauna, *J. Human Evol.* **29**(3):275-299.
Brain, C. K., 1981, *The Hunters or the Hunted*, Chicago University Press, Chicago.
Brown, J. S., Kotler, B. P., Smith, R. J., and Wirtz, W. O., 1988, The effects of owl predation on the foraging behavior of heteromyid rodents, *Oecologia* **76**:408-415.
Buckland, W., 1823, *Reliquiae diluvianae; or observations of the organic remains contained in caves, fissures and diluvial gravel, and on other geological phenomenon, attesting to the action of an universal deluge*, John Murray, London.
Chitty, D., 1996, *Do Lemmings Commit Suicide? Beautiful Hypotheses and Ugly Facts*, Oxford University Press, Oxford.
Coetzee, C. G., 1972, The identification of southern African small mammal remains in owl pellets, *Cimbebasia* **A(2)**:53-64.
Crandell, B. D., and Stahl, P. W., 1995, Human digestive effects on a micromammalian skeleton, *J. Archaeol. Sci.* **22**:789-797.
Dauphin, Y., Denys, C., and Kowalski, K., 1997, Analysis of accumulations of rodent remains: Role of chemical composition of skeletal elements, *Neues Jahrb. Geol. Paläontol. Abh.* **203**(3):295-315.
Dauphin, Y., Kowalski, C,. and Denys, C., 1994, Assemblage data and bone and teeth modifications as an aid to palaeoenvironmental interpretations of the open-air Pleistocene site of Tighenif (Algeria)., *Quat. Res.* **42**:340-349.
Denys, C., Fernandez-Jalvo, F. and Dauphin, Y., 1995, Experimental taphonomy: preliminary results of the digestion of micromammalian bones in the laboratory, *C. R. hebdomadaires Acad. Sci. Paris* **321**(2a):803-809.
Denys, C., Kowalski, K. and Dauphin, Y., 1992, Mechanical and chemical alterations of skeletal tissues in a recent Saharan accumulation of feaces from *Vulpes rueppelli* (Carnivora, Mammalia), *Acta Zool. Cracoviensia* **32**(2):265-283.
Dodson, P., and Wexlar, D., 1979, Taphonomic investigation of owl pellets, *Paleobiology* **5**:275-284.
Duke, G. E., Jegers, A. A., Loff, G., and Evanson, O. A., 1975, Gastric juice of some raptors, *Comp. Biochem. Physiol. A - Physiol.* **50**:649-656.
Dyczkowski, J., and Yalden, D. W., 1998, An estimate of the impact of predators on the British Field Voles *Microtus agrestis* population, *Mamm. Rev.* **28**(4):165-184.
Elton, C., 1966, *Animal Ecology*, Methuen & Co, London.
Fernandez-Jalvo, Y., 1995, Small mammal taphonomy at La Trinchera de Atapuerca (Burgos, Spain). A remarkable example of taphonomic criteria used for stratigraphic correlations and palaeoenvironmental interpretations, *Palaeogeogr. Palaeoclim. Palaeoecol.* **114**:167-195.
Fernandez-Jalvo, Y., 1996, Small mammal taphonomy and the Middle Pleistocene environments of Dolina, Northern Spain, *Quat. Int.* **33**:21-34.
Fernandez-Jalvo, Y., and Andrews, P., 1992, Small mammal taphonomy of Gran Dolina, Atapuerca (Burgos), Spain, *J. Archaeol. Sci.* **19**:407-428.
Fernandez-Jalvo, Y., Denys, C., Andrews, P., Williams, T., Dauphin, Y. and Humphrey, L., 1998, Taphonomy and Palaeoecology of Olduvai Bed-1 (Pleistocene, Tanzania), *J. Human Evol.* **34**:137-172.
Fisher, D. C., 1981, Crocodilian scatology, microvertebrate concentrations, and enamel-less teeth, *Paleobiology* **7**:262-275.
Geering, K., 1990, A taphonomic analysis of recent masked owl (*Tyto novaehollandiae castanops*) pellets from Tasmania, in: *Problem Solving in Taphonomy, Archaeological and Paleontological Studies for Europe, Africa and Oceania* (S. Solomon, I. Davidson, and D. Watson, eds.), Tempus, Queensland, pp. 135-143.
Glue, D. E., 1974, Food of the Barn owl in Britain and Ireland, *Bird Study* **21**:200-210.
Grimm, R. J., and Whitehouse, W. M., 1963, Pellet formation in a Great-horned owl. A roentgenigraphic study, *The Auk* **80**:301-306.
Grossman, M. L., and Hamlet, J., 1964, *Birds of Prey of the World*, Clarkson N. Potter, New York.
Hadly, E. A., 1999, Fidelity of terrestrial vertebrate fossils to a modern ecosystem, *Palaeogeogr. Palaeoclim. Palaeoecol.* **149**:398-409.
Hanney, P., 1963, Observations upon the food of the Barn owl (*Tyto alba*) in Southern Nyasaland, with a method of ascertaining population dynamics of rodent prey, *Ann. Mag. Nat. Hist.* **65**:305-313.
Harris, S., Morris, P., Wray, S., and Yalden, D., 1995, *A Review of British Mammals: Population Estimates and Conservation Status of British Mammals Other Than Cetaceans*, JNCC, Peterborough.
Haynes, G., 1980, Evidence of carnivore gnawing on Pleistocene and Recent mammalian bones, *Paleobiology* **6**:341-351.
Haynes, G., 1983, A guide for differentiating mammalian carnivore taxa responsible for gnaw damage to herbivore limb bones, *Paleobiology* **9**:164-172.

Henshilwood, C. S., 1997, Identifying the collector: evidence for human processing of the Cape dune mole-rat, *Bathyergus suillus*, from Blombos Cave, Southern Cape, South Africa, *J. Archaeol. Sci.* **24**:659-662.

Hoffman, R., 1988, The contribution of raptorial birds to patterning in small mammal assemblages, *Paleobiology* **14**:81-90.

Horwitz, L. K., and Goldberg, P., 1989, A study of Pleistocene and Holocene hyena coprolites, *J. Archaeol. Sci.* **16**:71-94.

Jacobs, L. L., 1985, Review of 'The Omo Micromammals' by H.B. Wesselman, *J. Vert. Paleontol.* **5**:281-283.

Korth, W. W., 1979, Taphonomy of microvertebrate fossil assemblages, *Annals Carnegie Mus.* **48**:235-285.

Kotler, B. P., Brown, J. S. and Hasson, O., 1991, Factors affecting gerbil foraging behavior and rates of owl predation, *Ecology* **72**:2249-2260.

Kowalski, K., 1990, Some problems with the taphonomy of small mammals, in: *International Symposium Evolution, Phylogeny and Biostratigraphy of Arvicolids (Rodentia, Mammalia)* (O. Fejfar and W. D. Heinrich, eds.), Geological Survey, Prague, pp. 285-295.

Kusmer, K. D., 1990, Taphonomy of owl pellet deposition, *J. Paleontol.* **64**:629-637.

Love, R. A., Webbon, C., Glue, D. and Harris, S., 2000, Changes in the food of British Barn Owls (*Tyto alba*) between 1974 and 1997, *Mamm. Rev.* **30**(2):107-129.

Mayhew, D. F., 1977, Avian predators as accumulators of fossil mammal material, *Boreas* **6**:25-31.

Mellett, J. S., 1974, Scatological origin of microvertebrate fossil accumulations, *Science* **185**:349-350.

Montgomery, W. I., 1975, On the relationship between sub-fossil and recent British Water voles, *Mamm. Rev.* **5**:23-29.

Murphey, P. C., Torick, L. L., Bray, E. S., Chandler, R., and Evanoff, E., 2001, Taphonomy, fauna and depositional environment of the *Omomys* Quarry, an unusual accumulation from the Bridger Formation (Middle Eocene) of Southwestern Wyoming (USA), in: *Eocene Biodiversity: Unusual Occurrences and Rarely Sampled Habitats* (G. F. Gunnell, ed.), Kluwer Academic/Plenum Publishers, New York, pp. 361-402.

Raczynski, J., and Ruprecht, A. C., 1974, The effects of digestion on the osteological composition of owl pellets, *Acta Ornithol.* **14**:1-12.

Rautenbach, I. L., 1978, Ecological distribution of the mammals of the Transvaal, *Annals Transvaal Mus.* **31**:131-157.

Saavedra, B., and Simonetti, J. A., 1998, Small mammal taphonomy: intraspecific bone assemblage comparison between South and North American barn owl *Tyto alba* populations, *J. Archaeol. Sci.* **25**:165-170.

Schmitt, D. N., and Juell, K. E., 1994, Towards the identification of coyote scatological faunal accumulations in archaeological contexts, *J. Archaeol. Sci.* **21**:249-262.

Simonetti, J. A., and Cornejo, L. E., 1991, Archaeological evidence of rodent consumption in central Chile, *Latin Am. Antiquity* **2**:92-96.

Stewart, K. M., Leblanc, L., Matthiesen, D. P., and West, J., 1999, Microfaunal remains from a modern east African raptor roost: patterning and implications for fossil bone scatters, *Paleobiology* **24**:483-503.

Taylor, I., 1994, *Barn Owls: Predator-Prey Relationships and Conservation*, Cambridge University Press, Cambridge.

Thornton, M. L., and Rasmussen, D. T., 2001, Taphonomic interpretation of Gnat-Out-of-Hell, an Early Uintan small mammal locality in the Unita Formation, Utah, in: *Eocene Biodiversity: Unusual Occurrences and Rarely Sampled Habitats* (G. F. Gunnell, ed.), Kluwer Academic/Plenum Publishers, New York, pp. 299-316.

Voorhies, M. R., 1969, Taphonomy and population dynamics of an early Pliocene vertebrate fauna, Knox County, Nebraska, *Univ. Wyoming Contrib. Geol., Spec. Paper* **1**:1-69.

Walton, A. H., 1990, Owl pellets and the fossil record, in: *Evolutionary Paleobiology of Behavior and Coevolution* (by A. J. Boucot), Elsevier, New York, pp. 233-241.

Wesselman, H. B., 1984, *The Omo Micromammals*, Contributions to Vertebrate Evolution 7, Karger, London.

Williams, J. P., 2001, Small mammal deposits in archaeology: a taphonomic investigation of *Tyto alba* (barn owl) nesting and roosting sites. Unpublished Ph.D. Thesis, University of Sheffield.

Yalden, D. W., 1999, *The History of British Mammals*, T & AD Poyser Natural History, London.

Chapter 15

Early Human Predation

RICHARD POTTS

1. Introduction 359
2. Interpreting Early Human Predation 360
3. Primate Background 361
4. Evolutionary Basis of Human Predatory Behavior...362
5. Ecological Overlap and Interaction between Hominins and Other Predator/Scavengers.........364
6. Predator-Prey Interactions during Human Evolutionary History 365
   6.1. Late Pliocene 365
   6.2. Early and Middle Pleistocene 368
   6.3. Late Pleistocene...369
   6.4. Holocene..370
7. Conclusion..371
   References..373

1. Introduction

From a phylogenetic perspective, human predatory behavior is an isolated phenomenon in which the propensity to capture and digest food from large vertebrate prey evolved in a minor clade of primates. From an ecological perspective, this phenomenon has had enormous repercussions for the present structure of terrestrial food webs and animal biological diversity.

All monkeys and apes, in which human phylogeny is embedded, have a predominantly plant diet, but most species also eat animals – insects, other invertebrates, or vertebrate tissues obtained by either active predation (hunting) or passive means (collecting of live animals). Human predation emerged in the context of an omnivorous diet, essentially an elaboration of behaviors present in apelike ancestors. The major evolutionary shift unique to humans, among all other primates, concerned the size of prey, specifically the capacity to exploit large mammals acquired by scavenging and hunting.

RICHARD POTTS • Human Origins Program, National Museum of Natural History, Smithsonian Institution, Washington, DC 20560-0112.

Predator-Prey Interactions in the Fossil Record, edited by Patricia H. Kelley, Michal Kowalewski, and Thor A. Hansen. Kluwer Academic/Plenum Publishers, New York, 2003.

The most carnivorous of all nonhuman primates today, the chimpanzee (*Pan troglodytes*), occasionally scavenges and hunts small (≤10 kg), mainly arboreal prey in forest or woodland environments (Stanford, 1998; Schoeniger et al., 2001). The fossil record shows, by contrast, that hominins of 2.5 to 1.8 million years ago (Ma) had begun to exploit terrestrial, savanna animals larger than 50 kg, probably relying on scavenging of the largest size classes (500 to >5000 kg). The manufacture of simple stone tools, useful in slicing meat and cracking bones, was associated with this transition. Modern *Homo sapiens* is capable of capturing and killing live prey of all sizes, including the largest land and marine animals.

The human fossil record, often considered poor compared with that of other organisms, is exceptionally rich in behavioral trace fossils. Archeological remains from well over 100,000 localities worldwide comprise many millions of hand-manufactured artifacts, typically stone tools. Fossilized remains of animal prey are often found associated with these artifacts. Much of what is known about the history of predatory behavior in human evolution comes from the interpretation of clusters of fossil animals and stone tools.

On the basis of current evidence, early human toolmakers began to gain access to large animal carcasses by about 2.5 million years ago (Ma) (de Heinzelin et al., 1999). This date also marks the oldest widely accepted occurrence of hominin stone flaking (Semaw et al., 1997). Major events in the history of human predation include complexly organized social hunting, cultural development of projectile implements (e.g., spears), and control over large animal populations, which eventually led to domestication.

Table 1 summarizes these and other markers in the evolutionary history of hominin carnivory and prey interaction, addressed further in the latter half of this chapter. The first half will consider the primate background of predatory behavior, the adaptive basis of human predation, and ecological interactions with large carnivorans (members of the Order Carnivora). The impact ascribed to predation in Pliocene and Pleistocene human evolution has been shaped by old assumptions and current debate; a short review of these issues is offered next.

2. Interpreting Early Human Predation

The association between Paleolithic tools and extinct fossil animals led scientists as early as the mid-19th century to speculate on the antiquity of human ancestry and the predatory heritage underlying human evolution. By the mid-20th century, elaborate and aggressive hunting was deemed to be the evolutionary hallmark of Pliocene *Australopithecus* and Pleistocene *Homo*; predation was considered a key innovation that distinguished even the oldest humans from apes (Dart, 1953; Ardrey, 1961). In the 1970s, growing awareness of the environmental, geological, and behavioral factors that govern fossil site formation ushered in an era of notable skepticism, which still exists, over whether Plio-Pleistocene hominins engaged in predation or even substantial carnivory (Brain, 1981; Binford, 1981; Potts, 1988; Shipman, 1986; Bunn and Ezzo, 1993).

This period of taphonomic research was triggered by two major research problems. First, Dart's (1953) concept of the "Osteodontokeratic Culture" led to questions whether the fragmented bones, teeth, and horn cores of large mammals found in *Australopithecus* cave sites were the predatory leftovers and cultural weapons of a "killer-ape." Brain (1981) showed that these South African caves, in fact, contained patterns of bone preservation typical of animals killed by nonhuman carnivores and reflected the inherent preservation biases of different skeletal elements. Brain's thesis that *Australopithecus* was more likely "the hunted" than "the hunters" underlined the fact that early humans were sympatric with large carnivorans; thus considerable predation risk attended any attempt by early hominins to invade a carnivorous niche.

The second problem focused on late Pliocene archeological sites in East Africa, particularly those of Olduvai Gorge, Tanzania (Leakey, 1971). These sites were originally assumed to preserve the remains of large prey dispatched by Oldowan toolmakers. The issues concerned (1) whether concentrations of mainly ungulate bones were indeed the product of hominin behavior; (2) if so, whether these remains were obtained by hunting or scavenging; and (3) whether sufficiently large amounts of meat were available to underwrite food sharing and the evolution of other behaviors typical of human hunter-gatherers (Isaac, 1978). These matters were scrutinized from a taphonomic perspective, which led to the discovery of microscopic methods for discerning stone tool butchery and percussion marks on bone – proving that early toolmakers were indeed the primary, but not sole, bone collectors active at several Pliocene archeological sites (Potts and Shipman, 1981; Bunn, 1981; Bunn and Kroll, 1986; Potts, 1988; Blumenschine and Salvaggio, 1988).

Although these methods, which remain the primary scientific advance from the 1980s research phase, proved that early humans had access to meat and bone marrow, no method has yet been devised to determine convincingly how that access occurred – by hunting, confrontational scavenging (driving predators from their kills), or more passive means (late scavenging). Thus the primary question – how regularly or habitually Plio-Pleistocene toolmakers engaged in predatory behavior – remains unresolved, an issue treated in greater depth later in this chapter.

3. Primate Background

The elaboration of predatory behavior in humans is often characterized as a shift from herbivory (in an ape ancestor) to omnivory or even carnivory. This scenario needs re-evaluation on the basis of current knowledge of primate diets and food acquisition behavior. Insectivory and consumption of other invertebrates are known widely in the order Primates (Harding and Teleki, 1981; McGrew, 2001). Systematic predation on vertebrates has been documented in capuchin monkeys (genus *Cebus*), baboons (*Papio*), and chimpanzees (*Pan troglodytes*), the closest genetic relative of humans (e.g., Rose, 1997, 2001; Stanford, 1998; Strum, 1981). For these predatory primates, prey animals are never larger than the predator's adult male body size. In primate populations where meat-eating is documented, vertebrate tissues make up a small component (2 to 5%) of the diet, although meat may contribute more than 700 kg to the diet of a wild chimpanzee community in some years (Stanford, 2001; Rose, 2001).

Hunting and meat-eating occur in chimpanzee populations scattered across low-latitude Africa. Scavenging is very rare. The best predictor of predatory behavior and success is the number of male participants (Boesch and Boesch, 1989; Mitani and Watts, 1999), although at one study site the presence of estrous females was strongly correlated with male chimps' decisions to hunt (Stanford, 1998). Although meat may provide nutritional benefits, chimpanzee hunting is not related to primary food (ripe fruit) scarcity but rather to social benefits, especially male social bonding and mating opportunities with females. That chimpanzees participating in a successful hunt will often share meat suggests that the male hunters solidify social alliances with other males, and it is this outcome that confers reproductive advantages in a species where male competition strongly impacts mating behavior (Mitani et al., 2002).

In every region where they live sympatrically, chimpanzees selectively hunt red colobus monkeys (*Procolobus badius*). Hunts of red colobus occur on average of 4 to 10 times per month, and success rates average over 50% across field study sites (Mitani et al., 2002). Predation strategies vary, however, from place to place. For example, in the Taï Forest (Côte d'Ivoire), chimps hunt cooperatively on a regular basis and initiate stalking behavior before the prey animals are in range; when successful, the chimpanzees almost always share the meat (Boesch, 1994); by contrast, at Gombe (Tanzania), hunting is opportunistic with little cooperation among males, and the participants are less disposed to share.

Although chimpanzee predation is often considered the ancestral condition from which human hunting evolved, this assumption is unfounded. The other three great apes (gorillas, bonobos, and orangutans) do not regularly hunt or eat vertebrate tissues. This includes the bonobo (*Pan paniscus*), which is as closely related to humans as chimpanzees. Thus while chimpanzee predation may offer reasonable hints as to possible factors underlying the elaboration of human predatory behavior, it is not necessarily a behavioral homology shared with the last human-chimp common ancestor and, therefore, with the earliest hominins.

4. Evolutionary Basis of Human Predatory Behavior

Although paleoanthropologists commonly talk about "the evolution of hunting," the elaboration of carnivory in humans arose from an intricate series of changes in anatomical, physiological, neurological, mental, and social characteristics. This amalgam of evolutionary shifts occurred over several million years, and made it possible for proficient large-mammal predation to emerge in a slow, bipedal, small-toothed descendant of an ape.

Anatomical changes included restructuring of the musculoskeletal system related to habitual bipedality (approximately 6 to 4 Ma), which freed the hands for manipulation. The latter was a necessary condition for toolmaking behavior, which was enhanced by lengthening the thumb, strengthening the metacarpals, and widening the distal finger pads, which occurred by ~2 Ma (Marzke, 1997). A later evolutionary change in body proportions involved lengthening and strengthening of the leg bones and muscles. This change occurred by 1.7 Ma, and enabled long-distance mobility and effective running behavior – prerequisites for long-distance search for carcasses and the pursuit of prey.

Physiological changes, which must have taken place by 2.5 or 2.0 Ma, enhanced the ability of the digestive system to process large quantities of raw meat and marrow. Carnivores tend to have shortened digestive tracts with smaller metabolic demands than do herbivores; since the mammalian gut is an energetically expensive organ system, it has been suggested that a shift to a more predatory existence would have allowed the redistribution of metabolic energy due to gut reduction. As the brain is by far the most expensive tissue, energy redistribution, then, was a precondition for encephalization – an idea known as the expensive-tissue hypothesis of brain expansion (Aiello and Wheeler, 1995). This intriguing hypothesis implies that dramatic brain size increase in human evolution could have been a secondary effect of increased carnivory rather than the result of direct selection for improved predatory behavior. A later influence on dietary physiology was the controlled use of fire, developed by about 400,000 years ago. Cooking enabled the initial breakdown of meat protein and detoxification of bacteria, creating a new physiological environment for the digestion of meat. Fire also eventually proved to be an effective tool in the control of game animals.

Neurological changes affected muscular control of the hand, including the ability to precisely flake and manipulate stone tools in cutting and crushing activities. Archeological remains suggest that such changes in neurological control had occurred by 2.5 Ma. Controlled aiming of thrown objects was another important neurological improvement and was certainly a condition for the development of well-designed throwing spears by 500 to 400 ka.

Change in human cognition enabled early humans to plan complex searches for animal food and to strategize how to dispatch large, solitary animals and game herds. The evolution of various aspects of human sociality also enhanced the sharing of large quantities of meat, including its exchange between neighboring groups of hunter-gatherers. Although the timing of these evolutionary shifts is difficult to know precisely, they were almost certainly in place over the past several hundred thousand years and accessible to all humans by the time *Homo sapiens* emerged, 200 to 100 ka.

All of these developments took place in the context of novel ecological relationships, which included confrontation with large carnivores (especially felids, hyenids, and canids), increased predation risk, improved predation avoidance, broad geographic and ecological dispersal of hominins, and the ability to respond to climatic shifts. The latter sometimes led to packing of human predators and prey populations into small areas (e.g., water sources), which appears to have contributed to late Pleistocene extinction of some herbivore species and Holocene domestication of others.

The question remains, what benefits accrued to early humans as they became more carnivorous and predatory? The adaptive advantages were likely both dietary and social, and included the sharing of large food packages that offset foraging uncertainty. Large quantities of muscle protein, nutritious organs, and fat were accessible by hunting or early scavenging of complete or nearly complete carcasses. In modern hunter-gatherers who live in low and mid-latitudes, meat makes up around 20 to 40% of the diet by weight, more than ten times as much meat per capita as in chimpanzees (Kaplan *et al.*, 2000; Lee and Devore, 1968). For Plio-Pleistocene ancestors, the acquisition of even a few bones late in the carcass consumption sequence offered fatty bone marrow, a substance that was probably critical during dry-season nutritional stress and would have offset the problem of "protein poisoning" from eating meat of lean African herbivores (Speth, 1989; Speth and Spielman, 1983).

Besides nutritional benefits, evidence from chimpanzees (noted earlier) and human hunter-gatherers points to strong social advantages of predatory behavior. Ethnographers have documented that hunting success and hunting reputation correlate statistically with positive weight gains, higher reproductive success, and greater social status and political advantages (Hawkes and Bliege Bird, 2002; Kelly, 1995). In these same foraging societies, meat is shared throughout the social group and even beyond to neighboring groups of relatives and acquaintances. Sharing is usually done in a manner not controlled by the hunter.

One of the oddest features of human behavior, associated with hunting activity, is the delayed consumption of prey (and gathered plant food). Rather than eating meat right away, or after a short wait at the prey kill site, humans take it elsewhere and distribute it among immediate relatives or more widely in the manner just described.

Although food sharing might have evolved via kin selection, modern humans tend to exhibit general reciprocity, the sort that emerges from mutualistic social interactions (Clutton-Brock, 2002), possibly connected with the buffering of uncertainty in risk-prone environments. This way of looking at meat-distributing behavior in humans makes sense in light of the intense variability of climate and adaptive settings confronted by the extant human lineage over its evolutionary history (Potts, 1996, 1998).

5. Ecological Overlap and Interaction between Hominins and Other Predator/Scavengers

Substantial predation risk and competition evidently accompanied the expansion of late Pliocene hominins into the large carnivore guild. Convincing evidence comes from fossil animal remains excavated in archeological sites of Bed I Olduvai Gorge, 1.85 to 1.78 Ma (Potts, 1988). In these excavations, damage caused by carnivore teeth and stone tools co-occur in the same bone assemblages, occasionally on the same bone specimens. Wherever the toolmakers transported and processed animal tissues, large carnivorans were attracted to the same places on the landscape – indicating hominin-carnivoran overlap in the use of space and carcasses. These sites were located some unknown distance from the original death sites, which are typically the places of greatest interference competition among African felids, canids, and hyenids. That the early Olduvai sites exhibit a strong signal of carnivoran activity, along with abundant butchery marks and hundreds to thousands of stone tools, offers reasonably solid confirmation of extensive competitive overlap among the carnivorous species.

Strong evidence of carnivore-carnivore predation exists in modern contexts and is taken to be a measure of interference competition among meat-eaters, usually due to the attraction of potential scavengers to a carcass (Van Valkenburgh, 2001). In the Bed I Olduvai sites, the fossil representation of Carnivora (mainly canids, viverrids, hyenids, and felids) is greater than expected (~1%) from modern and ancient African savannas. Carnivorans always comprise more than 1% of the macromammal MNI (minimum number of individuals), averaging 3% (n=6 bone assemblages) (Potts, 1988). Particularly striking is the stone-tool site of FLKNN level 3, where carnivorans comprise 21% of the MNI. In this same bone assemblage, three hominin bones exhibit clear bite marks from the teeth of a large carnivoran (on a parietal cranial bone) and a smaller one (on two foot bones, an astragalus and calcaneous, of *Homo habilis*). These finds seem to indicate the most direct expression of hominin-carnivoran competition in Bed I Olduvai. For some late Pliocene toolmakers, then, it is virtually certain that the initial propensity to exploit large animal carcasses was a risky enterprise, offset by the possibility of substantial dietary rewards.

The Plio-Pleistocene East African carnivore guild was, furthermore, structured differently from that of today. It included greater taxonomic and ecomorphic diversity – for example, the presence of a cursorial hyena (*Chasmoporthetes nitidula*) and sabertoothed cats (*Homotherium crenatidens, Megantereon cultridens,* and *Dinofelis* sp.), which possessed specialized meat-slicing dentitions. The presence of meat-slicing specialists and large bone-cracking hyaenids (e.g., *Pachycrocuta* sp.), which last appeared in Africa in the early Pleistocene, suggests that Plio-Pleistocene carnivoran biomass may have been substantially greater than present and that the milieu of predation, scavenging, and interference competition entailed higher predation rates, faster carcass turnover, and more confrontational scavenging opportunities than today (Lewis, 1997; Van Valkenburgh, 2001).

Existing knowledge of carnivore-carnivore interactions suggests that early toolmaking hominins faced strong selection pressure for strategies that reduced predation risk and strengthened competitive ability. One response included the removal of carcass parts by hominins from death sites and the incomplete (possibly rapid) processing of meat and marrow (Potts, 1988). This strategy would have reduced time spent in handling meaty tissues at carcasses and secondary bone accumulations. Other responses could have included increased body size in early African *H. erectus* (Walker and Leakey, 1993). A 50% or greater increase in stature (from approximately 4 feet tall in early *Homo* to 6 feet tall in African *H. erectus*), which occurred by 1.7 Ma, and a parallel enlargement of body mass would have reduced the predation risk inherent in confrontations with small and medium-sized canids, hyenids, or felids. It is thus tempting to relate this increase in hominin body size – the most dramatic in human evolutionary history – to predation pressure and risk, which initially was high if indeed hominin-carnivoran overlapping use of space and carcasses at earlier sites is indicative.

One of the most thoroughly studied mid-Pleistocene sites in East Africa – Olorgesailie, located in the southern Kenya rift valley – exhibits a surprisingly different relationship between hominins and carnivorans. By 1 Ma, for instance, substantial concentrations of stone tools and animal bones, which bear traces of human utilization, are largely devoid of carnivore gnawing and tooth marks. Carnivoran bones are extremely rare at these sites, and no other clue is present to suggest that canids, hyaenids, and felids interacted or overlapped with hominin toolmakers in similar ways to the situation described here for Olduvai, 800,000 years earlier (Potts, 1989; Potts *et al.*, 1999). One possibility is that mid-Pleistocene hominins and other predators had evolved ways of largely avoiding one another, although one possible factor – human control of fire – is not clearly evident at any archeological site (African or Eurasian) prior to about 400 ka.

While the competitive milieu evident at archeological sites had apparently changed by about 1 Ma, overlap between hominin and carnivoran meat-eaters persisted in Africa – evident, for example, at the site of Elandsfontein, South Africa (Klein, 1988; Klein and Cruz-Uribe, 1991). In Europe and Asia, middle and late Pleistocene hominins recurrently visited cave settings, which presented new opportunities for interaction with large carnivores. In the best studied examples, Neanderthal cave sites in southern Europe, caves were occupied alternately as denning sites by spotted hyenas and wolves, as hibernation sites by bears, and as ephemeral shelters by hominins, with little apparent temporal overlap (Stiner, 1994).

6. Predator-Prey Interactions during Human Evolutionary History

This section reviews evidence of human predation and scavenging through time. A summary of the key events is provided in Table 1.

6.1. Late Pliocene

Tool manufacture and use comprised the single most important means by which human ancestors modified their resource landscapes in ways conducive to carnivory and

TABLE 1. Evolutionary Record of Predation and Carnivory in Early Humans: Inferences and Evidence

Time Period (years ago)	Inferences	Nature of the Evidence
Late Miocene hominin-chimp last common ancestor, >6 million	• General faunivory (collecting/hunting small invertebrates & vertebrates)	• Faunivory found widely in higher primates, including all living ape species
Mio-Pliocene earliest hominins (*Orrorin, Ardipithecus*) 6 to 5 million	• Terrestrial bipedal locomotion emerged in Africa, freeing the hands for manipulating & carrying	• Proximal femur of *Orrorin tugenensis* appears to indicate habitual standing & walking
Early Pliocene hominins (*Ardipithecus, Australopithecus, Kenyanthropus*) 5 to 2.5 million	• Strong commitment to terrestrial bipedality by 4 Ma • Intentional stone flaking & toolmaking by 2.5 Ma; first known acquisition of meat & bone marrow from large mammals	• Skeletal morphology of *Australopithecus* • Oldest known stone tools & archeological sites, with associated faunal remains bearing tool cut & percussion marks
Plio-Pleistocene hominins (*Homo, Australopithecus, Paranthropus*) 2.5 to 0.8 million	• Small mammal predation continues from prior times • Exploitation of meat & marrow from diverse African bovid, giraffid, suid, hippopotamid, equid, rhinocerotid, primate, & elephant species; diverse use of carcass body parts of these large mammals, obtained probably by early scavenging, possibly some opportunistic hunting • Extensive competitive carnivoran-hominin overlap in the use of carcasses & bone collection sites, by ~2 Ma, in some places • First spread of hominin populations over Asia and SW Europe, with greater dependence on carnivory & possibly predation in temperate regions • In E. Africa, apparent relaxation of carnivoran-hominin competitive overlap by ~1.0 Ma, in some places	• Mammals of 15-50 kg found with stone tools • Animal bones of diverse species, body size, & skeletal representation, with low levels of stone tool cut & percussion marks, including meat-rich bones excavated from archeological sites • Evidence of gnawing & other damage typical of diverse-sized carnivorans found at archeological sites • Hominin fossil & archeological remains 1.75 Ma at Dmanisi (Caucasus); remains in E. Asia by at least 1.6 Ma, & NE Asia by ~1.4 Ma • Diminished carnivoran presence at archeological sites in S. Kenya
Middle Pleistocene hominins (*Homo* spp.) 800 to 128 thousand	• Continued scavenging & hunting of diverse animals, and (toward end of this period) earliest specialized exploitation of game herds, including driving of animals • Oldest projectile hunting (purposefully designed throwing spears) • Earliest definite evidence of control of fire, useful for cooking & habitat alteration, especially during late Pleistocene	• Animal bones from archeological sites • Javelin-like wooden spears preserved at European sites • First definite hearth-like lenses of ash and burnt sediment & bone

Time Period (**years ago**)	Inferences	Nature of the Evidence
Late Pleistocene/early Holocene hominins (*H. neanderthalensis, H. sapiens*) **128 to 5 thousand**	• Evidence of both diverse (broad spectrum) predation & specialized focus on single-species herds • Complex strategies of socially cooperative hunting • Oldest definite evidence of specialized predatory technologies, e.g., bone points used to catch fish • Oldest exploitation of marine mammals and shellfish • Potentially strong human contribution to prey animal extinctions ("overkill") • Shift from predation to control over the movement & reproduction of prey (food production revolution) • Intensified herding & development of pastoralism	• Animal bones from archeological sites • Archeological evidence of single-species kill sites • Bones & artifacts from archeological sites • Animal remains from archeological sites • Extinct megafaunal species found at sites in some regions • Extensive archeological evidence for animal domestication & the origin of food production • Evidence of intensified human & prey sedentism

predation. Dated at 2.5 Ma, the site of Bouri, Ethiopia, records the earliest known instance of hominin tool marks on animal bones and exploitation of large mammals (de Heinzelin *et al.*, 1999). The evidence consists of animal bones dispersed over an ancient landscape and cut and cracked in a manner typical of stone toolmakers. Almost all other late Pliocene archeological sites consist of bones and tools collected in small concentrated patches. By 2.3-2.1 Ma, hominin toolmakers exploited diverse herbivore species, especially ungulates such as bovids, and also diverse body sizes (<1 kg to >500 kg) and animal body parts (Kibunjia, 1994; Roche *et al.*, 1999; Plummer *et al.*, 1999). While most research has focused on large ungulates possibly acquired by scavenging, recent excavations at Kanjera, Kenya, document that, under certain circumstances, early toolmakers exploited small species (<50 kg) and immature individuals of larger ones, which is the first evidence of continuity between the predatory capabilities of nonhuman primates and late Pliocene hominins (Plummer *et al.*, 1999; Plummer and Stanford, 2000; Plummer *et al.*, 2001).

The largest, well preserved bone assemblage from any Plio-Pleistocene archeological site was excavated at FLK Zinj in Bed I Olduvai (1.8 Ma) (Leakey, 1971). Its study has produced wide debate regarding the predatory and carnivorous propensities of early human toolmakers. More than 60,000 fossilized bone specimens, including microfauna, were recovered from this site. Of these remains, approximately 3500 are assignable to large mammals, with an MNI of 45 representing at least 17 species. The large mammal assemblage is made up of diverse skeletal parts dominated by limb bone elements (Bunn and Kroll, 1986). The taphonomic formation of this bone concentration, found in a paleosol over an area of ~290 m^2, was complex and included the effects of hominin toolmakers (the dominant bone collector at this site), large and small carnivores, and other depositional and bone-damaging processes (Potts, 1988). The taxonomic and ecomorphic diversity of the fauna at FLK Zinj and other archeological sites in Bed I Olduvai indicate that carcass parts were introduced from wooded, grassy, grazing, browsing, high-mobility, and low-mobility habitats (Plummer and Bishop, 1994; Kappelman *et al.*, 1997). In a detailed analysis of tool cut marks on 172 bone specimens from FLK Zinj, Bunn and Kroll (1986) noted the predominance of marks on limb bone mid-shafts, which to these researchers implied a significant removal of muscle tissue. Thus Bunn and Kroll concluded that hominins had early access to carcasses, probably through hunting or confrontational scavenging. (See Bunn and Ezzo's, 1993, review of the hunting/scavenging debate.)

Questions remain, however, concerning the interpretation of hominin predation reflected at FLK Zinj. Shipman's (1986) analysis of the same bones, which applied stricter SEM criteria for identifying cut marks, recorded far less butchery damage. According to Shipman, the dominance of mid-shaft marks could indicate that hominins had relatively late access to carcasses in which the disarticulation of joints with stone tools was unnecessary (Shipman, 1986). Furthermore, Blumenschine's (1987) work in the Serengeti has shown that relatively leisurely scavenging opportunities may exist in certain seasons of high animal die-off, expecially in habitats avoided by bone-crunching hyenas (*Crocuta*). The problem with this "scavenger niche" interpretation is that bone-crunching carnivorans were active at FLK Zinj and other sites in Bed I Olduvai (Potts, 1988), and the type of habitat that offers high quality scavenging opportunities depends on rainfall, vegetation, and other factors that may not have applied to FLK Zinj (Tappen, 1995). Blumenschine's (1995) further study of the percussion-marked bones from FLK Zinj confirms that access to carcasses involved a complex interplay between hominins and carnivorans.

The degree to which Plio-Pleistocene humans were active predators, therefore, remains undetermined. Small, especially immature animals (<50 kg) were more likely obtained by hunting than by scavenging, since small prey are easily consumed to near entirety by a predator. The usual assumption is, furthermore, that the technological prowess of early toolmakers was inadequate to bring down animals larger than 500 kg. If this thinking is correct, the oldest known hominin exploiters of large mammals probably adopted a mixed hunting/scavenging strategy, in accord with most other mammalian carnivores, which display varying degrees of hunting and scavenging behavior in different circumstances. One thing is clear: certain populations of late Pliocene hominins acquired meat and marrow from diverse body parts derived from a wide diversity of mammalian species that represented a range of habitats and body sizes – all of which suggests that opportunism and behavioral flexibility characterized the earliest human intrusion into the large carnivore guild (Potts, 1988; Brantingham, 1998).

6.2. Early and Middle Pleistocene

Early humans of the genus *Homo* continued to gain access to a diverse range of mammalian taxa, body sizes, and skeletal parts. In the best-studied sequence at Olduvai Gorge, the body size of mammals associated with archeological sites significantly increased between 1.7 and 1.4 Ma (Leakey, 1971). Olduvai hominins of this period focused on skeletal regions richest in meat and marrow, particularly long bones, although the presence of tool cut marks on axial elements suggests that toolmakers exploited a wide range of carcass resources. Hominins also displayed varied strategies of acquiring carcasses, with a focus on early access probably by hunting and scavenging (Monahan, 1996).

Evolutionary enlargement of body size in hominins occurred near the time (ca. 1.75 Ma) of the earliest convincing evidence of *H. erectus'* spread into Asia, represented at the site of Dmanisi, Republic of Georgia (Gabunia *et al.*, 2000). Reliable carnivory may have been critical as populations dispersed into mid-latitude East Asia by 1.4 Ma (Zhu *et al.*, 2001) and eventually to western Europe by 800 ka (Carbonell *et al.*, 1995). A largely herbivorous primate, as it spread across diverse climatic and vegetation zones, would have faced the dangers posed by toxins from novel plants; a more carnivorous one could have reduced this risk by relying on meat and marrow, which have more uniform digestive qualities across ungulate species. It is apparent, however, that long-term inhabitants of some areas, such as the Levant of western Asia, depended on edible plants (Goren-Inbar et al., 2000). No fossil or archeological evidence exists that meat constituted the majority of any hominin species' diet prior to the late Pleistocene.

As early human populations grew or re-entered Europe around 500 ka, specialized hunting technology began to be made in the form of throwing spears. A puncture possibly due to projectile impact of an equid scapula from Boxgrove, England, ~500 ka (Roberts and Parfitt, 1999), and wooden spears designed like modern javelins from the site of Schöningen, Germany, ~400 ka (Thieme, 1997), suggest that planned hunts of large game had become part of the foraging repertoire of at least some middle Pleistocene hominin populations. Invention of the throwing spear initiated an entirely new set of possibilities for human predators, dispatching of prey from a distance of potentially many meters. However, what were once standard interpretations of proficient big-game hunting, such as the baboons of Olorgesailie, Kenya (~780 ka), and the elephants and horses of Torralba/Ambrona, Spain (~500 ka), have been criticized on taphonomic grounds and remain in a suspense category (Klein, 1999; Potts et al., 1999). Similarly, the view that *H. erectus* at Zhoukoudian, China (~500 ka), was a specialized hunter of deer and other ungulates is now considered uncertain pending further analysis of the remains or excavation of new sites (Klein, 1999; Binford and Ho, 1985).

6.3. Late Pleistocene

After 200 ka, hominins continued to exploit diverse animals and carcass resources, but hunters also began to concentrate on specific species of large herd animals. A key site is the cave of La Cotte de St. Brelade, located on Jersey of the Channel Islands (Scott, 1980, 1986). This site contains abundant Mousterian artifacts, typically associated in Europe with *Homo neanderthalensis*, and bone accumulations consisting of at least 20 mammoths and 5 woolly rhinoceros, dated 186 to 127 ka. Butchery marks indicate human involvement with the bone accumulation, vertical packing of the bones suggests a rapid accumulation, and the site marks a place where hunters apparently drove a small group of mammoths and rhinoceros over a steep rock face that dropped into a cavern (Scott, 1986; Mellars, 1996; Klein, 1999).

Stable isotope analyses of carbon and nitrogen in bone collagen of Neanderthal fossils from France, Belgium, and Croatia indicate that dietary protein was largely derived from animals rather than plants, similar to nonhuman carnivores from the same and other sites (Fizet et al., 1995; Boucherens et al., 1999; Richards et al., 2000). While scavenging opportunities were almost certainly available (Binford, 1989), the dietary importance of animal protein suggests that the Neanderthals were at times reliant on predation. According to a taphonomic study of Mousterian cave sites in Italy, Neanderthals relied increasingly over time on hunting, including the harvesting of prime adults (Stiner, 1994). Furthermore, butchery marks like those on ungulate remains have been documented on Neanderthal bones, and denote cannibalism and probably dietary stress related to short-term scarcity of animal protein (Defleur et al., 1999).

While faunal remains from many Neanderthal sites imply a continuation of general strategies of carcass procurement by hunting and scavenging, analyses of some western

European sites point to more specialized predation on particular species such as red deer (*Cervus elaphus*) and reindeer (*Rangifer tarandus*) (Chase, 1986). With the spread of *H. sapiens* to western Europe by around 40 ka, predation focused even more exclusively on herds of ungulates such as reindeer and horses (*Equus caballus*) (Chase, 1986; Olsen, 1989). Although much emphasis is placed on large game, hunting of small vertebrates continued through the late Pleistocene (Stiner *et al.*, 1999).

Tactical or intercept hunting by Middle Stone Age humans of Africa, possibly associated with the oldest populations of *H. sapiens* more than 100 ka, is indicated by catastrophic age profiles, including prime age adults, based on studies of animal dental remains from excavations in southern and eastern Africa (Marean, 1997; Klein, 1999; McBrearty and Brooks, 2000). Development of specialized technologies for capturing prey is also evident in Africa. For example, barbed bone points at the site of Katanda, D. R. Congo (75 to 90 ka), were associated with abundant large (>35 kg) catfish remains (*Clarius* sp.) – evidence that denotes a technology dedicated to catching aquatic resources (Brooks *et al.*, 1995; Yellen *et al.*, 1995). Predation on marine animals characterized human coastal populations of Africa after 40 ka, though sites possibly older than 110 ka do preserve marine shellfish and seals (McBrearty and Brooks, 2000). Marine animals are especially rich in Omega-3 fatty acids, which are concentrated only in the brain tissues of terrestrial mammals; on this basis, Broadhurst *et al.* (1998) suggest that access to marine animals was important in *H. sapiens*' brain evolution.

One reason why predators may narrow their choice of prey species is to adjust to diverse prey avoidance strategies, which make it difficult for a predator to capture more than one prey type efficiently (Pianka, 2000, p. 315). This reason, however, does not seem to explain specialized predation by late Pleistocene humans. Generalized predation strategies continued to be manifested at some archeological sites, where fossil remains from a diversity of prey are preserved. Moreover, late Pleistocene humans had evolved the ability to plan elaborate strategies of killing large numbers of animals at once and to organize ways of processing the meat, dispersing it, and storing it for later time. Complex planning, social coordination, and intensified exploitation of local resources were an integral part of human predation strategies by the time *H. sapiens* evolved, and led to the ability to ambush, drive, or corral large groups of single-species herds.

According to the "overkill hypothesis," human hunters were responsible for the extinction of a variety of late Pleistocene large prey species in North America and elsewhere (Martin, 1984). This view has been challenged on several grounds, and alternative explanations have been offered for late Pleistocene extinctions, including climate change, disruption of coevolved ecological relationships (e.g., Graham and Lundelius, 1984), and recovery of taxonomic equilibrium (stable diversity) following a peak period in species originations (Gingerich, 1977; Cifelli, 1981). The factors responsible for late Pleistocene faunal extinctions may have varied from one region or continent to another such that researchers now tend to invoke a combination of climatic and human causes: human predation had drastic effects in some regions because it occurred at a time of ecological stress induced by climate change (see Martin and Klein, 1984).

6.4. Holocene

As human populations became more sedentary toward the end of the Pleistocene and into the Holocene, their exploitation of landscapes and of faunal and floral resources dramatically intensified. This intensification is signaled by a greater density and distribution of archeological sites, their larger size, elaboration of shelters and other site features, and heightened concentration and processing of faunal remains. As a result, in several regions, unprecedented change occurred in the coevolutionary relationship between humans and animal prey. The ability to control the movement and killing of herd populations presented an opportunity to engage in planned selection and breeding of animals based on their appearance and nutritional properties. The process of domestication created morphological change in the prey species and increasing control over the biological diversity of entire landscapes. Humans promoted the survival of domesticated animals at the expense of wild prey and competing carnivores. Restructuring of prey and carnivore species biomass and diversity coincided with similar alterations in plant communities as human food production escalated; the combination is responsible for the development of human-dominated ecosystems (Potts and Behrensmeyer, 1992; Potts, 1996) (Figure 1). The evolutionary history of human predation thus provided crucial background to the large impact humans have had on terrestrial ecosystems globally.

7. Conclusion

1. During the late Pliocene, an important shift occurred in the carnivorous behavior of at least one early human lineage, which sought access to larger mammals (>15 kg up to >5000 kg). Since the ancestral diet probably included invertebrates and small vertebrates, the shift did not entail a transition from herbivory to carnivory. It did, however, involve an expansion in carnivorous behavior, accompanied by heightened carnivoran predation on early humans.

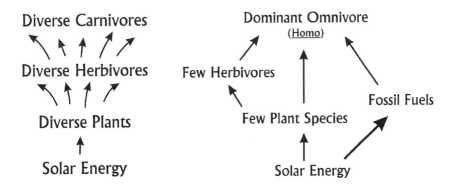

FIGURE 1. Simplified representation of trophic levels and biological diversity during the shift to human-dominated ecosystems. Since the end of the Permian, terrestrial ecosystems have been typified by a diverse array of carnivore species feeding on an even more diverse array of herbivores (left side). Since the origin of domestication and intensified food production (10 ka to present), humans have expanded the populations of a very few plant and herbivore species; have reduced the diversity of other species, including carnivores; and have situated themselves as the dominant meat- and plant-eating species (right side). In these human-dominated ecosystems, people have asserted an increasing degree of control on biological productivity (Potts and Behrensmeyer, 1992; Potts, 1996).

2. The scale and proficiency of large mammal predation by early humans are subjects of considerable debate. Although some studies suggest that the oldest toolmakers were passive scavengers who acquired animal tissues late in the carcass consumption sequence, most analysts note the presence of meat-rich skeletal elements that bear butchery marks at Pliocene archeological sites – indicative of early access by confrontational scavenging or hunting.
3. High taxonomic and ecological diversity of archeological faunal remains suggests that early human toolmakers were opportunistic in their acquisition of meat and fatty tissues. Convincing evidence of specialized hunting of particular big-game species is no older than about 180,000 years, and is related to the development of complex social planning of hunts, possibly symbolic communication, and highly developed forms of meat sharing and social reciprocity. While systematic exploitation of particular prey species was a phenomenon of the late Pleistocene, some forms of special hunting technology (e.g., throwing spears) developed by at least 400 to 500 ka.
4. In at least one instance – the major increase in body size from early *Homo* to early *Homo erectus* – predation pressure may have been a factor underlying anatomical change. This particular evolutionary change had important consequences for predator defense, mobility and geographic range expansion, and later developments in prey capture strategies.
5. Change in human predatory behavior over time was contingent on the evolution of various anatomical, physiological, cognitive, and social characteristics. Hunting and meat-eating offered both nutritional and social benefits. The ability to acquire large packages of animal food rich in protein and fat promoted the unusual delay in food consumption in which humans engage, and heightened the likelihood of sharing food over a wider social network. These behaviors spread the risk of resource and foraging uncertainty prevalent in dynamic Pleistocene environments. Fatty meat, marrow, and organs were probably important to consume as a means of buffering seasonal resource scarcity. Significant capturing of marine resources occurred only over the last 200 kyr, approximately the time of *Homo sapiens*' first appearance. Concentrations of specific lipids in marine foods may have assisted the evolution of brain functions and maturation.
6. Over the course of human evolution, a premium was placed on long-distance tracking of habitats and adaptability to novel settings. It is likely that predation and animal consumption proved critical in adjusting to Pleistocene environmental change. As archaic and modern populations spread into new geographic regions, consumption of animal tissues may have acted as a buffer against the lack of highly specific information about newly encountered plant species and their secondary compounds.
7. Control over the movement and breeding of animals has been the culmination of a mutualistic relationship that evolved between humans and certain species of large prey. Animal domestication resulted in microevolutionary change in both prey and predator (evolution of gracile bodies due to selective breeding of prey and increased sedentism in both prey and humans). It also provides a highly unusual example of how predator-prey interaction, even in a single predator species, may have significant consequences for ecological relationships and the structure of food webs.

8. Although fossil and archeological analysis has documented the broad narrative of early human predation, considerable uncertainties remain. The level of active predation, scavenging, and passive collection of animal protein and fatty tissues is a subject of ongoing dispute. The debate is important to paleoanthropologists because many ideas concerning early human social life, dependence on tools, and overall cognitive and behavioral evolution revolve around the complexity of the food search, especially the means of access to animal protein. This debate is unlikely to be resolved by framing it as an either/or issue – i.e., did early humans hunt or did they scavenge; did large animal carcasses mainly afford access to meat (muscle tissue) or to fat (e.g., bone marrow)? One of the challenges is to build stringent methods of determining the proportions of an archeological bone assemblage that reflect the diverse means of animal acquisition and nutritional resources potentially consumed by early human toolmakers. Independent of the archeological record, the degree of reliance on carnivory remains a key question throughout the time period of human evolution. Although the interpretation of diet from stable isotope data can be problematic, it may prove possible to extend this type of analysis, as already applied to Neanderthals, to the skeletal materials of earlier hominins (e.g., Sponheimer and Lee-Thorp, 1999). One result would be a firmer understanding of whether carnivory was ever as important as plant consumption in the lives of these earlier species. Finally, co-evolutionary relationships between hominin and carnivoran species have not yet been well documented. One important avenue is the comparative analysis of fossil remains recovered from precisely dated African, Asian, and European fossil sites, aimed at determining whether particular carnivoran species accompanied hominin populations as they spread, probably in multiple dispersal episodes, between regions and continents. Discovery of such associations would imply similar ecological opportunities available to both hominins and carnivorans, and tracing these associations through time would help to uncover the timing and conditions under which early hunters began to assert a significantly different and distinctly modern human influence on prey populations.

ACKNOWLEDGMENTS: Sincere thanks to Patricia Kelley for the invitation and encouragement to contribute to this volume. I acknowledge the Smithsonian Institution and National Science Foundation for support of research, extended over many years, at Olduvai Gorge, Olorgesailie, and Kanjera, mentioned in this review. My gratitude goes to Jennifer Clark for her assistance with the manuscript and preparation of Figure 1. This is a publication of the Smithsonian's Human Origins Program.

References

Aiello, L. C., and Wheeler, P., 1995, The expensive-tissue hypothesis: The brain and the digestive system in human and primate evolution, *Curr. Anthropol.* **36**:199-221.
Ardrey, R., 1961, *African Genesis.* Atheneum, New York.
Binford, L. R., 1981, *Bones: Ancient Men and Modern Myths,* Academic Press, New York.
Binford, L. R., 1989, Isolating the transition to cultural adaptations: An organizational approach, in: *The Emergence of Modern Humans* (E. Trinkaus, ed.), Cambridge University Press, Cambridge, pp. 18-41.
Binford, L. R., and Ho, C. K., 1985, Taphonomy at a distance: Zhoukoudian, "the cave home of Beijing Man"?, *Curr. Anthropol.* **26**:413-442.
Blumenschine, R. J., 1987, Characteristics of an early hominid scavenging niche, *Curr. Anthropol.* **28**:383-407.

Blumenschine, R. J., 1995, Percussion marks, tooth marks, and experimental determinations of the timing of hominid and carnivore access to long bones at FLK *Zinjanthropus*, Olduvai Gorge, Tanzania, *J. Human Evol.* **29**:21-52.

Blumenschine, R. J., and Salvaggio, M. M., 1988, Percussion marks on bone surfaces as a new diagnostic of hominid behavior, *Nature* **333**:763-765.

Boesch, C., 1994, Cooperative hunting in wild chimpanzees, *Animal Behav.* **48**: 653-667.

Boesch, C., and Boesch, H., 1989, Hunting behavior of wild chimpanzees in the Taï National Park, *Am. J. Phys. Anthropol.* **78**:547-573.

Boucherens, H., Billiou, D., Mariotti, A., Patou-Mathias, M., Otte, M., Bonjean, D., and Toussaint, M., 1999, *J. Archaeol. Sci.* **26**: 599-607.

Brain, C. K., 1981, *The Hunters or the Hunted? An Introduction to African Cave Taphonomy*, University of Chicago Press, Chicago.

Brantingham, P. J., 1998, Mobility, competition, and Plio-Pleistocene hominid foraging groups. *J. Archaeol. Meth. Theory* **5**:57-98.

Broadhurst, C. L., Cunnane, S. C., and Crawford, M. A., 1998, Rift Valley lake fish and shellfish provided brain-specific nutrition for early *Homo*, *British J. Nutrition* **79**:3-21.

Brooks, A. S., Helgren, D. M., Cramer, J. M., Franklin, A., Hornyak, W., Keating, J. M., Klein, R. G., Rink, W. J., Schwarcz,H. P., Smith, J. N. L., Stewart, K., Todd, N. E., Verniers, J., and Yellen, J. E., 1995, Dating and context of three Middle Stone Age sites with bone points in the upper Semliki Valley, Zaire, *Science* **268**:548-553.

Bunn, H. T., 1981, Archaeological evidence for meat-eating by Plio-Pleistocene hominids from Koobi Fora and Olduvai Gorge, *Nature* **291**:574-577.

Bunn, H. T., and Ezzo, J. A., 1993, Hunting and scavenging by Plio-Pleistocene hominids: nutritional constraints, archaeological patterns, and behavioral implications, *J. Archaeol. Sci.* **20**:365-398.

Bunn, H. T., and Kroll, E. M., 1986, Systematic butchery by Plio/Pleistocene hominids at Olduvai Gorge, Tanzania, *Curr. Anthropol.* **27**:431-452.

Carbonell, E., Bermúdez de Castro, J. M., Arsuaga, J. L., Diez, J. C., Rosas, A., Cuenca-Bescós, G., Salar, R., Mosquera, M., and Rodrígues, X. P., 1995, Lower Pleistocene hominids and artifacts from Atapuerca-TD6 (Spain), *Science* **269**:826-832.

Chase, P.G., 1986, The Hunters of Combe Grenal: Approaches to Middle Paleolithic Subsistence in Europe, *British Archaeol. Reports Int. Ser. 286*, Oxford.

Cifelli, R.,L., 1981, Patterns of evolution among the Artiodactyla and Perissodactyla (Mammalia), *Evolution* **35**:433-440.

Clutton-Brock, T., 2002, Breeding together: Kin selection and mutualism in cooperative vertebrates, *Science* **296**:69-75.

Dart, R., 1953, The predatory transition from ape to man, *Int. Anthropol. Linguistic Rev.* **1**:201-219.

Defleur, A., White, T., Valensi, P., Slimak, L., Crégut-Bonnoure, É., 1999, Neanderthal cannibalism at Moula-Guercy, Ardéche, France, *Science* **286**:128-131.

De Heinzelin, J., Clark, J. D.,White, T., Hart, W., Renne, P., WoldeGabriel, G., Beyene, Y., and Vrba. E., 1999, Environment and behavior of 2.5-million-year-old Bouri hominids, *Science* **284**:625-628.

Fizet, M., Mariotti, A., Bocherens, H., Lange-Badré, B., Vandermeersch, B., Borel, J., and Bellon, G., 1995, Effect of diet, physiology and climate on carbon and nitrogen stable isotopes of collagen in a Late Pleistocene anthropic palaeoecosystem: Marillac, Charente, France, *J. Archaeol. Sci.* **22**:67-79.

Gabunia, L., Vekua, A., Lordkipanidze, D., Swisher III, C. C., Ferring, R., Justus, A., Nioradze, M., Tvalchrelidze, M., Antón, S. C., Bosinski, T., Jöris, O., de Lumley, M.-A., Majsuradze, G., and. Mouskhelishvili A., 2000, Earliest Pleistocene hominid cranial remains from Dmanisi, Republic of Georgia: Taxonomy, geological setting, and age, *Science* **288**:1019-1025.

Gingerich, P. D., 1977. Patterns of evolution in the mammalian fossil record, in: *Patterns of Evolution as Illustrated by the Fossil Record* (A. Hallam, ed.), Elsevier, Amsterdam, pp. 469-500.

Goren-Inbar, N., Feibel, C. S., Verosub, K. L., Melamed, Y., Kislev, M. E., Tchernov, E., and Saragusti I., 2000, Pleistocene milestones on the Out-of-Africa corridor at Gesher Benot Ya'aqov, Israel, *Science* **289**:944-947.

Graham, R.W., and Lundelius, E. L., Jr., 1984, Coevolutionary disequilibrium and Pleistocene extinctions, in: *Quaternary Extinctions* (P.S. Martin and R.G. Klein, eds.), University of Arizona Press, Tucson, pp. 223-249.

Harding, R. S. O., and Teleki, G., eds., 1981, *Omnivorous Primates: Gathering and Hunting in Human Evolution*, Columbia University Press, New York.

Hawkes, K., and Bliege Bird, R., 2002, Showing off, handicap signaling, and the evolution of men's work, *Evol. Anthropol*, **11**:58-67.

Isaac, G. L., 1978, The food-sharing behavior of protohuman hominids, *Sci. Am.* **238**:90-108.

Kaplan, H., Hill, K., Lancaster, J., and Hurtado, A. M., 2000, A theory of human life history evolution: Diet, intelligence, and longevity, *Evol. Anthropol.* **9**:156-185.
Kappelman, J., Plummer, T., Bishop, L., Duncan, A., and Appleman, S., 1997, Bovids as indicators of Plio-Pleistocene paleoenvironments in East Africa, *J. Human Evol.* **32**:229-256.
Kelly, R. L., 1995, *The Foraging Spectrum: Diversity in Hunter-gatherer Lifeways*, Smithsonian Institution Press, Washington.
Kibunjia, M., 1994, Pliocene archaeological occurrences in the Lake Turkana basin, *J. Human Evol.* **27**:159-172.
Klein, R. G., 1988, The archaeological significance of animal bones from Acheulean sites in southern Africa, *African Archaeol. Rev.* **6**:3-25.
Klein, R. G., 1999, *The Human Career: Human Biological and Cultural Origins*, 2nd edition, University of Chicago Press, Chicago.
Klein, R. G., and Cruz-Uribe, K., 1991, The bovids from Elandsfontein, South Africa, and their implications for the age, palaeoenvironment, and origins of the site, *African Archaeol. Rev.* **9**:21-79.
Leakey, M. D., 1971, *Olduvai Gorge*, vol. 3, Cambridge University Press, London.
Lee, R. B., and DeVore, I., eds., 1968, *Man the Hunter*, Aldine, Chicago.
Lewis, M. E., 1997, Carnivoran paleoguilds of Africa: implications for hominid food procurement strategies, *J. Human Evol.* **32**:257-288.
Marean, C. W., 1997, Hunter-gatherer foraging strategies in tropical grasslands: Model-building and testing in the East African Middle and Later Stone Age, *J. Anthropol. Archaeol.* **16**:189-225.
Martin, P. S., 1984, Prehistoric overkill: the global model, in: *Quaternary Extinctions* (P.S. Martin and R.G. Klein, eds.), University of Arizona Press, Tucson, pp. 354-403.
Martin, P. S., and Klein, R. G., eds., 1984, *Quaternary Extinctions,* University of Arizona Press, Tucson.
Marzke, M. W., 1997, Precision grips, hand morphology, and tools, *Am. J. Phys. Anthropol.* **102**:91-110.
McBrearty, S., and A.S. Brooks. 2000. The revolution that wasn't: a new interpretation of the origin of modern human behavior, *J. Human Evol.* **39**:453-563.
McGrew, W. C., 2001, The other faunivory: Primate insectivory and early human diet, in: *Meat-eating and Human Evolution*. (C. B. Stanford and H. T. Bunn, eds.), Oxford University Press, Oxford, pp.160-178.
Mellars, P. A., 1996, *The Neanderthal Legacy: An Archaeological Perspective from Western Europe*, Princeton University Press, Princeton.
Mitani, J., and Watts D., 1999, Demographic influences on the hunting behavior of chimpanzees. *Am. J. Phys. Anthropol.* **109**:439-454.
Mitani, J. C., Watts, D. P., and Muller, M. N., 2002, Recent developments in the study of wild chimpanzee behavior, *Evol. Anthropol.* **11**:9-25.
Monahan, C. M., 1996, New zooarchaeological data from Bed II, Olduvai Gorge, Tanzania: implications for hominid behavior in the early Pleistocene, *J. Human Evol.* **31**:93-128.
Olsen, S. L., 1989, Solutré: A theoretical approach to the reconstruction of Upper Palaeolithic hunting strategies, *J. Human Evol.* **18**:295-328.
Pianka, E. R., 2000, *Evolutionary Ecology*, 6th edition, Addison Wesley Longman, San Francisco.
Plummer, T. W., and Bishop, L. C., 1994, Hominid paleoecology at Olduvai Gorge, Tanzania as indicated by antelope remains, *J. Human Evol.* **27**:47-76.
Plummer, T. W., and Stanford C. B., 2000, Analysis of a bone assemblage made by chimpanzees at Gombe National Park, Tanzania, *J. Human Evol.* **39**:345-365.
Plummer, T. W., Bishop, L. C., Ditchfield, P., and Hicks, J., 1999, Research on late Pliocene Oldowan sites at Kanjera South, Kenya, *J. Human Evol.* **36**:151-170.
Plummer, T., Ferraro, J., Ditchfield, P., Bishop, L., and Potts, R., 2001, Late Pliocene Oldowan excavations at Kanjera South, Kenya, *Antiquity* **75**:809-810.
Potts, R., 1988, *Early Hominid Activities at Olduvai*, Aldine de Gruyter, New York.
Potts, R., 1989, Olorgesailie: new excavations and findings in early and middle Pleistocene contexts, southern Kenya rift valley, *J. Human Evol.* **18**:477-484.
Potts, R., 1996, *Humanity's Descent: The Consequences of Ecological Instability*, Avon, New York.
Potts, R., 1998, Environmental hypotheses of hominin evolution, *Yearbook Phys. Anthropol.* **41**:93-136.
Potts, R., and Behrensmeyer, A. K., 1992, Late Cenozoic terrestrial ecosystems, in: *Terrestrial Ecosystems through Time* (A.K. Behrensmeyer, J.D. Damuth, W.A. DiMichele, R. Potts, H-D. Sues, and S.L. Wing, eds.), University of Chicago Press, Chicago, pp. 418-541.
Potts, R., and Shipman, P., 1981, Cutmarks made by stone tools on bones from Olduvai Gorge, Tanzania *Nature* **291**:577-580.
Potts, R., Behrensmeyer, A. K., and Ditchfield, P., 1999, Paleolandscape variation and Early Pleistocene hominid activities: Members 1 and 7, Olorgesailie Formation, Kenya, *J. Human Evol,* **37**:747-788.

Richards, M. P., Pettitt, P. B., Trinkaus, E., Smith, F. H., Paunovic, M., and Karavani, I., 2000, Neanderthal diet at Vindija and Neanderthal predation: The evidence of stable isotopes, *Proc. Nat. Acad. Sci.* **97**:7663-7666.

Roberts, M. B., and Parfitt, S. A., 1999, *Boxgrove: A Middle Pleistocene Hominid Site at Eartham Quarry, Boxgrove, West Sussex*, English Heritage, London.

Roche, H., Delagnes, A., Brugal, J.-P., Feibel, C., Kibunjia, M., Mourre, V., and Texier, P.-J., 1999, Early hominid stone tool production and technical skill 2.34 Myr ago in West Turkana, Kenya, *Nature* **399**:57-60.

Rose, L. M., 1997, Vertebrate predation and food-sharing in *Cebus* and *Pan*, *Int. J. Primatol.* **18**:727-765.

Rose, L. M., 2001, Meat and the early human diet: Insights from Neotropical primate studies, in: *Meat-eating and Human Evolution* (C. B. Stanford and H. T. Bunn, eds.), Oxford University Press, Oxford, pp.141-159.

Schoeniger, M. J., Bunn, H. T., Murray, S., Pickering, T., and Moore, J., 2001, Meat-eating by the fourth African ape, in: *Meat-eating and Human Evolution* (C. B. Stanford and H. T. Bunn, eds.), Oxford University Press, Oxford, pp.179-195.

Scott, K., 1980, Two hunting episodes of Middle Palaeolithic age at La Cotte de Saint-Brelade, Jersey (Channel Islands), *World Archaeol.* **12**:137-152.

Scott, K., 1986, The bone assemblages of layers 3 and 6, in: *La Cotte de St. Brelade, Jersey: Excavations by C.B.M. McBruney, 1961-1978*, Geo Books, Norwich, England, pp. 159-183.

Semaw, S., Renne, P., Harris, J. W. K., Feibel, C. S.,. Bernor, R. L., Fesseha, N., and Mowbray, K., 1997, 1.5-million-year-old stone tools from Gona, Ethiopia, *Nature* **385**:333-336.

Shipman, P., 1986, Scavenging and hunting in early hominids: theoretical framework and tests, *Am. Anthropol.* **88**:27-43.

Speth, J. D., 1989, Early hominid hunting and scavenging: the role of meat as an energy source, *J. Human Evol.* **18**:329-343.

Speth, J. D., and Spielman, K. A., 1983, Energy sources, protein metabolism and hunter-gatherer subsistence strategies, *J. Anthropol. Archaeol.* **2**:1-31.

Sponheimer, M., and Lee-Thorp, J., 1999, Isotopic evidence for the diet of an early hominid, *Australopithecus africanus*, *Science* **283**:368-370.

Stanford, C. B., 1998, *Chimpanzee and Red Colobus: The Ecology of Predator and Prey*, Harvard University Press, Cambridge.

Stanford, C. B., 2001, A comparison of social meat-foraging by chimpanzees and human foragers, in: *Meat-eating and Human Evolution* (C. B. Stanford and H. T. Bunn, eds.), Oxford University Press, Oxford, pp.122-140.

Stiner, M. C., 1994, *Honor Among Thieves: A Zooarchaeological Study of Neandertal Ecology*, Princeton University Press, Princeton.

Stiner, M. C., Munro, N. D., Surovell, T. A., Tchernov, E., and Bar-Yosef, O., 1999, Paleolithic population growth pulses evidenced by small animal exploitation, *Science* **283**:190-194.

Strum, S. C., 1981, Process and products of change: baboon predatory behavior at Gilgil, Kenya, in: *Omnivorous Primates: Gathering and Hunting in Human Evolution* (R. S. O. Harding and G. Teleki, eds.), Columbia University Press, New York, pp. 255-302.

Tappen, M., 1995, Savanna ecology and natural bone deposition, *Curr. Anthropol.* **36**:223-260.

Thieme, H., 1997, Lower Palaeolithic hunting spears from Germany, *Nature* **385**: 807-810.

Van Valkenburgh, B., 2001, The dog-eat-dog world of carnivores, in: *Meat-eating and Human Evolution* (C. B. Stanford and H. T. Bunn, eds.), Oxford University Press, Oxford, pp. 101-121.

Walker, A., and R. E. Leakey, eds., 1993, *The Nariokotome Homo erectus Skeleton*, Harvard University Press, Cambridge.

Yellen, J. E., Brooks, A. S., Cornelissen, E., Mehlman, M. H., and Stewart, K., 1995, A Middle Stone Age worked bone industry from Katanda, Upper Semliki Valley, Zaire, *Science* **268**:553-556.

Zhu R., Hoffman, K. A., Potts, R., Deng, C. L., Pan, Y. X.,. Guo, G., Shi, C. D., Guo, Z. T., Yuan, B. Y., Hou, Y. M., and Huang, W. W., 2001, Earliest presence of humans in northeast Asia, *Nature* **413**: 413-417.

II

Major Macroevolutionary Episodes in the History of Predation

Chapter 16

Origin and Early Evolution of Predators
The Ecotone Model and Early Evidence for Macropredation

MARK A. S. MCMENAMIN

---

1. Introduction 379
2. The First Bite 383
3. The Ecotone Model...385
   3.1. Estimating ö..389
   3.2. The Test of Time 391
   3.3. Assumptions of the Model..393
4. Summary...396
   References 398

---

1. Introduction

Our current knowledge on the origin and early evolution of large predators is summarized by Simon Conway Morris (1999, 153-154) as follows:

> ...for many years it was claimed that Cambrian marine communities were almost entirely free of predators . . . the seas were [thought to be] full of suspension-feeders gently swaying in the sea water and deposit feeders calmly digging their way through the sediment. This view is now seen to be far too idyllic, but the story of the rise of predators is still quite tentative. It does appear, however, that in contrast to Cambrian communities those of the Ediacaran were largely free of predators.

The origin of predation is thus lost in a realm of speculation about the remote Precambrian past. Predation certainly began at the microbial level, but the most that can be said about the earliest events is that either a larger microbe engulfed and digested a smaller microbe, or that a larger microbe was infected and killed by a smaller microbe, virus or prion acting as a molecular parasitoid.

MARK A. S. MCMENAMIN • Department of Earth and Environment, Mount Holyoke College, South Hadley, Massachusetts 01075.

Predator-Prey Interactions in the Fossil Record, edited by Patricia H. Kelley, Michał Kowalewski, and Thor A. Hansen. Kluwer Academic/Plenum Publishers, New York, 2003.

If I had to choose between the two scenarios, I would give priority to the latter, as there must have been numerous microscale genetic takeovers (involving microbial and molecular predators) associated with life's origin. One can even imagine a dramatic situation in which an early bacterium engulfs a tiny parasitoid (bacterial, viral, or proteinaceous), after which fate hangs in the balance as to whether the parasitoid will be metabolized by the host, or whether the host will be taken over and killed by the parasitoid. Should the contest end in stalemate, a chance opens for symbiogenesis, the fusion of relatively unrelated life forms to form a new composite (Khakhina, 1992). Indeed, the development of eukaryotic organelles such as chloroplasts (derived from ingested cyanobacteria) and mitochondria (derived from invading *Bdellovibrio* bacteria) are thought to have arisen in just this way (Margulis and Sagan, 2000), and as such may be considered circumstantial evidence for ancient predation. With the passage of time, the microbial predators, parasitoids and their prey become linked in an obligate symbiotic embrace.

Some form of predation, therefore, seems likely to have been associated with the origins of both the prokaryotes (i.e., early genetic takeovers) and the eukaryotes (i.e., the origins of organelles). This idea gives some sense of the extreme importance of predation for the development and advancement of life — predation involves "feedback" processes that lead to major new advances.

With the earliest autotrophic organisms, the fundamental trophic relationship was between the organism and its environment. With the advent of predation, life fed back on itself in such a way that the fundamentals were changed. No longer was the physical environment the sole source of mineral nutrients and other required compounds. One can scarcely imagine a more far-reaching ecological change.

Strata of the Neoproterozoic might be expected to provide the earliest fossil evidence for predators and predation, as the organisms of the Neoproterozoic were finally robust enough to have some chance of preservation outside of special taphonomic circumstances. Vermeij (1987, p. 361) inferred that the Proterozoic shelly organism *Cloudina* (a presumed early metazoan) was herald of the predator-prey escalation that began in the Cambrian. Bengtson and Yue Zhao (1992) claimed the earliest actual example of predation in supposed boreholes in Chinese cloudinids. This interpretation, however, has been challenged by Debrenne and Zhuravlev (1997), who argued that putative cloudinid boreholes are best interpreted as a result of microdolomite crystal formation with subsequent dissolution. A search continues for reliable ways to recognize evidence for predation in the fossil record (i.e., Lescinsky and Benninger, 1994).

A variety of techniques have been employed to assess possible instances of ancient predation, from tests of mechanical strength in modern shells (Moody and Steneck, 1993), to studies of intraspecific selective survival (Alexander, 1987), to field studies of durophagous predation (Moody, 2001), to examination of trace fossil evidence (Jensen, 1990; Conway Morris and Bengtson, 1994; Pickerill and Blissett, 1999), to identification of fossils showing healed damage (Vorwald, 1982; Conway Morris and Jenkins, 1985; McMenamin, 1987; McMenamin and McMenamin, 1990; Horny, 1997), to rare examples of fossil predators actually caught in the act of predation (Blake and Guensburg, 1992).

Fossils of early predators are perhaps the best clues we have to the origins of macropredation (i.e., predation in which both the predators and their prey are larger than a few millimeters in dimension). The first known fossil predators are protoconodont grasping spines such as *Protohertina* and *Mongolodus*. Having a global distribution, these early mouthpart elements are usually found as isolated conoidal sclerites. A recent report describes a naturally arranged, pyritized protoconodont apparatus (McIlroy and Szaniawski, 2000).

Considerable interest has been generated in the past decade in association with the discovery of large Cambrian predators such as *Anomalocaris* and *Laggania* (Whittington and Briggs, 1985; Briggs and Whittington, 1985; Conway Morris, 1985; Briggs, 1994; Collins, 1996). After a difficult early phase of interpretation – in which these taphonomically delicate creatures were subjected to the indignity of having their isolated bodyparts described as entire organisms (*Anomalocaris'* mouth was described as a jellyfish) – they were eventually recognized for what they were, a group of giant predators capable of inflicting serious damage to their Cambrian prey. The earliest evidence for predatory damage to a weakly mineralized organism (the Early Cambrian trilobite *Naraoia*) has been attributed to an anomalocarid by Nedin (1999).

Is there any evidence for macroscopic predation in the Precambrian? The oldest known gnathostomatous structure is the comb-like *Redkinia* from the Neoproterozoic of Ukraine (Fig. 1). Because of its flexible nature, however, *Redkinia* is thought to represent evidence for an early type of filter feeding rather than evidence for biting and crushing jaws (Burzin, 1999; Burzin and Gnilovskaya, 1999).

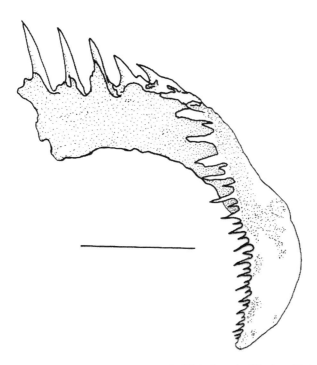

FIGURE 1. The Neoproterozoic gnathostome structure *Redkinia* from the Ukraine. Note the smaller teeth inset between the larger teeth, and the flexible nature of the entire structure. Scale bar 0.1 millimeter in length. After Burzin (1999).

A general view seems to be emerging that megascopic predation pressure (and even herbivorous grazing pressure) was remarkably low or nonexistent in the Neoproterozoic (Stanley, 1973; Nesis, 1995; Waggoner, 2000) and that the Ediacaran organisms lived in a comparatively peaceful "Garden of Ediacara" (McMenamin, 1998). Droser *et al.* (1998) argued that the preservation of Ediacarans and associated anactualistic sedimentary structures (i.e., structures that have no known equivalents in present-day environments) represent "a monument to the absence of benthic marine predators." This consensus is based primarily on negative evidence — no convincing trace fossils of predators, or specimens of damaged prey, are known from the Proterozoic. This allows us to infer a Proterozoic ecology of photoautotrophy, chemoautotropy, marine osmotrophy (passive absorption of dissolved nutrients; McMenamin, 1993), and filter feeding (Proterozoic sponges [Brasier *et al.*, 1997], *Redkinia*) that apparently gave way to a new marine ecosystem characterized by Cambrian predators and shelled animals.

Beginning at or just before the Cambrian with protoconodonts such as *Protohertzina*, a scourge of new predators led by anomalocarids swept through the marine biosphere. This plague of predators apparently resulted in the rapid (but not instantaneous; Jensen *et al.*, 1998) elimination of the apparently defenseless Ediacarans. A few Ediacaran survivors continued on into the Cambrian.

It is also reasonable to infer that prey animals of the Cambrian quickly developed strategies to avoid the new predators. Early predators have long been linked to the appearance of abundant shells in the Cambrian (Bergson, 1908; Evans, 1910). It has been suggested that early macropredators induced a major enhancement in the preservability of the marine biota as prey organisms acquired skeletons. Thus, the arrival of macropredators is deeply if indirectly inscribed in the rock record as the appearance of abundant shelly remains.

Other ideas have been proposed to explain the advent of biomineralized skeletons, including: removal of calcium (toxic in high concentrations) from soft tissue and its deposition as shell material; symbiotic acquisition of calcium-biomineralizing microbes; and biogeochemical attainment of threshold amounts of calcium in sea water sufficient for animals to form skeletons. Most of these alternate explanations, however, focus exclusively on the appearance of calcium carbonate skeletons, and we now know that many of the earliest animal skeletons were phosphatic. The changes of the Cambrian affected a diverse group of organisms with a wide range of body coverings, ranging from calcium carbonate and phosphatic skeletons to no skeletons at all. Any plausible hypothesis dealing with the onset of skeletonization must account for both the disparate chemical compositions of the early skeletons *and* Cambrian survival strategies involving behavioral shifts.

Survival strategies in the Early Cambrian included the following: elongate spines (Leighton, 2001); thick and ornamented shells (Gaudry, 1883; Culver, 1991; Leighton, 1999); deep burrowing (such as the domichnial burrow *Skolithos*; Conway Morris, 1999); rapid locomotion (*Pikaia* of the Burgess Shale); and presumed chemotoxic defense in both skeletonized (*Mickwitzia*; McMenamin, 1992*b*) and soft-bodied (*Aysheaia*; Conway Morris, 1999) animals. A search should be undertaken for chemotoxins preserved in Cambrian sediments.

2. The First Bite

Trace fossil evidence for predation on trilobites, and evidence suggesting that trilobites themselves were predators, has been presented since the mid-1970s (Alpert and Moore, 1975; Birkenmajer, 1977; Conway Morris and Jenkins, 1985; Jensen, 1990; Babcock, Ch. 3, in this volume). The most ancient evidence of such activity would be of considerable interest for any analysis of the origin of macropredation.

The Cambrian fossil record in the Cerro Rajón stratigraphic section of northwest Sonora, Mexico, begins with a trilobitoid trace fossil. This trace, *Rusophycus multilineatus* McMenamin, 2001, occurs in unit 3 of the La Ciénega Formation "in a green platy to fissile siltstone in which other trace fossils are rare" (McMenamin, 2001, p. 76). *Rusophycus multilineatus* is the rusophycid form of *Monomorphichnus multilineatus* Alpert, 1976. The latter form is known from the Deep Spring Formation, White-Inyo Mountains, California. Both ichnofossils are likely to have been made by the same type of arthropod, which must have resembled a fallotaspidoid trilobite.

FIGURE 2. *Rusophycus multilineatus* McMenamin, 2001 associated with *Planolites*. Hypichnial epirelief formed at the boundary between a thin clay layer and a sandy siltstone lamina. The larger ichnofossil has two lobes, with more prominent scratch marks visible on the right (representing the animal's left) lobe. Arrow indicates location of sinusoidal *Planolites* trace fossil, the maker of which is presumed to have been consumed as prey of the *Rusophycus multilineatus* tracemaker. This ichnofossil occurs in unit 3 of the La Ciénega Formation, Cerro Rajón area, northwestern Sonora, Mexico. Specimen IGM 3619 (deposited at the Instituto de Geología, Mexico City); width of image 2 cm.

The type specimen of *Rusophycus multilineatus* records an ancient predation event (Fig. 2). A slightly sinusoidal *Planolites* burrow occurs between the two lobes of *Rusophycus multilineatus*. The *Planolites* is positioned closer to and in part overlaps one of these lobes, and this lobe is excavated deeper into the substrate (as indicated by this hyporelief trace fossil), suggesting that the trilobite dug down deeper on this side in order to excavate and capture the *Planolites* tracemaker. As noted above, trace fossils are relatively rare in unit 3 of the La Ciénega Formation; thus it is unlikely that the close association of the two trace fossils is accidental.

Unless we accept the putative boreholes of *Cloudina*, the *Rusophycus multilineatus-Planolites* association depicted in Fig. 2 may stand as the oldest direct evidence for predation in the fossil record. A case can be made that the changing curvature (near the tip of the arrow in Fig. 2) of the *Planolites* trace as it approaches the center of the association represents an evasive maneuver on the part of the prey organism. *Planolites* is common in the Sonoran stratigraphic sequence, but this is the only specimen known that shows a progressive contraction in the radius of sinusoidal curvature at the end of the trace. Additional specimens are needed to more fully characterize the nature of this association.

The morphology of the *Rusophycus multilineatus-Planolites* association is quite similar to other Early Cambrian examples presented as evidence for trilobite predation (Jensen, 1990). Jensen (1990) described examples of *Rusophycus dispar* (possibly formed by olenellacean trilobites) associated with worm burrows. Jensen (1990) interpreted this association as evidence for trilobites attacking priapulids or other worm-like prey.

The main ichno-morphological difference between the *Rusophycus dispar* and *Rusophycus multilineatus* associations occurs in the relative position of the presumed prey burrows. In the *Rusophycus dispar* association the worm burrow runs approximately parallel to the long axis of the rusophycid trilobite trace, whereas in the *Rusophycus multilineatus* association the worm burrow is positioned approximately normal to the axis of the trilobitoid trace. This difference may be reasonably interpreted as due to variation in the behavior of different trilobite species as they homed in on their respective prey. The *Rusophycus multilineatus* predator was utilizing the scraping function of its right array of arthropod legs to exhume its prey, and using the opposed array as an anchor for mechanical leveraging. As might be expected for a later, more behaviorally advanced species, the *Rusophycus dispar* predator used an enhanced prey-capture protocol that utilized the scraping function of both left and right leg arrays. This would form a more efficient digging mechanism. Hence, this predator centered its body axis over the prey burrow, in contrast to the more lopsided approach favored by the *Rusophycus multilineatus* tracemaker.

The Sonoran trace fossil is of considerable biostratigraphic importance, as it is the eponymous form for the *Rusophycus multilineatus* assemblage. As this is currently the first Paleozoic assemblage in the Mexican sequence, the *Rusophycus multilineatus* assemblage serves to define the position of the Proterozoic-Cambrian boundary in Mexico. This represents a slight downward extension of the position of the boundary as defined in McMenamin (1996) and is upsection from the cloudinid occurrences represented by the *Sinotubulites* assemblage. It is remotely possible that all or part of the *Rusophycus multilineatus* assemblage could end up in the latest Neoproterozoic, depending on refinements in correlation of this interval and exact placement of the Proterozoic-Cambrian boundary. The Sonoran stratigraphic sequence has been proposed

Origin and Early Evolution of Predators

as the stratotype for the Proterozoic Lipalian Period and System (McMenamin, 1998, 2001). A sequence of assemblages crossing the Proterozoic-Cambrian boundary in the Sonoran sequence (McMenamin, 2001) is summarized in Figure 3.

The basal Cambrian position of the *Rusophycus multilineatus* assemblage in the Sonoran sequence, along with other evidence indicating the presence of predators in the Early Cambrian, supports arguments that the true distinction between marine life of the Proterozoic and that of the Cambrian is one of ecology. The ecological criterion that distinguishes these two times is the presence or absence of macropredators. One might even say that the primary distinguishing characteristic of Paleozoic rocks, as opposed to Precambrian rocks, is the presence of evidence for macropredation.

It is thus not merely coincidental that the earliest Cambrian trace fossil found in Sonora provides direct evidence for macropredation. Fossil assemblages both above and below the *Rusophycus multilineatus* occurrence currently support the interpretation of a basal Cambrian age for its assemblage (McMenamin, 2001). A compelling case can now be made for the sudden appearance of macropredators, and consequent ecological disruption, at or very near the beginning of the Cambrian.

3. The Ecotone Model

The concept of such a sudden overturn in the ecological composition of the marine biosphere (the ecological aspect of the "Cambrian Explosion") is an affront to any conventionally uniformitarian way of thinking about the earth sciences. It is known, however, that predators are able to diversify faster than the rest of the biota (Bambach and Kowalewski, 1999) and that, at certain times, catastrophic overkill events can occur (Martin, 1967; Anderson, 1989; Alroy, 2001).

Early Cambrian	*Elliptocephala* assemblage
	Nevadella assemblage
	Nevadia assemblage
	fallotaspidoid assemblage
	Rajonia ornata assemblage
	Rusophycus multilineatus assemblage
Late Precambrian	*Sinotubulites* assemblage
	Jacutophyton assemblage
	Evandavia assemblage
	Eomycetopsis assemblage

FIGURE 3. Sequence of assemblages spanning the Proterozoic-Paleozoic boundary in the Lipalian type section, Cerro Rajón, Sonora. The basal Cambrian boundary occurs at the base of the *Rusophycus multilineatus* assemblage (McMenamin, 2001).

Several mathematical models have been proposed to describe the dynamics of the Cambrian evolutionary radiation. For example, Hallam (1992, p. 131-132) evoked catastrophe theory (O'Shea, 1980, 1986) as a possible means of explaining the "dramatic radiation of the Metazoa in the early Cambrian." The primary motivation behind such modeling efforts is to arrive at an adequate explanation for the rapidity of the event. Although these mathematical models might at first appear difficult to test, with advances in knowledge our ability to test improves. Here I will elaborate a mathematical model (the ecotone model; McMenamin, 1992a) for the Cambrian event. The model demonstrates that the appearance of a single new predator species can set into motion a catastrophic paleoecological transformation. The sudden shift from one ecosystem (Proterozoic) to another (Cambrian) in the ecotone model is triggered by the rapidity with which the predation message passes through the marine biosphere.

The term "ecotone" was coined in 1905 by F. E. Clements, who defined it as a narrow transition zone or "line" between two qualitatively defined discrete habitats or ecosystem types. More recently, ecotones have been defined by Fortin (1994, p. 956) as "long narrow regions of high rates of change." As used by ecologists, "ecotone" bears a primarily geographic meaning. I have borrowed the term from ecologists and imparted to it a temporal connotation. The term is well suited to this purpose, as the transition from the Proterozoic to the Cambrian can be construed as a temporal gradient of comparatively rapid paleoecological change. Of particular interest is the appearance, during this interval, of evidence (in particular *Rusophycus multilineatus*) for early macropredators. It might be argued that predatorial cnidarians such as jellyfish and sea anemones were present in the Proterozoic, but this "impression" is based on a misinterpretation of the Ediacaran fossils (McMenamin, 1998). No convincing predatorial cnidarians are known from the Proterozoic.

The initiation of predator-prey relationships can be considered as a transmission of information from one species to another. Classic mathematical tools are available for such an analysis. The iterated growth equation, first introduced by P. F. Verhulst (1838), has been applied to describe the spread of a rumor (McCarty, 1976) and resembles the generating equation for the well-known bifurcation diagram of May (1976). A version of the Verhulst equation was employed to develop the ecotone model of Cambrian diversification (McMenamin, 1992a).

This model is useful for the study of the spread of information "units" or ecological "messages." The onset of macropredation is the ecological "message" of interest here, with the implication that the macropredatory habit can be acquired by triploblastic animals. The idea here is that an animal species can be permanently changed from a non-predatory species to a predatory species by being subjected to predation pressure. Such a change constitutes receipt of the ecological message.

Several assumptions will be required to model the spread of information through an ecosystem. These assumptions will be discussed in detail below, after a general presentation of the model.

Assume at the beginning of the model run a world filled with primary producers and first order consumers, and no (or a negligible number of) higher order consumers. Introduce one large predator to a biota free of megascopic predation. Assume this predator attacks two species per million years. Each first order consumer prey species develops aggressive behaviors that in turn allow it to attack, as a predator or parasite, two more species per million years. If $M(t)$ is the number of species attacked at time t,

Origin and Early Evolution of Predators

out of a population of first order consumer species V, then every million years each one of them will attack on average

$$2(1 - \frac{M(t)}{V}) \quad (1)$$

species not yet preyed upon. Consider the following differential equation:

$$\frac{dM(t)}{dt} = M(t)[2(1 - \frac{M(t)}{V})] \quad (2).$$

The variables are separable, and the equation can be rewritten:

$$\frac{dM(t)}{dt} = [\frac{2V - 2M(t)}{V}] \quad (3);$$

$$\frac{dM(t)}{dt} = M(t)\frac{2}{V}(V - M(t)) \quad (4);$$

$$\frac{dM(t)}{dt} = \frac{2}{V}M(t)(V - M(t)) \quad (5);$$

$$\frac{V}{M(t)(V - M(t))}\frac{dM(t)}{dt} = 2 \quad (6).$$

The solution to equation (6) will be (C a constant):

$$\int \frac{V}{M(t)(V - M(t))}\frac{dM(t)}{dt} dt = \int 2 dt = 2t + C \quad (7).$$

Such a differential equation is said to have separable variables (McCarty, 1976) because it can be written in the form:

$$P(y)y' = Q(x) \quad (8).$$

The solution will be of the form:

$$\int P(y)y' dx = \int Q(x) dx \quad (9).$$

where the indefinite integral $\int Q(x)dx$ represents a family of antiderivatives of functions whose derivative is $Q(x)$.

Observe that integrating the left-hand side of equation 7 yields:

$$\frac{V}{M(t)(V-M(t))} = \frac{1}{M(t)} + \frac{1}{V-M(t)} \quad (10).$$

Integrating both sides of equation (10)

$$\int \frac{V}{M(t)(V-M(t))} \frac{dM(t)}{dt} dt = \int \frac{\frac{dM(t)}{dt}}{M(t)} dt + \int \frac{\frac{dM(t)}{dt}}{V-M(t)} dt \quad (11)$$

gives

$$\int \frac{\frac{dM(t)}{dt}}{M(t)} dt + \int \frac{\frac{dM(t)}{dt}}{V-M(t)} dt = \ln|M(t)| - \ln|V-M(t)| + C \quad (12).$$

Since

$$0 \leq M(t) \leq V \quad (13)$$

the absolute value symbols may be dropped, yielding

$$\ln M(t) - \ln(V - M(t)) = 2t + C \quad (14).$$

Thus:

$$\ln\left(\frac{M(t)}{V-M(t)}\right) = 2t + C \quad (15);$$

$$\frac{M(t)}{V-M(t)} = e^{2t+C} = e^{2t}e^{C} = ke^{2t}, k > 0 \text{ and with } k = e^{C} \quad (16).$$

Solving for M as a function of t gives:

$$M(t) = ke^{2t}(V - M(t)) \quad (17);$$

$$M(t)(1 + ke^{2t}) = kVe^{2t} \quad (18);$$

$$M(t) = \frac{kVe^{2t}}{(1+ke^{2t})} = \frac{kVe^{2t}}{(1+ke^{2t})} \frac{e^{-2t}}{e^{-2t}} = \frac{kV}{e^{-2t}+k} \quad (19).$$

Note that

Origin and Early Evolution of Predators

as $t \to \infty$, $M(t) \to V$;

$$M(t) = \frac{kV}{k+e^{-2t}} \quad (20),$$

or, more generally,

$$M(t) = \frac{kV}{k+e^{-öt}} \quad (21).$$

where ö (odereisis) represents the number of "message deliveries" per unit time. Figure 4 shows a sample family of curves for equation (21) with ö varying from one to ten.

3.1. Estimating ö

In his study of diversification functions and the rate of evolution, Walker (1985) employed the logistic equation:

$$\frac{dN}{dt} = r\left(1 - \frac{N}{k}\right)N \quad (22).$$

Equation (22), like equation (2), is a variant of the Verhulst equation. In the former case r is the Malthusian parameter (intrinsic rate of diversification) and k is some limiting or equilibrium value (such as carrying capacity) of the habitat for the taxon in question. The Malthusian parameter is introduced into the ecotone model as ö in equation (21).

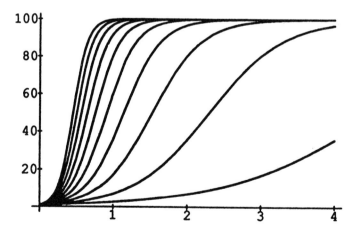

FIGURE 4. Ecological message deliveries per unit time. This figure shows a family of curves for equation (21) with ö varying from one to ten. The steepest curve on the far left denotes ö = 10; with a large number of "message deliveries" per unit time (ö), the curve rises steeply and function M(t) rapidly approaches the total (100%) population of first order consumer species (V). The ordinate represents the percentage attainment of "message delivery"; the abscissa can be taken to represent millions of years.

Walker (1985) was able to solve for *r* from the solution of equation (22):

$$r = \frac{1}{t}\ln\left[\frac{k-N(t_o)}{N(t_o)}\right] + \frac{1}{t}\ln\left[\frac{N(t)}{k-N(t)}\right] \quad (23).$$

Walker (1985) then searched for a way to simplify this equation. He noted that one cannot use $N=k$ since N never reaches or exceeds k. Nonetheless, it makes most sense to choose the largest value of N which makes equation (23) computable, and this would generally be the value $N(t')=k-1$, in other words, the population value that is one less than the equilibrium level or carrying capacity. Not only will this give the most accurate estimate, it will simplify Equation 2 to the form:

$$r = \frac{1}{t'}\ln\left[\frac{k-N(t_0)}{N(t_0)}\right] + \frac{1}{t'}\ln[k-1] \quad (24).$$

The expression $k-1$ is dimensionless. Walker (1985) assumed that the evolutionary radiation proceeded from a single ancestral species. Thus:

$$N(t_o)=1$$

and:

$$r = \frac{2}{t'}\ln[k-1] \quad (25).$$

Walker's (1985) attempt was the first time that the logistic model had been used to calculate an intrinsic rate of increase. Equation (25) can be adapted for the ecotone model (assuming a single, initial, bilateral, megascopic animal predator) as:

$$ö = \frac{2}{t'}\ln[V-1] \quad (26)$$

for $M(0) = 1$ and $M(t') = V - 1$.

Equation (26) would be applied in the following fashion. The unit of time in this case will be one million years. Say, for ease of calculation, that the time interval (t') during which the ecotone transformation occurs lasts only two million years. V represents the total population of animals liable to be affected by the ecotone transformation. In other words, these animals are the ones capable of changing from a peaceful neighbor to a predator or a parasite. The model does not require that the animal become an obligate predator; previous modes of feeding are still allowed. In other words, the model works if the new predators are omnivorous rather than being strictly carnivorous. Cases of reversion in which herbivorous prey become carnivorous predators, and then become herbivorous again, are ignored in this model, as the number predators, and then become herbivorous again, are ignored in this model, as the number of such cases could be expected to be small unless system-wide ambient predation pressure were to decrease for some reason, and we know that this was not the case.

Most animals near the Cambrian boundary were afflicted by large predators. Responses in prey animals varied from secretion of spiny scleritomes (Bengtson, 1994; for a modern variant see Eisner *et al.*, 1996) to secretion of chemotoxins (McMenamin, 1992*b*). The animals of variable V manifested a different response; a combination of aggressive behavior and newly acquired predatory (or, if they began as micropredators, macropredatory) feeding strategy.

With $t'=2$ and $V=100$ species:

$$\ddot{o} = (1)(\ln[99]) = 4.6 \qquad (27).$$

For $V=1000$ species:

$$\ddot{o} = (1)(\ln[999]) = 6.9 \qquad (28).$$

As will be shown below, an order of magnitude change in V requires only a relatively slight increase in $\ddot{o}$ to keep the ecotone interval constant. Stretching t' to twenty million years changes $\ddot{o}$ to 0.46 and 0.69 for $V = 100$ and $V = 1000$, respectively.

3.2. The Test of Time

Three quantitative descriptions of the Cambrian evolutionary event will now be tested. The first is the logistic hypothesis of Sepkoski (1992). This hypothesis ascribes the Cambrian radiation, as an explicit part of Sepkoski's model, to the exponential phase, following an early lag phase, of logistic equilibrium growth in early animal taxa. It differs from the ecotone model by relying on an intrinsic, within-taxon diversification to generate the Cambrian explosion.

Second is the cell type hypothesis of Valentine (1994). In this hypothesis, the Cambrian explosion owes to attainment of a threshold number of cell types (40-50) and ensuing radiations of several animal clades. Third is the ecotone hypothesis described above.

A test is provided by recalibration of the age of the beginning of the Tommotian stage of the Cambrian. Bowring *et al.* (1993) placed the point of "exponential" increase in Cambrian taxa at the base of the Tommotian, and calculated this horizon to be 530 million years in age. The Tommotian-Atdabanian interval of taxonomic increase lasted, according to Bowring *et al.*'s (1993) calculations, 5-6 million years. Calculations by Landing *et al.* (1998, their fig. 1) provide only 3-5 million years for the interval between the base of the Tommotian and the Atdabanian/Botomian boundary interval. The duration of this interval constitutes the test for the three mathematical descriptions of the Cambrian event.

Sepkoski (1992, p. 559) estimated the duration of the "exponential diversification" interval to be 15 million years. With the time available cut in third or less by the new recalibrations, the logistic/equilibrium hypothesis is considered here to be falsified by the fact that there is insufficient time for the exponential phase of logistic growth using Sepkoski's parameters. Even without the limitation imposed by the length of the exponential phase, the logistic model fails because a high per-taxon rate of diversification would have to be manifested in all clades simultaneously. It would be implausible to argue that biological factors intrinsic to the diversifying clades themselves would be sufficient to cause a simultaneous diversification, especially over such a brief time span. The Cambrian event is therefore more a step function than an S-curve.

The cell type hypothesis is considered here to be falsified by the geologic age recalibration for similar reasons. In the case of both the logistic and cell type models, the main difficulty is synchronizing the lag-to-exponential (or attainment of 40-50 cell types, respectively) transition to coincide exactly with the base of the Tommotian. For this to occur, Early Cambrian animal taxa would have to be viewed as taxonomic diversification time bombs, with all clades bearing synchronized, silently ticking fuses waiting to go off at exactly 530 million years before present.

Rising levels of oxygen (to pass beyond some biotically important threshold level) might be invoked to explain such simultaneous change, but this explanation is inadequate considering that active and presumably aerobic burrowers, known by their trace fossils, lived tens of millions of years before the Cambrian (McMenamin, 1996). Also, oxygen levels might very well have been higher at times during the Proterozoic than during the Early Cambrian (McMenamin and McMenamin, 1990). Therefore, the logistic and cell-type explanations are insufficient, as the evolutionary changes proposed by these hypotheses would have had to occur simultaneously in a variety of animal clades. This problem, serious before the recalibrations, becomes insurmountable when the Tommotian-Atdabanian interval is shrunk to 5 million years.

The ecotone model passes the test of a short Tommotian-Atdabanian interval. This model even predicts a rapid Cambrian event, for ecological message delivery (of, in this case, *threat of* and the *opportunities for* macropredation) can propagate quickly through a large group of taxa. The diversity of the Cambrian still must be generated (evolved), of course, but the ecotone model comes equipped with a potent evolutionary forcing mechanism — predation pressure itself. This is not a mere displacement of the problems encountered by the logistic and cell type models — high diversification rates become directly linked to escalating predation pressure (McMenamin, 1988), pressure that is applied in a rapid and pervasive fashion.

Another application of equation (26) will demonstrate the link between interval length and rate of message propagation. The overlapping consensus estimate (Bowring *et al.*, 1993; Landing *et al.*, 1998) of 5 million years will be used for t'. At $V = 1000$:

$$\ddot{o} = \left(\frac{2}{5}\right)(\ln[999]) = 2.76 \qquad (29).$$

At $V = 10,000$, the rate increases to:

$$\ddot{o} = \left(\frac{2}{5}\right)(\ln[9999]) = 3.68 \qquad (30).$$

For $V = 100,000$, $\ddot{o}$ rises only to 4.6 deliveries per million years. Thus, this model is quite robust to varying estimates of the numbers of individual taxa affected, and can model the spread of the required effect at a rate consistent with the latest geological data. In other words, assuming between 1,000 and 100,000 Neoproterozoic metazoan taxa bearing the potential for transformation to predators, a message delivery rate of between 2.7 and 4.6 messages per million years is required to transform the fauna to one rich in predators. For the higher rate (4.6), this would mean on average one message delivery per predator species per 217,000 years. This rate may even represent a conservative estimate, especially in light of indications that the transition from filter feeder to predator (in a case apparently lacking the accelerating influence of newly emergent "message-delivering" predators) can occur in as little as 7,000 years (Kelly-Borges, 1995).

The arguments above demonstrate that the ecotone model is consistent with a very rapid Cambrian diversification. It does not require that the diversification proceed at a high taxonomic level, but neither does it preclude the possibility of a high taxonomic level diversification at the Cambrian boundary.

3.3. Assumptions of the Model

Several assumptions are required by the ecotone model. These must be examined in detail, for it would be folly to construct a mathematical model consistent with geological data but at a variance with paleobiological data.

The first assumption is that the Neoproterozoic marine biota was largely free of megascopic predators, and was characterized instead by a variety of large autotrophic organisms. This "Garden of Ediacara" hypothesis (McMenamin, 1986, 1998) has gained support recently with the reports of the photosymbiotic fungiid coral *Leptoseris fragilis* photosynthesizing at 145 meters water depth (Schlichter and Fricke, 1991) and by calculations showing that symbionts in coral fix three to four times more carbon than photosynthetic plankton in the upwelling regions of all the oceans (Hawksworth, 1994). Micropredators were surely present in the Neoproterozoic, but the abundance of large, nonskeletogenous bodies (many with presumed autotrophic lifestyles) during this time argues forcefully against the presence of large, triploblastic animal predators. This is especially so considering the unusual preservation of Ediacarans and associated nonactualistic Proterozoic sedimentary structures. The first assumption thus appears to be valid.

The second assumption of the ecotone model is that attainment of a predator or parasite lifestyle is a viable response, by prey species which themselves are triploblastic animals, to the onset of megapredation. This assumption was challenged by Bengtson (1994, p. 425), who was unable to conceive "of any behavior pattern or selective advantage that would produce this effect." However, the switch from prey to parasite or predator has occurred repeatedly in the history of life. Vermeij (1987) suggests that swallowing whole prey is perhaps the most common initial type of predation, considering that it may only be a relatively small evolutionary step from trophic strategies of deposit feeding or particulate suspension feeding. Analogous change frequently occurs within the life of individual marine animals. Sebens and Koehl (1984) have documented the ontogenetic transformation of sea anemones from potential prey to predator.

The second assumption (i.e., predators can induce non-predators to become predators themselves) is speculative to the extent that such occurrences are hard to find among the modern biota. It is easy, however, to imagine selective pressures under which such a phenomenon might occur. It seems possible that such a transformation could occur quickly, requiring only a few or perhaps only one generation under the right circumstances. If so, it may therefore be possible to experimentally test the second assumption at a marine station specially prepared to conduct a relatively long term experiment of this nature.

Vacelet and Boury-Esnault (1995) provided a modern example of the required trophic transformation. They described how Mediterranean sponges belonging to genus *Asbestopluma* abandoned microphagous suspension feeding and evolved carnivory. These sponges capture crustaceans by means of filaments adorned with raised hook-shaped spicules.

The fact that Mediterranean *Asbestopluma*'s nearest relatives are deep water, open marine species raised questions about the origin of this shallow water sponge population. One possibility is that it originated from a sweepstake (i.e., low probability) colonization event from the North Sea, in which case the speciation event must have occurred rapidly, within the past 7,000 years (Kelly-Borges, 1995).

One could easily imagine a situation in which, once subjected to appropriate selection pressure, the filter-feeding gnathostomes of *Redkinia* (Fig. 1) are transformed into biting jaws. *Redkinia*, like *Asbestopluma*, thus could easily have acquired the ability to handle larger prey.

Early trilobites may have undergone the transformation from prey to predator. It is known that trilobite cuticular thickness and strength increases during the Early Cambrian. This trilobite cuticular strengthening could be plausibly attributed to carnivorous grazing pressure from predators such as *Anomalocaris* (McMenamin and McMenamin, 1990; Nedin, 1999; Babcock, Ch. 3, in this volume). At approximately the same time, trilobites began to use their skeleton to prey on other organisms. *Rusophycus multilineatus* from Mexico (Fig. 2) gives us direct evidence of this type of predation, right at the base of the Cambrian in the Sonoran stratigraphic sequence. Trilobite carnivory became more aggressive as time passed; the spiny gnathobases of the Middle Cambrian trilobite *Olenoides serratus* were utilized to tear and shred prey (Whittington, 1985; Babcock, Ch. 3, in this volume).

One may accept then that the second assumption is at least plausible. Virtually all undoubted heterotrophic Proterozoic animals, as known from their trace fossils, were bilateral grazers and deposit feeders, apparently generalized in their feeding preferences and with the potential to be transformed into predators without requiring a radical alteration of *baüplan*. The first large predator to appear (at or near the Cambrian boundary) came upon a marine biosphere filled with unprotected or weakly defended prey species. This predator induced a lifestyle shift to predation in numerous prey species, as battered and isolated survivor populations adjusted to the new conditions. Such circumstances were favorable to high rates of speciation. Other prey responses such as skeletons and escape behaviors also developed, but among the strategies most favored by natural selection were those that allowed the endangered prey to respond in a few generations with a minimum of morphologic change.

The only hard parts we would ordinarily expect to find in a Neoproterozoic grazing or deposit feeding animal would have been in the oral area — radular teeth, gnathostomes, or analogous structures. These would be among the first features co-opted by the organisms as defense against larger predators, and would therefore have had the highest potential for selective modification.

In an observation that is further consistent with the new radiometric dates, the interval of message spread (ended by recipient saturation) would be expected to be short since the species diversity of animals in the Neoproterozoic was not high relative to today. The predator-parasite transformation hypothesized here involved species of organisms already present in the marine biosphere, and could trigger a simultaneous and evolutionarily rapid change in unrelated animal lineages.

Fossil evidence suggests that most Neoproterozoic triploblasts existed at a similar level of behavioral development, and one might reasonably expect almost evenly matched predator-prey exchanges during the onset of macropredation. Prey lineages that survived these exchanges (in particular the trilobites) would be "preadapted" to become predators themselves, and to rapidly propagate the ecological message among similar organisms with propensities to respond in the same way. Indeed, the behavior switch confers a double selective advantage — what was once a source of mortality leads, perhaps even before a single new species has been generated, to a new and concentrated food resource. Such a system of rewards, along with other rewards associated with trophic exaptation of predator defenses, were no doubt needed to sustain an event as far-reaching and as sudden as the Cambrian diversification. More importantly, the ecotone transformation implies nearly simultaneous ecological change with potential to induce rapid evolutionary response in both the prey organisms and the predators themselves. In other words, the speed of the Cambrian event implies that it must have been, in some measure, a self-sustaining process.

Other factors could contribute to the self-sustainability of the process. As attested to by the presence of remoras (*Echeneis naucrates*) attached to sharks, predation is messy business. Macropredators in an aquatic setting scatter the environment with bits of shredded flesh. These fragments settle and are encountered by deposit and filter feeders as an unexpectedly high-quality food source. Thus, macropredators can condition an environment for ecotone transformation by exposing deposit and filter feeders to a meaty food source well before these low-order consumers have attacked, or been attacked by, something larger.

As noted at the beginning of this chapter, the predatory lifestyle, at least at a microbial scale, was present for hundreds of millions of years before the Cambrian. There may have been metazoan predators before the first shelly fossils, but these hypothetical early predators were too small or otherwise incapable of either inducing skeletonization in their prey or triggering significant sedimentological, ecological and/or evolutionary changes. Such transformation requires larger predators employing visual or other advanced sensory systems. Predation of this nature could induce evolutionary diversification, and might further encourage the prey to predator transformation by selecting for polymorphism in prey organisms (Bond and Kamil, 2002).

The third assumption underlying the calculation of $ö$ above is that the ecotone transformation was initiated by a single, exceptional predator species. Anomalocarids are now presumed to be monophyletic (Chen *et al.*, 1994), hence the possibility exists that the first vision-directed megascopic predator was an early anomalocarid. Early trilobites would also be candidates for the first megascopic predators. Like the anomalocarids, the earliest trilobites (and other early arthropods, for that matter) lacked a heavily mineralized skeleton.

The fourth and final assumption is the concept that, when it comes to predation, body size matters. Below a certain threshold size (say, one millimeter in greatest dimension), predators are unlikely to be capable of triggering the kinds of responses in larger prey organisms that have been discussed in this paper. Such small predators are too small to develop effective biting mouth parts, advanced visual systems and advanced cognition (I am excluding here cases of subsequent miniaturization). They must therefore be considered to be micropredators and are excluded from the analysis presented here. Of course, prey organisms may have attained large size in the first place as a means of avoiding attack by micropredators.

Viewed as an ecotone transformation, the Cambrian diversification involved a qualitative change in marine trophic relationships, and as such is unique and is of an importance comparable to that of the emergence of predation at the microbial scale. The Cambrian event must be set apart from all subsequent evolutionary radiations, which in the majority of cases (e.g., the Cenozoic radiation of Mammalia after extinction of the dinosaurs) represent occupation of ecological vacancies in an already structured ecosystem.

4. Summary

The record of escalating predation over time (Vermeij, 1987; Bambach and Kowalewski, 1999) begins at, or just before, the beginning of the Cambrian. Arguments (supported by evidence such as damaged prey, fossil predators, ichnofossil predation events, etc.) have been presented linking the origin of macroscopic predation to the decline of the Ediacaran biota, the skeletonization of Cambrian metazoans, and the diversification of Cambrian metazoans. Whether one prefers to see the Cambrian event as a skeletonization event (of already extant metazoan clades) or as an explosive evolutionary event (leading to Cambrian evolution of new higher taxa), sudden increase in predation pressure has been invoked to explain the nature of the Cambrian event. Other factors must have been involved as well, and these await elucidation by future research.

In this paper I have presented a mathematical model (compatible with both the skeletonization event view and the evolutionary diversification view) to explain how predation could so dramatically alter the marine biosphere during the brief Cambrian boundary interval. Recent radiometric date recalibrations, plus ichnofossil evidence showing that predators were active right at the base of the Cambrian, require the interval to be short considering the magnitude of ecological (and evolutionary) changes that took place at the time.

The ecotone model provides a plausible explanation of a large body of paleontological data that, owing to recent geologic date recalculations, cannot be adequately explained by other available models. Simply put, the ecotone model claims that introduction of a single macropredator species into the Proterozoic marine biosphere induced a paleoecological phase shift accomplished by predation-induced spread of the predatory lifestyle. The rate of ecological change predicted by the ecotone model is in close accord with the timing requirements of the radiometrically-calibrated geologic record.

The primary weakness of the model is its second assumption, namely, the contention that a presumably diverse group of medium to large sized early metazoans could all be relatively rapidly transformed from grazers and filter feeders to macropredators. This is a difficult premise to test as it requires profound changes to occur in poorly understood Neoproterozoic marine ecosystems. Suffice it to say for now that there is at least one example from the modern biota that supports the plausibility of the second assumption that predators can induce non-predators to become predators themselves. Other such examples may await discovery.

The implications of the ecotone model are twofold. First, it suggests that pronounced, ecosystem-wide ecological changes can occur with breathtaking speed. Second, it suggests that such change, in addition to being sudden, can be global and irreversible (or reversible only by eliminating the life forms involved).

This study suggests three new areas for research into the ecological dynamics of the Cambrian boundary interval. First, we need to determine the height of the systematic level at which the Cambrian diversification took place. It would also be helpful to determine, using high-resolution biostratigraphy, the Cambrian speciation rate. The ecotone model requires that a reasonable number of animal phyla be present before the Cambrian boundary, but makes no predictions about whether the ensuing metazoan diversification was a high taxonomic level radiation (new phyla and classes) or a low taxonomic level radiation (new orders, families and genera only). If the former were shown to be the case, the ecotone model could be helpful at analyzing the impetus for high-level taxonomic diversification.

Second, we need to consider the ecotone model from a Vernadskian perspective. Using a Vernadskian approach (Vernadsky, 1998, p. 67-69), it should be possible to quantify the velocity (meters per year) of the ecotone transition. Assuming a constant rate of ecotone spread from a single point of origin, we can estimate the velocity of global ecotone transformation. Taking the equatorial circumference of the earth to be 40,077 km, the speed for a transition time of $t' = 2$ million years would be $(40,077/(2)(2,000,000)) = 0.01$ km/year or approximately ten meters per year. For the consensus five million year value (Bowring *et al.*, 1993; Landing *et al.*, 1998), the speed drops to 4 meters per year. And for the twenty million-year interval it drops to one meter per year.

The best estimate therefore of what might be called the "speed of the Cambrian event" indicates that the ecotone transformation propagated across the face of the globe at an average rate of 4 meters per year. By means of comparison, this rate is 150 times faster than a typical value for a rate of tectonic plate motion (2.5 cm/yr), but roughly a thousand times slower than the spread of Christianity (4.6 km/yr, with it taking 400 years for the religion to reach Ireland). As correlations of the Cambrian boundary interval are refined to a precision of, say, 100,000 to 250,000 years (as should be possible in at least some stratigraphic sections with combined application of biostratigraphy, lithostratigraphy, magnetostratigraphy, isotope chemostratigraphy and new approaches to geochronology), it should be possible to trace the diachronous ecotone transformation back to its paleobiogeographic point of origin.

Finally, it should be possible to quantify macroscopic predation pressure in Vernadskian fashion. A full treatment of this subject is beyond the scope of this chapter, but it can already be said that future work will require new biogeometric parameters. A Vernadskian biogeometric parameter P (macropredation pressure intensity) could be defined in units of area, and would be analogous to what Vernadsky called kinetic geochemical energy ("pressure of life"; Vernadsky, 1998, p. 69). The value of P for the macroscopic part of the marine biosphere would be zero (or close to it) before the Cambrian, increase dramatically across the boundary, and increase again in the Albian as part of the rise in the abundance of Mesozoic predatory genera (Bambach and Kowalewski, 1999). This type of biogeometric parameterization (cf. McMenamin and Whiteside, 1999) is urgently needed to improve our understanding of global ecosystem dynamics, especially as we find ourselves in the midst of a new (anthropogenic), poorly understood ecotone transformation.

ACKNOWLEDGMENTS: Thanks to P. E. Cloud, J. W. Valentine, T. D. Walker, D. L. Schulte McMenamin, J. J. Sepkoski Jr., and an anonymous reviewer for enlightening discussion and comments, and to J. H. Stewart and J. Manuel Morales-Ramirez for assistance and comraderie in the field.

References

Alexander, R. R., 1987, Intraspecific selective survival within variably uniplicate Late Devonian brachiopods, *Lethaia* **20**:315-325.
Alpert, S., 1976, Trilobite and star-like trace fossils from the White-Inyo Mountains, California, *J. Paleontol.* **49**:661-669.
Alpert, S., and Moore, J. N., 1975, Lower Cambrian trace fossil evidence for predation on trilobites, *Lethaia* **8**:223-230.
Alroy, J., 2001, A multispecies overkill simulation of the end Pleistocene megafaunal mass extinction, *Science* **292**:1893-1896.
Anderson, A., 1989, Mechanics of overkill in the extinction of New Zealand moas, *J. Archaeol. Sci.* **16**:137-151.
Bambach, R. K., and Kowalewski, M., 1999, Diversity of predators compared to the records of prey-predator escalation—two tales of the history of predation, *Geol. Soc. Am. Abstr. Progr.* **31**:336.
Bengtson, S., 1994, The advent of animal skeletons, in: *Early Life on Earth* (S. Bengtson, ed.), Columbia University Press, New York, pp. 412-425.
Bengtson, S., and Yue Zhao, 1992, Predatorial borings in Late Precambrian mineralized exoskeletons, *Science* **257**:367-369.
Bergson, H., 1908, *L'Evolution créatrice*, Librairies Félix Alcan et Guillaumin Réunies, Paris.
Birkenmajer, K., 1977, Trace fossil evidence for predation on trilobites from Lower Cambrian of South Spitsbergen, *Norsk Polarinstitut Arsbok* **1976**:187-195.
Blake, D. B., and Guensburg, T. E., 1992, Caught in the act; a Late Ordovician asteroid and its pelecypod prey, *Geol. Soc. Am. Abstr. Progr.* **24**:6.
Bond, A. B., and Kamil, A. C., 2002, Visual predators select for crypticity and polymorphism in virtual prey, *Nature* **415**:609-613.
Bowring, S. A., Grotzinger, J. P., Isachsen, C. E., Knoll, A. H., Pelechaty, S. M., and Kolosov, P., 1993, Calibrating rates of Early Cambrian evolution, *Science* **261**:1293-1298.
Brasier, M., Green, O., and Shields, G., 1997, Ediacaran sponge spicule clusters from southwestern Mongolia and the origins of the Cambrian fauna, *Geology* **25**:303-306.
Briggs, D. E. G., 1994, Giant predators from the Cambrian, *Science* **264**:1283-1284.
Briggs, D. E. G., and Whittington, H. B., 1985, Terror of the trilobites. *Nat. Hist.* **94**:34-39.
Burzin, M., 1999, A mysterious world of Ediacaran organisms, *Science in Russia* **2**:22-28.
Burzin, M. B., and Gnilovskaya, M. B., 1999, Kakimi byli drevneishie zhibotnye, *Priroda* **11**:31-41.
Chen J., Ramsköld, L., and Gui-Qing, Z., 1994, Evidence for monophyly and arthropod affinity of Cambrian giant predators, *Science* **264**:1304-1308.
Collins, D., 1996, The "evolution" of *Anomalocaris* and its classification in the arthropod class Dinocarida (nov.) and order Radiodonta (nov.), *J. Paleontol.* **70**:280-293.
Conway Morris, S., 1985, Cambrian enigma, *Nature* **316**:677.
Conway Morris, S., 1999, *The Crucible of Creation*, Oxford University Press, Oxford.
Conway Morris, S., and Bengtson, S., 1994, Cambrian predators; possible evidence from boreholes, *J. Paleontol.* **68**:1-23.
Conway Morris, S., and Jenkins, R. J. F., 1985, Healed injuries in Early Cambrian trilobites from South Australia, *Alcheringa* **9**:167-177.
Culver, S. J., 1991, Early Cambrian foraminifera from West Africa, *Science* **254**:689-691.
Debrenne, F., and Zhuravlev, A. Yu., 1997, Cambrian food web: A brief review, *Geobios* **20**:181-188.
Droser, M. L., Jensen, S., and Gehling, J. G., 1998, The first grave robbers: Early Cambrian ichnofabric, *Geol. Soc. Am. Abstr. Progr.* **30**:233.
Eisner, T., Eisner, M., and Deyrup, M., 1996, Millipede defense: use of detachable bristles to entangle ants, *Proc. Nat. Acad. Sci. USA* **93**:10848-10851.
Evans, J. W., 1910, The sudden appearance of the Cambrian fauna, *11th Inter. Geol. Congr., Stockholm, 1910, Compte Rendu* **1**:543-546.

Fortin, M.-J., 1994, Edge detection algorithms for two-dimensional ecological data, *Ecology* **75**:956-965.
Gaudry, A., 1883, *Les Enchainements du Monde Animal Dans Les Temps Geologiques, Fossiles primaires*, Libraire F. Savy, Paris.
Hallam, A., 1992, *Great Geological Controversies*, Oxford University Press, Oxford.
Hawksworth, D. L., 1994, Strategies for living together, *Nature* **371**:570.
Horny, R. J., 1997, Shell breakage and repair in *Sinuitopsis neglecta* (Mollusca, Tergomya) from the Middle Ordovician of Bohemia, *Casopis Narodniho Muzea v Praze. Rada Prirodovedna* **166**:137-142.
Jensen, S., 1990, Predation by early Cambrian trilobites on infaunal worms—evidence from the Swedish Mickwitzia Sandstone, *Lethaia* **23**:29-42.
Jensen, S., Gehling, J. G., and Droser, M. L., 1998, Ediacara-type fossils in Cambrian sediments, *Nature* **393**:567-569.
Kelly-Borges, M., 1995, Sponges out of their depth, *Nature* **373**:284.
Khakhina, L. N., 1992, *Concepts of Symbiogenesis*, Yale University Press, New Haven.
Landing, E., Bowring, S. A., Davidek, K. L., Westrop, S. R., Geyer, G., and Heldmaier, W., 1998, Duration of the Early Cambrian: U-Pb ages of volcanic ashes from Avalon and Gondwana, *Can. J. Earth Sci.* **35**:329-338.
Leighton, L. R., 1999, Antipredatory function of brachiopod ornament, *Geol. Soc. Am. Abstr. Progr.* **31**:43.
Leighton, L. R., 2001, New example of Devonian predatory boreholes and the influence of brachiopod spines on predator success, *Palaeogeogr. Palaeoclim. Palaeoecol.* **165**:53-69.
Lescinsky, H. L., and Benninger, L., 1994, Pseudo-borings and predator traces; artifacts of pressure-dissolution in fossiliferous shales, *Palaios* **9**:599-604.
Margulis, L., and Sagan, D., 2000, *What is Life?*, University of California Press, Berkeley.
Martin, P. S., 1967, Pleistocene overkill, *Nat. Hist.* **76**:32-38.
May, R. M., 1976, Simple mathematical models with very complicated dynamics, *Nature* **261**:459-467.
McCarty, G., 1976, *Calculator calculus*, Page-Finklin Publishing Company, Palo Alto, California.
McIlroy, D., and Szaniawski, H., 2000, A Lower Cambrian protoconodont apparatus from the Placentian of southeastern Newfoundland, *Lethaia* **33**:95-102.
McMenamin, M. A. S., 1986, The Garden of Ediacara, *Palaios* **1**:178-182.
McMenamin, M. A. S., 1987, The fate of the Ediacaran fauna, the nature of conulariids, and the basal Paleozoic predator revolution, *Geol. Soc. Am. Abstr. Progr.* **19**:29.
McMenamin, M. A. S., 1988, Paleoecological feedback and the Vendian-Cambrian transition, *Trends Ecol. Evol.* **3**:205-208.
McMenamin, M. A. S., 1992a., The Cambrian transition as a time-transgressive ecotone, *Geol. Soc. Am. Abstr. Progr.* **24**:62.
McMenamin, M. A. S., 1992b, Two new species of the Cambrian genus *Mickwitzia*, *J. Paleontol.* **66**:173-182.
McMenamin, M. A. S., 1993, Osmotrophy in fossil protoctists and early animals, *Invert. Reprod. Develop.* **22**:301-304.
McMenamin, M. A. S., 1996, Ediacaran biota from Sonora, Mexico, *Proc. Nat. Acad. Sci. USA* **93**:4990-4993.
McMenamin, M. A. S., 1998, *The Garden of Ediacara*, Columbia University Press, New York.
McMenamin, M. A. S., ed., 2001, *Paleontology Sonora: Lipalian and Cambrian*, Meanma Press, South Hadley, Massachusetts.
McMenamin, M. A. S., and McMenamin, D. L. S., 1990, *The Emergence of Animals*, Columbia University Press, New York.
McMenamin, M. A. S., and Whiteside, J. H., 1999, Hypermarine upwelling and a new Vernadskian metric, *J. Biosph. Sci.* **1**, http://www.mtholyoke.edu/courses/mmcmenam/journal.html.
Moody, K. E., 2001, Patterns of Predation on juvenile blue crabs in lower Chesapeake Bay: size, habitat and seasonality, in: *Proceedings of the Blue Crab Mortality Symposium* (V. Guillory, H. Perry, and S. Vanderkooy, eds.), Gulf States Mar. Fish. Commis. Pub. No. **90**, Ocean Springs, Mississippi, pp. 84-92.
Moody, K. E., and Steneck, R. S., 1993, Mechanisms of predation among large decapod crustaceans of the Gulf of Maine Coast: functional vs. phylogenetic patterns, *J. Exper. Mar. Biol. Ecol.* **168**:111-124.
Nedin, C., 1999, *Anomalocaris* predation on nonmineralized and mineralized trilobites, *Geology* **27**:987-990.
Nesis, K. N., 1995, Khishchniki na zare zhizni, *Priroda* **8**:60-62
O'Shea, D., 1980, *An Exposition of Catastrophe Theory and its Applications to Phase Transitions*, Queen's Papers in Pure and Applied Mathematics Number 47, Queen's University, Kingston, Ontario.
O'Shea, D., 1986, Elementary catastrophes, phase transitions and singularities, *Math. Model.* **7**:397-411.
Pickerill, R. K., and Blissett, D., 1999, A predatory *Rusophycus* burrow from the Cambrian of southern New Brunswick, eastern Canada, *Atlantic Geol.* **35**:179-183.
Schlichter, D., and Fricke, H. W., 1991, Mechanisms of amplification of photosynthetically active radiation in the symbiotic deep-water coral *Leptoseris fragilis*, *Hydrobiologia* **216/217**:389-394.

Sebens, K. P., and Koehl, M. A. R., 1984, Predation on zooplankton by the benthic anthozoans *Alcyonium siderium* (Alcyonacea) and *Metridium senile* (Actiniaria) in the New England subtidal, *Mar. Biol.* **81**:255-271.

Sepkoski, J. J., 1992, Proterozoic-Early Cambrian diversification of metazoans and metaphytes, in: *The Proterozoic Biosphere* (J. W. Schopf and C. Klein, eds.), Cambridge University Press, Cambridge, pp. 553-561.

Stanley, S. M., 1973, An ecological theory for the sudden origin of multicellular life in the late Precambrian, *Proc. Nat. Acad. Sci. USA* **70**:1486-1489.

Vacelet, J., and Boury-Esnault, N., 1995, Carnivorous sponges, *Nature* **373**:333-335.

Valentine, J. W., 1994, The Cambrian explosion, in: *Early Life on Earth* (S. Bengtson, ed.), Columbia University Press, New York, pp. 401-411.

Verhulst, P. F., 1838, Notice sur la loi que la population suit dans son accroissement, *Coresp. Math. Phys.* **10**:113-121.

Vermeij, G. J., 1987, *Evolution and Escalation: An Ecological History of Life*, Princeton University Press, Princeton, New Jersey.

Vernadsky, V. I., 1998, *The Biosphere*, Springer-Verlag, New York.

Vorwald, G. R., 1982, Healed injuries in trilobites; evidence for a large Cambrian predator, *Geol. Soc. Am. Abstr. Progr.* **14**:639.

Waggoner, B. M., 2000, Rewriting the rulebook; community ecology through the Neoproterozoic-Cambrian transition, *Geol. Soc. Am. Abstr. Progr.* **32**:41.

Walker, T. D., 1985, Diversification functions and the rate of taxonomic evolution, in: *Phanerozoic Diversity Patterns* (J. W. Valentine, ed.), Princeton University Press, Princeton, New Jersey, pp. 311-334.

Whittington, H. B., 1985, *The Burgess Shale*, Yale University Press, New Haven.

Whittington, H. B., and Briggs, D. E. G., 1985, The largest Cambrian animal, *Anomalocaris*, Burgess Shale, British Columbia, *Trans. R. Soc. London, Series B, Biol. Sci.* **309**:569-609.

Chapter 17

Durophagous Predation in Paleozoic Marine Benthic Assemblages

CARLTON E. BRETT

1. Introduction 401
2. Paleozoic Predators...402
   2.1. Arthropods...403
   2.2. Cephalopods 404
   2.3. Gnathostomes 405
   2.4. Shell Drilling Gastropods 406
3. Trace Fossil Evidence of Predatory Attack..406
   3.1. Coprolites..406
   3.2. Bite Marks...407
   3.3. Predatory Drill Holes...409
4. Possible Consequences of Predator Escalation 418
   4.1. Shell Strengthening and Sculpture...419
   4.2. Spinosity..420
   4.3. Platyceratid Targeting?..425
5. Discussion: Tests of the Mid Paleozoic Escalation Hypothesis and a Preliminary Model of Escalation 427
6. Summary...428
   References 429

1. Introduction

Fossil evidence of biotic interactions is of considerable importance in understanding broad-scale evolutionary patterns and has long attracted the interest of paleontologists (Clarke, 1921; Moodie, 1923; for exhaustive compendium see Boucot, 1990). Although it is notoriously difficult to document with fossils, predation is ubiquitous in modern communities and often of fundamental importance in controlling the abundance and diversity of organisms in marine environments (e.g., Paine, 1969).

CARLTON E. BRETT • Department of Geology, University of Cincinnati, Cincinnati, Ohio 45221.

Predator-Prey Interactions in the Fossil Record, edited by Patricia H. Kelley, Michał Kowalewski, and Thor A. Hansen. Kluwer Academic/Plenum Publishers, New York, 2003.

Predator-prey interactions were probably also of fundamental importance in shaping and directing long term trends of evolutionary adaptation. Vermeij (1977, 1983, 1987) and others have amassed considerable evidence for the intensification of predation pressure on hard-shelled marine organisms during the middle Mesozoic Era. The so-called "Mesozoic Marine Revolution" involved an arms race between durophagous, shell-crushing, predators and their prey. The arms race was triggered by a new "arsenal" of crushing teeth, claws, and drills (radulae). Concurrently, especially among gastropods, prey shells evolved a series of spines, thickened lips, slit-like apertures. These morphological trends may have been mitigated, at least in part, by intensified predation (Vermeij, 1977, 1987). But this was an intensification of existing interactions and recent studies have yielded evidence for predation as early as the late Proterozoic (Bengtson and Zhao, 1992). By the Early Cambrian animals up to a meter in length were preying upon marine shelly benthos (Babcock and Robison, 1989; Babcock, 1993; Conway Morris and Bengtson, 1994). Indeed, the first wave of predation may have been of fundamental importance in triggering the acquisition of hard skeletons by numerous taxa of organisms in over ten phyla during the Cambrian explosion. The rapid development of sclerites, valves, and armor seen in the Early Cambrian may well have been driven by biting organisms, such as anomalocaridids. This critical phase in the rise of predation is reviewed by Babcock (Ch. 3, in this volume), who proposes to call this interval the "Early Paleozoic Revolution".

The vast numbers of skeletonized benthic invertebrates of Paleozoic seas, including brachiopods, molluscs, trilobites, and echinoderms, provided a major potential food resource for predators that could penetrate the skeletal armor. Signor and Brett (1984) postulated that evolution of varied durophagous (shell crushing) predators in the middle Paleozoic produced an intensification of selection pressure similar to, but of lesser magnitude than, the Mesozoic marine revolution. A variety of studies, since the mid-1980s, have tested this notion and supplied further data on durophagous predatory interactions in the Paleozoic Era (Alexander 1986a,b; Boucot, 1990; Aronson, 1991; Kowalewski et al., 1998; Baumiller et al., 1999; Leighton, 2001). Moreover, new information indicates that the effects of this middle Paleozoic intensification of predation go beyond shell crushing and include biting, peeling, and drilling strategies.

In the present paper, I review and update both predatory and defensive aspects of this putative middle Paleozoic predatory radiation. First, I consider the types of durophagous predators that existed in the Paleozoic and briefly review their evolutionary history (for more details see Brett and Walker, in press). Second, I review direct trace fossil evidence for shell breakage and penetration of prey shells for both types of attack during the Cambrian to Permian. Third, I will consider the predicted evolutionary consequences of intensified attack on hard-shelled prey, vis-a-vis the empirical record of changes in skeletal morphology. In the final section, I will use a number of specific case studies to discuss preliminary tests for escalation in response to predation pressure.

2. Paleozoic Predators

Instances of predators preserved in direct association with their victims or prey organisms preserved in stomach contents of predators provide compelling evidence, but are understandably so rare as to be of little consequence. Still, some striking examples have come to light and they provide important information on predator-prey interactions.

More significant is the body fossil record of organisms whose morphology or phylogenetic relationship indicates a durophagous carnivorous habit (Signor and Brett, 1984; Brett and Walker, in press; Figs. 1, 2).

2.1. Arthropods

Survey of the fossil record of arthropods and their relatives indicates the existence of certain groups of probable predators as early as the Cambrian. Among the oldest were the anomalocaridids, a widespread Cambrian group with arthropod affinities, in which a circular slicing oral ring indicates a capacity for predation at least on soft or chitinous bodied organisms (see Babcock, Ch. 3, in this volume). Conway Morris and Jenkins (1985) illustrated bite marks on Cambrian trilobites that might be the work of these large predators.

Other durophagous arthropods include phyllocarids, which possessed molariform calcareous mandibles. Phyllocarids appeared in the Late Ordovician and show a diversity peak in the Devonian (see Signor and Brett, 1984). Decapod crustaceans with claws for prey crushing appeared in the Devonian and diversified in the later Paleozoic, but they were mainly small and uncommon.

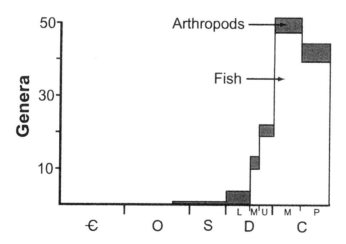

FIGURE 1. Plot of relative abundance of durophagous predators, excluding cephalopods. A majority of these predators were fishes. From Signor and Brett (1984).

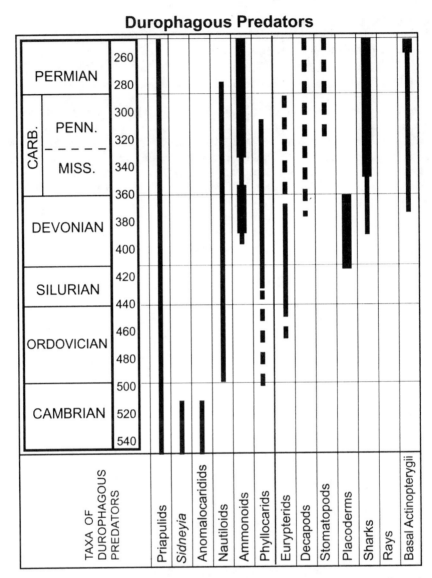

FIGURE 2. Ranges of several durophagous predators in the Paleozoic Era. Thick lines indicate abundant occurrence; thin line: present but relatively uncommon; dashed line: present but rare.

2.2. Cephalopods

Nautiloid cephalopods also appeared in the Cambrian, although they only attained abundance and large size in the Ordovician. Early forms may have been bottom-feeding scavengers (Bandel, 1985). Nautiloids, presumably with chitinous beaks, based on uniformitarian reconstructions, were abundant in marine benthic assemblages from the Early Ordovician onward, but calcified rhyncholites are virtually unknown prior to the Mesozoic (see Vermeij, 1983). Alexander (1986a) reported a possible calcified rhyncholite embedded in a shell of the brachiopod *Rafinesquina* from the Late Ordovician; he further suggested that nautiloids were a major predator of Ordovician brachiopods. However, Lescinsky and Benninger (1994) argued that this putative rhyncholite fossil is actually a shell fragment, which had become embedded in the brachiopod as a result of pressure solution. Possible gut contents found in the body chambers of nautiloids from the Upper Ordovician (Cincinnatian) contain abundant fragments of trilobite exoskeletons (C. E. Brett pers. observ., 2001). Possible crop residues from large nautiloids contain ammonoids and other shells (Zangerl *et al.*, 1969).

2.3. Gnathostomes

Earliest well-known gnathostome fishes are Silurian acanthodians, although possible acanthodian spines and chondrichthyan (shark) denticles are known from the Middle Ordovician (Benton, 1997). These fishes and their later Paleozoic descendants possessed sharp teeth adapted for predation on soft-to-chitinous invertebrates and other fishes. But the earliest major radiation of durophagous fishes undoubtedly occurred in the Early to Middle Devonian. Varied placoderms, including especially the arthrodires, with sharp shearing gnathal plates, rhenanids with blunt crushing plates, and a ray-like, benthic mode of life, and ptyctodonts with hypermineralized tritors evolved during the Devonian. These may have been important crushers of hard-shelled prey, although their remains are rare in most invertebrate-rich assemblages. Other groups, though not strictly durophagous, such as cladodont sharks may have had an impact on invertebrate prey.

Virtually all of the placoderm taxa became extinct by the end of the Devonian (Moy-Thomas and Miles, 1971). They were replaced by varied sharks, including the successful hybodontids and ctenacanthoids, some of which developed broadened teeth, and also by varied holocephalans, many of which were durophagous (Moy-Thomas and Miles, 1971; Mapes and Benstock, 1988). These fishes remained rather common through the end of the Paleozoic. Holocephalans (chimaeroids), such as the helodontoids, cochliodontoids, and petalodontids, possessed autostylic (fused) skulls and hypermineralized broad crushing dentition analogous to that of earlier ptyctodonts. Certain Carboniferous chimaeras, such as *Helodus*, have been implicated as producers of distinct crush marks in Carboniferous brachiopod and bivalve shells (Brunton, 1966; Boyd and Newell, 1972; Alexander, 1981).

Stomach contents and coprolites of late Paleozoic holocephalans, sharks and other fish are known to contain fragmented invertebrate remains (Zangerl and Richardson, 1963; Zangerl *et al.*, 1969; Moy Thomas and Miles, 1971). For example, specimens of the Permian petalodont *Janassa* have been found with brachiopods and crinoid ossicles preserved as probable gut contents (Malzahn, 1968).

Durophagous holocephalans show a five-fold increase in taxonomic richness in the Carboniferous relative to the Devonian (Mapes and Benstock, 1988). However, durophagous holocephalans also underwent a major decline in the Late Pennsylvanian and Permian (Mapes and Benstock, 1988).

Overall, while large predators were present from Cambrian times onward, the number of marine hard-shelled crushing organisms showed an abrupt increase in the Middle Devonian. These included cephalopods, crustaceans, and several fish groups.

2.4. Shell-Drilling Gastropods

It is clear that shell drilling was important from the late Proterozoic/Cambrian onward (Bengtson and Zhao, 1992; Conway Morris and Bengtson, 1994). Circular bore holes become more numerous in the later Paleozoic. Until recently it was quite unclear what organisms were responsible for these holes, but recent discoveries suggest that platyceratid gastropods were among the culprits. Baumiller (1990) recorded a spectacular example of a drill hole in the tegmen of a Mississippian crinoid beneath the shell of an attached platyceratid gastropod. In a similar example, Baumiller *et al.* (1999) reported a platyceratid apparently in situ on a drilled brachiopod shell. While these examples may record parasitism, rather than true predation, they are highly important in proving the capacity for radular drilling among platyceratids and implicating them as possible shell boring predators.

The record of platyceratid gastropods spans Middle Ordovician to Late Permian (Bowsher, 1955); during this interval more specialized members of the group were clearly commensal/parasitic on pelmatozoan echinoderms. However, others, notably *Cyclonema* (Ordovician-Silurian) and *Naticonema* (Ordovician-Devonian), retained unspecialized shells and may have been facultatively free-living scavengers and predators. The occurrence of *Platyceras* on a Mississippian brachiopod cited above may indicate that certain members of this more modified genus also did not live on crinoids, but rather were parasites or predators on brachiopods and other low-tier benthos.

As with durophagous predation, there is evidence for intensification of shell drilling in the Devonian (Kowalewski *et al.*, 1998), although the fossil record of platyceratids shows relatively little increase in diversity or abundance during this time (Bowsher, 1955).

3. Trace Fossil Evidence of Predatory Attack

Traces provide both direct and indirect evidence for predation. There is growing evidence of various sources for this type of interaction.

3.1. Coprolites

One line of evidence comes from the record of coprolites or petrified feces. A compilation of references on coprolites by Häntschel (1968) indicates a general rise in frequency of shell-bearing coprolites during the late Paleozoic. Only about 25 reports of pre-Devonian coprolites were found; most of these are small and phosphatic, lacking shell fragments. Conversely a larger number of shell-bearing coprolites are known from the late Paleozoic. Kloc (1987) reported pyritic coprolites from the Upper Devonian of New York containing remains of dozens of juvenile goniatites. Zangerl and Richardson (1963) and Zangerl *et al.* (1969) illustrate a number of shell fragment-bearing coprolites from the Mississippian and Pennsylvanian. Such evidence is anecdotal, but suggestive of an increase in a hard-shelled diet during the middle Paleozoic.

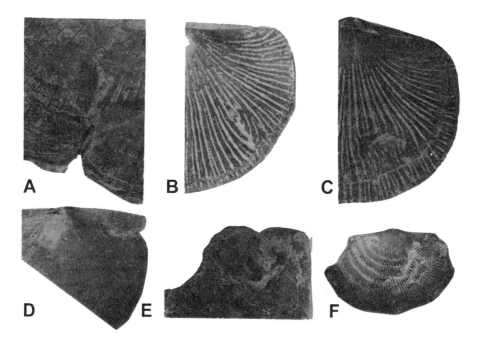

FIGURE 3. Healed breakage in Late Ordovician brachiopods; Richmond Group, near Richmond, Indiana. (A), (D) clefts in *Rafinesquina alternata*; X 2.0 and X1.5. (B), (C) healed cleft and divot in shells of *Plaesiomys subquadrata*, both views X2.5. (E) Embayed fracture in *Rafinesquina*; X2. (F) embayed fracture in *Leptaena richmondensis*; X2.5. Adapted from figures in Alexander (1986a); see that paper for details of locations and repositories.

3.2. Bite Marks

A variety of clefts and divots are preserved in fossil shells of brachiopods, arthropods, and mollusks (Fig. 3). These traces may provide evidence for predation if mechanical causes, such as storm wave smashing, can be ruled out, which may be difficult. However, as pointed out by Vermeij (1987), mechanical fracturing only represents a small fraction (<10%) of shell damage in many marine environments.

Bite marks usually cannot be ascribed to a particular predator (Table 1). Exceptions include several instances of circular punctures in Carboniferous goniatites observed by Mapes and Hansen (1984). The size and spacing of these holes implicates the associated cladodontid shark *Symmorium* as a probable predator (see Hansen and Mapes, 1990). Similar, repetitive punctures were also observed in Mississippian brachiopods and ascribed to sharks (Alexander, 1981).

Clefts, divots, and larger predation scars have been identified in brachiopods as old as the Late Ordovician and are considered to be marks of durophagous predators, probably large nautiloids (Alexander, 1986a,b).

Healed skeletal damage has been observed in many Paleozoic organisms. For example, several authors observed regenerated arms in crinoids (Springer, 1920; Signor and Brett, 1984; Ubaghs, 1978; Schneider, 1988).

Alexander (1981, 1986a,b) distinguished between lethal and sublethal (healed) shell damage in assemblages of brachiopods from Ordovician to Pennsylvanian age.

Sublethal fractures and punctures prove to be more common in certain morphotypes of brachiopods, such as strophomenides. Alexander (1986a) documented an increased proportion of so-called "lethal fractures" in the late Paleozoic, together with a decline in repaired damage. This evidence suggests an increase in the effectiveness of predatory attacks.

Well-preserved shells of Paleozoic gastropods display scalloped fractures of the outer lip that resemble marks made by predatory crustaceans that "peel" gastropod shells to reach the body of the snail (Vermeij et al., 1981; Schindel et al., 1982; Ebbestad and Peel, 1997; Ebbestad, 1998; Table 1). The oldest instance of healed lip-peeling type of fracture is from the Middle Ordovician (Peel, 1984); most case studies of Ordovician and Silurian gastropods suggest that lip-peeling was rare (Peel, 1984; Schindel et al., 1982). For example, Upper Ordovician gastropods were recently found to have low frequencies (~7%) of shell repair (Ebbestad and Peel, 1997) in association with architectural strengthening, such as narrow apertures and increased ornamentation.

Shell repair was also found to be present, with frequencies of ~10% in well preserved Middle Devonian gastropods (Brett and Cottrell, 1982); this suggests a frequency similar to that reported by Schindel et al. (1982) who observed 10-20% frequency of repaired lip peeling in Pennsylvanian gastropods. Such data, though somewhat anecdotal, suggest intensified frequency (or decreased effectiveness) of predatory bite marks in the late Paleozoic beginning with the Ordovician.

Moreover, extensive study of Carboniferous gastropod shells suggests frequencies of shell repair similar to those recorded for snails of comparable size in the Mesozoic (Schindel et al., 1982). This does not necessarily imply that the frequencies of attacks were comparable between the Paleozoic and Mesozoic: an increased rate of predation might also be coupled with a greater frequency of lethal predation in which shells were destroyed. Five species of Mississippian benthic and nektonic ammonoids from a single locality were found to exhibit varying degrees of shell repair that did not correlate with shell morphology in either species (Bond and Saunders, 1989). Bond and Saunders also noted that thick-shelled species sustained higher frequencies of non-lethal injury than did thin-shelled species, suggesting the thick-shelled species were more difficult to attack. Interpretation of healed shell breakage is not unambiguous because it monitors not only predatory attack but also the ability of prey to withstand attack. Nonetheless, the middle Paleozoic appears to mark a time of intensification of this type of interaction.

TABLE 1.. Instances of Shell Breakage and Repair Among Paleozoic Fossil and Recent Gastropods

Age	Type	Frequency (%)	References
Ordovician	lip-peeling	rare (<<1)	Steel and Sinclair, 1971
Silurian	lip-peeling	rare (<<1)	Peel, 1984
Devonian	lip-peeling spire breakage	10-30	Brett and Cottrell, 1982
Carboniferous (Pennsylvanian)	lip-peeling spire breakage	15-25	Schindel et al., 1982
Recent	lip-peeling spire breakage	20-96	Vermeij et al., 1980

3.3. Predatory Drill Holes

Several taxa of marine organisms have evolved the ability to bore shells using mechanical and/or chemical processes. Modern shell-drilling organisms include one mesogastropod and several neogastropod families, octopod cephalopods, a few flatworms, sipunculids, and polychaetes (Carriker and Yochelson, 1968; Bromley, 1981; Kabat, 1990). There is definitive evidence that certain now-extinct taxa, including other lineages of gastropods, independently evolved the ability to drill shells for predation and/or parasitism (Baumiller, 1990).

Unlike shell-crushing predation, which may commonly destroy evidence of successful attack, boreholes do not generally break down shells and can thus provide evidence for attacks on living organisms. However, boreholes may weaken shells and bias the record toward lowered observed frequencies of drilling (Roy *et al.*, 1994).

A major problem is to distinguish non-predatory domichnial borings (*Trypanites*) from those of predatory drillers (*Oichnus* of Bromley, 1981; Table 2). Certain circular borings, which were once ascribed a predatory origin (e.g., borings in Ordovician dalmanellid brachiopods noted by Bucher, 1938), have been shown to be post-mortem substrate borings. Richards and Shabica (1969) showed that in some cases these holes penetrated through two or more valves that had been stacked upon each other. Clearly these holes were made after death of the brachiopods and represent simple dwelling sites referable to *Trypanites*.

Carriker and Yochelson (1968) and Kowalewski *et al.* (1998) discussed several diagnostic criteria for predatory/parasitic drilling, including: a) borehole tapers inward from exterior of shell, b) drilling is confined to a single valve, c) there is only one complete borehole per shell, d) holes occur preferentially in certain positions on shells, and e) certain species of (prey) shells are selectively bored (Table 2). Modern gastropod borings are commonly perpendicular to shell surfaces and have chamfers or bevels; incomplete boreholes show a centrally elevated boss (Figs. 4, 5). On the basis of these criteria, a number of circular boreholes in Paleozoic skeletons appear to represent parasitic or predatory attack on live organisms.

The second issue, not readily resolved, is the distinction between parasitic and predatory boreholes. In some cases, parasitism is clearly indicated (see Brett, 1978, for discussion). These include: (a) instances of multiple complete and incomplete boreholes, (b) strictly incomplete boreholes that do not penetrate through skeletons, (c) pits that are demonstrably embedment structures formed by growth of the host's skeleton around an epibiontic organism, and (d) instances in which there is evidence from attachment scars that the driller occupied a site on a host for prolonged period of time.

TABLE 2. Features of Domicile vs. Predatory Boreholes

Domicile borings	Predator borings
holes commonly do not penetrate through shell	holes normally pass through shells
commonly multiple in single shell	generally one complete hole per shell
may originate from exterior *or* interior	originate from exterior
no preferred position	preferred position
slight or no host shell preference	preference for certain species
examples: algae, annelids, fungi, bivalves, sponges, barnacles, bryozoans, "worms"	examples: naticid and muricid gastropods, cephalopods, Turbellaria

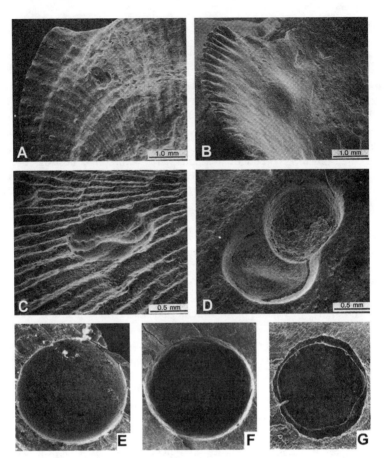

FIGURE 4. Gastropod-like boreholes in Paleozoic brachiopods. A-C, Middle Devonian Hamilton Group; A, B, Incomplete bore holes showing "healing" by brachiopod. Exterior and interior views of the brachiopod *Rhipidomella vanuxemi*; Middle Devonian Hamilton Group, western New York; (A) exterior view of incomplete bore hole; (B) interior view of shell showing blister or "pearl" where the brachiopod has responded to the presence of the incomplete hole; (C) incomplete hole in *Rhipidomella* showing central boss; (D) two overlapping sediment filled bore holes on *Parazyga*; upper hole is complete; note chamfer well displayed on lower hole. E-G, borings in Mississippian brachiopods; Edwardsville Formation, Missouri; (E) incomplete "type A" borehole in *Rugosochonetes planumbona*; X 40; (F) complete "type B" borehole in *Composita globosa*, X13; (G) incomplete and "healed" type A hole in *Cleiothyridina parvirostra*; note shelf of secreted material that represents an early phase of hole closure by the brachiopod. Figures A-D from Smith *et al.* (1985). Figures E-G from Ausich and Gurrola (1979).

Table 3 lists examples of Paleozoic circular boreholes that were either predatory or parasitic in nature. Note that circular, penetrative borings range from 0.05 mm to nearly 4 mm in diameter and occur on shells from the Early Cambrian onward. At least four distinct types of borings occur in Paleozoic shells and these almost certainly record the work of different types of organisms (see Ausich and Gurrola, 1979; Fig. 4; Table 4). These are distinguished on the basis of size, position, substrate type, and presence/absence of a chamfer or bevel (cylindrical or conical boreholes).

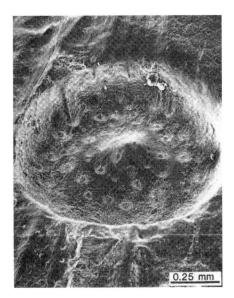

FIGURE 5. Incomplete borehole in *Douvillina* (Middle Devonian, Hamilton Group, western New York) showing central raised boss; note truncated pseudopunctae of shell.

1. Type 1. The first category comprises very minute pits, less than 0.05-0.2 mm in diameter, and lacking a chamfer. It is not clear what the range of this type of boring is, but it includes minute pits seen in Cambrian *Moburgella* (Conway Morris and Bengtson, 1984). Typical examples were illustrated in a paper by Chatterton and Whitehead (1987) from phosphatic inarticulates (lingulates) from the Silurian of Oklahoma. Despite their small size these holes fit all criteria (e.g., restriction to a single species, stereotyped positioning, single complete hole per shell) of a prey-species selective small predator, probably a gastropod.

2. Type 2. A second type of boring consists of cylindrical holes generally 0.5 to 1.5 mm in diameter, which occur in articulate brachiopods (Fig. 4E, G). Ausich and Gurrola (1979) referred to this morphology as "Type A" borings. The oldest examples include borings in the small orthid *Dicoelosia* from the Ordovician to Lower Devonian (Rohr, 1976). Sheehan and Lesperance (1978) identified similar holes, showed that these borings were stereotyped in their position, and suggested that they were produced by small gastropods. However, shells may exhibit multiple complete borings that exhibit a preferred distribution on parts of the brachiopod shell that would have been in contact with the substrate. For example, a specimen of the Middle Devonian spiriferid *Cyrtina* was found with four small boreholes in the interarea, which was certainly pressed against the substrate in life.

The producers of these holes display a distinct preference for particular brachiopod species. Ausich and Gurrola (1979) concluded that these holes were produced by infaunal parasites, possibly polychaete worms. This group ranges from Ordovician to at least Pennsylvanian. Leighton (2001) illustrated this type of borehole from Mississippian brachiopods and showed that they preferentially attacked the shell from below.

TABLE 3. Reports of Probable Predatory Boreholes in Paleozoic Shells

Age	Reference	Characteristic
Early Cambrian	Bengtson, 1968	0.10 mm; in phosphatic shells of *Mobergella*
Late Cambrian	Miller and Sundberg, 1984	0.05-.20 mm; in phosphatic inarticulate brachiopods
Middle Ordovician	Cameron, 1968	0.4 - 2.0 mm; rarely chamfered; in 3 species of articulates
Silurian	Rohr, 1976	0.32 mm; in one articulate brachiopod (*Dicoelosia*)
	Chatterton and Whitehead, 1987	0.06-0.13 mm; in phosphatic inarticulates
	Brett (unpubl.)	0.35-0.85 mm; in 3 species of articulate brachiopods
Early Devonian	Sheehan and Lespe rance, 1978	0.78-1.17 mm; in 1 species of articulate brachiopod
Middle Devonian	Brett (unpubl.)	(Type A) 0.30-0.80 mm; in 12 species of articulate brachiopods
	Smith *et al.*, 1985	(Type B) 0.2-3.08 mm; beveled boss; 10 species of articulates
	Leighton, 2001	(Type A) <1mm; in 3 species of articulate brachiopods
Late Devonian	Rodriguez and Gutschick, 1970	small (<1mm); in 1 species of articulate brachiopod
Mississippian	Ausich and Gurrola, 1979	(Type A) 0.1 - 1.6 mm; cylindrical in 11 species of articulate brachiopods
		(Type B) 1.16 -3.2mm; beveled with boss; in 6 species of articulate brachiopods
	Brunton, 1966	0.1-1.0 mm; cylindrical; in articulate brachiopods
Pennsylvanian	Hoffmeister *et al.*, 2001	(Type A) <0.2 mm; cylindrical to beveled, 1 species of articulate brachiopod
Permian	Kowalewski *et al.*, 2000	(Type B) beveled; in articulate brachiopods and bivalves

Ausich and Gurrola (1979) found a relatively high frequency of "Type A" boreholes (Type 2 herein) in Mississippian brachiopods from Indiana with drilling frequencies of about 20%, 22%, 36.5%, and 44% in *Cleiothyridina, Athyris, Rugosochonetes,* and *Rhipidomella,* respectively.

3. *Type 3.* A third category of borings, "Type B" boreholes of Ausich and Gurrola (1979), includes holes that are conical, typically larger: 1-3 mm in diameter, display a chamfer, and in incomplete boreholes possess a central raised knob or boss (Figs. 4C, 5). These holes, thus, closely resemble drillings of modern naticid gastropods. Such boreholes were first reported by Buehler (1969) from Middle Devonian brachiopods of New York State.

Smith *et al.* (1985) also described the chamfered type of borehole in Middle Devonian brachiopods. Normally, only a single complete borehole was present in a given brachiopod. Rarely, two or three pits were observed on a single shell, but with only one of these penetrating the shell. The occurrence of rare "pearls" beneath incomplete boreholes from the Devonian Hamilton Group shows that these holes were produced in live brachiopods that reacted to repair the potential damage (Smith *et al.*, 1985; Fig. 4B, G).

These Devonian borings show a distinct stereotypy in position, occurring most frequently near the umbones of the shells (Smith *et al.*, 1985; Fig. 6), as is typical of modern predatory gastropods. Similar preferential positioning of boreholes is observed in Upper Devonian atrypides and productids (Leighton, 2001) as well as in Mississippian spiriferides (Ausich and Gurrola, 1979).

Smith *et al.* (1985) noted a distinctly preferred occurrence of boreholes in certain brachiopod genera, notably *Athyris, Rhipidomella, Pseudoatrypa,* and strophomenids (*Protodouvillina, Pholidostrophia*), in several Middle Devonian samples, whereas rhynchonellids and terebratulids appear to have been avoided (Table 5). It is notable, therefore that the same families/genera were preferentially bored in later assemblages. For example, Leighton (2001) found *Douvillina* and *Pseudoatrypa* to have been drilled in the Late Devonian. He found stereotypy in positioning of boreholes and a decided preference for one of the valves (Fig. 6B-D). Leighton also suggested that most of the brachiopods were attacked from below, leaving open the possibility of infaunal parasites as the producers of these holes. *Athyris, Cleiothyridina,* and *Composita*, all athyrid brachiopods, were the most frequently drilled, with Type 3 (or Type B), boring in Ausich and Gurrola's (1979) Mississippian samples. These studies imply that the preference of predators for particular clades of prey existed for a period of tens of millions of years.

New evidence from Ordovician rocks suggests that the predatory drilling behavior may have evolved earlier. Some of the Ordovician boreholes described by earlier authors are certainly *Trypanites domichnia* made in post-mortem shells (Carriker and Yochelson, 1968; Richards and Shabica, 1969), but others appear to have been made by predators.

Kaplan and Baumiller (2000) reconsidered the occurrence of abundant boreholes on *Onniella meeki* from the Cincinnatian Waynesville Formation in Indiana. They considered taphonomic bias, noted as had previous authors that the convex pedicle valves were more subject to fragmentation when packed together and demonstrated this experimentally. They further recognized biased valve ratios and developed a model to restore the relative frequency of boreholes on specific valves. In the context of their model, they concluded that a small but significant percentage of the holes in these shells were the work of valve- and site-selective predators.

TABLE 4. Classification and Characteristics of Early and Middle Paleozoic Shell Borings of Three Types

	Type 1	Type 2	Type 3	Type 4
Size	0.05 - 0.20 mm	0.1 - 1.5 mm	1.5 - 3.5 mm	<0.2 mm
Shape	cylindrical	cylindrical	parabolic	parabolic
Morphology of floor	flat, rarely bossed	flat	commonly bossed	flat
Bevel	in some	no	in some	in some
Species preference	yes	yes	yes	yes
Position/valve preference	yes	minor (on lower valve)	yes (on exposed valve)	yes (valve preference)
Multiple borings	rare	present	rare, generally one per shell	rare, generally one per shell
Substrate	Chitinophospate	$CaCO_3$	$CaCO_3$	$CaCO_3$
Known range	Mid-Cambrian – Silurian?	Mid-Ordovician - Carboniferous	Mid-Devonian - Permian	Pennsylvanian - Permian
Interpretation	undetermined predator	infaunal parasite	predaceous gastropod	predaceous gastropod

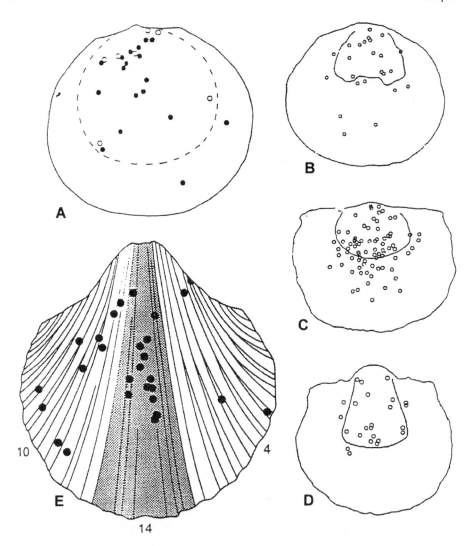

FIGURE 6. Borehole position plots on Paleozoic brachiopods showing stereotypy of position on valves over meat-rich muscle fields. (A) on *Rhipidomella* from the Middle Devonian Hamilton Group, New York; (B-D) on Upper Devonian brachiopods from Iowa; (B) *Pseudoatrypa*, (C) *Douvillina*; (D) *Devonoproductus*; (E) on *Spirifer* from Mississippian Fort Payne Formation, Missouri, showing preference for sulcus region; A from Smith *et al.* (1985); B-D from Leighton (2001); E from Baumiller *et al.* (1999).

Unpublished data suggest that at least some Type 3 boreholes in Cincinnatian brachiopods were made in live shells (S. Felton, pers. comm., 2001). Rare *Platystrophia* and other large brachiopods, including articulated specimens, from the Cincinnatian show single large (up to 3 mm diameter), parabolic holes that lie very consistently in or near the center of the sulcus of the pedical valve. The specific occurrence of these holes on a few species of brachiopods and their occurrence in stereotypic positions on the valves strongly suggests a predatory/parasitic attack on live hosts.

Survey of large collections of Silurian brachiopods failed to turn up evidence for Type A boreholes and is notable that this type of borings does not appear to be common prior to Middle Devonian time. Hence, Type 3 "naticid-like" chamfered borings may have evolved more than once, even during the Paleozoic.

Type 3 boreholes appear to have been moderately common in certain species of brachiopods in the Devonian to Mississippian, although never seemingly as abundant as Type 2 holes. Smith et al. (1985) report frequencies of occurrence of 7% boring of this type in a population of 298 Devonian *Athyris* brachiopods and Ausich and Gurrola (1979) similarly record up to 7% of this morphotype of boring in a population of 247 Mississippian *Composita globosa*, which also showed about 11% of Type 2 borings. It is notable that all brachiopods reported by Ausich and Gurrola to have Type B (Type 3) boreholes also have a similar or larger number of Type A (Type 2, herein) holes. The converse was not found to be true; several species with Type 2 boreholes had few or no Type 3 holes.

4. Type 4. Large, parabolic borings appear to have been relatively rare in the Pennsylvanian-Permian (Kowalewski et al., 1998) However, Hoffmeister et al. (2001a) report very small borings resembling Ausich and Gurrola's (1979) "Type B boreholes" (Type 3 herein) in the small Pennsylvanian brachiopod *Cardiarina cordata*. I tentatively assign these to a new category, Type 4, because of their tiny size with respect to most Type 3 boreholes.

Chamfered boreholes are also noted in silicified shells of both brachiopods and bivalves from the Permian of the Glass Mountains of West Texas (Hoffmeister et al., 2001b). The authors documented frequencies of up to 2.3% drilling in brachiopods and up to 4.3% in bivalves.

Type 3 and 4 boreholes appear to have been made by genuinely predatory organisms. Their morphology closely resembles that of modern predatory naticid gastropods. Smith et al. (1985) argued that predatory behavior did not evolve in archeogastropods and suggested that these holes were made by an unknown, soft-bodied predator. However, subsequent discoveries indicate that gastropods may have been responsible.

As noted above, Baumiller (1990, 1996) and Baumiller et al. (1999) have found direct evidence that platyceratid archeogastropods were capable of radular drilling, and attacked both pelmatozoan echinoderms and brachiopods. In the case of the echinoderms it is clear that the interaction was not predatory, but probably parasitic, as the gastropod resided on the crinoid or blastoid theca for extended periods of time (Baumiller 1993, 1996). To date, the oldest reported boreholes in pelmatozoans are Baumiller's (1996) examples of Middle Devonian (Eifelian) blastoids. However, recently S. Felton (pers. comm., 1999) has found Type 3 boreholes at the bases of anal sacs of *Dendrocrinus* from the Upper Ordovician Cincinnatian of Ohio. These boreholes are associated with the commensal/parasitic platyceratid gastropod *Cyclonema*. This suggests that some of the oldest platyceratids were already capable of radular drilling.

As noted, in the case of the brachiopods it is more likely that the relationship involved true predation (but see Baumiller et al., 1999, for a discussion of possible parasitism).

Naticonema, a non-specialized platyceratid gastropod, is found in close association with the bored Middle Devonian brachiopods and may have been a predatory driller. Likewise, *Cyclonema* is a common associate of Type A boreholes in Late Ordovician brachiopods.

TABLE 5. Brachiopods with gastropod-like boreholes from several units of the Middle Devonian Hamilton Group; Numbered stratigraphic units are 1) Centerfield Limestone; 2) Wanakah Shale; 3) Jaycox Shale; 4) Kashong Shale; 5) Windom Shale-Fall Brook bed; 6) Windom Shale-Taunton beds. Asterisks indicate species that were significantly preferred; crosses show species that were avoided. From Smith *et al.* (1984).

Species	Number bored/total by stratigraphic unit					
	1	2	3	4	5	6
Athyris spiriferoides	3/14*	21/298	2/34	2/81	6/39	4/26
Cupulorostrum spp.				0/252		0/3
Cyrtina hamiltonensis	1/87			0/1	2/21	
Delthyris cf. *D. sculptilus*	0/9		0/8	0/19		
Devonochonetes coronatus	0/1			0/5		
Elita fimbriata		0/1	0/4	0/2	0/2	
Mediospirifer audaculus	0/3	0/9	1/56	0/7	3/17	0/12
Megastrophia concava		0/2	1/6			
Meristella cf. *M. haskinsi*	1/11		0/8	2/6		
Mucrospirifer mucronatus	0/3	0/7	0/56+	0/31	0/7	0/1
Nucleospira concinna	1/15	0/4	0/6	5/180		1/2
Parazyga hirsuta	1/3		4/41		3/9	
Pentamerella pavilionesis	0/6					
Pholidostrophia nacrea	4/18*	0/4	8/18*			
Protodouvillina inequistriata	1/8	1/1	6/21*	0/2		0/5
Pseudoatrypa cf. *P. devoniana*	2/82				5/86	1/109
Rhipidomella spp.	5/47	3/15	8/84		7/27*	0/6
Schuchertella arctostriata		0/1?		0/2		
Spinatrypa spinosa					5/50	0/9
Spinocyrtia granulosa	0/2	0/3	0/6		0/1	0/11
Strophodonta demissa	0/8	0/2	0/6			
Tropidoleptus carinatus		0/2	0/100+	0/150		1/3

*Species that were significantly preferred. Chi-square values (and ratios) for units 1 through 6, respectively: 24.0 (19/319), 5.6 (25/349), 80.4 (30/469), 15.8 (10/835), 17.0 (32/276), and 19.8 (7/190) (28). +Species that were avoided.

It is notable that a few small specimens of *Naticonema* and rare, small specimens of *Cyclonema* from the Cincinnatian also show Type 3 boreholes (C. Brett, unpubl. data; S. Felton, pers. observ., 2001). The latter observations prompt comparisons with modern naticids, which preferentially attack juveniles of conspecific or at least con-familial naticids (Kabat, 1990). It is certainly possible that cannibalism was established by Ordovician time. This behavior, coupled with other evidence, highlights the probably opportunistic behavior of platyceratids.

Kowalewski et al. (1998) compiled a database on boreholes through the Phanerozoic based on some 85 publications. This compilation shows a distinct peak of boreholes – representing mainly Type 2 (or Type A of Ausich and Gurrola, 1979) borings – in the Devonian to Carboniferous (Fig. 7). Frequencies of borings in preferred species of Devonian – Mississippian brachiopods are variable, but average about 5 – 6% (Smith et al., 1985; Leighton, 2001; Table 5), generally lower than for gastropod borings in Cenozoic bivalves (~20%) (Kowalewski et al., 1998). This parallels other trends in Paleozoic predation and supports the notion of a middle Paleozoic intensification of predation.

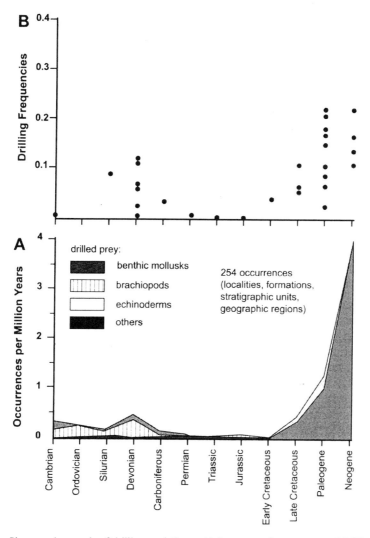

FIGURE 7. Phanerozoic records of drilling predation. A) frequency of occurrences of drilling in single localities through time; B) drilling frequency per million years. Note minor spike in the Devonian. Adapted from Kowalewski et al. (1998).

Initially, it was thought that the frequency of drilling declined in the Late Paleozoic (Kowalewski et al., 1998). However, this was for assemblage-based data. For individual species data, the frequency of drilling can exceed 40% per species examined. For example, Ausich and Gurrola (1979) report overall drilling frequencies (Types A and B boreholes combined) as high as 44% for one species of Mississippian brachiopod, which is similar to Brunton's (1966) findings. Additionally, Hoffmeister et al. (2001a) report a drilling frequency (Type C drill holes) of 32% in the Pennsylvanian brachiopods from New Mexico. Furthermore, the drilling predator (or parasite) showed specific prey preference (i.e., preferring *Cardiarina* over other brachiopods), valve preference (i.e., pedicle valves are twice as likely to be drilled than brachial valves), and drill hole site selectivity indicating sophistication in drilling of Late Paleozoic brachiopod prey (Hoffmeister et al., 2001a). As noted above, Hoffmeister et al. (2001b) also reported drilling frequencies of 2.3% and 4.3%, for silicified brachiopods and bivalves, respectively, from the Permian of West Texas.

In a similar study, however, the brachiopod *Composita* had little evidence of drilling predation (or parasitism), with very low frequencies reported throughout the Carboniferous Period (Hoffmeister et al., 2002). Thus, drilling frequencies can be greatly affected by a preferred prey, or conversely, by non-preferred prey that have low drilling frequencies or none at all. Therefore, more species per locality should be examined when drilling frequencies are compared through the Phanerozoic. As Hoffmeister et al. (2002) acknowledge, there are still serious gaps in our knowledge concerning drilling predation in the Paleozoic, especially with regard to spatial and temporal distributions of drilled prey.

4. Possible Consequences of Predator Escalation

The foregoing examples serve to illustrate that by early middle Paleozoic time, shelled benthic invertebrates were beset by a variety of predators and parasites. There is also evidence that selective pressure from shell-crushing, biting, and drilling predators was markedly increased during Devonian time.

This leads to several predictions regarding possible morphological adaptation of the potential prey: a) prey should make behavioral modifications to avoid predators; b) shells should show increased thickening and other mechanical strengthening; and c) prey should show evidence of predator-deterrent spines.

Increased infaunal behavior of brachiopods and several groups of bivalves has been documented by Thayer (1983). Although this life habit may have evolved for various reasons, including stabilization on soft, fluid substrates, it may also have conferred some selective advantage to sedentary organisms in a regime of intensified predation. Thus, an increase in proportion of endobenthic life habits may be partially a response to predation pressure. However, this pattern requires much more documentation. This tendency became even more pronounced in the Mesozoic with the advent of siphonate bivalves. However, it is also notable that the most successful late Paleozoic brachiopods, the productids and chonetids, were quasi-infaunal.

Signor and Brett (1984) explored several other morphological features that serve to strengthen invertebrate skeletons or make them more difficult to attack. These features are surveyed in the following sections.

4.1. Shell Strengthening and Sculpture

The presence of an open umbilicus renders molluscan shells easier to crush (Vermeij, 1983). Therefore, one might predict a decline in umbilicate forms in the face of increasing predation pressure. In a sample of some 60 genera of bellerophontids, Signor and Brett (1984) found a substantial decline in umbilicate form, beginning in the Silurian Period (Fig. 8). They also noted a distinctive decrease in proportion of genera with disjunct coiling in the late Paleozoic and an increase in the percentage of shells with sculpture.

Ribbing and fluting may render shells more resistant to crushing; again Signor and Brett (1984) found an increased incidence of sculpture in post-Silurian nautiloids and bellerophontids. Similarly, Alexander (1986b) found a decline in the proportion of brachiopods with smooth shells lacking ribs and an increase the proportion of species with coarse ribs during the late Paleozoic. Such adaptations could aid in the resistance to shell breaking and crushing predators. In addition, Bordeaux and Brett (1990) observed that strongly ribbed brachiopods, such as rhynchonellids, also have a very low frequency of borings in Devonian samples. Possibly, the strong corrugated shells were harder to penetrate than other types of brachiopods.

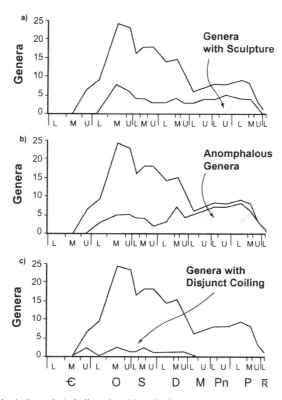

FIGURE 8. Morphological trends in bellerophontid mollusks. Upper curve shows total number of genera; cross-hatched areas show proportion of total genera showing indicated feature; top graph shows presence of sculpture; middle graph proportion of anomphalous genera (lacking umbilicus); lower graph, genera with disjunct coiling. Note loss of genera with disjunct coiling and increase in proportion of genera with sculpture and lacking an umbilicus. From Signor and Brett (1984).

Late Paleozoic crinoids show a tendency to small size and highly compact, thick-walled calyxes (Signor and Brett, 1984; Fig. 11a, herein). Certain Permian forms, such as members of the disparid Allagecrinacea and the cladid Codiacrinacea are extraordinarily thick plated in relation to their calyx diameters. It is notable that an increase in the proportion of these compact crinoids coincides with a decrease in spinosity (Fig. 11, see also below) suggesting that these morphologies may have been correlated and related, at least indirectly to durophagous predation. However, other explanations for thick walled cups exist (e.g., relationship to warm temperatures and saturation of calcium carbonate).

4.2. Spinosity

The last and perhaps less ambiguous morphologic trend predicted by Vermeij (1977) and Signor and Brett (1984) is an increase in the frequency of taxa with spines; spines may also be predicted to increase in length and sharpness. Increased spinosity might serve to deter both durophagous and drilling predators.

The rise to dominance of the productid strophomenides in later Paleozoic led to a substantial increase in the proportion of brachiopods that possessed spines on both the pedicle and brachial valves (Signor and Brett, 1984; Fig. 9). Most likely, these spines originated as a stabilizing "rooting" spines for these endobenthic brachiopods (Rudwick, 1970); however, these spines also appear to have deterred boring predators (Leighton, 2001) and perhaps shell crushers. *Devonoproductus* had a much lower frequency of completed boreholes than either contemporaneous atrypids or *Douvillina*. If so, increased proportion of productides could have been partially the result of increased durophagous predation and drilling predation or parasitism. This suggests that, regardless of their origin, the presence of spines may have prevented drilling predators from penetrating the shell.

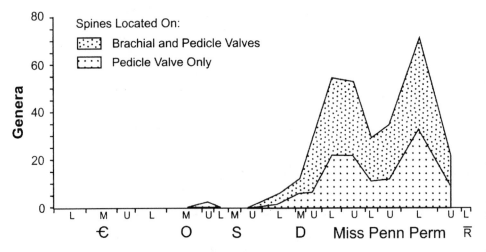

FIGURE 9. Brachiopod genera, primarily productids, with spines on the pedicle or both valves. Both show consistent trends. From Signor and Brett (1984).

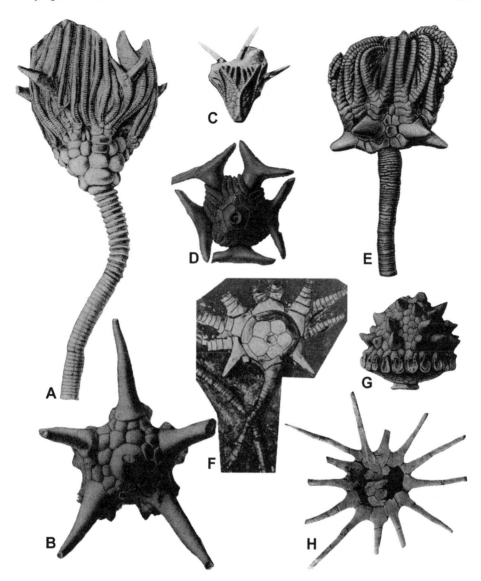

FIGURE 10. Spinosity in Paleozoic crinoids. From Signor and Brett (1984). (A), (B) *Dorycrinus mississippiensis* Roemer; Mississippian camerate crinoid with stout tegminal spines; note that spines protrude between arms in A; Keokuk Limestone; Iowa; both views approximately X1. (C) *Thamnocrinus springeri* Goldring; Middle Devonian camerate crinoid with tegminal spines; Moscow Formation, western New York; X1. (D) *Pterotocrinus bifurcatus* Wetherby; note sharp "wing plates" between arms; Mississippian; Kentucky; (E) *Wachsmuthicrinus spinifer* Hall; Mississippian flexible crinoid with stout spines on axillaries of the arms; Burlington Limestone; Burlington, Iowa; X1. (F) *Idosocrinus bispinosus* Wright; Lower Carboniferous, Thornton Burn, Scotland, X1. (G) *Batocrinus icosodactylus* Wachsmuth and Springer; tegmen studded with spines; Mississippian Burlington, Iowa; X1. (H) *Eiromocrinus grossus* Strimple and Watkins; Pennsylvanian cladid crinoid with long spines on top of anal sac; Millsap Lake Fm., Hood Co., Texas; X0.5. Figures A, B, D, G from Wachsmuth and Springer (1897); C from Goldring, (1923); F from Wright (1954); H from Strimple and Watkins (1969).

Few Paleozoic gastropods were spiny – the net proportion of genera with spines barely exceeds 5%. Table 6 shows Paleozoic gastropods for each period that exhibit spine development based on compilation of the Treatise on Invertebrate Paleontology (Knight et al., 1960). Spinose gastropods are unknown from the early Paleozoic and first appeared in the Silurian. It is also noteworthy that all of the spinose forms are inferred, based on shell morphology and other direct evidence, to have been relatively sedentary. Most noteworthy are the platyceratids. Several species of spinose platyceratids appear in the Devonian. In this case, direct evidence indicates that these forms were sedentary, as they are preserved attached to the tegmens of crinoids. The platyceratids are inferred to have been commensalistic (coprophagous) or possibly parasitic on crinoids (Bowsher, 1955; Rollins and Brezinski, 1988; Baumiller, 1990; Boucot, 1990). Such a permanently sessile life style may have rendered these gastropods particularly vulnerable to predatory attack, and spines may have provided a major adaptive advantage.

Trilobites showed pleural and genal spines, probably as adaptations for stabilization on soft substrates; enrollment also evolved in several groups during the Ordovician. An abrupt increase of species with spines along the axial lobe and protruding forward from the cephalon occurs within the Early to Middle Devonian. The well-known and highly diverse trilobites from the Emsian – Eifelian of Morocco and North America show a high frequency of spinose genera in several lineages.

Kloc (1993, 1997) has also documented the occurrence of possible camouflage strategies in the Early Devonian selenopeltid trilobite *Dicranurus*. Large cephalic spines are commonly bored and encrusted by bryozoans and algae and may have served to camouflage the trilobite from predators that relied on vision, such as cephalopods and fish.

Among echinoderms increased mobility of spines may also be observed. Echinoids, some asteroids and ophiuroids, a few edrioasteroids and at least one genus of crinoid evolved mobile, articulated spines that can be swiveled into positions to increase protective function. Only in crinoids, however, is the record complete enough to document trends (Signor and Brett, 1984).

Spines on the calyces and tegmens of crinoids could have served as a deterrent to would-be predators. Therefore, it is significant that virtually no crinoids possessed spinose calyces prior to the middle Silurian, when *Calliocrinus* displayed large tegminal spines (Signor and Brett, 1984). Both camerate and cladid crinoids show a substantial increase in the proportion of spinose genera commencing in Early Devonian time (Figs. 10, 11). The proportion of spinose genera increases to a maximum in middle Mississippian time and then declines in the late Paleozoic, in concert with the decline of camerate crinoids during the Chesterian crisis identified by Ausich et al. (1990). A number of different families of camerates and cladids (and a few of flexibles) developed spinose characters independently during the Devonian – Mississippian interval (Fig. 10).

Spinosity occurs in two distinct forms. Many post–Silurian cladid crinoids and some camerates developed spines on the elongated anal sac or anal chimney (Fig. 10H). This region may have been particularly vulnerable to predation as it contained large extensions of the viscera, and possibly gonads (see Lane, 1984). Other crinoids, primarily Devonian – Mississippian camerates, but also a few late Paleozoic cladids, developed elongate spines on the calyx. A few genera developed spinose plates on the axillaries of the arms. *Arthroacantha*, a very common and widespread Devonian camerate, possessed articulated spines, as in echinoids, on the calyx, as well as spines on the arms (Fig. 12). Within this genus there is also a trend toward increasing spine length into the Late Devonian (G. C. McIntosh, pers. comm., 2001).

Most of the crinoids that evolved spines possessed large, long columns and relatively thin plated calyces, seemingly making them more vulnerable to predation. However, paradoxically, some larger crinoids, including most flexibles, did not evolve spinose calyces. One can only speculate that these taxa may have possessed alternate predator deterrent strategies such as toxic chemicals.

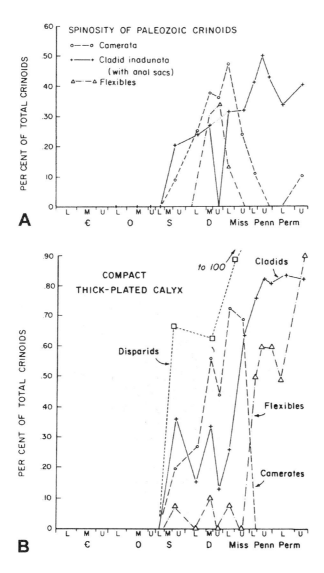

FIGURE 11. Morphological trends in Paleozoic crinoids. (A) percentage of spinose genera; (B) percentage of thick walled cups. Modified from Signor and Brett (1984).

TABLE 6. Occurrence of Spines in Paleozoic Gastropod Genera; Based on Data from Treatise on Invertebrate Paleontology (Knight et al., 1966)

Age	N	# Spines	% Spines
Middle Cambrian	2	0	0
Late Cambrian	15	0	0
Early Ordovician	45	0	0
Middle Ordovician	57	0	0
Late Ordovician	58	0	0
Early Silurian	48	0	0
Late Silurian	96	1	0.9
Early Devonian	81	2	2.5
Middle Devonian	63	3	4.8
Late Devonian	47	4	4.2
Mississippian	77	4	5.2
Early Pennsylvanian	74	4	5.4
Middle Pennsylvanian	80	1	1.2
Late Pennsylvanian	81	2	2.5
Permian	66	1	1.5

FIGURE 12. Stylized drawing of the spinose Middle Devonian crinoid *Arthroacantha carpenteri* with attached *Platyceras* gastropod; based on specimens from the Silica Shale; note jointed spines on calyx plates and stout spines on axillaries of arms forming a protective shroud around the tegmen and commensal gastropod. Modified from Kesling and Chilman (1975).

4.3. Platyceratid Targeting?

There is at least one other important correlate of the spinosity of crinoids that involves coevolution with commensal/coprophagous to parasitic platyceratid gastropods. One of the most intriguing correlates with spinosity is that for Devonian and Mississippian camerates. Virtually all spinose taxa were common hosts to platyceratid gastropods. The common feature of host crinoids was a large flat tegmen, which provided an easily accessible site for the gastropods. Indeed, it has been argued that the development of long anal chimneys and sacs may have been a deterrent to gastropod infestation (Meyer and Ausich, 1983).

Interestingly, the subset of most spinose crinoids also includes some of the most frequently platyceratid infested forms. Notably, *Arthroacantha*, the most specialized of spiny crinoids (Fig. 12), was heavily infested. In populations of *A. carpenteri* from the Silica Shale of Ohio, up to 70% of individuals have attached gastropods; an even higher proportion was found in the Devonian Arkona Shale in Ontario populations (unpubl. data G.C. McIntosh). Even more intriguing is the fact that some of the platyceratids infesting *Arthroacantha* are among the most spinose of all Paleozoic gastropods. Common hosts (e.g., *Gilbertsocrinus* and *Acanthocrinus*) were similarly highly spinose. Only a few of the non-spinose crinoids, notably the flexible taxocrinids, were common hosts to platyceratids.

Thus, beginning in the Middle Devonian, a majority of crinoids that served as platyceratid hosts were spinose and nearly all spinose crinoids are known to have been hosts (Table 7). Conversely, in pre-Devonian time *none* of the common host echinoderms (e.g., *Dimerocrinites*, *Ptychocrinus*, *Macrostylocrinus,* and the rhombiferan cystoid *Caryocrinites*) were spinose (Table 7).

Why did several different lineages of platyceratid-infested crinoids suddenly and selectively develop spinosity in the Devonian? It might be suggested that an increased spinosity in crinoids was a response to platyceratid attachment, especially if the gastropods were parasitic. However, the fact that platyceratids coevolved with crinoid hosts for over 50 million years before the appearance of spines in the hosts casts considerable doubt on this interpretation. Moreover, it is clear that spines on crinoid calyces were utterly ineffective in keeping platyceratid gastropods off the tegmens. Altruism on the part of crinoid hosts or platyceratids is most unlikely, especially in view of possible harmful effects of the snails on crinoids.

The explanation of this paradoxical situation seems to involve the rather dramatic increase in predation pressure during the Devonian. In view of the direct evidence for predatory attack on Devonian gastropods, it is likely that platyceratids were an attractive food source. As already noted, the evolution of spines in platyceratids and other sedentary gastropods suggests that they were particularly prone to predation. Thus, the presence of platyceratids on crinoids might invite predatory attack, even if the crinoids themselves were not attractive prey (see Meyer, 1985). In the process of attacking the gastropod the crinoid would also have been damaged if not partially ingested. In short, given the high frequency of infestation, the platyceratids may have "targeted" crinoid hosts for predator attacks and *indirectly* triggered an adaptive response on the part of the host crinoids. In the crinoids that frequently harbored platyceratids there may have been a greatly increased predation pressure that made the chance appearance of spines or protuberances a key adaptive advantage.

TABLE 7. Occurrences of platyceratid gastropods in well preserved pelmatozoan echinoderm faunas from the Early Silurian (Wenlock) and Middle Devonian (Givetian) of eastern North America giving characteristics of the hosts. Symbols: + feature is present; - feature is absent; N= *Naticonema*; P= *Platyceras*; Ps= spinose *Platyceras* (mainly *P. dumosum*); * major host crinoid (>25% of specimens possess platyceratids). Note absence of spines on host crinoids and platyceratids in Silurian as compared to presence of spines in nearly all host crinoids and some of the gastropods in the Middle Devonian.

Age/Genus	Gastropod	Large/Flat Tegmen	Spines	Location of Spines/Jointing
Silurian (Rochester Shale, Waldron Shale)				
Caryocrinites	N, P	+	-	-
Dimerocrinites	N	+	-	-
Lyriocrinus	P, N	+	-	-
Macrostylocrinus	P, N	+	-	-
Periechocrinites	P, N	+	-	-
Other crinoids (N=21)	-	small tegmen	-	-
Total N = 26 genera				
Devonian (Silica Shale, Arkona Shale, Hamilton Group)				
Acanthocrinus	P*	+	+	calyxc unjointed
Aorocrinus	Ps	+	+	arms
Arthroacantha	Ps*	+	+	calyxc unjointed
Corocrinus	P*	+	-	-
Dolatocrinus	P	+	-	-
Genneaocrinus	P*	+	+	calyxc unjointed
Gilbertsocrinus	P*	+	+	calyxc unjointed
Megistocrinus	P (rare)	+	+	tegmen
Taxocrinus	N*	-	-	-
other crinoids (N=26)	-	Small tegmen	-	No spines except on anal sacs
Total N = 35 genera				

Indirect targeting of this sort may be compared to the effect of hermit crabs on gastropod shell evolution. Vermeij (1987) has pointed out that by providing an accessory food supply they may have promoted intensified predation, which fed back into evolution of greater resistance to predation on the part of the mollusks.

5. Discussion: Tests of the Mid Paleozoic Escalation Hypothesis and a Preliminary Model of Escalation

Since the publication of the mid Paleozoic escalation hypothesis (Signor and Brett, 1984), a number of studies have been undertaken as partial tests of this hypothesis. In general, these studies have provided further support for the notion of increased predation pressure beginning in the Silurian-Devonian. However, they also suggest that the pattern of escalation may not be continuous or linear, and that, in fact, prey may adopt alternative evolutionary strategies depending upon the intensity of predatory pressure.

Leighton (2001) provided a partial test of the mid Paleozoic escalation hypothesis. He suggested that if spines in brachiopods serve as an anti-predatory strategy there should be evidence that more spinose forms had lower rates of successful predation. In his survey of borehole frequency, Leighton did observe that the spinose productids had a lower rate of successful predator attack. This suggests that spinosity may have been an effective deterrent to drilling predation as well as shell crushing.

Aronson (1991) argued that if predation pressure had been a significant factor in crinoid communities, a major decline in crinoid thickets would be observed in fossil assemblages through time, from pre-Devonian to Carboniferous time intervals. This prediction came in part from evidence that stalked crinoids migrated offshore in the face of the Mesozoic marine revolution of predators (Meyer and Macurda, 1977). Aronson made corrections for differences in rock volume of various ages and durations of particular intervals to predict an expected proportion of dense crinoid assemblages for each age. He found that the observed proportion of dense crinoid thickets did not differ significantly from the predicted proportion. Thus, crinoid thickets do not show a decline during the post-Devonian interval. This provides negative evidence for the escalation hypothesis and might suggest that predation pressure was not, in fact, a major factor in controlling crinoid density. Alternatively, Aronson suggests that the general lack of reefs in the Mississippian caused a decline in specialized reef-dwelling fish predators. This inference is questionable as most known Devonian gnathostome fish fossils are not associated with reefs, but rather with open marine settings.

One might also suggest that crinoids were able to adjust to the increased pressure of grazing by sharks, holocephalans and other fishes, *up to a point*. The rise in spinosity, plate thickness, and other strategies (e.g., biochemical defenses) may have been temporarily effective in preventing decimation by predators. Thus, for an interval in the Devonian and Mississippian crinoids maintained dense thickets and high diversity. However, ultimately these "strategies" may have been ineffective, as suggested by a decline of high diversity crinoid thickets and the decrease in spinosity in the late Paleozoic (Signor and Brett, 1984).

A second study, by Waters and Maples (1991), provided positive evidence of the impact of predation on crinoids in the Late Mississippian. Waters and Maples documented a major faunal reorganization event that occurred near the Genevievian-Gasperian stage boundary in eastern North America. Long-stemmed and relatively large camerates were replaced by smaller camerates and blastoids with robust, crush-resistant calyces. The authors argued that this turnover may reflect predation pressure: crinoids took refuge in small size and stem length reduction. Ironically, the Gasperian crinoids show *lower* spinosity than do earlier Mississippian crinoids, the opposite of what might be predicted by the escalation hypothesis (see also Fig. 11). This trend was already noted by Signor and Brett (1984); all crinoid subclasses exhibit *decreased* spinosity in the late Paleozoic following an Early Mississippian high. Waters and Maples suggest that predators were able to "keep up" with the armaments of their prey and that spinose plates became ineffective as a defensive strategy; smaller size and greater compactness of the calyx may then have been more effective strategies.

Studies such as those of Aronson (1991) and Waters and Maples (1991) suggest that prey escalation may follow a stepwise, but non-linear pattern. In early phases of predator escalation, prey with strengthened or spinose skeletons may have an adaptive advantage; such would be the case with brachiopods and crinoids in the Devonian-Mississippian. During this phase, prey may show a trend toward increasing size or effectiveness of armor and population densities may temporarily remain stable or even increase, as with Early Mississippian crinoids. In a second phase, durophagous predators become larger and better adapted to cope with prey defenses and armaments, such as spines become ineffective; prey may take temporary evolutionary refuge in small size, camouflage and other strategies and may show a reversal in trends toward armaments, such as spines. Such a pattern may be reflected especially in Pennsylvanian-Permian crinoids. At this point predation pressure may begin to make serious inroads into prey population densities. Subsequently, if durophagous predators become still more numerous, eurytopic, and/or effective at shell crushing, prey may either be eliminated altogether or display major habitat shifts. This phase is exemplified by widespread infaunalization of mollusks in the Mesozoic and the shift of stalked crinoids into deep oceanic refugia (Macurda and Meyer, 1977).

Finally, it might be noted that the escalation of predator-prey interactions probably did not develop gradually and progressively. Boucot (1990) pointed out that predator-prey interactions, as with other types of complex organism coactions, seem to have developed rapidly and early and remained similar for long periods. Certain predator-prey links can be observed to persist for extended periods; these include drilling predators and certain families of their preferred prey.

Boucot (1990), Brett and Baird (1995) and Brett et al. (1996) argued that marine organisms exhibit ecological as well as morphological stasis. The persistence of both prey and predator species through blocks of stability (ecological evolutionary subunits) suggests that these sorts of interactions were established early on during these blocks and remained fixed, although the notion has yet to be rigorously tested. If so, the process of escalation, as with the rest of evolution was not continuous, but highly episodic. During major faunal overturn and restructuring events that characterize the ends of EE subunits predator-prey interactions are abruptly intensified (or possibly decreased) and escalation attains a new and metastable equilibrium. However, the way in which these episodes of change operated is poorly understood and the pattern itself requires much more detailed documentation.

6. Summary

There is a growing body of evidence that predation on hard-shelled marine organisms intensified during the middle Paleozoic. The direct fossil record of potential predators shows a substantial increase in durophagous shell crushing predators, including cephalopods, phyllocarids, and several lineages of gnathostome fishes.

Trace fossil evidence provides a strong case for the existence of predatory attacks on shelled organisms as early as the late Neoproterozoic. Coprolite evidence is weak, but suggestive of an increase in shell crushing predation during the middle Paleozoic. Brachiopod and molluscan shells and trilobite exoskeletons show evidence of healed bite marks and peeling from the Cambrian onward, but with an increased frequency in the Devonian. Drill holes attributable to predation, with stereotypical position and prey-species preference, are found in brachiopods (Cambrian onward) and mollusks (Ordovician onward); boreholes also show increased frequency in the middle Paleozoic. Present evidence suggests that certain of these boreholes are attributable to gastropods, most probably members of the family Platyceratidae.

Hard-shelled organisms responded to crushing and drilling predation by evolving a variety of thicker more spinose skeletons. Few early Paleozoic organisms exhibit long projecting spines, but several groups, including brachiopods, gastropods, trilobites, and crinoids show an abrupt increase in spinosity, beginning in the Silurian- Devonian. Such spines seem to have been effective, at least in reducing effective drilling and durophagous predation, up to a point. Groups that may have been specifically targeted for predation, including platyceratid gastropod-infested crinoids and sedentary epifaunal forms appear to have preferentially developed anti-predatory architecture. However, there is some evidence that spines may have become ineffectual against predators with large jaws. Late Paleozoic forms may have taken refuge in smaller size and resistant thick walled skeletons, and, eventually, in habitat shifts.

In a broad view, the rise of predation pressure and the skeletal response appear to have been progressive in the middle-late Paleozoic. However, escalation of predator–prey interactions may have developed in incremental steps associated with periods of abrupt biotic reorganization. These episodes interrupted much longer times of ecological stasis.

ACKNOWLEDGMENTS: Thanks to Richard Bambach, Tom Baumiller, Michał Kowalewski, and Sally Walker, who provided reviews of this paper and suggested important improvements. Evelyn Mohalski aided in preparation of figures.

References

Alexander, R. R., 1981, Predation scars preserved in Chesterian brachiopods: probable culprits and evolutionary consequences for the articulates, *J. Paleontol.* **55**:192-203.
Alexander, R. R., 1986a, Resistance to and repair of shell breakage induced by durophages in Late Ordovician brachiopods, *J. Paleontol.* **60**:273-285.
Alexander, R. R., 1986b, Frequency of sublethal shell-breakage in articulates through geologic time, in: *Les Brachiopodes Fossiles et Actuels* (P. R. Racheboeuf and C. Emig, eds.), Université de Bretagne Occidentale. Biostratigraphie du Paleozoique, pp. 159-166.
Aronson, R. B., 1991, Escalating predation on crinoids in the Devonian: negative community-level evidence, *Lethaia* **24**:123-128.
Ausich, W. I., and Gurrola, R. A., 1979, Two boring organisms in a Lower Mississippian community of southern Indiana, *J. Paleontol.* **53**:335-344.
Ausich, W. I., Kammer, T. W., and Baumiller, T. K., 1994, Demise of the middle Paleozoic crinoid fauna: a single extinction event or rapid faunal turnover, *Paleobiology* **20**:345-361.
Babcock, L. E., 1993, Trilobite malformations and the fossil record of behavioral symmetry, *J. Paleontol.* **67**:217-229.
Babcock, L. E., and Robison, R. A., 1989, Preferences of Paleozoic predators, *Nature* **337**:695-696.
Bandel, K., 1985, Cephalopod morphology and function, in: *Mollusks: Notes for a Short Course* (D. J. Bottjer, C. S. Hickman, and P. D. Ward, eds.), University of Tennessee Department of Geological Sciences, Studies in Geology **13**, pp. 190-201.
Baumiller, T. K., 1990, Non-predatory drilling of Mississippian crinoids by platyceratid gastropods, *Palaeontology* **33**:743-748.
Baumiller, T. K., 1993, Boreholes in Devonian blastoids and their implications for boring by platyceratids, *Lethaia* **26**:41-47.
Baumiller, T. K., 1996, Boreholes in the Middle Devonian blastoid *Heteroschisma* and their implications for gastropod drilling. *Palaeogeogr. Palaeoclimat. Palaeoecol.* **123**:343-351.

Baumiller, T. K., Leighton, L. R., and Thompson, D. L., 1999, Boreholes in Mississippian spiriferide brachiopods and their implications for Paleozoic gastropod drilling, *Palaeogeogr. Palaeoclimatol. Palaeoecol.* **147**:283-289.
Bengtson, S., and Zhao, Y., 1992, Predatorial borings in late Precambrian mineralized exoskeletons, *Science* **257**:367-369.
Benton, M. J., 1997, *Vertebrate Palaeontology*, 2nd ed., Chapman and Hall, London.
Bishop, G. A., 1975, Traces of predation, in: *The Study of Trace Fossils* (R. Frey, ed.), Springer-Verlag, New York, Heidelberg, Berlin, pp.267-281.
Bond, P. N., and Saunders, W. B., 1989, Sublethal shell injury and shell repair in Upper Mississippian ammonoids, *Paleobiology* **15**:414-428.
Bordeaux, Y. L., and Brett, C. E., 1990, Substrate specific associations on Middle Devonian brachiopods: implications for paleobiology, *Hist. Biol.* **4**:203-220.
Boucot, A. J., 1990, *Evolutionary Paleobiology of Behavior and Coevolution*, Elsevier, Amsterdam, 725 pp.
Bowsher, A. L., 1955, Origin and adaptation of platyceratid gastropods, *Univ. Kansas Paleont. Contrib., Mollusca* **5**:1-11.
Boyd, D.W., and Newell, N. D., 1972, Taphonomy and diagenesis of a Permian fossil assemblage from Wyoming, *J. Paleontol.* **46**:1-17.
Brett, C. E., and Baird, G. C., 1995, Coordinated stasis and evolutionary ecology of Silurian to Middle Devonian marine biotas in the Appalachian basin, in: *New Approaches to Speciation in the Fossil Record* (D. Erwin and R. Anstey, eds.), Columbia University Press, New York, pp. 285-315.
Brett, C. E., and Cottrell, J. F., 1982, Substrate specificity of the Devonian tabulate coral *Pleurodictyum*, *Lethaia* **15**:248-263.
Brett, C. E., Ivany, L. C., and, Schopf, K. M. 1996, Coordinated stasis: An overview, *Palaeogeogr. Palaeoclimatol. Palaeoecol.* **127**:1-20.
Brett, C. E., and Walker, S. E. 2002. Predators and predation in Paleozoic marine environments, in: *The Fossil Record of Predation* (M. Kowalewski and P. H. Kelley, eds.), Special Papers in Paleontology, **8**, in press.
Bromley, R. G., 1981, Concepts in ichnotaxonomy illustrated by small round holes in shells, *Acta Geol. Hisp.* **16**:55-64.
Brunton, H., 1966, Predation and shell damage in a Visean brachiopod fauna, *Palaeontology* **9**:355-359.
Bucher, W. H., 1938, A shell-boring gastropod in a *Dalmanella* bed of upper Cincinnatian age, *Am. J. Sci.* **36**:1-7.
Buehler, E. J., 1969, Cylindrical borings in Devonian shell, *J. Paleontol.* **43**:1291.
Cameron, B., 1967, Oldest carnivorous gastropod borings found in Trentonian (Middle Ordovician) brachiopods, *J. Paleontol.* **41**:147-150.
Carriker, M. J., and Yochelson, E. L., 1968, Recent gastropod boreholes and Ordovician cylindrical borings, *U.S. Geol. Surv. Prof. Pap.* **B593**:1-26.
Chatterton, B., and Whitehead, H. L., 1987, Predatory borings in inarticulate brachiopods *Artiotrema* from the Silurian of Oklahoma, *Lethaia* **20**:67-74.
Clarke, J. M., 1921, Organic dependence and disease: their origin and significance, *New York State Museum Bulletin* **221/222**:1-113.
Conway Morris, S., and Bengtson, S., 1994, Cambrian predators: possible evidence from boreholes, *J. Paleontol.* **68**:1-23.
Conway Morris, S., and Jenkins, R. J. F., 1985, Healed injuries in Early Cambrian trilobites from South Australia, *Alcheringa* **9**:167-177.
Ebbestad, J. O. R., 1998, Multiple attempted predation in the Middle Ordovician gastropod *Bucania gracillima*, *Geol. Fören. Stockholm Forhand.* **120**:27-33.
Ebbestad, J. O. R., and Peel, J. S., 1997, Attempted predation and shell repair in Middle and Upper Ordovician gastropods from Sweden, *J. Paleontol.* **7**:1007-1019.
Goldring, W. 1923, Devonian Crinoids of New York. N.Y. State Mus. Bull.
Hansen, M. C., and Mapes, R. H., 1990, A predator-prey relationship between sharks and cephalopods in the late Paleozoic, in: *Evolutionary Paleobiology of Behavior and Coevolution* (by A. J. Boucot), Elsevier, Amsterdam, pp.189-195.
Häntschel, W., 1968, Coprolites: An Annotated Bibliography, *Geol. Soc. Amer. Mem.* **108**:1-132.
Hoffmeister, A. P., Kowalewski, M., Bambach, R. K., and Baumiller, T. K., 2001a, Intense drilling predation on the brachiopod *Cardiarina cordata* (Cooper, 1956) from the Pennsylvanian of New Mexico, *Geol. Soc. Am. Abstr. Progr.* **33**:A248.
Hoffmeister, A. P., Kowalewski, M., Bambach, R. K., and Baumiller, T. K., 2001b, Evidence for predatory drilling in late Paleozoic brachiopods and bivalve mollusks from West Texas, *N. Am. Paleont. Conv., Paleobios*, **21**(suppl. 2):66-67.

Hoffmeister, A. P., Kowalewski, M., Bambach, R. K., and Baumiller, T. K., 2002, A boring history of drilling predation on the Paleozoic brachiopod *Composita*, *Geol. Soc. Am. Abstr. Progr.* **34**:A116.
Kabat, A. R., 1990, Predatory ecology of naticid gastropods with a review of shell boring predation, *Malacologia* **32**:155-103.
Kaplan, P., and Baumiller, T. K., 2000, Taphonomic inferences on boring habit in the Richmondian *Onniella meeki* Epibole, *Palaios* **15**:499-510.
Kesling, R. V., and Chilman, R. B., 1975, Strata and megafossils of the Middle Devonian Silica Formation, *Univ. Mich. Mus. Paleont. Pap. Paleont.* **8**:1-408.
Knight, J. B., Cox, L. R., Keen, A. M, Batten, R. L., Yochelson, E. L., and Robertson, R., 1960, Systematic descriptions, in: *Treatise on Invertebrate Paleontology, Pt. I, Mollusca 1* (R. C. Moore, ed.), Geological Society of America and University of. Kansas, Boulder, Colorado, and Lawrence, Kansas, pp. I161-I331.
Kowalewski, M., Dulai, A., and Fürsich, F. T., 1998, A fossil record full of holes: the Phanerozoic history of drilling predation, *Geology* **26**:1091-1094.
Kowalewski, M., Simões, M. G., Torello, F. F., Mello, L. H. C., and Gilardi, R. P., 2000, Drill holes in shells of Permian benthic invertebrates, *J. Paleontol.* **74**:532-543.
Kloc, G. J., 1987, Coprolites containing ammonoids from the Devonian of New York, *Geol. Soc. Am. Abstr. Progr.* **19**:23.
Kloc, G. J., 1992, Spine function in the odontopleurid trilobites *Leonaspis* and *Dicranurus* from the Devonian of Oklahoma, *N. Amer. Paleont. Conv. Abstr. Progr., Paleont. Soc. Spec. Publ.* **6**:167.
Kloc, G. J., 1993, Epibionts on *Selenopeltine* (Odontopleuriae) Trilobites, *Geol. Soc. Am. Abstr. Prog.* **25A**:103.
Kloc, G. J., 1997, Epibionts on *Dicranurus* and some related genera, Second Intl. Trilobite Conf., Brock University, St Catharines, Ontario, Abstracts **28**.
Lane, N. G., 1984, Predation and survival among inadunate crinoids, *Paleobiology* **10**:453-458.
Lescinsky, H. L., and Benninger, L., 1994, Pseudo-borings and predator traces: Artifacts of pressure dissolution in fossiliferous shales, *Palaios* **9**:599-604.
Leighton, L. R., 2001, New examples of Devonian predatory boreholes and the influence of brachiopod spines on predator success, *Palaeogeogr. Palaeoclimatol. Palaeoecol.* **165**:71-91.
Malzahn, E., 1968, Uber neue Funde von *Janassa bituminosa* (Schloth.) im neiderrheinisschen Zechstein, *Geol. Jahrb.* **85**:67-96.
Mapes, R. H., and Benstock, E. J., 1988, Color pattern on the Carboniferous bivalve *Streblochondria*? Newell, *J. Paleontol.* **62**:439-441.
Mapes, R. H., and Hansen. M. C., 1984, Pennsylvanian shark-cephalopod predation: a case study, *Lethaia* **17**:175-183.
Meyer, D. L., 1985, Evolutionary implications of predation on Recent comatulid crinoids, *Paleobiology* **11**:154-164.
Meyer, D. L., and Ausich, W. I., 1983, Biotic interactions among Recent crinoid and among fossil crinoids, in: *Biotic Interactions in Recent and Fossil Benthic Communities* (M. J. S. Tevesz and P. L. McCall, eds.), Plenum Press, New York, pp. 377-427.
Meyer, D. L., and Macurda, Jr., D. B, 1977, Adaptive radiation of the comatulid crinoids, *Paleobiology* **3**:74-82.
Miller, R. H., and Sundberg, F. A., 1984, Boring Late Cambrian organisms, *Lethaia* **17**:185-190.
Moodie, R. L., 1923, *Paleopathology: An Introduction to the Study of Ancient Evidence of Disease*, University of Chicago Press, Chicago, 567 pp.
Moy-Thomas, J. A., and Miles, R. S., 1971, Palaeozoic Fishes. W. B. Saunders Co., Philadelphia, 259 pp.
Paine, R. T., 1969, A note on trophic complexity and community stability, *Am. Nat.* **103**:91-93.
Peel, J. S., 1984, Attempted predation and shell repair in *Euomphalopterus* (Gastropoda) from the Silurian of Gotland, *Lethaia* **32**:163-168.
Richards, R. P., and Shabica, C. W., 1969, Cylindrical living burrows in Ordovician dalmanellid brachiopod shells, *J. Paleontol.* **43**:838-841.
Rodriguez, R. P., and Gutschick, R. C., 1970, Late Devonian – Early Mississippian ichnofossils from western Montana and northern Utah, in: *Trace Fossils* (J. P. Crimes and J.C. Harper, eds.), *Geol. J., Spec. Issue* **3**:407-438.
Rohr, D. M., 1976, Silurian predator borings in the brachiopod *Dicaelosia* from the Canadian Arctic, *J. Paleontol.* **65**:687-688.
Rollins, H. B., and Brezinski, D. K., 1988, Reinterpretation of crinoid-platyceratid interaction, *Lethaia* **21**:189-292.
Roy, K., Miller, D. J. and LaBarbera, M., 1994, Taphonomic bias in analyses of drilling predation: effects of gastropod drill holes on bivalve shell strength, *Palaios* **9**:413-421.
Rudwick, M. J. S., 1970, *Living and Fossil Brachiopods*, Hutchinson University Library, London, 199 pp.

Schindel, D. E., Vermeij, G. J., and Zipser, E., 1981, Frequencies of repaired shell, fractures among the Pennsylvanian gastropods of north-central Texas, *J. Paleontol.* **56**:729-740.

Schneider, J. A., 1988. Frequency of arm regeneration of comatulids in relation to life habit, in: *Proceedings of the 6th International Echinoderm Conference* (R. D. Burke, P. V. Mladenov, P. Lambert, and R. L. Parsley, eds.), Victoria, British Columbia, pp. 97-102.

Sheehan, P. M., and Lesperance, P. J., 1978, Effect of predation on population dynamics of a Devonian brachiopod, *J. Paleontol.* **52**:812-817.

Signor, P. W., III, and Brett, C. E., 1984, The mid-Paleozoic precursor to the Mesozoic marine revolution, *Paleobiology* **10**:229-245.

Smith, S. A., Thayer C. W., and Brett, C. E., 1985, Predation in the Paleozoic: gastropod-like drill holes in Devonian brachiopods, *Science* **230**:1033-1037.

Springer, F., 1920, The Crinoidea Flexibilia, *Smithsonian Inst. Pub.* **2871**:1-238.

Steel, M. H., and Sinclair, G. W., 1971, A Middle Ordovician fauna from Braeside, Ottawa Valley, Ontario, *Geol. Surv. Can., Bull.* **211**:1-96.

Strimple, H. L., and Watkins, W. T., 1969, Carboniferous crinoids of Texas with stratigraphic implications. *Paleont. Amer.* **6**:141-275.

Thayer, C. W., 1983, Sediment-mediated biological disturbance and the evolution of the marine benthos, in: *Biotic Interactions in Recent and Fossil Benthic Communities* (M. J. S. Tevesz and P. J. McCall, eds.), Plenum Press, New York, pp. 479-595.

Ubaghs, G., 1978, Skeletal morphology of fossil crinoids, in: *Treatise on Invertebrate Paleontology, Pt. 1, Mollusca 1* (R. C. Moore and C. Teichert, eds.), Geological Society of America and University of. Kansas, Boulder, Colorado, and Lawrence, Kansas, pp. T58-T216.

Vermeij, G. J., 1977, The Mesozoic marine revolution: Evidence from snails, predators, and grazers, *Paleobiology* **3**:245-258.

Vermeij, G., 1983. Shell breaking predation through time, in: *Biotic Interactions in Recent and Fossil Benthic Communities* (M. J. S. Tevesz and P. L. McCall, eds.), Plenum Press, New York, pp. 649-669.

Vermeij, G. J., 1987, *Evolution and Escalation: An Ecological History of Life*, Princeton University Press, Princeton, New Jersey, 527 pp.

Vermeij, G. J., Zipser, E., and Dudley, E. C., 1980, Predation in time and space: Peeling and drilling in terebrid gastropods, *Paleobiology* **6**:352-364.

Vermeij, G. J., Schindel, D. E., and Zipser, E., 1981, Predation through geological time: evidence from gastropod shell repair, *Science* **214**:1024-1026.

Wachsmuth, C., and Springer, F., 1897, The North American Crinoidea Camerata, *Harvard Mus. Compar. Zool., Mem.*, **20**:1-897.

Waters, J. W., and Maples, C., 1991, Mississippian pelmatozoan community reorganization: a predation-mediated faunal change, *Paleobiology* **17**:400-410.

Wright, J., 1954, *Idosocrinus* gen. nov. and other crinoids from Thornton Burn, East Lothian, *Geol. Mag.* **91**:167-170.

Zangerl, R., and Richardson, E., 1963, Paleoecological history of two Pennsylvanian black shales, *Fieldiana Geol. Mem.* **4**:1-352.

Zangerl, R., Woodland, B. G., Richardson, E. S., and Zachry, D. L., 1969, Early diagenetic phenomena in the Fayetteville Black Shale (Mississippian) in Arkansas, *Sed. Geol.* **3**:87-120.

Chapter 18

The Mesozoic Marine Revolution

ELIZABETH M. HARPER

1. Introduction 433
2. Establishing a Chronology for the MMR..435
   2.1. Predation by Simple Ingestion..435
   2.2. Predation by Prying..436
   2.3. Predation by Breakage..437
   2.4. Drilling Predation...438
   2.5. Summary of Temporal and Spatial Patterns...439
3. Prey Responses to Increasing Threat of the MMR 440
   3.1. Restriction to Refugia...440
   3.2. Life Habit Changes...443
   3.3. Morphological Defenses...445
   3.4. Behavioral and Physiological Changes..447
4. Discussion and Future Directions 447
   References 451

1. Introduction

The modern oceans teem with animals which kill others to live, from killer whales that form pods of several individuals in co-ordinated attacks on their quarry (Pitman *et al.*, 2001) to the drilling activities of tiny predatory foraminifers (Hallock *et al.*, 1998). Most authors believe that predator-prey interactions, in tandem with competition, are key factors in controlling structure in modern communities. Classic work by Connell (1970) and Paine (1974) showed how predation in rocky shore communities prevented domination by major space occupiers and thus promoted overall diversity. In such situations taxa with antipredatory adaptations will be at an advantage and, if predation has been similarly important over geological time, we should anticipate that it has been an important agent of natural selection. Indeed, it is often suggested that the first appearance of shelled organisms in the "Cambrian explosion" might be due, in part, to the rise of predators (Conway Morris, 2001).

ELIZABETH M. HARPER • Department of Earth Sciences, Cambridge University, Cambridge CB2 3EQ, United Kingdom.

Predator-Prey Interactions in the Fossil Record, edited by Patricia H. Kelley, Michal Kowalewski, and Thor A. Hansen. Kluwer Academic/Plenum Publishers, New York, 2003.

There is clear evidence from the fossil record of predator-prey interactions throughout the entire Phanerozoic; indeed predatory drillholes may even be present in the very earliest shelly fossils (Bengtson and Zhao, 1992). It is also apparent that a range of predatory methods were in use before the close of the Paleozoic, for example hyolithids and brachiopods swallowed whole within the guts of Middle Cambrian Burgess Shale priapulid worms (Conway Morris, 1977), drillholes in Cambrian brachiopods (e.g. Conway Morris and Bengtson, 1994), repaired crushing damage in brachiopods and gastropods (Alexander, 1981; Ebbestad and Peel, 1997) and possible evidence of Ordovician asteroids prying apart bivalve prey (Blake and Guensburg, 1994). It is increasingly evident that over the course of the Paleozoic there was a marked increase in the number of groups of crushing predators, notably during the Devonian (Signor and Brett, 1984), and there is also some evidence that levels of drilling predation increased at this time (Kowalewski *et al.*, 1998; Harper, submitted). This "mid-Paleozoic marine revolution" had a profound impact on the evolution of various prey taxa (Brett, Ch. 17, in this volume).

Further major changes in the levels of predation occurred post-Paleozoic when a marked increase in the number and diversity of predatory taxa occurred in concert with changes in levels of bioturbation (Thayer, 1983) and grazing (Steneck, 1983). This phenomenon, popularly known as the Mesozoic Marine Revolution (MMR), has been linked to a general restructuring of shallow marine communities into those that are recognizable today and the rise of specific morphological adaptations that are considered primarily defensive.

Although Papp *et al.* (1947) first noted post-Paleozoic changes in gastropod morphology and Stanley (1977) noted a radiation of burrowing bivalves in the Mesozoic, which they explained by increasing levels of predation pressure, our understanding of the MMR has largely been fuelled by the energies of Geerat Vermeij, principally from a series of articles (Vermeij, 1977, 1978, 1982, 1987) that documented a wide array of evidence from different taxa. Vermeij's central thesis is that biological hazards, i.e. predation and competition, are the prime agents of natural selection and that the radiation in durophagous predators over the last 540 million years has driven an escalating "arms race" between predators and prey. He predicted that increasing levels of predation and competition would result in the evolution of an increasing variety of defensive adaptations (morphological and behavioral) and that the primary selection pressure on organisms would be the evasion of their own predators.

Despite being the best known of the marine revolutions, many questions remain about the MMR. For example, when exactly did it occur? Can the increase in predation pressure be viewed as a rising crescendo over the last 250 million years or have there been definite events within the MMR? Was it experienced in both benthonic and planktonic ecosystems? Were its effects felt equally at all latitudes? How confident can we be that rising predation pressure, particularly in response to specific predatory groups, was responsible for the changes seen?

This chapter examines some of these questions. First, I review the methods by which we chart escalating predation pressure and consider the timetable of the MMR. I then consider critically some of the changes in prey abundance, life habits and morphologies attributed to the MMR. The final section points to areas where future work might be profitable in order to establish a greater understanding of the MMR.

2. Establishing a Chronology for the MMR

Different predatory taxa use different methods to attack and subjugate their prey and prey organisms require different defensive adaptations if they are to escape these differing methods of attack. For example, adaptations that confer mechanical strength on a shell may be an effective defense against crushing predators, but may offer no assistance in warding off predators that chemically drill through the shell. In order to demonstrate that the rise of particular predatory groups caused specific responses among their prey we must have very clear evidence that there is a plausible temporal and geographical coincidence between the onset of the presumed threat and the response.

Information on the history of predation is largely acquired through indirect methods; there are few examples of predators fossilized literally "in the act." Fundamental evidence comes from our knowledge of modern predator-prey interactions. The onset of a particular predatory threat may be identified from the first appearance of taxa that are known to feed in a given way today, the recognition of particular functional morphologies, for example claws or jaws, and the recognition of diagnostic patterns of injury in prey items. There are clearly problems with this approach: some taxa may have evolved particular methods of predation relatively late in their evolutionary history, structures associated with predation may in fact have evolved for another function that merely pre-adapted the organism to use it offensively, and many predators leave no trace on their prey or leave evidence that may be indistinguishable from taphonomic damage. Assessment of past predation levels requires quantitative analysis of the proportion of injured prey items as either indices of the number of individuals killed or the frequency of unsuccessful attacks. Since few injuries that can be recognized are so diagnostic that they can be assigned with great confidence to particular predatory taxa, the history of predation is perhaps best explored in terms of predatory method.

Vermeij (1987) identified five principal methods by which predators attack molluscan prey but only four of these (whole animal ingestion, prying, breakage and drilling) are wholly distinct. His fifth category, transport, involves taking the prey item to a new location before employing one of the other techniques. The principal four categories, with only slight modification, can be used to consider predators that feed on a range of prey taxa, although given the great multitude of predatory methods which have evolved (see Taylor et al. [1980] and Taylor [1998] for the range of predatory gastropod activities, for example) this approach is necessarily rather simplistic. The following sections provide a brief account of the major taxa that use each predatory method, their geological history and the ease of identifying their activities in the fossil record. First appearance data for each family are taken from Benton (1993).

2.1. Predation by Simple Ingestion

Vermeij (1987) identified a wide range of modern predatory groups (worms, fish, gastropods and some starfish) that feed on molluscan prey by whole animal ingestion. Here, I use an extended definition to also include taxa that feed simply by biting off parts of the prey, for example cymatiid gastropods feeding on ascidians (Laxton, 1971). The simplicity of the method of attack is widely taken to indicate its primitiveness, and indeed there is rare evidence of the ability in Middle Cambrian priapulid worms from the Burgess Shale (Conway Morris, 1977). However, the number of predatory taxa that feed by simple ingestion and their importance is virtually impossible to assess from the fossil record. Simple ingestors require few morphological adaptations, which are mostly confined to the softparts, usually leave no recognizable trace on their prey, and many of the modern groups that feed in this manner (for example worms and some starfish) have extremely poor fossil records. However, a number of modern gastropods, for example tonnids, olivids, cymatiids and conids, feed by this method on a variety of prey items, including holothurians, molluscs, worms and fish. Many have evolved sophisticated methods, such as conids injecting and immobilizing their fish prey with venom (Terlau et al., 1996). They all appear for the first time in the Early Cretaceous, which may indicate a surge in level of this type of predation at that time.

2.2. Predation by Prying

Large prey items that are made up of more than one skeletal element, such as bivalves, brachiopods and barnacles, may be successfully tackled by predators that are able to pry open the skeletal parts and feed through the opening created. The chief predators in modern oceans that feed in this way are the extra-oral feeding starfish, which use their suckered tube feet to pull apart the valves of their prey before inserting their eversible stomach lobes between them. Their importance can be gauged by the severe damage that they inflict on commercial shellfisheries. Other pryers include melongenid gastropods, which use the lip of their own shell to wedge apart the valves of prey bivalves (Nielsen, 1975) and rapanine muricid gastropods, which use labral spines on the shell lip to pry apart barnacle plates (Vermeij and Kool, 1994), while a number of non-chelate crustaceans, e.g. slipper lobsters, use flattened dactyli to open oysters (Lau, 1987). Modern octopods use their suckered arms to pull apart their prey (e.g. McQuaid, 1994). Again the evolutionary history and past importance of these predatory methods is difficult to evaluate. Most prey items that have been pried apart show no morphological damage, and even where "accidental breakage" occurs, it is indistinguishable from taphonomic damage (Harper, 1994*a*). It is also difficult to infer the ability from the predator morphology. There is a dispute over when asteroids gained the ability to feed extra-orally. Gale (1987) claimed that it did not arise until the Jurassic with the appearance of the Asteriidae, citing critical changes in the musculature of the mouthframe and the suckered tube feet to have occurred at that time. By contrast Blake (1987) and Blake and Guensburg (1992) argued that some Paleozoic taxa also fed extra-orally, a point that is seemingly supported by a single specimen of *Promopalaeaster dyeri* from the Ordovician apparently straddling a large bivalve (Blake and Guensburg, 1994). Dietl and Alexander (1998) have used an ingenious way of establishing when the buccinoid gastropod genus *Busycon* began wedging open its bivalve prey. They noted that modern melongenids not infrequently break the lip of their own shells while attacking their prey and that these breaks are repaired. In a survey of shell repair in Miocene and Recent gastropods from New Jersey, they noted a low frequency of repaired apertural breaks in *Busycon,* which they interpreted as indicating that wedging predation either had not evolved or had not been perfected by the Miocene, at least in the eastern USA, despite the fact that melongenids have a fossil record dating back to the Albian. Labral spines appear to be a Neogene trait within the rapanine gastropods that evolved at least five times (Vermeij and Kool, 1994).

2.3. Predation by Breakage

Other predators gain access to the flesh of shelly prey by breaking them apart. Within the vertebrates, modern jawed fish of a variety of groups, including sharks, rays and teleosts, are considered important predators. Paleozoic placoderms were probably important durophages (Signor and Brett, 1984) but were extinct by the Devonian, after which a number of elasmobranchs (e.g. "pavement toothed" sharks and holocephalans) may have become important in their place (Brett, Ch. 17, in this volume). Among the bony fish, Mesozoic pycnodontiform and semionotiform fish both appeared in the record in the Early Triassic and persisted until the Eocene and Late Cretaceous respectively, but it is clear that the major radiation of predatory teleosts has been a Cenozoic event. Cartilaginous fish believed to be predatory include the hybodontid sharks, which first appeared in the Late Devonian and further radiated in the Triassic, and rays and other sharks in the Jurassic and Late Cretaceous. Breaking activity in vertebrate predators may be inferred from tooth morphology, gastric residues and coprolites (Speden, 1969), and their victims may be identified by repaired breakages, although it is generally not possible to relate styles of damage to particular foes. An exception to this is the report of distinctive bitemarks in a Callovian ammonite attributed to a semionotid fish (Martill, 1990). There were a number of large marine reptiles during the Triassic, including chelonids, placodonts, nothosaurs, pachypleurosaurs, plesiosaurs and ichthyosaurs, the latter two continuing into the Jurassic. Although there is little direct evidence of their diets, the well-developed jaws of these Mesozoic reptiles suggest that they must have been voracious predators. Gut contents of ichthyosaurs and marine crocodiles indicate that they fed on belemnites (Pollard, 1990; Martill, 1986) and there is a suggestion that mosasaurs preyed on ammonites (Kauffman and Kesling, 1960), although the evidence for this is debatable (Kase et al., 1994) (see also Mapes and Chaffin, Ch. 7, in this volume for a discussion of this debate). There is a clear need for further work to consider the diets of these Mesozoic reptiles and to search for ways of identifying their prey.

The second major taxon to use breaking techniques extensively are the decapod crustaceans, in particular the crabs (Taylor, 1981). Lau (1987) reviewed the range of decapod feeding strategies, from outright crushing of shelled prey to chipping and peeling vulnerable apertures. The known fossil record of crustacean families that are active crushing predators today is shown in Figure 1. There are two radiations, one in the Mesozoic (when the Homariidae, Palinuridae, Nephropidae, Sculidae, Paguridae, Calappidae, Xanthidae and Carpiliidae make their first appearance, most of them in the Early Cretaceous), and a Paleogene burst (Portunidae, Cancridae, Grapsidae and Ocypodidae all being recognized for the first time) in the Eocene. However, Vermeij (1982) warned that claw-like appendages may have first evolved as defensive traits before they were used offensively and large specialized crushing chelae do not appear until the Cenozoic (Vermeij, 1987). As with fish damage, successful crushing predation is difficult to identify in the fossil record because of the difficulty of discriminating between it and post-mortem damage, in particular for predators that simply "nutcracker" their prey. However, some crustaceans do make distinctive patterns. For example, stomatopods make characteristic holes that have been recognized in Plio-Pleistocene prey from Florida (Geary et al., 1991) and gastropods which have been unsuccessfully peeled by crabs (e.g. calappids) may show characteristic repair patterns. Monks (2000) interpreted repaired apertural breakages in heteromorph ammonites as due to failed crab predation and Allmon *et al.* (1990) used repaired peeling damage to quantify predation levels in turritellids.

It is also widely suggested that cephalopods may have been important crushing predators in the past: Alexander (1981) suggested that nautiloids were responsible for damage in Ordovician brachiopods. Certainly in modern seas octopods are voracious predators of a range of prey (Ambrose, 1986) and one of the predatory methods they use is to crush prey with their powerful beaks; modern *Nautilus* is capable of shearing through chicken bone (Alexander, 1986). Mesozoic seas apparently teemed with a variety of cephalopods (ammonites, nautiloids, coleoids), some of which may also have been able to feed in this way.

2.4. Drilling Predation

Modern marine predators that make neat drillholes in their prey, either to feed through or through which to inject a toxin, are found in a wide range of taxa, e.g. a variety of worms, several groups of gastropods and octopods (Bromley, 1981; Kabat, 1990; Ponder and Taylor, 1992; Morton and Chan, 1997). Many of these have a significant impact on the mortality of their prey species. The oyster drills *Urosalpinx cinerea* and *Thais haemastoma* are a serious menace to ostreiculture (White and Wilson, 1996) and Saunders *et al.* (1991) discovered that 50% of dead *Nautilus* shells from Papua New Guinea had been drilled by octopods. The presence of drillholes in shelly prey is perhaps one of the easiest forms of predation to recognize in the fossil record and there is a huge amount of ethological information that can be wrung from these traces, for example prey size preferences, stereotypy in attack, and success and failure rates (see Kelley and Hansen, Ch. 5, in this volume). However, the study of fossil drillholes and their makers is also most vexing.

The first problem is the difficulty of establishing with any certainty which taxon was responsible for drilling which holes; few drillers make holes that are totally unlike those drilled by other taxa (Bromley, 1981); exceptions are the distinctive tear-shaped holes which are *sometimes* drilled by octopods (Bromley, 1993) or the jagged holes in echinoid tests made by cassid gastropods (McNamara, 1994).

A second problem is that it is not possible to infer whether or not a particular predator had the ability to drill from its hardparts; for example drilling by muricid and naticid gastropods is effected largely by secretions from the accessory boring organ located either within the foot or the proboscis tissue respectively (Carriker, 1981). Furthermore, several of the taxa that are known to drill today, e.g. octopods and worms, have extremely poor fossil records and it seems not unlikely, given the number of different extant gastropods that can drill, that some extinct gastropods had also evolved the ability. There is, therefore, a major challenge in trying to study the evolution of drilling predators. Drillholes are very common in bivalves and gastropods from Cretaceous and Cenozoic shallow water assemblages (e.g. Taylor, 1970; Taylor *et al.*, 1983). Such traces co-occur with a number of muricid and naticid gastropod taxa and there can be little doubt that the vast majority of these holes were drilled by these predators. The oldest known muricids and naticids, along with their presumed drillholes, appear in the Albian (Taylor *et al.*, 1983). Cassid gastropods first occur in the Early Cretaceous (Santonian/Coniacian) but the first cassid-like holes found in echinoids are from the Eocene (Sohl, 1969). The first recognized octopod is known from the Carboniferous (Kluessendorf and Doyle, 2000) and yet the oldest recognized octopod drillholes are of Pliocene age, although doubtless older examples exist unrecognized (Bromley, 1993; Harper, 2002). There have been numerous reports of gastropod-like drillholes in Paleozoic prey, chiefly in brachiopods (although many of these are equivocal), and fewer reports from the Mesozoic (Kowalewski et al., 1998), perhaps suggesting that drilling was not a major selection pressure until the Albian appearance of muricids and naticids. However, where drillholes do occur in Triassic and Jurassic prey they appear to account for a significant level of mortality in their favored prey species (e.g. Newton, 1983; Fürsich and Jablonski, 1984; Harper et al., 1998).

2.5. Summary of Temporal and Spatial Patterns

Figure 1 summarizes the above information by showing the first appearance data for the most important predatory taxa. It clearly supports the contention that the number of predatory taxa has increased markedly over the last 250 million years. However, there are a number of uncertainties, which make it difficult to recognize the full pattern, not least that associated with the timing of the appearance of extra-oral feeding in asteroids and our uncertainty as to whether or not cephalopod groups were major crushing predators in the Mesozoic. Nevertheless, three episodes within the MMR can be identified when there were significant radiations in predatory taxa: (i) Triassic appearance of crushing fish and homaridean lobsters, (ii) Early Cretaceous, which saw a massive radiation of predatory gastropods using a wide range of techniques (Taylor et al., 1980), and (iii) early Cenozoic radiation of the teleosts and brachyuran crabs.

Although there appears to be good evidence that predation pressure in general has increased, there have been few attempts to gauge levels of predation on particular clades over long periods of geological time. Because of the difficulty of recognizing most forms of predation, those studies that have been conducted have dealt with either drilling or apertural peeling. Different stories have emerged. Vermeij et al. (1980) and Allmon et al. (1990) studied drilling and peeling predation in terebrid and turritellid gastropods and found that there were no significant differences in frequency of either from the Late Cretaceous to the present day. Harper (1994b) investigated drilling predation on corbulid bivalves and found highly variable levels of predation between faunas but no significant differences between geological periods. In contrast, Vermeij (1987) provided data that show that conid gastropods suffered a significant increase in peeling predation from the Eocene onwards.

What of the geographical controls on predation pressure? In modern faunas it is well known that taxonomic diversity increases towards the tropics (e.g. Stehli, 1968) and it is reasonable to suggest that there are more predatory taxa at low latitudes. This may be illustrated also within some predatory taxa, for example temperate rocky shores tend to support only one drilling muricid (e.g. *Nucella lapillus* in western Europe and *Urosalpinx cinerea* in North America), which is a key predator of epifaunal bivalves and barnacles. In contrast, in tropical localities several muricid taxa may co-exist (Taylor, 1980). Other predatory taxa, however, are today more diverse and important at higher latitudes. Taylor et al. (1980) showed that buccinid gastropods were most important at high latitudes. Vermeij (1990) suggested that extra-oral feeding asteroids were more common in temperate latitudes and are not an important selective force in the tropics. However, Harper (1994a) reported on *Coscinasterias acutispina* on a Hong Kong shore where it is a voracious predator of molluscs and barnacles, suggesting that there may be insufficient observations and published data on tropical asteroids. Another way of studying the latitudinal control on predation pressure is to consider the ratio of carnivorous to non-carnivorous taxa at different latitudes. To date there have been few such studies, most focussed on terrestrial clades (e.g. beetles and birds) and have argued for broadly constant ratios at all latitudes, perhaps with a slight decrease in boreal regions (Jeffries and Lawton, 1985; Warren and Gaston, 1992). However, in an intriguing recent paper, Valentine *et al.* (2002) have uncovered high values for the ratio of carnivorous/non-carnivorous gastropods in tropical and arctic regions in the northeastern Pacific with much lower values in the temperate zones.

Sadly, there have been few attempts to establish the geographical distribution of predators in the geological past. This is particularly unfortunate because the detailed analysis of predatory gastropods undertaken by Taylor *et al.* (1980) indicated that there were marked differences between the pattern observed today and that of the Late Cretaceous when diversity was higher at high latitudes, a finding which is interestingly rather similar to the pattern uncovered by Valentine *et al.* (2002). There is a clear need for studies on the fossil distribution of other predatory taxa to be undertaken.

3. Prey Responses to Increasing Threat of the MMR

Prey response to the increasing threat of predation of the MMR may be divided into three broad categories: restriction to refugia where predation pressure is less intense, exploration of new niches, and the evolution of morphological defensive adaptations. In order to link putative defensive adaptations with the MMR we need to formulate testable hypotheses to avoid the charge of telling "adaptive fairy tales" (Gould and Vrba, 1982). In some cases it is possible to test experimentally whether or not a particular trait is truly defensive in the manner postulated. Other supporting evidence can be gained where a particular adaptation has been acquired polyphyletically over a relatively short interval of geological time that coincides with the onset of a particular predatory method or appearance of a specific predatory taxon in an area. Most of the following examples come from investigations of the Mollusca, Echinodermata and Brachiopoda. These taxa have excellent fossil records, which assists greatly with understanding the timing of particular events. In the case of the molluscs and echinoderms, much is known about their Recent predators and there is the opportunity to carry out "evolutionary experiments" on modern representatives.

3.1. Restriction to Refugia

Unquestionably there was a marked change in the taxa that dominated shallow water benthic communities during the Mesozoic. Suspension-feeding echinoderms (e.g. stalked crinoids and ophiuroids) and brachiopods were abundant and diverse in these environments during the Paleozoic and early Mesozoic, but more recently they have been restricted to deeper waters, driven, some suggest, by the increased predation pressure of the MMR (Stanley, 1977; Jablonski and Bottjer, 1990). The best evidence to support increased predation pressure as the driving force for these offshore migrations has been amassed for the case of echinoderms, in particular for the ophiuroids (Aronson, 1994). Ophiuroids formed dense beds in Paleozoic and early Mesozoic shallow-water deposits and the frequency of these brittlestar beds drops off sharply during the early

Mesozoic Marine Revolution

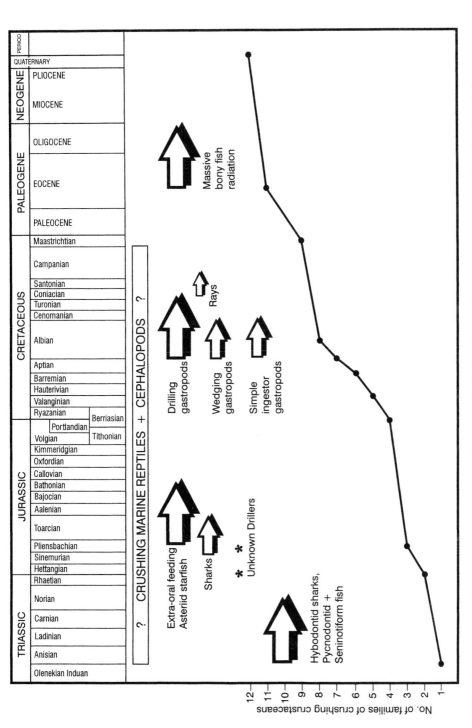

FIGURE 1. First appearances of major predatory taxa. Line graph represents the familial diversity of crustaceans known to be capable of crushing shelly prey. Arrows mark major radiation events of other important taxa, and the asterisks show the isolated occurrences of significant drillholes produced by unknown predators. All first appearance data are derived from Benton (1993).

Mesozoic (Aronson and Sues, 1987; Aronson, 1994). Since the Mesozoic, dense populations of brittlestars are restricted to more challenging environments such as deeper and colder waters where predation levels are low (Aronson and Sues, 1987; Aronson *et al.*, 1997). The hypothesis that this sharp change in the distribution of brittlestar beds was due to increased predation pressure in shallow water was directly testable. In a comparison of the frequency of sublethal predation in Carboniferous, Jurassic, and Recent ophiuroids, Aronson (1987) demonstrated that, although less than 2% of individuals from fossil localities provided evidence for arm regeneration following injury, over 60 % of individuals collected live from the Caribbean did bear signs of damage. Furthermore, a series of experiments where he tethered ophiuroids in the field revealed that they suffered high levels of attack; known predators include fish, crabs and starfish. It has also been suggested that another suspension-feeding echinoderm group, the stalked crinoids, were similarly displaced into deeper waters by the MMR (e.g., Bottjer and Jablonski, 1988; Aronson, 1994). Interestingly, although stalked crinoids appear to show morphological evidence of increased predation pressure during the Devonian (Meyer and Macurda, 1977; Signor and Brett, 1984), there was no decline in their communities over that period (Aronson, 1991).

In the echinoids, McNamara (1994) demonstrated that predation pressure from cassid gastropods may have driven some Neogene Australian echinoids into deeper water. Noting that cassids first appear in the Australian fossil record during the late Oligocene, McNamara showed that, within a lineage of the genus *Lovenia*, three species, which succeeded one another from the early to late Miocene, showed a statistically significant decrease in mortality (28.8% to 8.3%) from cassid drillholes and also that there was a decline in the success rate (i.e. more incomplete drillholes) in the younger species. Furthermore he noted a decrease in mortality due to cassids within a single species sampled at 1-meter intervals up the lithological sequence studied. The three species of *Lovenia* inhabited progressively finer sediments, which McNamara interpreted as evidence that these echinoids had been driven into deeper water to escape the cassids.

There are also similar notions that brachiopods, which had enjoyed high levels of diversity and high abundance in Paleozoic shallow-water communities but now are highly restricted in deep, cold waters or cryptic niches in the tropics, were forced into these refugia by increased predation pressure (Stanley, 1977; Donovan and Gale, 1990). However, attractive though these ideas might be, the case has never been supported satisfactorily. There is an interesting conflict between these suggestions and the repeated assertion that brachiopods are (and were) not favored prey items, particularly in comparison with bivalves, which offer a higher flesh yield, and even the notion that their flesh may be unpalatable (Thayer, 1985; Thayer and Allmon, 1990). There are a number of reports of predation damage on Paleozoic brachiopods (e.g. Brunton, 1966; Conway Morris and Bengtson, 1994; Leighton, Ch. 8, in this volume), but the post-Paleozoic records are few, although drillholes are not uncommon in some Mesozoic brachiopods (Harper and Wharton, 2000). Moreover, it is known that a range of modern predators (gastropods, fish, starfish) do eat brachiopods. Further work is required to demonstrate whether or not predation levels in the Mesozoic and Cenozoic were sufficient to cause the distribution of brachiopods today.

The intertidal zone provides another probable refuge for vulnerable prey species. The dual problems of desiccation and thermal stress severely limit the ability of predatory crabs and sea-stars to operate high in the intertidal zone, whereas mussels, oysters, littorinid and limpet gastropods and barnacles have evolved a number of strategies, for example the ability to tightly seal their valves or plates, which has allowed them to thrive there. However, it is difficult to demonstrate whether the invasion of the intertidal zone was predator-driven, since fossil shorelines are seldom recognized due to their small areas and their ephemeral nature. Certainly a rich rocky shore fauna of bivalves, brachiopods and serpulid worms was in place by the Late Cretaceous (Surlyk and Christensen, 1974).

3.2. Life Habit Changes

Several taxonomic groups appear to have undergone post-Paleozoic adaptive radiations in which taxonomic diversification has been accompanied by new successful clades adopting a greater diversity of life habits. It is possible that some of the new, more derived life habits may have some defensive value and, therefore, their evolution may be linked to the MMR.

The vast majority of predatory groups are either epifaunal or nektonic; few of them have the ability to search deep within the sediment for their prey and it has been suggested that one form of defense against predators is to adopt a deep infaunal life habit. Vermeij (1987) has shown that during the early Mesozoic there was a sharp increase in the number of different taxa (including bivalves, gastropods, echinoids, crustaceans) that had the ability to burrow deeply (>10 cm). Both bivalve and echinoderm workers have sought to demonstrate a link between the adoption of a deep infaunal existence and the MMR.

Most authors suggest that bivalves were primitively infaunal, living in shallow burrows with their posterior commissures more or less at the sediment-water interface (e.g. Morton, 1996). Such shallow burrowing bivalves today fall easy victim to predators that can locate them in life position, or when they are exhumed by water movement. Deeper burrowing, as evidenced by the possession of a pallial sinus, appears to be a largely post-Paleozoic phenomenon, although a sinus is present in the Ordovician *Lyrodesma* and a number of upper Paleozoic anomalodesmatans (Runnegar, 1974). The ability to burrower deeper into the substrate is chiefly dependent on fusion of the mantle to produce long siphons in order to maintain contact with the seawater above, and also the possession of an active digging foot. During the Mesozoic there was a proliferation of superfamilies and families containing siphonate burrowers (Stanley, 1968; Skelton *et al.*, 1990). Stanley's work suggested that there were only two clades of siphonate burrowers, the majority of siphonate families belonging to a large heterodont clade, with a smaller clade comprising only a handful of anomalodesmatan families (e.g. Pandoroidea and Clavagelloidea). Whereas the siphonate anomalodesmatans can be shown to have Paleozoic origins within the Pholadomyidae (Runnegar, 1974), Stanley believed that the siphonate heterodonts were all part of a single spectacular radiation from an astartid-like ancestor early in the Mesozoic. Later, he attributed this radiation directly to the rise of durophagous predators at that time (Stanley, 1977).

It now seems that the radiation of deeper burrowing bivalves may be more complicated; even within Triassic veneroids siphon formation was polyphyletic, there being at least three clades (Skelton *et al.*, 1990). The multiple evolution of siphons within a relatively short geological interval perhaps supports the idea of a response to an external selection pressure that emerged at that time. Recently, McRoberts (2001) has undertaken a study of the generic diversity of Triassic bivalves and observed that different extinction rates between in- and epifaunal genera produced an overall increase in the former during the period. However, he argued that predation pressure was not responsible; potential Triassic molluscivores (ichthyosaurs, placodonts and gastropod-like drillers) were neither abundant nor widespread enough to have been a credible selection pressure. Instead, he urged that environmental stresses (e.g. oxygen, salinity and temperature) and competition be further considered as possible causes of the trend observed (see Discussion).

Modern regular echinoids have a variety of predators, including fish, asteroids and other echinoderms, against which they possess a variety of morphological defenses (e.g. robust spines, toxic secretions) and cryptic behaviors (Smith, 1984). Irregular echinoids evolved from regular echinoids in the Early Jurassic, enabled by a range of morphological adaptations of the tube feet, flattening of the test and migration of the periproct detailed by Kier (1974). In general these infaunal echinoids are less well armored than their epifaunal counterparts and it has been suggested by Stanley (1977), Kier (1982) and Vermeij (1987) that the adoption of the deep burrowing habit might be a direct result of the increase in predation pressure.

These hypotheses that suggest that the onslaught of durophagous predators literally drove some bivalves and echinoids to seek safety deep within the sediment are attractive. However, it should be noted that they are based only on the coincidence of timing between the evolution of the habit and the MMR. As noted by Smith (1984), reduced predation may have been merely a welcome secondary benefit from an adaptation that resulted from a different selection pressure. It should also be noted that life below the sediment has not remained a perfect haven, free from predation. Other taxa have subsequently become important infaunal predators, for example naticid and cassid gastropods have become major predators of infaunal bivalves and echinoids respectively (e.g., Kelley and Hansen, 1993; McNamara, 1994). If a deep infaunal life habit is an important defense, it is perhaps surprising that the brachiopods have not also escaped from their enemies in this way. Only the lingulids burrow and even then not deeply. This apparent failure is probably explained by the limited flexibility of the brachiopod bauplan: inability to form extensive mantle fusion and the lack of an equivalent to the bivalve foot.

Other life habit changes that occurred at the start of the Mesozoic may have been defensive within the epifaunal Bivalvia (Harper and Skelton, 1993). Paleozoic epifaunal bivalves were almost exclusively epibyssate, although examples are known to have cemented (e.g. Pseudomonotidae) and bored into hard substrata (e.g. *Coralliodomus*) (Newell and Boyd, 1970; Pojeta and Palmer, 1976). In a study of familial diversity in the class, however, Skelton *et al.* (1990) identified a steady post-Paleozoic decline in epibyssate families and a concomitant increase in derived epifaunal life habits. This decline appears to start in the earliest Mesozoic and again is characterized by the polyphyletic acquisition of a new habit in large numbers of taxa. For each of the new habits a byssate ancestry can be demonstrated and most retain a byssally attached phase early in their ontogeny. Harper (1991) identified around 20 clades of cementing bivalves, of which the most important groups, e.g. the oysters and spondylids, first appeared in the Triassic and Jurassic. At first sight it seems counterintuitive that rigid cementation to a hard substratum might confer more protection against predators than the more flexible attachment of the byssate ancestors of these clades. However, experiments demonstrate that crabs and starfish are less able to eat artificially cemented mussels than those that are byssally attached (Harper, 1991). The basis for this selection appears to be an inability to manipulate the cemented prey, thus decreasing their profitability compared to the byssate mussels. Interestingly, similar experiments using muricid gastropod predators, which do not manipulate their prey, showed no statistical difference between numbers of cemented and byssate prey taken (Harper and Skelton, 1993). It is also striking that boring is a polyphyletic habit within the bivalves, appearing in nine post-Paleozoic clades, with the most important (Lithophagidae, Hiatellidae and Gastrochaenidae) first appearing in either the Triassic or Jurassic (Vermeij, 1987). It seems likely that occupying a boring within a hard substrate might provide good protection from crushing fish or crustacean predators, although Harper and Skelton (1993) have summarized experimental data and examples from the literature which show that both starfish and gastropods remain capable of feeding on borers.

An increase in the numbers of boring and cementing taxa during the Mesozoic is not restricted to the bivalves. A study of changes in hardground faunas from the Ordovician to the Cretaceous has revealed parallel changes in a number of encrusting and endolithic taxa, which Palmer (1982) attributed in part to the MMR increase in predation pressure.

3.3. Morphological Defenses

Defensive adaptations can only evolve by natural selection if there is some variation in prey morphology and behavior and if predation is not always successful (Vermeij, 1982). There is much evidence that modern predators feed "optimally," i.e., they exploit prey in such a way as to maximize the net energetic yield by selecting items that cost less (in terms of either energy or time) to locate and handle (e.g. Hughes and Elner, 1979; McQuaid, 1994). Therefore, prey that are more difficult either to find or to handle gain an advantage.

A great many hard-part modifications in a range of taxa have been interpreted as defensive adaptations. Possible morphological defenses and the timing of their first appearance have perhaps received most attention in the gastropods. Initial work by Papp et al. (1947) and Vermeij (1977) indicated that gastropods displaying what appear to be weaker morphologies (e.g. open, planispiral coiling) were much more common in Paleozoic and early Mesozoic faunas than they were later in the Phanerozoic, while high-spired taxa, and those with narrow or toothed apertures, which restrict entry by chelae, have increased in importance over that time interval. In a more exhaustive study, Vermeij (1987) charted the distribution of these traits from faunal surveys of shallow warm-water deposits across the Phanerozoic. His results were striking, reinforcing the earlier findings. He picked out the Triassic and Jurassic as the major changeover point but showed the proportion of gastropod taxa displaying these defensive traits rising more or less continuously towards the present day. A notable exception appears to be numbers of high-spired gastropod subfamilies, which, after a spectacular increase in the early Mesozoic, have been in decline since the mid-Cretaceous. But were these morphological changes truly defensive? Vermeij (1987) also collected data on the frequency of repaired breakages and the incidence of incomplete drill holes in a range of Recent gastropods as a measure of defensive effectiveness. Although morphological defenses against drilling predators proved difficult to identify, these data supported his contention that taxa with high spires, narrow apertures, thickened lips and increased shell ornamentation are all more capable of withstanding crushing attack. Additionally, Palmer (1979) has produced experimental data which show that the artificial removal of spines from muricid gastropods makes them more vulnerable to successful attack by crushing predators.

Bertness (1981) showed experimentally using modern gastropods that those with narrow apertures and thicker shells were more resistant to crushing predation. Appleton and Palmer (1988) demonstrated that gastropods, which are exposed to water in which crushed conspecifics and their crab predators have been kept, respond by producing thicker shells. It is important to note that not all morphological traits can be linked to enhanced protection against predation. Allmon *et al.* (1990) found that there was no correlation between various aspects of the morphology of turritellid gastropods (e.g. strength of shell sculpture and apertural form) and the level of drilling or peeling predation sustained.

It is not only possible to show that the proportion of various traits has changed within Phanerozoic gastropod faunas, but also to show morphological changes and to chart their relative success within particular clades. In a study of aporrhaid gastropods, Roy (1994) demonstrated two major phases of increase in taxonomic diversity. In the first, during the Middle Jurassic, two distinct morphological groups emerged, one with narrow apertures and elaborate lip ornament, the other apparently more sturdy with less elaborate apertures. The second, post-Albian, radiation produced an increase in the taxonomic diversity of the latter category. Roy hypothesized that these most robust forms may have arisen as a consequence of increasing crushing predation; however, his preliminary survey of shell repair in the two groups failed to substantiate this hypothesis.

Bivalved organisms (i.e. bivalves and brachiopods) have a very poor capacity for rapid and effective repair if damaged, with the result that even unsuccessful predation attempts may ultimately prove fatal (Vermeij, 1983; Alexander and Dietl, 2001). As a result, bivalves are generally thought to show fewer morphological defenses than do gastropods on the grounds that their most effective defense is avoidance of capture in the first place. However, a number of probable adaptations may be identified. Vermeij (1987) suggested that traits such as tight closure, radial ribs, occlusion of permanent gapes, and the possession of flexible or overlapping margins and spiny ornaments might all be defensive. These hypotheses have not been tested, nor their temporal occurrence investigated, as rigorously as his ideas on gastropod defense. Nevertheless, Vermeij and Veil (1978) demonstrated that there was a significant equatorward decrease in the number of Recent bivalve taxa with permanent shell gapes, which presumably correlates with the increased numbers of predators in low latitudes and the need to evade them. Similarly, Harper and Skelton (1993) identified an equatorward increase in the number of Recent epifaunal bivalve taxa with spines. Stone (1998) further studied the evolution of spines and flanges in epifaunal bivalves. Experiments that examined the vulnerability of naturally smooth-shelled mussels in comparison with those with artificially added spines found that the latter suffered a statistically lower mortality rate when attacked by drilling muricid gastropods but that spines had no defensive value against extra-oral feeding starfish. Although some late Paleozoic bivalves had the ability to produce highly ornamented shells, for example the pseudomonotids, Stone found that the majority of spinose groups had emerged since the beginning of the Jurassic. In a study of naticid predation on Miocene bivalves, Kelley (1991) demonstrated that the valves of five prey taxa gradually thickened and that this was accompanied by a decrease in drilling predation. Other traits that may also increase drilling time in bivalve prey include inhibiting the dissolving action of the ABO secretion, e.g. by possessing conchiolin sheets within the shells of corbulids (Lewy and Samtleben, 1979). However, taphonomic losses and changes make it difficult to assess the timing of first appearances of these sheets (Harper, 1994b; Kardon, 1998).

Ward (1981) noted that many Late Cretaceous ammonites developed heavy ribbing and spinose conchs and that such extravagant ornamentation was likely to have been deleterious to their hydrodynamic function, suggesting therefore some adaptive function. He suggested that these may have been defensive against the (largely unknown) predators of the ammonites, or to strengthen the thin shells during combat with their own prey. These are tempting scenarios, but much more data need to be collected on rates of repair in ammonites of varying morphologies; unfortunately this task is hampered by the fact that most are preserved as internal molds.

The temporal appearance of morphological defenses has not been studied as intensively in non-molluscan groups. Nevertheless, it is known that barnacles have become progressively more highly calcified since the Triassic (see Vermeij, 1987). Post-Paleozoic regular echinoids have changed their spines in a number of different ways, which may be interpreted as defensive (Kier, 1974). Cidaroids acquired large, robust spines that presumably fend off their predators, whereas most non-cidaroids showed an increase in the number of spines and a reduction in their lengths, combined with an increase in the number of tube feet, which Kier suggested assisted in the attachment of camouflaging shell debris.

3.4. Behavioral and physiological changes

A great many modern prey taxa employ behavioral or physiological defenses. Predators that sense their prey by visual means may be avoided by becoming nocturnal (e.g., the echinoid *Echinoneus cyclostomus*, which hides during the day [Rose, 1978]), or by being camouflaged by epibionts (which may also confuse the metabolic cues released by the prey). Experimental removal of fouling organisms from the shells of bivalves increases their vulnerability (Vance, 1978). Limpets have "mushrooming" and running responses to avoid starfish predation (Branch, 1979), and an increasing number of intertidal mussel species are known to bind their would-be muricid gastropod predators using their byssal threads (Day et al., 1991). Some echinoids deliver poison from their hypodermic-like spines (Smith, 1984) and, as noted above, it has been suggested that modern articulate brachiopods are also in some way toxic (Thayer, 1985). Such behavioral and physiological defenses are virtually impossible to track in the fossil record and therefore there can be no real test as to whether they too showed some increase due to the MMR. One possible line of inquiry, however, could be to study the incidence of complete but non-fatal drillholes in particular taxa. Kitchell et al. (1986) noted that during experiments on living terebrids several individuals were completely drilled (several times) but did not die. The repeated finding of multiple complete but non-fatal drillholes suggests that the prey had some active escape response or that perhaps the flesh was in some way distasteful. Multiple complete drillholes in other prey taxa may signal that they may possess this or other behavioral or physiological defenses.

4. Discussion and Future Directions

Nearly 25 years after the first detailed account of the MMR, there is mounting evidence to support Vermeij's hypothesis that, over the course of the Phanerozoic, individual prey organisms have been exposed to increasingly greater risk of predation and that they have responded with a concomitant increase in defensive adaptations. Three marked episodes of rapid diversification of predatory taxa have been identified: the "Cambrian explosion," the mid-Paleozoic marine revolution and the MMR. The MMR put "a modern stamp" on late Mesozoic and Cenozoic communities and saw predation pressure on major invertebrate groups increasing to its present level (Taylor, 1981). Much of the evidence gathered to support the MMR concerns animals with hard parts, principally molluscs and echinoderms, but doubtless it had a similar effect on the evolutionary histories of wholly soft-bodied organisms, although this would be difficult to document from the fossil record. The details of the MMR are, however, incompletely known and a number of areas remain to be addressed.

It remains remarkably unclear what drives periods of predator-prey escalation. The early MMR was probably shaped by the profound biological reorganization which took place after the Permian mass extinctions as vacated niches were refilled by different organisms. Vermeij (1995) linked biological revolutions to abiotic changes in the global environment. Specifically commenting on the late Mesozoic radiation of predators, he cited massive submarine volcanism that has been recorded at that time as having increased seawater temperatures and nutrient levels. These factors, combined with the associated transgression, which created large areas of shallow epeiric seas, would, he argued, have provided the enhanced level of resources necessary to allow organisms to acquire more energy-demanding life habits or traits. For example, the production of massive shells is energetically expensive and is easier in warmer water, where calcium carbonate is less soluble.

Our understanding of the evolutionary history of different predator groups needs much more refining. Identifying whether a particular predator group was active or not appears to be dogged by an uncertainty principle; either one can infer from functional morphology that a predator might have fed in a particular way (e.g. a clawed crab) but yet not be able with great confidence to attribute particular damage to it, or be able to recognize clear predatory damage (e.g. simple predatory drillholes) and yet not be able to determine which taxon might have caused it. As noted by Donovan and Gale (1990), our knowledge of modern predators and their prey is driven largely by our knowledge of commercially important groups; hence we know far more about predators of bivalves than we do about those of brachiopods. Obviously there are some predatory taxa, for example various soft-bodied worms with low preservation potentials, about which we are never likely to know much. However, other potential predatory taxa, such as the Mesozoic marine reptiles, may have been important but about which information about prey types and attack methods are rather scanty. To date most studies have focussed on the effects of predators on benthic prey; little appears to be known about those on planktonic and nektonic organisms.

Modern techniques of phylogenetic reconstruction using molecular and morphological data sets have led to the publication of robust phylogenies of a wide range of taxonomic groups. These studies provide the opportunity to test a number of hypotheses associated with the MMR and one fruitful area to consider is the way in which predatory taxa have diversified. Taylor (1998) used this approach to examine the evolutionary diversification of some major caenogastropod predators and to consider whether certain key innovations were responsible. One of the taxa studied, the extraordinarily diverse Conoidea, are well known for their ability to attack a range of prey using hypodermic darts, modified from radular teeth, to deliver rapid-acting toxins.

Using a morphological phylogeny of the Conoidea, Taylor showed that even many of the least derived conoidean species produced venom and that hypodermic teeth have evolved independently in six different clades. Interestingly, only two of these clades, the Terebridae and Conidae, have shown marked diversification. Both have independently acquired a radular caecum, a quiver-like structure in which multiple hypodermic teeth can be stored and used consecutively in an attack. Taylor considered this to be the key innovation in the marked diversification of the Conoidea.

Details of the precise timing of the appearance of various predatory groups and methods are not known completely, and yet such information is essential if we are to link them to particular prey responses. First appearances of taxa known today to attack prey in a given way may be a grossly inaccurate way of tackling the problem, particularly if predatory traits are often exaptive (Vermeij, 1982). It is also important to establish how effective different predators were over geological time. At present, our only quantitative surveys of predation level concern either drilling predation or apertural peeling. Further effort is required to study the prey remains of modern predators, in particular crushers, with a view to being able to identify such remains in the fossil record and to be able to establish rates of predation. Seemingly intractable problems may be tackled by novel methods, such as the recognition of wedging predation by studying lip damage on the predator itself (Dietl and Alexander, 1998). Other problems may be resolved by looking at more appropriate prey organisms. For example, there is an apparent mismatch between theories that invoke an increase in predation pressure from the very beginning of the Mesozoic (e.g. Stanley, 1968, 1977) and the observations that the most marked radiation of predatory groups (gastropods, bony fish, crabs) did not occur until significantly later, during the Early Cretaceous. If this discrepancy really exists then it seriously undermines many of the hypotheses discussed above and, indeed, McRoberts (2001) has recently questioned whether predation pressure on Triassic bivalves was really sufficient to cause the radiation in infaunal taxa, although he did note that the influence of Triassic predators might be underestimated because they did not leave recognizable traces. Potentially important predatory groups did emerge during the Triassic (Fig. 1), but most would have employed unspecialized crushing methods that seldom leave distinctive patterns of damage, and bivalves may be a particularly unfortunate group to use to study levels of crushing predation because of their poor capacity for repair (Vermeij, 1983; Alexander and Dietl, 2001). Alexander (1986) noted that brachiopods, for example rhynchonellids and terebratulids, which showed little or no repair may signal either that attacks on them were highly successful or that they were virtually immune from this type of predation. Gastropods, with their enhanced ability to repair shells, are perhaps a better indicator of levels of molluscivorous predation. Although Triassic data are scarce, some gastropod taxa from the St. Cassian Formation (Italy) do show high frequencies of shell repair, suggesting that crushing predators were present and active (Vermeij, 1987). Furthermore there is evidence that in at least some Triassic localities, where shell preservation is exceptionally good, significant numbers of some bivalve taxa had been drilled (Newton, 1983; Fürsich and Jablonski, 1984), which may well indicate that drilling, by unknown predators, may have been more widespread than commonly acknowledged (Harper et al., 1998). Much more data are required on predation levels from a variety of Triassic and Jurassic localities.

The mechanisms of predator-prey escalation demand some attention. Although the idea of evolutionary arms races between specific predators and prey is attractive there is little evidence of such coevolution between marine taxa. There are two important factors to consider. First, the majority of modern predators have very catholic diets; they are capable of attacking a diverse number of prey taxa and tend to feed opportunistically such that if high predation rates on one prey item lead to its being encountered less frequently, the predator will switch to a more abundant prey type. Second, most prey taxa are preyed on by a number of different predators that often employ very different strategies. For example the mussel *Mytilus edulis* is attacked by crushing crabs, drilling muricids and prying starfish (Kitching et al., 1959). There are no universal defensive adaptations and some traits, for example the cemented habit in bivalves, may be effective against some predators and yet potentially leave the prey more exposed to others (Harper and Skelton, 1993). Instances where a prey species is preyed upon by only one taxon, where tight coevolution between predator and prey might occur, are rare. One possible example is found in tonnid gastropods, which have become specialists at attacking holothurians, whose toxic secretions are unpalatable to many would-be predators (Morton, 1991). In the few instances where long term prey responses to predation pressure have been examined, Kelley (1991) found evidence of increased shell thickness in response to naticid predation, whereas Allmon et al. (1990) found no increase in morphological adaptation in turritellines but did accept that there may have been (undetectable) behavioral or physiological responses. Vermeij (1982, 1987) argued that the evolution of predators was more likely to be driven by the need to avoid their own predators than to become more effective predators themselves.

The evolution of prey defenses must involve compromise between the need to defend against different predatory methods and the need to balance the energetic requirements involved with those defenses and their other physiological activities. For example, most molluscs have abandoned the primitive nacreous shell microstructure despite the fact that it is mechanically superior to more derived arrangements (Taylor, 1973) and would presumably have offered a more effective defense against the multitude of crushing predators. However, nacreous shells are more expensive to produce and are slower to repair when damaged (Palmer, 1983). There is also the constraint of the prey organism's bauplan: not all prey are capable of acquiring a particular defense no matter how effective it might be (Harper and Skelton, 1993). For example, artificial spines added to mussel shells were shown to be beneficial in deterring a range of predators (Stone, 1998) but no spiny mytilids are known, probably because the thickened periostracum that characterizes the group is incapable of describing such intricate structures during shell formation (Harper, 1997).

Further information is required on the different preadaptations and constraints that have controlled the differing responses to increasing predation pressure. There is also a need to elucidate whether adaptations spread by intraspecific variation or by increased taxonomic diversity in particular clades, as argued by Roy (1994) for aporrhaid gastropods possessing stronger shells. Another interesting use of the plethora of robust phylogenies now emerging in the literature will be to compare the diversification patterns of sister taxa, one of which possesses a particular trait that is perceived as a defensive adaptation. Roy (1996) conducted such a study comparing the stromboidean gastropod families Aporrhaidae and Strombidae. The latter family possesses a distinctive operculum that assists individuals in a leaping escape response. He showed that the strombids have undergone a progressive increase in taxonomic diversity since the beginning of the Cenozoic and that they have tended to replace the aporrhaids at lower latitudes where presumably predation pressure has been more intense. Most studies are rather generalized and deal with higher taxa and there are relatively few that attempt to combine studies on the morphological variation in a specific clade over long geological periods and large geographic areas with quantitative analyses of predation levels and physical changes in the environment.

References

Alexander, R. R., 1981, Predation scars preserved in Chesterian brachiopods: probable culprits and evolutionary consequences for the articulates, *J. Paleontol.* **55**:192-203.
Alexander, R. R., 1986, Resistance to and repair of shell breakage induced by durophages in Late Ordovician brachiopods, *J. Paleontol.* **60**:273-285.
Alexander, R. R., and Dietl, G. P., 2001, Shell repair frequencies in New Jersey bivalves: a recent baseline for tests of escalation with Tertiary, Mid-Atlantic congeners, *Palaios* **16**:354-371.
Allmon, W. D., Nieh, J. C., and Norris, R. D., 1990, Drilling and peeling of turritelline gastropods since the Late Cretaceous, *Palaeontology* **33**:595-611.
Ambrose, R. F., 1986, Effects of octopus predation on motile invertebrates in a rocky subtidal community, *Mar. Ecol. Prog. Ser.* **30**:261-273.
Appleton, R. D., and Palmer, A. R., 1988, Water-borne stimuli released by predatory crabs and damaged prey produce more predation-resistant shells in a marine gastropod, *Proc. Natl. Acad. Sci., USA* **85**:4387-4391.
Aronson, R. B., 1987, Predation on fossil and Recent ophiuroids, *Paleobiology* **13**:187-192.
Aronson, R. B., 1991, Escalating predation on crinoids in the Devonian: negative community-level evidence, *Lethaia* **24**:123-128.
Aronson, R. B., 1994, Scale-independent biological interactions in the marine environment, *Oceanogr. Mar. Biol. Ann. Rev.* **32**:435-460.
Aronson, R. B., Blake, D. B. and Oji, T., 1997, Retrograde community structure in the late Eocene of Antarctica, *Geology* **25**:903-906.
Aronson, R. B., and Sues, H.–D., 1987, The paleoecological significance of an anachronistic ophiuroid community, in: *Predation: Direct and Indirect Impacts on Aquatic Communities* (W. C. Kerfoot and A. Sih, eds.), University Press of New England, Hanover, NH, pp. 355-366.
Bengtson, S., and Zhao, Y., 1992, Predatorial borings in Late Precambrian mineralized exoskeletons, *Science* **257**:367-360.
Benton, M. J. (ed.), 1993, *The Fossil Record 2*, Chapman and Hall, London, 845 pp.
Bertness, M.D., 1981, Crab shell-crushing predation and gastropod architectural defense, *J. Exper. Mar. Biol. Ecol.* **50**:213-230.
Blake, D. B., 1987, A classification and phylogeny of post-Palaeozoic sea stars (Asteroidea: Echinodermata), *J. Nat. Hist.* **21**:481-528.
Blake, D. B. and Guensburg, T. E., 1992, Predatory asteroids and the fate of brachiopods – a comment, *Lethaia* **23**:429-430.
Blake, D. B., and Guensburg, T. E., 1994, Predation by the Ordovician asteroid *Promopalaeaster* on a pelecypod, *Lethaia* **27**:235-239.
Bottjer, D. J., and Jablonski, D., 1988, Paleoenvironmental patterns in the evolution of post-Paleozoic marine invertebrates, *Palaios* **3**:540-560.
Branch, G. M., 1979, Aggression by limpets against invertebrate predators, *Anim. Behav.* **27**:408-410.
Bromley, R. G., 1981, Concepts in ichnotaxonomy illustrated by small round holes in shells, *Acta Geol. Hisp.* **16**:55-64.
Bromley, R. G., 1993, Predation habits of octopus past and present and a new ichnospecies, *Oichnus ovalis*, *Bull. Geol. Soc., Denmark* **40**:167-173.
Brunton, H., 1966, Predation and shell damage in a Visean brachiopod fauna, *Palaeontology* **9**: 355-359.
Carriker, M. R., 1981, Shell penetration and feeding by naticacean and muricacean predatory gastropods: a synthesis, *Malacologia* **20**:403-422.
Connell, J. H., 1970, A predator-prey system in the marine intertidal region. 1. *Balanus glandula* and several predatory species of *Thais*, *Ecol. Monographs* **40**:49-78.
Conway Morris, S., 1977, Fossil priapulid worms, *Spec. Papers Palaeont.* **20**:1-95.
Conway Morris, S., 2001, Significance of early shells, in: *Palaeobiology II* (D.E.G. Briggs and P.R. Crowther, eds.), Blackwell Science, Oxford, pp. 31-40.
Conway Morris, S., and Bengtson, S., 1994, Cambrian predators: possible evidence from boreholes, *J. Paleontol.* **68**:1-23.
Day, R. W., Barkai, A., and Wickens, P.A., 1991, Trapping of three drilling whelks by two species of mussel, *J. Exper. Mar. Biol. Ecol.* **149**:109-122.

Dietl, G. P. and Alexander, R. R., 1998, Shell repair frequencies in whelks and moon snails from Delaware and southern New Jersey, *Malacologia* **39**:151-165.

Donovan, S. K. and Gale, A. S., 1990, Predatory asteroids and the decline of the articulate brachiopods, *Lethaia* **23**:77-86.

Ebbestad, J. O. R., and Peel, J. S., 1997, Attempted predation and shell repair in Middle and Upper Ordovician gastropods from Sweden, *J. Paleontol.* **71**:1007-1019.

Fürsich, F. T., and Jablonski, D., 1984, Late Triassic naticid drillholes: carnivorous gastropods gain a major adaptation but fail to radiate, *Science* **224**:78-80.

Gale, A. S., 1987, Phylogeny and classification of the Asteroidea (Echinodermata), *Zool. J. Linn. Soc.* **89**:107-132.

Geary, D. W., Allmon, W. D., and Reaka-Kudla, M. L., 1991, Stomatopod predation on fossil gastropods from the Plio-Pleistocene of Florida, *J. Paleontol.* **65**:355-360.

Gould, S. J., and Vrba, E. S., 1982, Exaptation - a missing term in the science of form, *Paleobiology* **8**:4-15.

Hallock, P., Talge, H. K., Williams, D. E., and Harney, J. N., 1998, Borings in *Amphistegina* (Foraminiferida): evidence of predation by *Floresina amphiphaga* (Foraminiferida), *Hist. Biol.* **13**:73-76.

Harper, E. M., 1991, The role of predation in the evolution of the cemented habit in bivalves, *Palaeontology* **34**:455-460.

Harper, E., 1994a, Molluscivory by the asteroid *Coscinasterias acutispina* (Stimpson), in: *The Malacofauna of Hong Kong and southern China III* (B. Morton, ed.), Hong Kong, Hong Kong University Press, pp. 339-355.

Harper, E. M., 1994b, Are conchiolin sheets in corbulid bivalves primarily defensive?, *Palaeontology* **37**:551-578.

Harper, E. M., 1997, The molluscan periostracum: an important constraint in bivalve evolution, *Palaeontology* **40**:71-97.

Harper, E. M., 2002, Plio-Pleistocene octopod drilling behavior in scallops from Florida, *Palaios* **17**:292-296.

Harper, E. M., Forsythe, G. T. W., and Palmer, T., 1998, Taphonomy and the Mesozoic Marine Revolution: preservation state masks the importance of boring predators, *Palaios* **13**:352-360.

Harper, E. M., and Skelton, P. W., 1993, The Mesozoic Marine Revolution and epifaunal bivalves, *Scripta Geol., Spec. Issue* **2**:127-153.

Harper, E. M., and Wharton, D. S., 2000, Boring predation and Mesozoic articulate brachiopods, *Palaeogeogr. Palaeoclimatol, Palaeoecol.* **158**:15-24.

Hughes, R. N., and Elner, R.W., 1979, Tactics of a predator, *Carcinus maenus* and morphological responses of the prey, *Nucella lapillus*, *J. Anim. Ecol.* **48**:65-78.

Jablonski, D. and Bottjer, D.J., 1990, Onshore-offshore trends in marine invertebrate evolution, in: *Causes of Evolution: A Paleontological Perspective* (R. M. Ross and W. D. Allmon, eds.), University of Chicago Press, Chicago, pp. 21-75.

Jeffries, M. J., and Lawton, J. H., 1985, Predator-prey ratios in communities of freshwater invertebrates: the role of enemy free space, *Freshw. Biol.* **15**:105-112.

Kabat, A. R., 1990, Predatory ecology of naticid gastropods with a review of shell boring, *Malacologia* **32**:155-193.

Kardon, G., 1998, Evidence from the fossil record of an antipredatory exaptation: conchiolin layers in corbulid bivalves, *Evolution* **52**: 68-79.

Kase, T., Shigeta, Y. and Futakami, M., 1994, Limpet home depressions in Cretaceous ammonites, *Lethaia* **27**:49-58.

Kauffman, E. G., and Kesling, R., 1960, An Upper Cretaceous ammonite bitten by a mosasaur, *Contr. Mus. Paleontol. Univ. Mich.* **15**:193-248.

Kelley, P. H., 1991, The effect of predation intensity on rate of evolution of five Miocene bivalves, *Hist. Biol.* **5**:65-78.

Kelley, P. H., and Hansen, T. A., 1993, Evolution of the naticid gastropod predator-prey system: an evaluation of the hypothesis of escalation, *Palaios* **8**:358-375.

Kier, P. M., 1974, Evolutionary trends and their significance in post-Paleozoic echinoids, *J. Paleontol., Paleontol. Soc. Mem.* **5**:1-95.

Kier, P. M., 1982, Rapid evolution in echinoids, *Palaeontology* **25**:1-9.

Kitchell, J. A., Boggs, C. H., Rice, J. A., Kitchell, J. F., Hoffman, A., and Martinell, A., 1986, Anomalies in naticid predatory behavior: a critique and experimental observations, *Malacologia* **27**:291-298.

Kitching, J. A., Sloan, N., and Ebling, F.J., 1959, The ecology of Lough Ine VIII. Mussels and their predators, *J. Anim. Ecol.* **28**:331-341.

Kluessendorf, J., and Doyle, P., 2000, *Pohlsepia mazonensis,* an early 'octopus' from the Carboniferous of Illinois, USA, *Palaeontology* **43**:919-926.

Kowalewski, M., Dulai, A., and Fürsich, F. T., 1998, A fossil record full of holes: The Phanerozoic history of drilling predation, *Geology* **26**:1091-1094.
Lau, C. J.,1987, Feeding behaviour of the Hawaiian slipper lobster (*Scyllarides squammosus*) with a review of decapod crustacean feeding tactics on molluscan prey, *Bull. Mar. Sci.* **41**:378-391.
Laxton, J. H., 1971, Feeding in some Australasian Cymatiidae (Gastropoda: Prosobranchia). *Zool. Jour. Linn. Soc.* **50**:1-9.
Lewy, Z., and Samtleben, C., 1979, Functional morphology and palaeontological significance of the conchiolin layers in corbulid bivalves, *Lethaia* **12**:341-351.
Martill, D. M., 1986, The diet of *Metriorhynchus*, a Mesozoic marine crocodile, *Neues Jahrb. Geol. Paläont. Mh.* **1986**:621-625.
Martill, D. M., 1990, Predation on *Kosmoceras* by semionotid fish in the Middle Jurassic Lower Oxford Clay of England, *Palaeontology* **33**:739-742.
McNamara, K. J., 1994 The significance of gastropod predation to patterns of evolution and extinction in Australian Tertiary echinoids, in: *Echinoderms through Time* (B. David, A. Guile, J.-P. Féral, and M. Roux, eds.), A. A. Balkema, Rotterdam, pp. 785-793.
McQuaid, C., 1994, Feeding behaviour and selection of bivalve prey by *Octopus vulgaris* Cuvier, *J. Exper. Mar. Biol. Ecol.* **177**:187-202.
McRoberts, C. A., 2001, Triassic bivalves and the initial marine Mesozoic revolution: a role for predators?, *Geology* **29**:359-362.
Meyer, C. A., and Macurda, D. B., 1977, Adaptive radiation of the comatulid crinoids, *Paleobiology* **3**: 74-82.
Monks, N., 2000, Mid-Cretaceous heteromorph ammonite shell damage, *J. Moll. Stud.* **66**:283-285.
Morton, B., 1991, Aspects of predation by *Tonna zonatum* ((Prosobranchia: Tonnoidea) feeding on holothurians in Hong Kong, *J. Moll. Stud.* **57**:11-19.
Morton, B., 1996, The evolutionary history of the Bivalvia, in: *Origin and Evolutionary Radiation of the Mollusca,* (J.D. Taylor, ed.), Oxford University Press, Oxford, pp. 337-356.
Morton, B., and Chan, K., 1997, The first report of shell boring predation by a member of the Nassariidae (Gastropoda), *J. Moll. Stud.* **63**:476-478.
Newell, N. D., and Boyd, D. W., 1970, Oyster-like Permian Bivalvia, *Bull. Amer. Mus. Nat. Hist.* **143**:219-282.
Newton, C. R., 1983, Triassic origin of shell-boring gastropods, *Geol. Soc. Am. Abstr. Progr.* **15**:652-653.
Nielsen, C., 1975, Observations on *Buccinum undatum* L. attacking bivalves and on prey responses with a short review of attack methods of other prosobranchs, *Ophelia* **13**:87-108.
Paine, R. T., 1974, Intertidal community structure. Experimental studies on the relationship between a dominant competitor and its principal predator, *Oecologia* **15**:93-120.
Palmer, A. R., 1979, Fish predation and the evolution of gastropod shell sculpture: experimental and geographical evidence, *Evolution* **33**:697-713.
Palmer, A. R., 1983, Relative cost of producing skeletal organic matrix versus calcification: evidence from marine gastropods, *Mar. Biol.* **57**:287-292.
Palmer, T. J., 1982, Cambrian to Cretaceous changes in hardground communities, *Lethaia* **15**:309-323.
Papp, A., Zapfe, H., Bachmayer, F., and Tauber, A.F., 1947, Lebenspuren maner Krebse, *Königl. Akad. Wissensch. Wien, Mathem. Naturwiss. Klass. Sitzber.* **155**:281-317.
Pitman, R. L., Balance, L. T., Mesnick, S. I,. and Chivers, S. J., 2001, Killer whale predation on sperm whales; observations and implications, *Mar. Mamm. Sci.* **17**:494-507.
Pojeta, J., and Palmer, T. J., 1976, The origin of rock boring mytilacean pelecypods, *Alcheringa* **1**:167-179.
Pollard, J. E., 1990, Evidence for diet, in: *Palaeobiology: A Synthesis* (D. E. G. Briggs and P. R. Crowther, eds.),. Blackwell, Oxford, pp. 362-367.
Ponder, W. F. and Taylor, J. D., 1992, Predatory shell drilling by two species of *Austroginella* (Gastropoda: Marginellidae), *Jour. Zool., London* **228**:317-328.
Rose, E. P. F., 1978, Some observations on the Recent holectypoid echinoid *Echinoneus cyclostomus* and their palaeoecological significance, *Thalass. Jugoslav.* **12**:299-306.
Roy, K., 1994, Effects of the Mesozoic Marine Revolution on the taxonomic, morphologic, and biogeographic evolution of a group: aporrhaid gastropods during the Mesozoic, *Paleobiology* **20**:274-296.
Roy, K., 1996, The roles of mass extinction and biotic interaction in large-scale replacements: a reexamination using the fossil record of stromboidean gastropods, *Paleobiology* **22**:436-452.
Runnegar, B., 1974, Evolutionary history of the bivalve subclass Anomalodesmata, *J. Paleontol.* **48**:904-939.
Saunders, W. B., Knight, R. L., and Bond, P. N., 1991, *Octopus* predation on *Nautilus:* evidence from Papua New Guinea, *Bull. Mar. Sci.* **49**:280-287.
Signor, P. W., and Brett, C. E., 1984, The mid-Paleozoic precursor to the Mesozoic marine revolution, *Paleobiology* **10**:229-245.

Skelton, P. W., Crame, J. A., Morris, N. J., and Harper, E. M., 1990, Adaptive divergence and taxonomic radiation in post-Palaeozoic bivalves, in: *Major Evolutionary Radiations. The Systematics Association Special Volume 42* (P. D. Taylor and G. P. Larwood, eds.), Clarendon Press, Oxford, pp. 91-117.
Smith, A. B., 1984, *Echinoid Palaeobiology,* George Allen and Unwin, London, 190 pp.
Sohl, N. F., 1969, The fossil record of shell boring by snails, *Am. Zool.* **9**:725-734.
Speden, I.G., 1969, Notes on New Zealand fossil Molluscs – 2 Predation on the New Zealand Cretaceous species of *Inoceramus* (Bivalvia), *N. Z. J. Geol. Geophys.* **14**:56-70.
Stanley, S. M., 1968, Post-Paleozoic adaptive radiation of infaunal bivalve molluscs - a consequence of mantle fusion and siphon formation, *J. Paleontol.* **42**:214-229.
Stanley, S. M., 1977, Trends, rates, and patterns of evolution in the Bivalvia, in: *Patterns of Evolution, As Illustrated by the Fossil Record* (A. Hallam, ed.), Elsevier, Amsterdam, pp. 209-250.
Stehli, F. G., 1968, Taxonomic diversity gradients in pole location: the Recent model, in: *Evolution and Environment* (E.T. Drake, ed.), Yale University Press, New Haven, pp. 163-227.
Steneck, R. S., 1983, Escalating herbivory and resulting adaptive trends in calcareous algal crusts, *Paleobiology* **9**:44-61.
Stone, H. M. I., 1998, On predator deterrence by pronounced shell ornament in epifaunal bivalves, *Palaeontology* **41**:1051-1068.
Surlyk, F., and Christensen, W. K., 1974, Epifaunal zonation on an Upper Cretaceous rocky coast, *Geology* **2**:529-534.
Taylor, J. D., 1970, Feeding habits of predatory gastropods in a Tertiary (Eocene) molluscan assemblage from the Paris Basin, *Palaeontology* **13**:254-60.
Taylor, J. D., 1973, The structural evolution of the bivalve shell, *Palaeontology* **16**:519-534.
Taylor, J. D., 1980, Diets and habitats of shallow water predatory gastropods around Tolo Channel, Hong Kong, in: *The Malacofauna of Hong Kong. The Proceedings of the First International Workshop on the Malacofauna of Hong Kong and Southern China, 1977* (B. Morton, ed.), Hong Kong University Press, Hong Kong, pp. 163-180.
Taylor, J. D., 1981, The evolution of predators in the Late Cretaceous and their ecological significance, in: *The Evolving Biosphere* (P.L. Forey, ed.), British Museum (Natural History) and Cambridge University Press, pp. 229-240.
Taylor, J. D., 1998, Understanding biodiversity: adaptive radiations of predatory marine gastropods, in: *The Marine Biology of the South China Sea* (B. Morton, ed.), Hong Kong University Press, Hong Kong, pp. 187-206.
Taylor, J. D., Cleevely, R. J., and Morris, N. J., 1983, Predatory gastropods and their activities in the Blackdown Greensand (Albian) of England, *Palaeontology* **26**:521-533.
Taylor, J. D., Cleevely, R. J., and Taylor, C. N., 1980, Food specialization and the evolution of predatory prosobranch gastropods, *Palaeontology* **23**:375-409.
Terlau, H., Shoon, K., Grilley, M., Stocker, M., Stuhmer, W., and Olivera, B., 1996, Strategy for rapid immobilization of prey by a fish-hunting marine snail, *Nature* **381**:148-151.
Thayer, C. W., 1983, Sediment-mediated biological disturbance and the evolution of the marine benthos, in: *Biotic Interactions in Recent and Fossil Benthic Communities* (M. J. S. Tevesz and P. L. McCall, eds.), Plenum, New York, pp. 479-595.
Thayer, C. W., 1985, Brachiopods versus mussels: competition, predation and palatability, *Science* **228**:1527-1528.
Thayer, C. W., and Allmon, R., 1990, Unpalatable thecideid brachiopods from Palau: ecological and evolutionary implications, in: *Brachiopods through Time* (L. MacKinnon and D. Campbell, eds.), Balkema, Rotterdam, pp. 253-260.
Valentine, J. W., Roy, K., and Jablonski, D., 2002, Carnivore/non-carnivore ratios in northeastern Pacific marine gastropods, *Mar. Ecol. Prog. Ser.* **228**:153-163.
Vance, R. R., 1978, A mutualistic interaction between a sessile marine clam and its epibionts, *Ecology* **59**:679-685.
Vermeij, G. J., 1977, The Mesozoic marine revolution: evidence from snails, predators and grazers, *Paleobiology* **3**:245-258.
Vermeij, G. J., 1978, *Biogeography and Adaptation: Patterns of Marine Life,* Harvard University Press, Cambridge, MA, 332 pp.
Vermeij, G. J., 1982, Unsuccessful predation and evolution, *Am. Nat.* **120**:701-720.
Vermeij, G. J., 1983, Traces and trends in predation, with special reference to bivalved animals, *Palaeontology* **26**:455-465.
Vermeij, G. J., 1987, *Evolution and Escalation: An Ecological History of Life,* Princeton University Press, Princeton, NJ, 527 pp.
Vermeij, G. J., 1990, Asteroids and articulates: is there a causal link?, *Lethaia* **23**:431-432.

Vermeij, G. J., 1995, Economics, volcanoes, and Phanerozoic revolutions, *Paleobiology* **21**:125-152.
Vermeij, G. J., and Kool, S.P., 1994, Evolution of labral spines in *Acanthais*, new genus, and other rapanine muricid gastropods, *Veliger* **37**:414-424.
Vermeij, G. J., and Veil, J. A., 1978, A latitudinal pattern in bivalve shell gaping, *Malacologia* **17**:57-61.
Vermeij, G. J., Zipser, E., and Dudley, E. C., 1980, Predation in time and space: peeling and drilling in terebrid gastropods, *Paleobiology* **6**:352-364.
Ward, P., 1981, Shell sculpture as a defensive adaptation in ammonoids, *Paleobiology* **7**: 96-100.
Warren, P.H., and Gaston, K.J., 1992, Predator-prey ratios: a special case of a general pattern?, *Phil. Trans. R. Soc. Lond., Ser B,* **338**:113-130.
White, M. E., and Wilson, E. A., 1996, Predators, pests, and competitors, in: *The Eastern Oyster* Crassostrea virginica (V. S. Kennedy, R. I. E. Newell, and A. F. Eble, eds.), Maryland Sea Grant College, Maryland, pp. 559-579.

Index

Ammonoid cephalopods
    predation on, 180, 182–197, 201–208, 220, 405, 408, 437
    predation traces on, 182–187, 189, 192–197, 201–208, 220, 408, 437
    as predators, 9, 143, 195, 220, 319, 438
    temporal trends in shell armor, 206–207, 447
Anomalocarid arthropods, 64–66, 68, 84, 143, 381, 395, 402–403
Antipredatory traits
    in bivalve molluscs, 132, 144–145, 149–155, 418, 443–446
    in brachiopods, 221–222, 225–230, 418–420, 427, 429, 447
    in bryozoans, 242, 250, 252–257
    in cephalopods, 206–207, 419
    in crinoids, 263–270, 273–274, 420–429
    in dinosaurs, 337
    in echinoids, 422, 444, 447
    in gastropods, 132, 144–149, 151–155, 419, 422, 424, 429, 445–446, 450
    in ostracods, 106
    in reef-builders, 33–34, 37–43, 45–49
    in small mammals, 355
    in trilobites, 70–75, 422, 429
Archaeocyathids, 46–49
Armor, 39, 132, 142, 145, 154, 170, 207, 226, 280, 296, 326, 402, 428, 444; *see also* Antipredatory traits
Arms races, 50, 269, 296, 322, 402, 434, 449; *see also* Coevolution; Escalation
Arthropods, 40, 44, 56–57, 65–67, 69, 71, 73, 75, 84, 144, 196–197, 204–205, 207, 220–221, 395, 401, 403, 407; *see also* Anomalocarid arthropods; Barnacles; Crustaceans; Eurypterids; Ostracods; Phyllocarid arthropods; Trilobites
Asterozoan echinoderms
    asteroid, 9, 12, 66, 144, 148–149, 216, 219, 242, 285, 422, 434–436, 439, 441, 444; *see also* Sea stars
    ophiuroid, 9, 66, 283, 422, 440, 442

Barnacles, 44, 94, 104, 115, 292, 409, 436, 439, 443, 447
Bathymetric gradient, 122, 169, 263, 268, 275, 296; *see also* Gradients in predation
Bioerosion, 25, 35, 44, 50, 245, 247
Biomineralization, 56, 66–67, 70, 74, 83–84, 382

Bioturbation, 34, 36, 39, 41, 44, 47, 180, 189, 195, 321, 434
Bipedality
    in dinosaurs, 326, 331, 337
    in humans, 362, 366
    in rodents, 355
Birds, as predators, 8, 93, 144, 158–159, 280, 286–288, 297, 336–337, 341–356, 440
Bivalve molluscs; *see also* Pelecypod molluscs
    antipredatory adaptations of, 129, 132, 144–145, 149–155, 166, 226, 443, 446, 450
    breakage of, 141–145, 151–152, 154–170, 226, 405, 449
    drilling on, 94, 96, 99, 103, 113–120, 122–133, 412, 415, 417–418, 438–439
    predation on, 94, 96, 98, 103–104, 113–120, 122–133, 141–145, 151–152, 154–170, 226, 405, 412, 415, 417–418, 434, 436, 438–439, 443–446
    predation traces on, 113–120, 126–131, 142–145, 151–170, 405, 412, 415, 417, 438
    spatial trends in predation on, 122, 169–170
    temporal trends in predation on, 122–133, 166–168, 439–445
Bone damage
    bite marks, 312–313, 334, 364
    breakage, 343–347, 350, 353, 356
    butchery, 361, 364, 368–369, 372
    cooking, 344, 363, 366
    cut marks, 344, 367–368
    digestion, 346–350, 353
    gnawing, 343–344, 365–366
    tooth marks, 317, 318, 328, 334, 343–344, 365
Borings, 26–27, 38, 42, 44, 145, 152, 184, 380, 384, 418–419; *see also* Drill holes; Drilling predation
    as domiciles, 25, 42, 44, 115, 145, 409, 445
    parasitic, 21, 25, 61, 218, 406, 409–411, 413, 415
    post–mortem, 61, 70, 109, 196, 409, 413
    predatory, 10, 13–14, 21, 23–24, 56–57, 59, 61, 69–70, 81, 95–106, 108, 114–115, 117, 127, 131, 141, 179–181, 196, 217–219, 229, 245–247, 249, 252, 257, 406, 409–415, 417, 420, 427, 429

Brachiopods, 68, 74, 187, 193–194, 280, 296, 402, 440, 443
  antipredatory adaptations of, 70, 84, 169, 218, 221–229, 418–420, 427, 444, 447
  breakage of, 157, 216, 219–221, 226–229, 407–408, 434, 438, 442, 446
  drilling on, 113, 123, 216–219, 221–229, 406, 409–418, 429, 434, 439, 442
  predation on, 68, 113, 123, 143, 215–230, 405–408, 410–418, 427, 429, 434, 436, 438–440, 442, 446, 448–449
  predation traces on, 113, 143, 157, 219–221, 223–229, 407–418, 434
  spatial trends in predation on, 169, 221, 230
  temporal trends in morphology of, 221, 224–227, 229, 429
  temporal trends in predation on, 221, 224–229, 408, 417–418
Breakage, 14–15, 18–19, 21, 23, 27, 40, 42, 46, 60–61, 64–66, 68–69, 117, 141–145, 148, 151–155, 157–170, 178–183, 188–189, 194–195, 197, 201, 208, 240, 245, 247, 256, 271–272, 287–289, 314–315, 328, 343–347, 350, 353, 356, 402, 407–408, 419, 433, 435–437, 445; *see also* Bone damage; Crushing predation; Durophagy
Bryozoans
  antipredatory adaptations of, 40, 240, 242–245, 250–257
  grazing on, 242–244, 246–247, 256
  growth habits of, 21, 40, 46–47, 242–244, 251–252, 256–257
  predation on, 21, 115, 239–257
  predation traces on, 243–249, 256–257
  regeneration of, 240, 247–250
  temporal trends in morphology of, 250–257
Burrowing, 56, 80, 145, 150–151, 155, 168, 195–196, 216, 382, 434, 443–444

Cambrian Period, predation during, 44, 55–57, 63–69, 74, 76, 80, 83–84, 143, 221, 319, 379–385, 402–406, 410–413, 428–429, 433–435; *see also* Cambrian explosion
Cambrian explosion
  ecotone model for, 385–397
  role of predation in, 57, 66–67, 70, 83–84, 382, 385–386, 390–391, 393–397, 402, 433, 448
Cannibalism, 11, 117, 132, 179, 181, 183, 328, 369, 416
Carboniferous Period, predation during, 143, 197–206, 221, 245, 405, 407–408, 417–418; *see also* Mississippian Period; Pennsylvanian Period
Carnivory, 16, 19, 28, 35, 44, 56, 64, 69, 74–75, 80–81, 186, 196, 303–312, 314–315, 343–349, 351, 353, 356, 359–369, 371–373, 390, 394, 403, 440; *see also* Predation
Cenozoic Era, predation during, 123–133, 144, 154–155, 166–169, 247–249, 252, 256–257, 265, 291–297, 320, 417, 437–439, 442; *see also* Eocene Epoch; Miocene Epoch; Neogene Period; Oligocene Epoch; Paleocene Epoch; Paleogene Period; Pleistocene Epoch; Pliocene Epoch; Recent; Tertiary Period
Cephalopod molluscs; *see also* Ammonoid cephalopods; Nautiloid cephalopods; Octopod cephalopods
  predation on, 178–187, 190–208
  predation traces on, 180–186, 190–206
  as predators, 64, 66, 73, 144, 179, 181, 220, 404–405, 409, 428, 438–439
Chordates, origin of predation in, 304–311
Cnidaria, 34, 40, 46, 65, 243, 386; *see also* Corals
Coevolution, 27, 46, 50, 73, 121–122, 126, 128–129, 131–132, 371, 425, 449–450; *see also* Arms races
  definition of, 1–2, 121
Commensalism, 242, 251, 288, 406, 415, 422, 424–425
Competition, 34, 37–38, 41–43, 47, 121, 243, 327, 355, 362, 364–366, 371, 433–434, 444
Conodonts, 187, 309–310, 318
Coprolites, 57, 68–69, 186–187, 304, 312–316, 318–321, 334, 343, 405–406, 437; *see also* Pellets; Scats
Corals; *see also* Cnidaria
  antipredatory traits of, 34, 37–43, 45–50
  predation on, 33–50
Cretaceous Period, predation during, 9, 13, 23, 26, 44, 94–95, 97, 100–102, 104, 122–131, 133, 143–144, 154, 160, 162, 166–169, 183–184, 186–187, 206, 229, 248–249, 256–257, 265, 291–296, 317, 327–337, 436–441, 445, 447, 449
Crinoid echinoderms
  antipredatory adaptations in, 264–270, 273–274, 420–425, 427, 429
  bathymetric patterns of, 47, 263–265, 268–269, 275, 427, 440, 442
  parasitism on, 113, 268, 283, 406, 415, 422, 425
  predation on, 263–275, 405–406
  predation traces on, 265, 267, 271–272
  regeneration in, 265–275, 407
Crocodilians, 327, 332, 336, 347, 437
Crushing predation, 117–118, 151–152, 155, 165, 219, 226, 247, 256, 283, 409, 419, 428, 434–435, 437–439, 446, 449–450; *see also* Breakage; Durophagy
Crustaceans, 17–18, 21, 40, 144, 269, 295, 394, 443
  crabs, 11, 13–14, 35, 38, 115, 120, 141–142, 144, 148, 151, 160–161, 165, 186, 196–197, 206, 266, 285, 294, 314, 426, 437, 439, 442, 444, 449–450

Crustaceans (*continued*)
  lobsters, 144, 160, 206, 284–285, 436, 439
  as predators, 9, 14, 35, 66, 75, 158, 205, 280, 284, 288–289, 294, 297, 403–405, 408, 436–437, 441, 445
  shrimp, 8, 11, 13, 40, 266
  stomatopods, 144, 158, 162, 437
Devonian Period, predation during, 44, 66, 77, 80–81, 123, 166, 196, 217–229, 245, 254, 264, 269, 275, 309, 312, 316–318, 403–406, 408, 410, 419, 425–429, 434, 437, 442
Digestion, 15–19, 39, 67, 77, 242, 311–316, 321, 328, 335, 343, 346–350, 356, 363, 379
  effects on prey morphology, 15, 312–316, 321, 346–350
Dinosaurs
  evidence of predation in, 327–330
  feeding adaptations of, 329–337
  ornithischian, 326, 328–329, 337
  as predators, 327–337
  prey defenses of, 337
  sauropod, 326, 328–329, 337
  scavenging by, 327–328, 330, 333, 337
  theropod, 325–337
  trackways of, 328–329
  tyrannosaurid, 325, 327–328, 330–331, 333–335, 337
Diversity, 45, 352, 364, 443, 446, 448, 450
  of assemblages accumulated by predators, 351–353, 367, 372
  effects of predation on, 37–38, 288, 359, 371, 433, 442, 444
  of predators, 220–222, 403–406, 434, 439–441
Drill holes, 94–109, 434, 448; *see also* Borings; Drilling predation
  beveled, 13–14, 23, 95, 115, 217–218, 410, 412–413
  cylindrical, 13, 95, 115, 217–218, 281, 411–413
  failed, 99–102, 119–121, 129, 132
  incomplete, 14, 99–102, 119–121, 129–132, 409–412
  morphology of, 13, 23, 95, 115, 217, 281–282, 284, 291, 295, 409–413, 415, 438, 445
  multiple, 13–14, 23, 99–102, 120–121, 130–132, 249, 284, 409–413, 447
  nonfunctional, 99, 119–120, 447
  site selectivity of, 103–104, 118, 126–128, 131–132, 208, 229, 249, 281, 293, 409, 411–414, 418
  size selectivity of, 104–106, 119, 128, 130–132
  taphonomy of, 26, 116–117, 161, 289–290, 295
Drilling predation, 420, 427–429
  on bivalves, 103, 113–120, 123–133, 417, 438–439, 446, 449

Drilling predation (*continued*)
  on brachiopods, 113, 123, 216–219, 221–227, 229, 406, 409–418, 420, 434, 439
  on bryozoans, 247–249
  by cassid gastropods, 281–284, 292–293, 295, 438–442
  on cephalopods, 178–181, 196, 438
  on cloudinids, 81, 84, 380
  cost-benefit analysis of, 119, 129
  on crinoids, 406
  on echinoids, 280–284, 291–297, 438, 442
  by edge drilling, 115, 132
  on foraminifera, 10, 13, 94, 114
  by gastropods, 13, 23, 93–107, 113–120, 122–133, 280–284, 291–293, 296–297, 406, 409, 411, 415–418, 438, 441
  on gastropods, 113–120, 123–133, 438–439, 446–447
  by muricid gastropods, 13, 93–102, 104–105, 114–120, 126–128, 409, 438–439, 446
  by naticid gastropods, 13, 23, 93–97, 99–103, 105–106, 114–120, 123–133, 409, 416, 438–439
  by nematodes, 13, 23, 113
  by octopods, 113, 178–181, 196, 438
  on ostracodes, 93–109, 114
  process of, 94–95, 99, 114, 281, 438
  trends in, 122–133, 225–226, 229, 293, 402, 406, 417–418, 434, 439, 442, 446, 449
Durophagy, 44, 70, 83, 141–145, 245, 264, 270, 290–291, 403, 420, 434, 437, 443–444
  by arthropods, 27, 64–67, 69, 143–144, 158, 160, 182, 196, 205, 220, 266, 284–285, 294, 403, 437
  by birds, 144, 158–159, 286–287
  on bivalves, 141–170
  on brachiopods, 143, 216, 219–221, 226, 228–229, 407
  on bryozoans, 245
  by cephalopods, 66, 143, 179, 181, 187, 205, 216, 220, 404–405, 438
  on cephalopods, 177–210, 404–408, 437
  on crinoids, 264
  by echinoderms, 267, 285
  on echinoids, 284
  by fish, 66, 143, 158–159, 181–182, 186, 196, 203–205, 216, 219, 266, 286, 294, 405, 407, 437
  on foraminifera, 11–14
  by gastropods, 144, 158, 160, 290
  on gastropods, 141–170, 408, 438
  by mammals, 144, 286–287
  by marine reptiles, 143, 184, 286, 437
  peeling method of, 141, 146–147, 160, 162, 165–166, 196–197, 402, 408, 428, 437–439, 446, 449

Durophagy (*continued*)
   trends in, 62, 166–170, 183, 206–207, 221, 226–229, 419, 428, 439
   on trilobites, 65, 69

Echinoderms, 305, 309, 406, 425–426, 440; *see also* Asterozoan echinoderms; Crinoid echinoderms; Echinoid echinoderms; Pelmatozoan echinoderms
   antipredatory adaptations of, 84, 422
   blastoid, 113, 264, 415, 427
   holothurian, 12, 26, 40, 44, 247, 283, 436, 450
   predation on, 19, 21, 68, 218, 415
   as predators, 38, 66, 144, 247, 285, 297
   predation traces, 285
Echinoid echinoderms
   antipredatory adaptations of, 422, 443–444, 447
   parasitism of, 283–284, 288, 291, 294–297
   predation on, 280–283, 284–297, 442–444
   predatison traces on, 281–283, 288–297, 438
   as predators, 9, 12, 16, 26, 28, 93, 144, 243–244, 247, 256–257, 280
   and reefs, 35–36, 44, 297
   temporal trends in drilling of, 291–293
Eocene Epoch, predation during, 45, 50, 96, 122–125, 127–131, 167, 229, 245, 247, 270, 275, 292, 312, 318–319, 343–344, 353, 437–439, 441
Escalation; *see also* Arms races; Coevolution, 34, 50, 66, 83, 128, 144–145, 155, 161, 163, 165–166, 168, 170, 322, 380, 402, 418, 448–450
   definition of, 1, 121
   hypothesis of, 121–128, 131–133, 142, 166, 183, 207–208, 280, 296, 426–429
Eurypterids, 66, 76, 143, 205, 220
Extinction, 207, 448
   and escalation, 121, 124, 127, 131, 133, 170, 229
   and predation, 65–66, 98, 124, 127, 131, 133, 363, 367, 370
   selectivity of, 47–48, 50, 124, 131, 221, 225, 444

Fishes
   coprolites of, 311–316, 318–322
   diversification of, 44, 50, 66, 83, 221, 245, 403–405, 437, 439, 441, 449
   mode of predation in, 144, 159, 181, 204–205, 286, 304, 317, 435, 437
   morphology of feeding adaptations of, 317
   origin of predation by, 305–311
   predation by, 8–9, 15, 22, 35–38, 44, 64–66, 83, 93, 142–144, 147, 159, 178, 181–182, 186, 190, 192–193, 195–196, 203–205, 221, 240–245, 265–266, 274–275, 280, 285–289, 294–295, 297, 303, 311–323, 345, 403, 405, 427–428, 437, 442, 444–445

Fishes (*continued*)
   predation on, 312–314, 316–319, 321, 328, 335–336, 342, 436
   and reefs, 34–38, 40–42, 44, 50, 427
   sharks, 143, 181, 186–187, 190, 192–193, 195–196, 203–205, 208, 219, 245, 247, 264, 269, 271, 304, 315, 317–319, 395, 405, 407, 427, 437
   stomach contents of, 93, 247, 264, 312–316, 319–320, 405
   traces of predation by, 15, 22, 36–37, 159, 181–182, 186, 190, 192, 195, 243, 294, 313–320, 437
Flatworms, 9, 113, 244, 409
Foraminifera
   benthic, 8–17, 19–27
   feeding processes of, 16–21, 28
   planktonic, 8, 11, 13, 15–19, 21, 24–28
   predation on, 8–16, 22–27, 69, 94, 114
   as predators, 16–22, 28

Gastropod molluscs; *see also* Drill holes; Drilling predation
   antipredatory adaptations of, 132, 145–149, 152–155, 422, 424–426, 434, 443, 445–447, 449–450
   drilling by, 10, 13, 23, 25, 27, 93–107, 113–120, 122–133, 196, 218–219, 245, 247, 280–284, 288–295, 406, 409, 411, 415–418, 438–439, 441–442, 446–447, 450
   grazing by, 35, 243–244, 246–248
   parasitism by, 268, 283–284, 288–291, 293–295, 297, 406, 409–410, 415, 425–426
   predation on, 18, 113–117, 120, 123–133, 142–149, 154–155, 159–170, 425, 439, 446–447
   predation traces on, 113–117, 120, 126–133, 142–145, 148, 154–155, 159–170, 408, 437–439, 445, 447, 449
   as predators, 9–10, 12–13, 15, 23, 25–28, 35–36, 93–107, 113–120, 122–133, 144, 152–153, 158, 160, 168, 218–219, 243, 245, 247, 280–283, 288–294, 297, 406, 409, 411, 415–418, 435–436, 438–442, 444, 446–450
   spatial variation in predation in, 122, 168–170, 439–440
   temporal trends in predation in, 123–133, 155, 166–168, 293, 408, 417–418, 439
Gradients in predation
   bathymetric, 122, 169–170, 256–257, 263
   latitudinal, 122, 169, 221
Gut contents, 8, 11, 56, 67, 69, 75, 80–81, 242, 257, 265, 267, 309, 312–316, 320, 328, 335, 405, 437; *see also* Stomach contents

Holocene Epoch: *see* Recent
Human predation, 280, 347; *see also* Bone damage; Mammals
  and cooking, 344, 363, 366
  hunting strategies of, 363, 366–370, 372
  impact on ecosystems of, 359, 370–371
  overkill hypothesis of, 367, 370
  sociality and, 362–364, 367, 370, 372
  tool–making technology, 360–370, 372
Hyoliths, 84

Ichnofossils, 115, 383, 396; *see also* Predation traces; Trace fossils
Infaunalization, 56; *see also* Burrowing
  as response to predation, 296, 428
Injury, 43, 57–58, 60–65, 74, 142, 157, 163–164, 166, 181–183, 187, 190, 208, 245, 265–266, 268, 275, 328, 330, 408, 435, 442; *see also* Breakage; Drilling predation; Durophagy; Regeneration; Repair scars; Sublethal predation

Jurassic Period, predation during, 23, 25, 45, 50, 114, 122–123, 127–128, 143–144, 183–184, 187, 229, 249, 251, 270, 327, 332–333, 337, 436–437, 439, 441–442, 444–446, 449

Latitudinal gradient: *see* Gradients in predation

Malformations, 57–62, 182, 295
Mammals 328, 336; *see also* Bone damage; Human predation; Small mammals
  predation by humans on, 359–360, 362–373
  predation on, 328, 336, 341–355, 361–362, 364–365, 368
  as predators of invertebrates, 280, 286–288, 297
Marine reptiles, 184, 320, 437, 448
  ichthyosaurs, 437, 444
  mosasaurs, 144, 184–185, 317, 437
  placodonts, 143, 437, 444
  turtles, 143, 181, 280, 286, 288, 297, 437
Marine revolutions; *see also* Cambrian explosion; Mesozoic marine revolution
  early Paleozoic, 56–57, 66, 83, 402
  mid–Paleozoic, 44, 66, 83, 143, 207, 221, 230, 270, 402–429, 434, 448
Mesozoic Era; predation during, 44–45, 49, 83, 113–114, 133, 143–144, 154, 183, 186, 206–207, 220, 229–230, 256, 264, 293, 320, 327, 332–333, 336, 397, 402, 408, 418, 427–428, 434, 437–439, 442–445, 448–450; *see also* Cretaceous Period; Jurassic Period; Mesozoic marine revolution; Triassic Period
Mesozoic marine revolution, 83, 206, 220, 230, 264, 397, 402, 434–450
  prey responses to, 440, 442–447
  timing of, 435–439, 441

Miocene Epoch, predation during, 23–24, 44–45, 50, 96, 103, 106, 122–132, 144, 154, 166–168, 245, 247, 292–294, 316, 343, 353, 366, 436, 442, 446
Mississippian Period, predation during, 143, 182–183, 186, 218–221, 225, 228, 264, 269–273, 275, 406–408, 410–415, 417–418, 425, 427–428; *see also* Carboniferous Period
Molluscs, 21, 25, 35–36, 44, 84, 93, 113–114, 116, 122–127, 131, 141–145, 147, 150–151, 155, 158, 160, 164–167, 182, 419, 428, 435–436, 439–440, 444, 447–450; *see also* Ammonoid cephalopods; Bivalve molluscs; Cephalopod molluscs; Gastropod molluscs; Nautiloid cephalopods; Octopod cephalopods; Scaphopod molluscs

Nautiloid cephalopods, 64, 143, 187–191, 404, 419
Nautilus, 178–182, 184–186, 194–195, 200, 205–208, 438
  predation on, 180, 182, 190–191, 193–201, 203–208
  as predators, 64, 143, 186, 216, 220–221, 319, 404–405, 407, 438
Nematodes, 9, 13–14, 19, 21, 23, 27, 69, 84, 113, 244, 253
Neogene Period, predation during, 123, 125, 127–133, 166, 249, 337, 442; *see also* Miocene Epoch; Pliocene Epoch
Neoproterozoic, predation during, 55, 66, 81, 84, 380–382, 384, 393–396, 428; *see also* Precambrian
Nonlethal predation, 142; *see also* sublethal predation

Octopod cephalopods, as predators, 113, 115–116, 144, 178–181, 196, 217, 409, 436, 438
Oligocene Epoch, predation during, 45, 94, 123–132, 167, 245, 292–293, 343
Ordovician Period, predation during, 59, 64–66, 69, 72, 76–84, 115, 143–144, 157, 166, 216, 219–221, 225, 228, 403–408, 411–416, 422, 429, 434, 436, 438
  radiation during, 83, 73
Origin of predation, 304–311, 379–397
Ornamentation
  as antipredatory trait, 106, 148, 168, 206–207, 408, 445, 447
  in bivalves, 116, 157
  in brachiopods, 218–219, 221–227
  in cephalopods, 187, 189, 200, 447
  in gastropods, 148, 408, 445
  in ostracods, 103, 106
Ostracods, 18, 93–108, 114–115, 319

Paleocene Epoch, predation during, 94–100, 102–107, 123–125, 127–130, 167, 169, 249, 252, 292–293, 441

Paleogene Period, predation during, 123, 125, 127–128, 131–133, 167, 437, 441; *see also* Eocene Epoch; Oligocene Epoch; Paleocene Epoch

Paleozoic Era, predation during, 24, 49–50, 63, 66–67, 73, 81, 83, 113, 123, 143–144, 166, 169, 182, 207–208, 216–222, 225–226, 228–230, 256, 269–270, 275, 316, 318, 321, 384–385, 402–410, 413–415, 417–420, 426–429, 434, 436–437, 439, 442, 448; *see also* Cambrian Period; Carboniferous Period; Devonian Period; Mississippian Period; Ordovician Period; Pennsylvanian Period; Permian Period; Silurian Period

Parasitism, 21, 25, 60–62, 75, 79, 113, 116, 182–183, 216–218, 240, 268, 279–281, 283–284, 288, 291, 293–297, 304, 386, 390, 393–394, 406, 409–411, 413–415, 420, 422, 425

Pelecypod molluscs, 116, 124, 182, 215–216, 226, 229; *see also* Bivalve molluscs

Pellets, 343, 345–347, 349, 352; *see also* Coprolites

Pelmatozoan echinoderms, 216, 218, 406, 415, 426; *see also* Crinoid echinoderms; Echinoderms

Pennsylvanian Period, predation during, 22, 24, 160, 166, 168, 186, 195, 220–221, 228–229, 274–275, 312, 404–408, 412–413, 415, 418, 428; *see also* Carboniferous Period

Permian Period, predation during, 24, 66, 122–123, 143, 219, 228, 245, 247, 264, 275, 319, 402, 404–405, 412–413, 415, 417–418, 420, 428

Phyllocarid arthropods, 65–66, 143–144, 196, 204–205, 220, 403–404, 428

Pleistocene Epoch, predation during, 23, 37, 122–125, 127–130, 132, 145, 161, 167, 248, 343, 360–361, 363–370, 372, 437

Pliocene Epoch, predation during, 23, 122–125, 128–130, 132, 167, 170, 293, 343, 360–361, 363–364, 366–368, 371–372, 437, 439, 441

Polychaetes, 9, 11–12, 15, 18, 21, 35–36, 44, 73, 104, 116, 253, 409, 411

Porifera, 35, 44

Precambrian, predation during, 57, 113, 381, 385; *see also* Neoproterozoic; Proterozoic Eon

Predation
  by arthropods, 27, 64–67, 69, 75–81, 143–144, 158, 160, 182, 196, 205, 220, 266, 284–285, 294, 383–385, 403, 436–437
  on arthropods, 57–70
  by birds, 8, 93, 144, 158–159, 280, 286–288, 297, 336–337, 341–356, 440

Predation (*continued*)
  on bivalve molluscs, 94, 96, 98, 103–104, 113–120, 122–133, 141–145, 151–152, 154–170, 226, 405, 412, 415, 417–418, 434, 436, 438–439, 443–446
  on brachiopods, 68, 113, 123, 143, 215–230, 405–418, 420, 427, 429, 434, 436, 438–440, 442, 446, 448–449
  on bryozoans, 21, 115, 239–257
  by cephalopods, 64, 66, 73, 143–144, 179, 181, 187, 205, 216, 220, 404–405, 409, 436, 438–439
  on cephalopods, 177–208, 408, 437
  on crinoids, 263–275, 405–406
  by dinosaurs, 325–337
  by echinoderms, 38, 66, 144, 247, 267, 285, 297, 436
  on echinoderms, 19, 21, 68, 218, 279–294, 415
  by fish, 8–9, 15, 22, 35–38, 44, 64–66, 83, 93, 142–144, 147, 158–159, 178, 181–182, 186, 190, 192–193, 195–196, 203–205, 216, 219, 221, 240–245, 265–266, 274–275, 280, 285–289, 294–295, 297, 303, 311–323, 345, 403, 405, 407, 427–428, 437, 442, 444–445
  on fish, 312–314, 316–319, 321, 328, 335–336, 342, 436
  by foraminifera, 16–22, 28
  on foraminifera, 8–16, 21–27, 69, 94, 114
  by gastropods, 9–10, 12–13, 15, 23, 25–28, 35–36, 93–107, 113–120, 122–133, 144, 152–153, 158, 160, 168, 218–219, 243, 245, 247, 280–283, 288–294, 297, 406, 409, 411, 415–418, 435–436, 438–442, 444, 446–450
  on gastropods, 18, 113–117, 120, 123–133, 142–149, 154–155, 159–170, 425, 439, 446–447
  by humans, 359–373
  by mammals, 144, 280, 286–288, 297
  on mammals, 328, 336, 341–355, 361–362, 364–365, 368
  by marine reptiles, 143, 184, 286, 437
  on marine reptiles, 317
  opportunistic, 181–182, 195, 205, 207, 351–352
  optimal foraging theory of, 119, 129, 163
  on ostracods, 93–109, 114
  on reef-building organisms, 34–37, 44–45
  sublethal, 14–15, 22, 42, 56–57, 59, 61–66, 81–82, 99–102, 119–121, 129–132, 141–148, 155, 157–168, 169–170, 178–183, 188, 190, 206, 208, 219–220, 227–229, 246–250, 267, 270, 273–274, 279–280, 285–286, 294, 407–412, 434, 436–438, 442, 445–447, 449–450

Predation frequency, 125–126, 187, 189, 197–198, 200–203, 206, 274; *see also* Predation intensity
Predation intensity, 154, 221–222, 229, 239–240, 242, 267, 275, 297; *see also* Predation frequency; Predation pressure
 calculation of, 116
 for crushing predation, 142, 161, 163–164, 226, 273
 for drilling predation, 106, 292–293
Predation pressure, 15, 25–27, 38, 40, 48, 57, 66–67, 70, 73–74, 78, 84, 103, 108, 147, 155, 169–170, 225, 252, 263–265, 268–270, 280, 288, 365, 372, 382, 386, 391–392, 396–397, 402, 418–419, 425, 427–429, 434, 439–440, 442, 444–445, 448–450
Predation traces, 46, 289; *see also* Ichnofossils; Trace fossils
Predator efficiency, 130, 229
Predatory behavior, 64–65, 72, 293, 343
 of anomalocarids, 64
 of asteroids, 144, 216, 285, 434–436
 of cephalopods, 179, 181
 of crabs, 148, 151, 160, 165, 285, 437, 444
 of dinosaurs, 327–337
 of drilling gastropods, 114, 120
 of fish, 311, 316–317, 321
 of foraminifera, 17–21
 of humans, 359–372
 of stomatopods, 144, 158
 of trilobites, 56, 75–77, 80, 394
Prey effectiveness, 120, 129–130, 160–161
Primates, 344, 353, 359–361, 366–367, 369; *see also* Human predation
Proterozoic Eon, predation during, 380–382, 386; *see also* Neoproterozoic; Precambrian
Pterosaurs, 328

Recent, predation during, 114, 247, 343, 350, 354, 363, 367, 370
Reefs, 33–50, 243, 245, 265, 269, 273, 280, 285, 297, 427; *see also* Cnidaria; Corals
Refuge from predation, 8, 84
 nocturnalism as, 38, 447
 size as, 43, 94, 105, 119, 164–165, 229, 427–429
 spatial, 11, 37–39, 45, 49, 225, 229, 263, 268–269, 428–429, 440, 442
Regeneration, 37–38, 42–43, 46, 48, 60–61, 245, 249–250, 264–268, 270–275, 442; *see also* Repair scars
Repair scars, 15, 22, 42, 57, 119–120, 141–148, 155, 179–180, 183, 188, 190, 206, 208, 219–220, 246–250, 274, 279, 285, 294, 407–408, 412, 434, 436–438, 445–447, 449–450; *see also* Durophagy; Prey effectiveness; Regeneration; Sublethal predation
 calculation of frequency of, 160–165, 228

Repair scars (*continued*)
 geometry of, 61, 157–160, 181–182, 286, 294, 407, 437
 relation to predation intensity of, 142, 267
 spatial variation in, 169–170
 temporal trends in, 62–65, 166–168, 227–229, 270, 273, 408
Rodents, 355; *see also* Small mammals

Scaphopod molluscs, 9–11, 13–15, 25–28, 114
Scats, 343–344, 346–347, 353; *see also* Coprolites
Scavenging, 21, 60, 67–69, 75–76, 115, 180–181, 186, 189, 191, 193, 195–197, 205, 285, 307, 311, 313–315, 318, 320, 326–328, 330–333, 337, 344, 359–361, 363–369, 372–373, 404, 406
Sea stars, 144, 243, 266, 285; *see also* Asterozoan echinoderms
Sea urchins, 12, 44, 144, 244, 267, 280, 284–288; *see also* Echinoid echinoderms
Selective predation, 9–11, 126, 318
 of prey size, 104–106, 119, 128–132, 164–165, 197–202
 of prey species, 9, 15, 23, 26, 28, 104, 115, 119, 164, 200, 203, 242, 286
 of site on prey shell, 62–64, 103–104, 118, 126–128, 131–132, 164–166, 218, 229, 249, 267, 271, 281, 284, 293, 409, 411–414, 418
 of valve, 119
Silurian Period, predation during, 80, 143, 166, 221, 224–225, 319, 404–406, 408, 411–413, 415, 427, 429
Size–frequency distributions
 for lethal breakage, 197–203
 for lethal drilling, 283
 for sublethal breakage, 164–165
Small mammals; *see also* Birds; Coprolites; Pellets; Scats
 antipredatory adaptations of, 355
 bone modification of, 343–350
 evidence of predation on, 343–350
 population fluctuations of, 353–354
 predation on, 341–356, 366
 seasonal variation in predation, 350
Small shelly fossils, 84, 395, 434
Sociality: *See* Human predation
Stereotypy, 99, 103, 118–119, 126–127, 131–133, 165–166, 217–218, 229, 332, 411–414, 428, 438; *see also* Selective predation
Stomach contents, 93, 143, 164, 178, 186, 247, 264, 319, 402, 405; *see also* Gut contents
Sublethal predation: *See* Predation

Taphonomy, 26–27, 46, 62, 65, 117, 289, 293, 295, 321, 354, 413
Tertiary Period, predation during, 44, 124, 126, 344
Tool use, 360–368, 372–373; *see also* Human predation
Toxicity, 21, 40, 229, 242

Trace fossils, 36, 56–57, 61, 80, 82–83, 320–322, 328, 337, 360, 380, 382–385, 392, 394, 402, 406–418, 428; *see also* Ichnofossils; Predation traces

Trends, 402, 417; *see also* Ammonoid cephalopods; Antipredatory adaptations; Bathymetric gradient; Bivalve molluscs; Brachiopods; Bryozoans; Drilling predation; Durophagy; Echinoid echinoderms; Gastropod molluscs; Gradients in predation; Repair scars; Trilobites
    morphological, 34, 83, 132, 145–155, 224–229, 250–257, 269–270, 296, 394, 402, 419–423, 429
    paleoecologic, 268–269, 275, 296, 418
    in predation through time, 64–65, 123–133, 163, 166–170, 224–229, 297, 402, 417–418, 427–428

Triassic Period, predation during, 44, 114, 143–144, 166, 183, 245, 247, 327–328, 331–333, 337, 437, 439, 441, 443–445, 449

Trilobites
    antipredatory adaptations of, 70–75, 83–84, 422, 429
    predation on, 56–70, 81–84, 381, 383, 394, 402–403, 405, 428
    predation traces on, 57–65
    as predators, 56, 75–82, 84, 205, 383–384, 394–395
    temporal trends in predation, 64–65

Turtles: *See* Marine reptiles

Whole animal ingestion, 141, 144, 181, 240, 433, 435–436